PERSPECTIVES IN CONDENSED MATTER PHYSICS
A Critical Reprint Series

Condensed Matter Physics is certainly one of the scientific disciplines presently characterized by a high rate of growth, both qualitatively and quantitatively. As a matter of fact, being updated on several topics is getting harder and harder, especially for junior scientists. Thus, the requirement of providing the readers with a reliable guide into the forest of printed matter, while recovering in the original form some fundamental papers suggested us to edit critical selections on appealing subjects.

In particular, the present Series is conceived to fill a cultural and professional gap between University graduate studies and current research frontiers. To this end each volume provides the reader with a critical selection of reprinted papers on a specific topic, preceded by an introduction setting the historical view and the state of art. The choice of reprints and the perspective given in the introduction is left to the expert who edits the volume, under the full responsibility of the Editorial Board of the Series. Thus, even though an organic approach to each subject is pursued, some important papers may be omitted just because they lie outside the editor's goal.

The Editorial Board

PERSPECTIVES IN CONDENSED MATTER PHYSICS
A Critical Reprint Series: Volume 3

SIMULATION APPROACH TO SOLIDS

edited by

Marco Ronchetti
Gianni Jacucci

KLUWER ACADEMIC PUBLISHERS

Jaca Book

prima edizione
gennaio 1991

copertina e grafica
Ufficio grafico Jaca Book

ISBN 88-16-96005-1 (Jaca book)
ISBN 0-7923-0383-0 (Kluwer)

per informazioni sulle opere pubblicate e in programma
ci si può rivolgere a Editoriale Jaca Book spa
via Rovani, 7, 20123 Milano, telefono 4988927

FOREWORD

This volume concentrates on the simulation of crystalline solids. The existence of recent complete account of the use of computer simulation in condensed matter[1,2,3,4] has freed us from the general task of attempting a comprehensive review of the field. As a result we followed more heartily our inclination in the choice of papers to be included. The imaginary path we have chosen in describing advances in computer simulation work is one bringing from statistical mechanics and solid state physics to materials science.

The volume addresses a fascinating field. We are satisfied with this selection of papers. However we must point out that we had problems in squeezing the material we wished to include, in order to fit the prescribed volume size and number of papers.

We believe that the following four are the main reasons for this difficulty:

1. Computer simulation research in solids has been productive of outstanding results in *diverse areas* over the last twenty years;

2. Almost each topic (atomic diffusion, lattice vibrations, melting etc.) grants a non trivial and hardly negligible extension in the area of *surface physics*;

3. Papers in the area are seldom *self-contained*: a new method is invented and presented in a first work along with a test case, results for relevant systems often come only in later works;

4. Methods are often involved and their presentation requires the introduction of a variety of conceptual frameworks and interdisciplinary languages, a material hard to exclude and requiring plenty of space to be didactically useful.

We have done our best in an attempt to meet publisher specifications. In many cases we had to take drastic decisions, much to our regret, preferring shorter versions to more complete papers, or singling out a paper from a temporal sequence, in the strive not to leave out any major area of research. Nontheless, many interesting papers that in our opinion play a well definite role in the development of this science, were to be expelled from the privileged bunch. More papers of a given research group are often included, in order to favour coherence in presentation. No attempt is made to represent all existing schools in any one subject.

Table of Contents

M. Ronchetti and G. Jacucci, Simulation in solids

1. INTRODUCTION

In this chapter we will briefly outline the history of computer simulation, introduce the technique of Molecular Dynamics and discuss the attitude of researchers towards simulation. In section 2, a short presentation of the collected paper will be given. Aim of this presentation is to briefly introduce the reader to the fields and to give a few additional references, without attempting any exhaustive review. Among these remarks, some are more extended then others, in particular those regarding topics for which we felt that it is difficult to find elsewhere an introduction to the main ideas. Finally, in section 3 future perspectives of computer simulation are briefly discussed.

1.1 History

Newton mechanics allows in principle to determine the time evolution of any physical system for which quantum effects are not relevant. However, such calculation would require the simultaneous solution of a huge number of differential equations even for a tiny fragment of matter: an obviously impossible task in practice. Statistical mechanics was therefore born in nineteenth century, with the goal of evaluating the average values of physical quantities of complex systems.

Early computers allowed for the first time in history to approach the typical problems of statistical mechanics by brute force, at least for systems with hundred to thousand degrees of freedom. The first calculations were mainly concerned with statistical mechanics issues: Fermi, Pasta and Ulam[5] approached the problem of the reversibility paradox: the microscopic description of a classical many-body system is based on *reversible* Newton equations, but the macroscopic description of the same system (i.e. its thermodynamics) is governed by *irreversible* laws. The system chosen by Fermi (a one-dimensional anharmonic chain) turned out to be too simple, lacking Lyapunov instability: as a consequence, the laws of statistical mechanics could not be recovered.

Metropolis et al.[6] were the pioneers of condensed matter investigation via computer simulation. Their early work started what is now known as "Metropolis Monte Carlo". The method is called "Monte Carlo" (MC) because it allows an exploration of configuration space by throwing dices: random numbers in fact provide the power, and the steering wheel is a simple rule which locally samples in prevalence the regions of configuration space where Boltzmann weight is highest. The procedure allows to collect statistics in an efficient way, and therefore it provides a technique to calculate averages of physical quantities. The method is concerned with configuration space, rather than with the whole phase space: it is therefore possible to gather information only on the so-called static properties. No clue is available on time-dependent quantities: therefore informations on dynamic properties can not be obtained in this way.

A different method to explore phase space follows the trajectories of representative points by solving explicitly the equations of motion. When applied to atoms and molecules the method is called Molecular Dynamics (MD). The first MD calculation in two and three dimensions was concerned with the very same problem posed by Fermi: Alder and Wainwrigth[7] studied the motion of 100 hard spheres with equal speed and different velocities. They were able to prove that momentum distribution for such a system converges rapidly to equilibrium. Thermodynamic properties calculated in this way are consistent with those calculated by MC.

The first attempt to investigate materials dates to 1960, when Vineyard[8] and coworkers used MD to investigate the dynamics of radiation damage. Their system was based on a short-range repulsive interatomic potential, plus a constant inward force responsible for the cohesion of the crystal. A single atom was endowed with a high kinetic energy, and the effect on the crystal was studies. Simulation was used as a very powerful microscope:

in the paper several "photographs" of the atomic trajectories are shown: for the first time a window over the very small was open. Through that window many researcher looked and still look, to gain qualitative insight on microscopic processes besides quantitative evaluation of thermodynamic properties.

Anees Rahman initiated in 1964 the study of systems with continous potentials[9]: he used 864 particles in periodic boundary conditions to simulate liquid argon. It was probably surprising to find that few particles (a very small collection, if compared with Avogadro number) can reproduce thermodynamic properties of real systems. It is exactly this fortunate fact which makes computer simulation of condensed matter possible.

Since then many researchers followed the route opened by these vanguards, and many reports based on these techniques have been published. Some important papers have been reprinted in a volume[1], and a recent review of the field can be found in the proceedings of a school[2].

1.2 Simulation of equilibrium solids

Computer simulation is most useful when it is difficult to use traditional analytic techniques to solve problems: for this reason many computational studies deal with liquids and amorphous systems. Solids are rather well described by traditional techniques of lattice dynamics. However, many physical phenomena can be correctly described only by taking into account anharmonic terms in the potentials, which in turn requires to use approximate theories. By MD the dynamics of the system can be exactly solved: therefore the computer is useful also to model fundamental phenomena of chemical physics of solids.

Applications of the technique are limited by the maximum length which can be investigated: in general only phenomena with a characteristic length of 10-50 Å can be reproduced. On such a length scale, and with periodic boundary conditions it was necessary to constrain the focus on perfect or almost perfect crystals. In spite of these restrictions, the effort constituted the first attempt toward materials science, meaning a physics of materials. First applications were mostly concerned with atomic systems. Soon the technique was extended to deal with molecular solids[10]. Research spanned over several topics both in the bulk and on surfaces: structural stability and transformations, calculation of elastic properties, vibrations (phonons), structural defects, atomic diffusion and superionic conductors, melting and crystallization. In the present book the reader will find a collection of articles which cover all these arguments, plus some papers on the application of recent techniques of quantum simulation to solids.

1.3 Technique

Molecular Dynamics, in its most straightforward realization, is a very simple technique. Given the interaction potential and an initial configuration of N particles at time t, the resulting forces acting on each atom are calculated. Newton equations of motion are then numerically solved for a small time interval Δt under the assumption of constant force, obtaining the system configuration at time $t+\Delta t$. In the limit $\Delta t \rightarrow 0$ the solution is exact. The procedure can be iterated *ad infinitum*, and the evolution of the system can therefore be followed.

In its simplest and most used form, MD is based on the equations of motion derived by the classical Lagrangian

$$\mathcal{L} = \frac{1}{2} \sum_{i}^{N} m_i \dot{\mathbf{r}}_i^2 - \sum_{i>j}^{N} U(\mathbf{r}_{ij})$$

where m_i and r_i are mass and position of the i-th particle, and U(r) is the interaction potential. A simple discretization of the corresponding equations of motion, first used by Verlet[11], is obtained by expanding in

Taylor series the position at time t+Δt and t−Δt :

$$\mathbf{r}(t+\Delta t) = \mathbf{r}(t) + \dot{\mathbf{r}}(t)\,\Delta t + \frac{1}{2}\ddot{\mathbf{r}}(t)\,\Delta t^2 + O(3)$$

$$\mathbf{r}(t-\Delta t) = \mathbf{r}(t) - \dot{\mathbf{r}}(t)\,\Delta t + \frac{1}{2}\ddot{\mathbf{r}}(t)\,\Delta t^2 - O(3)$$

By summing the two terms and rearranging, one has

$$\mathbf{r}(t+\Delta t) = 2\,\mathbf{r}(t) - \mathbf{r}(t-\Delta t) + \ddot{\mathbf{r}}(t)\,\Delta t^2 + O(4)$$

and by subtracting

$$\dot{\mathbf{r}}(t) = \frac{\mathbf{r}(t+\Delta t) - \mathbf{r}(t-\Delta t)}{2\,\Delta t} + O(3)$$

Therefore, given the coordinates $\mathbf{r}_i$ at time t and at time t−Δt, it is possible to calculate the configuration at time t+Δt, and the velocities at time t. By using the energy equipartition theorem one can obtain the temperature of the system.

MD can be used to calculate time averages of physical quantities of interest, or as a powerful microscope to examine in detail local configurations and typical atomic trajectories. By adding a dissipative term to the equations of motion the method can also be used to find local minima in the potential energy (see e.g. R7 and R8). MD can be also be a tool for studying non-equilibrium phenomena.

The typical number of particles used in such simulations is of the order of 1000: if they are contained in a cube, almost half of them are on the cube's faces. Since most often the interest is focussed on bulk properties, it is necessary to impose periodic boundary conditions to avoid surface effects. This fact is not without consequences: while averages taken on a system in free boundary conditions correspond to a statistical ensemble where the pressure P is fixed to zero, (trivial) periodic boundary conditions imply that the volume V is constant. More precisely, as pointed out by Ray and Rahman[12], the corresponding statistical ensemble is NEh, meaning that the conserved quantities are the number of particles N, the energy E and the dynamical variable h (a matrix such that $\det h = V$) which describes shape and size of the MD cell. For many years MD calculations have been performed only in this setting.

In the last ten years much work has been devoted to the generalization of the method to different statistical ensembles. Such generalization are obtained by writing an *ad hoc* Lagrangian, and then showing that the equation of motion obtained from it generate the trajectories of the desired statistical ensemble. The first extension led to constant pressure MD[13]. For such system the Lagrangian is

$$\mathcal{L} = \frac{1}{2}\sum_{i}^{N} m_i \dot{\mathbf{r}}_i^2 - \sum_{i>j}^{N} U(r_{ij}) + \frac{1}{2} W \dot{V}^2 - PV$$

where W is a coupling parameter which can be though as the mass of a piston, and the volume V becomes an additional dynamical variable.

Today it is possible to study systems at constant temperature, strain, volume or pressure. Further extensions of the method make possible the use of MD to treat systems with frozen internal degrees of freedom, like rigid molecules[14].

MD does not allow to calculate some quantities like free energy differences which can instead be obtained, in certain cases, by MC. On the other hand, MD provides also information on dynamic properties which are unachievable by MC. As far as sampling of configuration space is concerned, a comparison[15] between MD and MC shows that the two techniques have a comparable efficiency.

Further details of the MD technique can be found in the original papers, collected in this book and in the already mentioned selection of reprints[1]. Moreover, an excellent and comprehensive introductions to the tricks and the art of simulation has recently appeared[3], and other very helpful books are available[4].

1.4 Philosophy

It took some time until the dignity of computational physics was recognized. As recently as in 1980 on the columns of the authoritative french journal "La Recherche", the Nobel laureate P.W.Anderson[16] spoke about the syndrome "de la machine enchantée". Such syndrome, in his view, is the illusion that it is possible to explain a physical phenomenon by simply solving a given number of fundamental equations which describe a single particle each. He argued that macroscopic phenomena cannot be analyzed as a simple "addition" of individual atomic microphenomena, and concluded that "la physique 'calculante' n'a donc aucune chance ... d'aboutir à un résultat d'un interêt quelconque, puisque sa règle consiste à individualiser les atomes." We believe that some truth can be found in these words, in that a whole range of phenomena are simply too complex to be represented starting from the atomic composition, although in principle this could be done (we shall come back to this point in section 3). On the other hand, it sounds ungenerous (and simply not true) to conclude that nothing has been learned from the use of computer to model physical phenomena.

The results obtained by computer simulation prove that in fact this approach can give important contribution to the understanding of physical phenomena. The best example is probably the theory of simple liquids, where the computational approach allowed to discover entirely new phenomena, like the solid-liquid transition for hard spheres, and the non-exponential decay of the velocity autocorrelation function. It also allowed to discriminate between well founded theories and interesting but incorrect ideas. In fact, simulation can provide *exact* results to problems which could otherwise only be solved by approximate methods, or which could otherwise be intractable. Of course it cannot solve all classes of problems: it should address those question which depend on a large but not huge number of degrees of freedom, and which are complicated enough that it is not *a priori* obvious that certain approximations actually work.

Computer simulation can be used with different philosophies in mind, which can be exemplified by two extreme cases. In the first approach (which we will call "experimental") a model is assumed neglecting whether it is accurate for reproducing some real material or phenomenon. The calculation gives the exact consequences of that model, without approximations. The prediction of theories can then be checked against the exact results, so as to control the validity of the theory and the importance of the approximations done by the theory itself. Such simulations are useful also in Mathematical Physics, where it is sometimes useful to know if a conjecture is likely to be true before trying to demonstrate it.

The second approach instead consists in actually trying to reproduce experimental results starting from a model, and check the exact predictions of the model, as obtained from the calculation, against the experimental results. Simulation is hence in this conception a theoretical tool which is used to measure the validity of models, and to predict new facts. It is probably for this dual nature of simulation that computational physicists tend to be considered as experimentalists from theoretician, and as theoreticians from the experimentalists!

These two views of computer simulation emerge from the original papers: often authors speak, even in paper titles, of "computer experiments". The meaning is clearly expressed in Verlet's words[11]: "The *exact* machine computation relative to classical fluids have several aims: it is possible to realize *experiments* in which the intermolecular forces are known; approximate theories can thus be unambiguously tested and some guidelines are provided to build such theory whenever they do not exist".

Frequently such experiments are extended by playing with the system: one can introduce an *ad hoc* perturbation and then examine the response of the system to a perturbation. An example of such extended experiments is the already mentioned paper by Vineyard on radiation damage[8]. In the present collection of reprints such approach can be found for instance in a paper by Parrinello and Rahman[R2] where the interatomic potential is suddenly switched from argon to rubidium in the middle of the simulation, to show how the structure of the system is sensitive to the form of the interaction.

An example of "theoretical" attitude is well expressed in a paper by A. Rahman[17] on the properties of water: " The overall philosophy behind the technique of MD, applied to a phenomenon as complex as liquid water, should be the achievement of reasonably good agreement with the experimental values of several static and dynamic properties. Given such agreement, if calculations, in addition, provide evidence of subtle, hitherto unsuspected, properties, it can be considered worthwhile to direct experimental effort towards such properties.".

These two approaches are not in contrast, being rather complementary. Their blend gives the unique flavor which makes computer simulation a fascinating discipline.

2. THE PAPERS

2.1 Crystal Structure and Polymorphic Transformations

Should all crystals be BCC? This question was raised by Alexander and McTague in a well known paper[18]. Based on the Landau theory of second order phase transitions, they argued that, as a matter of principle, nucleation of Body-Center Cubic (BCC) crystalline phase should be common to all materials. A.Rahman used MD as an obvious tool to answer such questions. This was the beginning of a series of applications of MD as a good technique for checking the relative stability of phases, which was fruitful of physical results and technical improvements of the method. BCC nucleation had in fact been observed by Hsu and Rahman in a MD study of 500 particles interacting via a pair potential suited for liquid rubidium[19]. The same authors[R1] extended this study by monitoring the nucleation process starting from the same liquid configuration and using four different potentials: rubidium, rubidium truncated, Lennard-Jones and Lennard-Jones truncated. Only in the first case they observed BCC nucleation: in all other cases the outcome was a Face-Centered Cubic (FCC) crystal.

The use of traditional NEh-MD for calculations as the one reported above implied long times in which the system lived in a glassy, metastable state. Periodic boundary conditions and fixed box can in fact impose conditions which hinder the realization of the stable state. If, for example, the equilibrium phase was incommensurate with the side of the simulation box, it would be impossible to grow a true crystal during the simulation. The use of constant pressure MD[13] can help to reproduce more realistic conditions, and in fact some relevant study was made possible by this technique, like the investigation of pressure-induced transformation from crystalline into amorphous ice[20], or the study of molecular systems[21]. However any transformation requiring anisotropic expansions of the box walls is still be forbidden, and internal stresses accumulate and cannot be released. Such conditions are realized, for instance, in transitions between two crystalline structures, where the shape of the corresponding elementary cells are different (e.g. in a transition from orthorombic to monoclinic). This problem was solved by Rahman and Parrinello[R2] by generalizing the idea which lead to the constant pressure MD (CPMD). While volume fluctuations are allowed in Andersen's CPMD , the ratios between the edges of the simulation cell remain constant, Rahman-Parrinello method allows instead to change the shape of the cell: therefore periodic boundary conditions do not introduce any artificial stress into the system. The first application of this method was concerned again with the transition from BCC to FCC when the potential was switched from rubidium to Lennard-Jones.

A further refinement of the technique allows to exert arbitrary external stresses on the system[R3]. This extension was achieved introducing into the Lagrangian new dynamical variables corresponding to the stress tensor components. The new method completed the evolution of MD as a tool for studying structural transitions in simple solids.

Among the applications of this method, the reader will find in this collection of reprints a paper on high pressure nitrogen[R4] and one on high pressure helium[R5]. In this last paper the method was extended by Nosé and Klein to more complex solids: namely to molecular crystals.

Relevant work on structure of molecular crystals includes an extended study on one of the best known plastic crystals: adamantana[22,23]. Adamantana ($C_{10}H_{16}$), which has a highly symmetric crystalline structure, transforms on heating to a high temperature plastic phase in which the molecules are dynamically orientationally disordered. The amount of orientational disorder may vary significantly when two different plastic crystals are compared: molecular orientations may change by rotational jumps between several equilibrium positions or vary continously by rotational diffusion.

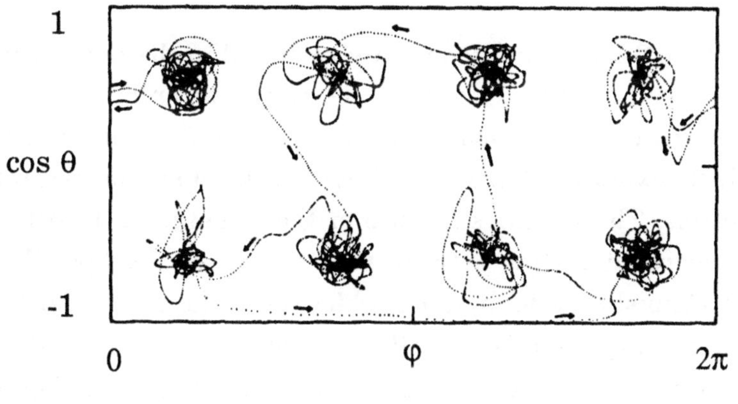

Fig. 1 - Time evolution of bond orientation in adamantana (from Ref.23)

Other studies of structural stability by MD have been concerned with surfaces. An example of such applications is a recent work[R8,24] on gold surface reconstruction. Besides the usual two-body interaction, in this work the authors successfully introduced a many-body force called "glue" to take into account the electronic cohesion.

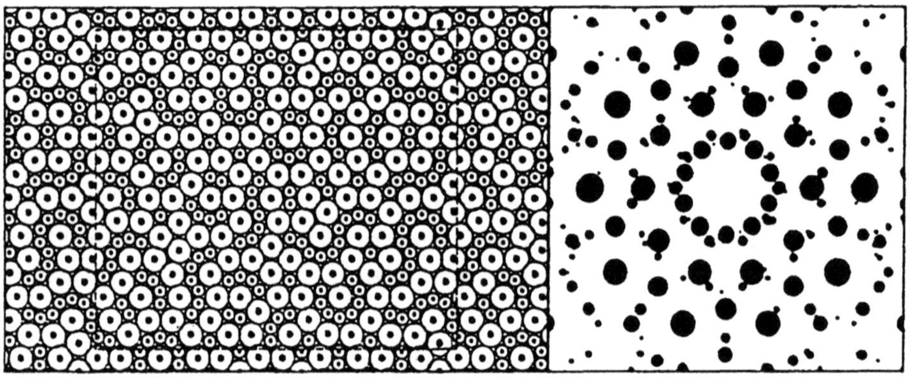

Fig.2 Quasicrystalline configuration (left) and its diffraction pattern (right). (From Ref. 26)

For a long time MD has been used mostly as an aid in the investigation of already rather well studied system. Recently the technique has been applied also to newly discovered materials: quasicrystals[25], a new phase of the matter which exhibits sharp diffraction peaks and non-crystallographic symmetries. The atomic structure of these materials is still unknown: it is therefore interesting to try to show that it is indeed possible to find reasonable atomistic model systems which show the main structural features of observed quasicrystals. Investigations by MD have shown[26] that such atomistic models exist and that they can in fact be stable. Figure 2 shows one such configuration. Other studies could establish the behavior of quasicrystalline materials under homogeneous stress[R7].

The idea behind the study of stability by MD is that, given enough time, some driving force will lead the system to its stable configuration. This is certainly true provided that internal stresses can be relaxed, and in the limit $t \rightarrow \infty$. For any finite time instead there is the danger that the system remains trapped in some metastable state. A better way to check the relative stability is therefore the direct calculation of the free energy difference. While this task is trivial at zero temperature, at finite temperature the entropic contribution makes it rather complex. However, a technique has been developed[27] to evaluate the free energy difference between two not too different systems. The application of this method has made possible to evaluate the relative stability of FCC and BCC structures for model systems at high temperatures.[R6]

2.2 Phonons and Elastic Constant

The dynamical structure function $S(\mathbf{q},\omega)$ which can be experimentally obtained by neutron scattering contains important informations on the dynamics of the system. For instance, its peaks give informations about the phonons in the system, and its dispersion curve yields the sound velocity. Van Hove has shown[28] that the $S(\mathbf{q},\omega)$ is related to the classical space-time correlation function $G(\mathbf{r},t)$ defined as

$$G(\mathbf{r},t) = \frac{1}{N}\left\langle \sum_{i,j} \delta\left(\mathbf{r} - \mathbf{R}_i(t) + \mathbf{R}_j(0)\right) \right\rangle$$

where N is the number of particles, brackets indicate statistical average and $\mathbf{R}_i(t)$ gives the position of particle i at time t. In particular, $S(\mathbf{q},\omega)$ can be obtained as space-time Fourier transform of $G(\mathbf{r},t)$. (A didactic introduction to these topics can be found in a recent book by Cusack[29]). Since all the informations needed to calculate the space-time correlation function are available from MD, it is possible to study phonons by computer simulation. The first calculations of $S(\mathbf{q},\omega)$ by MD were concerned with liquids[30]. Hansen and Klein[31,R11] presented the first applications to rare-gas solids.

In a paper[R12] on NaCl the predictions of the quasiharmonic approximation (which allows harmonic frequencies to change with volume) and of the one-phonon approximation (which only keeps lowest order term in an expansion of $S(\mathbf{q},\omega)$) were checked against exact results obtained by simulation at four different temperatures, giving a detailed analysis of anharmonic effects.

As examples of phonon calculations in molecular systems and on surfaces, the reader will find in the present collection a reprint on the orientationally disordered solid β–N_2[R13] and a paper on surface phonons in Xe overlayers on an Ag surface[R14].

Calculation of elastic constants were made possible by the constant stress MD due to Rahman and Parrinello: fluctuations of the elastic strain in such ensemble yield in fact the elastic modulii, as shown in R9. The paper is important because it defines the method. However, calculations based on its straightforward application are affected by serious convergence problems. A refined method[R10] improves efficiency by using the constant entropy MD only to let the system choose the average shape of the simulation cell: further calculations are performed with a traditional fixed-cell MD. This procedure allowed Rahman et al. to accurately calculate for the first time elastic constants of a solid. The main results for this kind of calculation are on sodium[32] and on crystalline[33] and amorphous[34] silicon. The problem of the calculation of elastic constants by MD is discussed in depth in a long paper by J.R.Ray[35].

2.3 Lattice Defects and Atomic Diffusion

Atomic diffusion in crystalline solids is substained by lattice defects. Point defects are generally mobile and erratically traverse the lattice, leaving behind a trail of displaced atoms. The effect is an atomic mean square displacement which increases with time, and a related diffusion coefficient D ($<R^2>=6Dt$). The statistical mechanical formulation of the diffusion problem traditionally breaks into two parts:

a) the concentration of thermally generated point defects, and the associated free energy of formation;

b) a rate theory description of the elementary atomic jump process.

More complete treatments do address the role of dynamic correlations between subsequent jump events. Below Debye temperature, thermal disorder in crystals can be subjected to analytic treatment (Lattice Dynamics). This task still requires the solution of an N-body problem. However, it can be carried out on modern machines for N~1000 by standard computer routines performing function minimization and matrix diagonalization. At temperatures at which diffusion is important, anharmonic contributions to lattice defect thermodynamics are large. Machine calculations such as MC and MD are exact in this respect. In addition, MD is the primary tool to investigate the dynamics. Trajectories of the representative point in 6-N dimensional phase space can be generated and analyzed. Jump frequencies can be measured in a simple way. The actual path taken in the jump process can be monitored. All questions connected with crystal memory and persistence in the direction of motion of the defect, multiple jumps and isotope effects, are open to direct observation.

The first article we have included[R15] is one of Bennett's milestone contributions[27,36] to the applications of computer simulation technique to the field of point defect diffusion. R16 is an example (taken from metals) of the richness of results to be harvested on atomic diffusion with MD-based methods.

R17, R18 and R19 are remarkable examples of proficous use of a theoretical framework to complement, support and interpret simulation data. R20 exemplifies surface point defect diffusion cases, while R21 addresses a somewhat different case of the dynamics of disorder, taken from molecular solids.

Reviews of part of the field are available elsewhere[37].

2.4 Motion of Ions in Superionic Conductors

Superionic conductors are interesting materials since, in spite of being solid, they exhibit a rather strong ionic conduction. Moreover, they are of technological importance. Both because of their intrinsic disorder and because of their strong anharmonicity it is difficult to study them with analytic theories. They are therefore a perfect subject for computer simulation, and in fact the contribution given by computer simulation to the understanding of the mechanisms responsible for this behavior has been very important, and they are still subject of active investigations. The first simulations of superionic conductors did not deal with AgI, the most representative element of this class, being it easier to obtain reasonable potentials for other compounds. Once again the precursor was Anees Rahman[R22]: his "feasibility study", a calculation on CaF_2 dates back to 1976. This first paper showed evidence for diffusion of F^- ions, while no diffusion was observed for Ca^+ ones.

Further investigations on the same system performed by Jacucci and Rahman[R23] were important in clarifying the mechanisms of ionic conductivity. An explanation often put forward for the behavior of these materials was that the ionic conductivity arises from the melting of one of the sublattices, while the other maintains its solid state. Simulation allowed instead (through a detailed analysis of F^- ions motion) to discover that diffusion occurred via jumps between sites of a simple cubic lattice. Time of flight was negligible with respect to the time spent on the lattice sites, and successive jumps were generally uncorrelated in space and time. Lattice defects were identified as important actors in the diffusion mechanism. The role of defects has been confirmed in recent years in an extended study[38, R27] on the same compound. Fig. 3, from a didactic paper by Gillan[39], shows the diffusive motion of F^- (small circles) while the cations sit on lattice sites forming their Debye-Waller cloud. The hopping behavior of the anions was confirmed by calculations performed on $SrCl_2$[40] and on PbF_2[41].

Simulations of AgI began[R24] right after the successful work on CaF_2, achieving good agreement with the experiments.

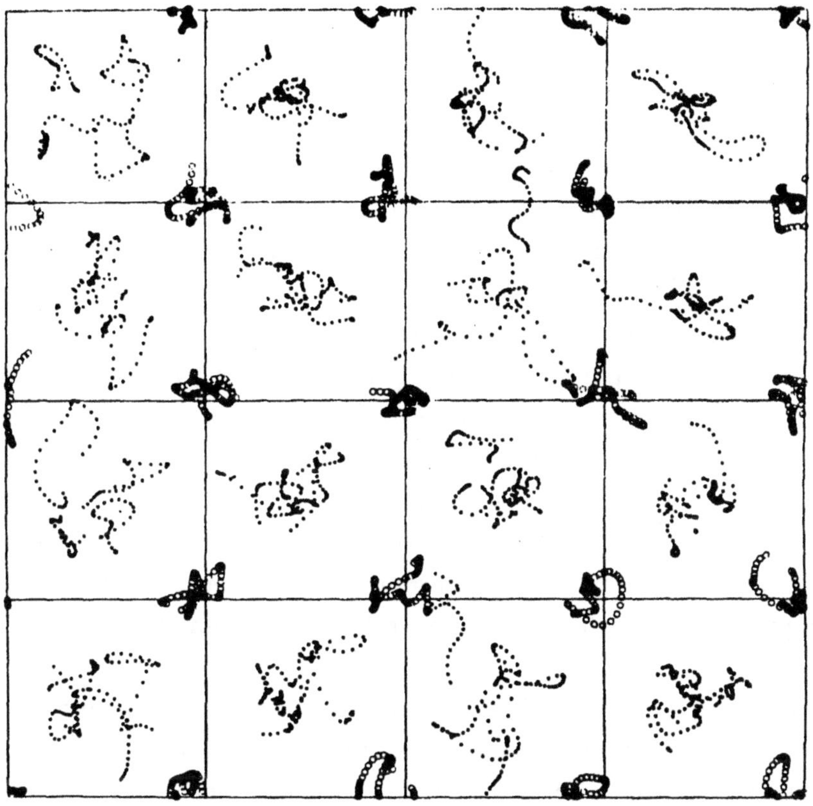

Fig. 3 Diffusion of atoms in CaF_2 (From Ref. 39)

Recent calculation on AgI[R28] could confirm the validity of the jump diffusion Chudley-Elliot model[42] and clarify the relation between Ag^+ diffusive motion and the spectrum of inelastic neutron scattering.

Structural properties of AgI were studied with the constant pressure MD. It was possible to reproduce the $\alpha \leftrightarrow \beta$ (wurtzite) phase transition[R25].

The $\beta \rightarrow \alpha$ transition was investigated[R26] also for Li_2SO_4. In the superionic α–phase this system revealed an interesting behavior: at large enough temperature cation hopping is presented together with rotations of the $SO_4^{=}$ group which lead to orientational disorder.

2.5 Melting and Crystallization

How does melting occur? Is it initiated in the bulk, maybe favoured by defects like vacancies? Is it the effect of an increased density of dislocations? Or is it rather initiated on surfaces? These question are still unsettled, and theoretical and experimental work is in progress. The answer might not be unique: there are in fact indications that different systems behave in qualitatively different ways. As an example, different behaviors have been observed in argon[43] and gold[R30]: while in fact surface-nucleated melting has been found in the first, no such phenomenon is observed on well-packed reconstructed (1 1 1) surface in the second. Moreover, different behaviors have also been observed on unequivalent surfaces of the same system[R31]. Problems in this area are complex, and it is difficult to build a comprehensive theory. For this reason, the role of simulation becomes crucial, since it is the only effective theoretical tool available: it may then happen to find a paper in which experimental and theoretical (simulation) work are presented together, as in the work by Landman et al.[R32].

Similarly complex is solidification: much work has been done for many years on the glass transition, i.e. on the freezing of liquids into a metastable amorphous solid as opposed to the formation of stable, ordered solids (crystals and maybe quasicrystals). We do not include any paper on glasses in this collection, the main focus being on crystalline solids. Crystallization was simulated by several authors (see references in R29), using the traditional fixed-box MD. Variations in mean pressure and temperature are not welcome in such studies: is is therefore important to use a different statistical ensemble: NPT (i.e. constant temperture and pressure), as done by Nosé and Yonezawa[R29] on a Lennard-Jones system.

As for melting, also in the case of crystallization one can observe different behaviors in different materials. Systems with isotropic potentials (as metals or rare gases) and systems with anisotropic bonds show different behaviors, crystal formation being much easier for the first class (and in fact in nature glasses are always associated with the second category, although in recent years technology has been able to produce metallic glasses, e.g. by ultrafast cooling). Such a difference is well presented in two papers by the Schneider et al. on epitaxial growth: the first[R33] being concerned with Lennard-Jones systems, and the second[R34] with silicon.

2.6 Quantum Simulations

Although traditional classical simulation achieved a remarkable success in calculating properties of condensed matter, extension of the simulation method to quantum mechanisms is important. In first place it allows to treat systems where quantum effects are dominant, as He and H. Moreover, the explicit treatment of electrons allows to treat from first principles those systems for which the approximation of pairwise additive potentials breaks down. The main techniques are the Ab-initio Molecular Dynamics method, developed by Roberto Car and Michele Parrinello[44], and the Quantum Monte Carlo methods. Although the main focus of this book is on MD, it is useful to give space also these methods, due to their importance.

2.6.1 *Ab-initio Molecular Dynamics*

The main assumption of MD is that atoms behave like classical particles which follow Newton equations of motion. Moreover, it is assumed that electronic motion is much more rapid than the motion of ions, so that the Born-Oppenheimer approximation holds, and the forces between atoms can therefore be obtained from a simple energy surface. In the simplest case, empirical 2-body potentials are sufficient to reproduce with good accuracy the properties of matter. Most computer simulations rely on such simple recipe. The most critical of these approximations turns out to be the description of the interatomic potential as superposition of two-body functions. More accurate results can be obtained in some cases by using more complex three-body potentials. Such solution is much more time-expensive, since the time needed to calculate the forces grows as N^3 rather than as N^2 with the number of particles N. Moreover it is difficult to obtain good three-body potentials. For these reasons simulations based on such potentials are not frequent in literature.

A more accurate approach would be to derive the interatomic potential directly from first principles, i.e. from the electronic ground state. This can be done by using Density Functional (DF) methods, which for given simple structures are extremely successful in obtaining accurate energy surfaces.

Let $\{\mathbf{R}\}$ denote the configuration of a multiatomic system, and let $\phi(\{\mathbf{R}\})$ be the classical Born-Oppenheimer potential energy surface for such system. The problem of finding $\phi(\{\mathbf{R}\})$ can be solved in the frame of (Local) DF theory by minimizing the total energy functional $E(n(\mathbf{r}),\{\mathbf{R}\})$ with respect with the occupied orthonormal single-particle orbitals ψ that represents the electronic charge density $n(\mathbf{r})=\sum_i \psi_i$.

The total energy functional E is the sum of five terms: the electrostatic interaction energy between the nuclei, the single particle (quantum) kinetic energy of the electrons, the interaction energy of the electrons with the external nuclear potential, the average electrostatic interaction energy between the electrons, and the exchange and correlation energy of the electrons. The first four terms can be exactly calculated: to deal with the last, which contains all the information on the many-body Fermi character of the problem, the so-called Local Density Approximation is used. It consists of assuming that the sought term is given by \I(„ **dr** n(**r**) $E_{xc}(n(\mathbf{r})))$ where $E_{xc}(n(\mathbf{r}))$ is the exchange and correlation energy per particle of a homogeneous electron gas with density n, which is known with sufficient accuracy. Once the functional E is known, the minimum problem of determining $\phi(\{\mathbf{R}\})$ is mapped onto the task of finding the solution of a set of self-consistent single-particle Schrödinger equations. At this point $\phi(\{\mathbf{R}\})$ could be used into the classical Newton equations, and the motion of the atoms can be calculated for a time step Δt, resulting in the new configuration $\{\mathbf{R'}\}$. The whole procedure should then be repeated to get the new potential $\phi(\{\mathbf{R'}\})$. DF calculations can in principle be coupled with other simulation techniques. One such example is reprint R37, where DF calculations together with quasiharmonic lattice dynamics allow to calculate the free energy formation of lattice vacancy in silicon. Unfortunately however, DF calculations are very computer-time intensive on themselves: it is therefore unthinkable to perform a separate self-consistent electronic minimization at every MD step. The application of DF to MD simulations seemed therefore to be unfeasible.

The approach known as "Car and Parrinello Method" (CP) overcomes this difficulty and efficiently combines MD and DF methods. In this approach the total energy E(n(**r**),{**R**}) is treated as the potential energy of a fictitious classical dynamic system consisting of nuclear plus electronic degrees of freedom: R and ψ. The corresponding equations of motion are

$$\mu\ddot{\psi}(\rho,\tau) = -\frac{\delta E}{2\delta\psi^*(r,t)} + \sum_i \lambda_{ij}\ \psi_i(r,t)$$

$$M\ddot{\mathbf{R}} = -\frac{\partial\phi}{\partial\mathbf{R}}$$

where μ is an adjustable parameter, M are the nuclear masses, and λ_{ij} are Lagrangian multipliers used to impose the orthonormality between the wavefunctions. These equations can be integrated numerically even for relatively complex systems.

So obtained nuclear trajectories do not coincide, in general, with those generated by the Newton equation of motion. However, one can choose the parameter μ so as to make the electronic dynamics sufficiently faster than the nuclear. In such conditions there is little mixing between the fast electronic and the slow nuclear degrees of freedom: thermal equilibration requires a very long time. As a consequence the nuclear trajectories deviate only very slowly from the Born-Oppenheimer surface, on which initially they lie. This "classical adiabatic dynamics" allows simulation of the motion of the nuclei without the need of a separate self-consistent electronic minimization at every step, overcoming in this way the main difficulty of coupling DF and MD.

A check that classical adiabatic dynamics is actually followed can be performed by constantly checking the temperature of the system. Unlike classical MD simulations, here one can define two different temperatures: an electronic temperature T_c, related to the average (classical) kinetic energy of the electrons

$$<K_c> = < \int dr\, \mu\, \dot{\psi}^2 >$$

and a nuclear temperature T related to the average nuclear kinetic energy

$$<K_i> = \frac{1}{2} < \sum M\dot{R}^2 >.$$

Under the conditions of classical adiabatic dynamics one must have $T_c << T$: in this way the electronic degrees of freedom simply oscillate around a minimum of $E(n(\mathbf{r}),\{\mathbf{R}\})$, while continually readjusting to the new minimum as the nuclei evolve. The temperature T corresponds to the physical temperature of the system, allowing therefore, as in classical MD, to simulate thermal treatments as annealing and quenching by simply varying the particle velocities.

The advantage of the CP scheme is not only in solving the problem of the interatomic potential, but also in allowing a more detailed information about the simulated system. Since both the nuclear trajectories and the electronic charge densities are generated, one can, for instance, follow directly the evolution of the chemical bonds resulting from atomic motion. The knowledge of electronic charge densities also allows to compute consistently electronic properties, which could obviously not obtained by classical MD simulations.

The CP method is very expensive in terms of computer time and memory: the investigation of a system with one hundred atoms currently requires one 20 seconds per step on a CRAY-YMP supercomputer, meaning that a full calculation easily needs more than 100 hours. However the method makes possible an investigation from first principles, which allows very accurate calculations, and gives important insight into materials of technological interest. Two important examples of such calculations can be found in this book: an investigation of structural, dynamic and electronic properties of amorphous silicon[R35] and a study of silicon microclusters[R36]. Many other applications of the technique, mostly due to the Trieste group, have followed the first ones: among these calculations on the structure of selenium and silicon clusters[45], proton diffusion in crystalline silicon[46] and structural characterization of amorphous Si and C[47].

2.6.2 *Quantum Monte Carlo*

The classic Monte Carlo system, known as Metropolis Monte Carlo, is based on a simple idea: the Boltzmann weight is used to guide random sampling through the relevant regions of configuration space. Statistical averages are then obtained by averaging over the collection of states which have been visited. A typical Metropolis MC program is very simple (can be less than hundred lines of code) and rather efficient, although more sophisticated and efficient algorithm have been devised and used.

The idea behind the Quantum Monte Carlo method is at the same time very simple and rather old, being suggested by Fermi in the late 40's. A good introduction to the technique can be found in a review paper by Lester and Hammond[48]. Here we shall briefly outline the method. The main point is the observation that the Schrödinger equation becomes a diffusion equation when written in imaginary time: its solution therefore can be found by generating a random walk. By mapping the time t into a variable s ($s=it/\hbar$), where i is the imaginary constant) the Schrödinger equation becomes

$$\frac{\partial\psi(\mathbf{r},s)}{\partial s} = D\nabla_{\mathbf{r}}^2\,\psi(\mathbf{r},s) - (V(\mathbf{r}) - E_T)\,\psi(\mathbf{r},s)$$

where $D=\hbar^2/2m$ plays the role of a diffusion constant, V is the potential, and E_T is an energy offset which is useful in the problem. When considering only the first member on the right hand side, the equation is an ordinary diffusion equation. If instead only the second term is considered on the right hand side, then the equation becomes a first-order rate, with rate constant ($V(\mathbf{r})-E_T$). The solution can therefore be sought as the combination of a diffusion and a branching process. The simulation proceeds by randomly moving in space a set of representative points (called walkers or psips). Then, after each diffusion step, branching occurs and walkers are either replicated or killed according with the rate constant. In this way the density of walkers is increased in regions where V is smaller than E_T, and decreased elsewhere. In the course of simulation the number of walkers can vary dramatically, according to E_T. This value is continously adjusted during the simulation, hopefully converging to the energy eigenvalue of the ground state. In the infinite time limit, the system reaches a stationary limit, and the density of walkers gives a representation of the wavefunction, achieving the full knowledge of the system.

A method to decrease fluctuations and to reduce the statistical error consists in using a trial function $\phi(\mathbf{r})$. By multiplying the Schrödinger equation by $\phi(\mathbf{r})$ and readjusting in terms of $\varphi(\mathbf{r},s)= \psi(\mathbf{r},s)\phi(\mathbf{r})$, one gets:

$$\frac{\partial\varphi(\mathbf{r},s)}{\partial s} = D\nabla^2_{\mathbf{r}}\, \varphi(\mathbf{r},s) - D\nabla_{\mathbf{r}} \left(\varphi(\mathbf{r},s)\, F(\mathbf{r})\right) + \left(E_T - E_L(\mathbf{r})\right) \varphi(\mathbf{r},s)$$

where $E_L(\mathbf{r}) = \phi^{-1} H\phi$ is the local energy and $F(\mathbf{r})=\nabla_{\mathbf{r}} \ln | \phi(\mathbf{r}) |^2$ plays the role of a driving force. Like before, also in this case one has a diffusion equation with a branching term. The diffusion equation is slightly more complicated, having a driving force term (as in the Schmoluchowsky equation) which drives walkers faster away from the unfavorable regions, so that convergence is improved (if the trial function is reasonably good).

In this form QMC method are useful to simulate boson systems. Treating fermions is more difficult, since ψ will also have negative regions, where density of walkers would be negative. The problem can be solved by using a trial function ϕ which has the same sign as ψ. In this case the product $\varphi= \psi\phi$ is always positive and the method can be applied again. The condition is implemented by never allowing a walkers to cross nodes of the trial function. This method, called fixed-node approximation, works well in the cases where the node locations are known (for instance because of symmetry considerations). Unfortunately, this is usually not the case for many-body systems.

QMC has been successfully used mostly for chemical calculation of atoms and molecules[48]. An early application to solids was a study of He[49]. A relevant QMC calculation of a Fermi system (solid H at high pressure) is reprinted here[R38]. Additional information on the method (including also Variational MC and Green's function MC) can be found in this reprint.

2.6.3 *Path-Integral Monte Carlo*

A different quantum method which also makes use of Monte Carlo is based on path integrals. In Feynman's formulation of quantum mechanics a quantum particle of mass m is mapped into a chain of N representative points linked with harmonic springs. The spring constant is equal to $mN(k_BT/\hbar)^2$, where k_B is the Boltzmann constant and T is the temperature of the system. Low temperatures correspond therefore to loose springs: the representative points spread around like the wave function of the system. A quantum system can therefore be simulated by using classical chains: sampling can be performed by either classic Monte Carlo or Molecular Dynamics. In this case though MD does not provide any real dynamic information, being only a mean to explore the relevant portion of configuration space. Of course also this kind of quantum simulation is much heavier than classic simulations, because each quantum particle is represented by N classical points. The isomorphism between the quantum statistical system and the classic chain is exact in the limit $N\rightarrow\infty$, but in practice very good results can be obtained with finite N.

Although most application of path integrals simulation are concerned with liquids[50], important work concern solids, in particular light elements (as H in metals[51,R40] and He[52,R39]) for which quantum effects are important.

3. THE CRYSTAL BALL: FUTURE DEVELOPMENTS

Computer technology is still progressing, and faster and faster machines will be available in near future. Computer speeds range at present from one MegaFlop (one million FLoating point OPerations per seconds) of a microVAX to the 200 MegaFlops of a CRAY. GigaFlops (one billion flops) machines will be soon available, and TeraFlops computers (1000 billions Flops) have been announced for a near future. Several attempts have been done to build dedicated machines, i.e. machines having a hardware architecture specifically designed to reflect software architecture of a specific problem. Although this approach is appealing, we feel that the results are so far not outstanding, since in the time needed to design and build these machines, there is usually a development of general purpose computers which allows to achieve similar results on the specific problem, while maintaining flexibility. We therefore believe that general-purpose supercomputers will be the machines of choice for state-of-the-art simulation also in future. General-purpose parallel architectures will make possible large speed gains, and the development of algorithms for simulation on these machines has started[53].

Graphics is likely to become a very useful tool for simulation. Sometimes, in fact, it is much easier to understand qualitative behavior of complex systems by inspecting visual representations of relevant parameters, rather than trying to directly extract informations from the numbers. Fast graphic workstations are now available, and new software standards (e.g. X-Window) allow to visualize in real time on a local graphic workstation the results of a calculation performed on a remote mainframe.

The gap between computational power of small, inexpensive, single user machines (workstations and fast personal computers) and top-of-the-line, expensive and very fast supercomputers has been reducing over the last ten years. This trend seems not to be over. While however we should expect that the main research effort will continue to be done on medium and large computers, there should be an important fall-out on didactics, using the very same algorithms for building "conceptual laboratories". It already is possible to visually follow the dynamical evolution of a two-dimensional system of up to 100 particles in real time on a personal computer like an Apple Macintosh based on a Motorola 68030 processor, which costs today less than $3000: we can expect this computational power to be available for less than $1000 within the next three years. Molecular Dynamics plus graphics can therefore become an easy available, wonderful tool for teaching some physics, from the foundation of mechanics to those of statistical mechanics and solid state.

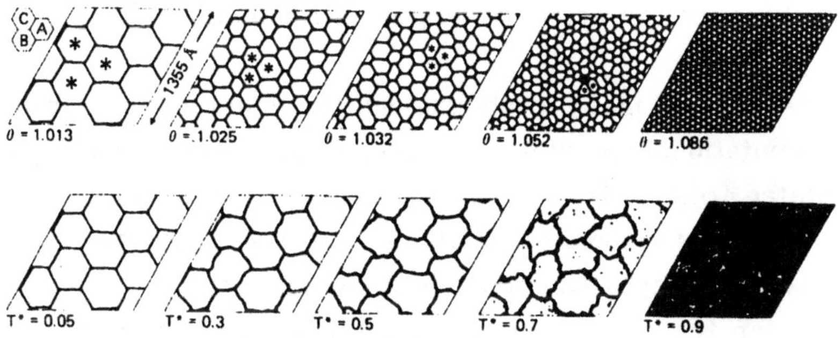

Fig 4. Kripton domain formation on a graphite surface (from Ref. 54)

In spite of these progresses, will computers ever become Anderson's "machine enchantée", which allows to approach every kind of physical phenomenon starting from an entirely microscopic description? At first sight one would foresee a dramatic improvement in the ability of computer simulation to treat more and more complex systems, and in fact some very heavy calculation was already made possible in the last few years by the increased computational power. An example is provided by the work of F.Abraham et al.[54] who studied the incommensurate phase of kripton on graphite using more than 100'000 atoms, achieving a system size comparable with the dimension of experimentally studied samples (which was possibible since the problem was quasi-two dimensional).

Unfortunately it is easy to realize that too optimistic expectations do not correspond to reality. Let's in fact assume that time needed to perform a simulation have linear dependence in the number of particles (in reality it is slightly worse than linear for traditional MD with neighbors tables, and definitely worse for ab-initio MD). In this approximation, each factor 1000 in speed allows to increase the linear size of a three-dimensional system by a factor 10, going from today's 40-100 Å to 1000 Å with teraflops machines. Such size is not enough to treat on an atomistic basis non homogeneous systems (micro-crystallites). Even worse is the situation close to phase transitions, where the characteristic lengths of fluctuations diverge.

There are of corse many problems which would benefit from an increased computational speed: for instance, all problems which require a longer simulated time. A factor 1000 in computer speed would allow to reach simulated times of the order of 10^{-6} sec on systems of approximately 1000 particles. Such improvement would allow to reach the experimental cooling rates for the production of metallic glasses. Moreover, even relatively small gain in speed would be very useful for ab-initio MD, allowing to grow from a few tenth of particles to system sizes comparable to those which are today achieved in an ordinary MD simulation.

However, in spite of all the possible advances in computer speed, it seems that computer simulation in its atomistic form will not be able to tackle a vast class of phenomena: those connected with materials science. There are fields which, due to their spatial or statistic complications, need computer modelling: the range between 1μ and 100μ is inaccessible to statistical mechanics. The physics of this range, which cannot be described on a microscopic basis, and cannot be represented by statistical mechanics is now known as mesoscopic physics. Its complications are intrinsic, and the continuum representation breaks down. The very same phenomena mentioned by Anderson (catalysis, corrosion) and others (like formation and migration of dislocations, grain boundaries, microfractures) are those which need new modelling and testing. Computer simulation, in a new form, is exactly what can help solving the problems. We have learned in fact by now that simulation, in the double function of testing theories and testing models, can give an invaluable support to the advance in understanding the world. We expect therefore that in the near future new techniques will be developed to bridge the gap between microscopic and macroscopic, giving birth to a computer simulation of materials at the *mesoscopic* level, as opposed to macroscopic and microscopic levels.

Such techniques will be different from both macroscopic and microscopic approaches. Comparing the methodology of numerical modelling of continuous bodies (as a field in applied mathematics and engineering) and that of atomistic modelling of condensed matter (as a field in solid state physics and statistical mechanics), differences in empirical input and founding equations are apparent.

In finite element treatments of deformations of continuous bodies, the constitutive properties of materials (e.g. stress vs. strain plots) are inputted into continuity and conservation equations. Mechanical properties (elastic and plastic behavior, fluid flow), thermal properties (heat flows) and their coupling can thus be computed. Complications can be introduced in the models by geometrical and surface effects as well as by the presence of regions characterized by different constitutive properties.

In atomistic modelling of matter, on the other hand, atomic interactions derived either from fitting of experimental behavior or from ab initio calculations are used in the classical equations of motion of particle assemblies. The resulting equation of state yields mechanical and thermodynamic properties of the material under study. Transport coefficients can also be computed by this method. Some complications as interfaces and board can be introduced at the atomistic level.

However atomistic modelling of materials at the micron scale is not practical, while the homogeneous continuum model is often inapplicable in real materials much below the millimeter scale. Furthermore, microstructure is often too complex to be described with the introduction in the Finite Elements model of multiple different spatial regions. As a result, materials science and engineering lack well established computer modelling and simulation methods precisely at the length scale where interesting chemical physics phenomena often occur, i.e. at the mesoscopic level. Surface oxidation, powder compaction and sintering, plasma spraying deposition of coatings, are further examples of important processes related to materials where mesoscopic phenomena appear. The properties of these materials are irreducible to equivalent homogeneous models using statistical properties and distribution functions. This is related to nonlocal features of the network of pores, cracks, channels, that are essential to any useful description.

Computer simulation at the mesoscopic level involves drops and grains as actors. Their mechanical and thermal properties (e.g. splatting, solidification, heat transfer) become an empirical input. The successive happening of stochastic events influenced by steric effects and complemented by qualitative rules of behavior (helping to decide the outcome of individual events) are the constituents of a procedure replacing continuum equations and particle equations of motion in mesoscopic simulation. First cases in which procedural rules are included in simulation programs can already be found, for instance in a program modelling the fabrication process of plasma sprayed coatings described by J.H.Harding et al.[55]. Examples of such rules are:

- the curvature of solid splats and the magnitude of their surface stresses is made to depend on the surface temperature during deposition;
- incoming drops can or cannot fill voids upon splashing, depending on the form of the local surface structure made up of previous splats;
- adhesion or crack-forming can result from different impact conditions with the deposition surface.

Application of such rule-oriented simulations requires help also from Artificial Intelligence disciplines, as expert systems, learning algorithms and qualitative physics. Also on this front, some investigation has started[56].

The hope is therefore that important developments will in future concern processes that occur at the mesoscopic scale, opening a whole new area to computer simulation.

REPRINTED ARTICLES

R1 C.S.Hsu and A.Rahman, J.Chem Phys **71**, 4974-4986 (1979).

R2 M.Parrinello and A.Rahman, Phys.Rev.Lett. **45**, 1196-1199 (1980).

R3 M.Parrinnello and A.Rahman, J.Appl Phys **52**, 7182-7190 (1981).

R4 S.Nosé and M.L.Klein, Phys.Rev.Lett. **50**, 1207-1210 (1983).

R5 D.Levesque, J.J.Weis and M.L.Klein, Phys.Rev.Lett. **51**, 670-673 (1983).

R6 A.Rahman and G.Jacucci, Nuovo Cimento **4D**, 357-381 (1984)

R7 F.Nori, M.Ronchetti and V.Elser, Phys.Rev.Lett. **61**, 2774-2777 (1988)

R8 F.Ercolessi, E.Tosatti and M.Parrinello, Phys.Rev.Lett. **57**, 719-722 (1986).

R9 M.Parrinello and A.Rahman, J.Chem.Phys **76**, 2662-2666 (1982)

R10 J.R.Ray, M.C.Moody and A.Rahman, Phys.Rev.B **32**,733-735 (1985)

R11 J.P.Hansen and M.L.Klein, Phys.Rev.B **13**, 878-887 (1976)

R12 E.R.Cowley, G.Jacucci, M.L.Klein and I.R.McDonald, Phys.Rev.B **14**, 1758-1769 (1976)

R13 M.L.Klein, D.Levesque and J.J.Weis, J.Chem.Phys. **74**, 2566-2568 (1981)

R14 G.G.Cardini, S.F.O'Shea, M.Marchese and M.L.Klein, Phys.Rev.B **32**, 4261-4263 (1985)

R15 C.H.Bennett Thin Solid Films **25**, 65-70 (1975)

R16 A.Da Fano and G.Jacucci, Phys.Rev.Lett. **39**, 950-953 (1977)

R17 G.De Lorenzi and G.Jacucci, Phys.Rev.B **33**, 1993-1996 (1986)

R18 G.De Lorenzi, G.Jacucci and C.P.Flynn Phys.Rev.B **36**, 9461-9468 (1987)

R19 M.Marchese, G.Jacucci and C.P.Flynn Phys.Rev.B **36**, 9469-9481 (1987),

R20 G.De Lorenzi, G.Jacucci and V.Pontikis, Surf.Science **116**, 391-413 (1982).

R21 J.P.Ryckaert, M.L.Klein and I.R.McDonald, Phys.Rev.Lett. **58**, 698-701 (1987)

R22 A.Rahman, J.Chem.Phys.**65**, 4845-4848 (1976)

R23 G.Jacucci and A.Rahman, J. Chem.Phys.**69**, 4117-4125 (1978).

R24 P.Vashishta and A.Rahman, Phys.Rev.Lett. **40**, 1337-1340 (1978)

R25 M.Parrinello, A.Rahman and P.Vashishta, Phys.Rev.Lett. **50**,1073-1076 (1983).

R26 R.W.Impey, M.L.Klein and I.R.McDonald, J.Chem.Phys.**82**, 4690-4698 (1985).

R27 M.J.Gillan.in *"Transport Structure Relaxation in Fast ion and Mixed Conductors Proceedings of the 6th Risø International Symposium on Metallurgy and Materials Science"*, Risø National Laboratory, Roskilde, Denmark, September 1985, p.461-466.

R28 G.L.Chiarotti, G.Jacucci and A.Rahman, Phys.Rev.Lett. **57**, 2395-2398 (1986).

R29 S.Nosé and F.Yonezawa, J.Chem.Phys.**84**,1803-1814 (1986).

R30 P.Carnevali, F.Ercolessi, E.Tosatti, Phys.Rev.B **36** 6701-6704 (1987)

R31 F.F.Abraham and J.Q.Broughton, Phys.Rev.Lett. **56**, 734-737 (1986)

R32 U.Landman, W.D.Lüdtke, R.N.Barnett, C.L.Cleveland, M.W. Ribarsky, E. Arnold, S.Ramesh, H.Baungart, A.Martinez and B. Khan, Phys.Rev.Lett. **56**, 155-158 (1986).

R33 M.Schneider, I.K.Schuller and A.Rahman, Phys.Rev.B **36**, 1340-1343 (1987).

R34 M.Schneider, I.K.Schuller and A.Rahman, in *Interfaces, Superlattices and Thin Films* Mat.Res.Soc.Symp.Proc. **77**, 91-98, Pittsburgh (1987)

R35 R.Car and M.Parrinello, Phys.Rev.Lett. **60**, 204-207 (1988).

R36 P.Ballone, W.Andreoni.R.Car and M.Parrinello, Phys.Rev.Lett. **60**, 271-274 (1988).

R37 G.B.Bachelet, G.Jacucci, R.Car and M.Parrinello, in *The Physics of Semiconductors,* World Scientific (1987), p.801-804

R38 D.M.Ceperley and B.J.Alder, Phys.Rev.B **36**, 2092-2106 (1987)

R39 D.M.Ceperley and G.Jacucci Phys.Rev.Lett. **58**, 1648-1651 (1987).

R40 M.J.Gillan, Phys.Rev.Lett. **58**, 563-566 (1987)

OTHER REFERENCES

1 G.Ciccotti, D.Frenkel and I.R.McDonald eds., *Simulation of Liquid and Solids, Molecular Dynamics and Monte Carlo methods in Statistical Mechanics*, North Holland 1987

2 W.G. Hoover and G.Ciccotti eds., *Molecular Dynamics Simulation of Statistical Mechanics Systems. Proceedings of the International School of Physics E.Fermi* North-Holland 1986

3 M.P.Allen and D.J.Tildesley *Computer Simulation of Liquids*, Clarendon Press 1987;

4 J.P.Hansen and I.R.MacDonald, *Theory of Simple Liquids*, Academic Press, London 1976;

W.G. Hoover, *Molecular Dynamics*, Lecture Notes in Physics 258, Springer Verlag Berlin 1986;

5 E.Fermi, J.Pasta and S.Ulam, Los Alamos Report LA-1940 (1955), later published in *E.Fermi: Collected Papers* (Chicago, Ill.) 1965

6 N.Metropolis, A.W.Rosenbluth, M.N.Rosenbluth, A.H.Teller and E. Teller, J.Chem. Phys.. **21**,1087 (1953)

7 B.J.Alder and T.E.Wainwrigth, J.Chem.Phys. **27**, 1207 (1957)

8 J.B.Gibson, A.N.Goland, M.Milgram and G.H.Vineyard, Phys.Rev. **120**,1229 (1960)

9 A.Rahman, Phys.Rev. **136**, A405 (1964)

10 For a review see M.L.Klein, Ann.Rev.Phys.Chem. **36**, 525 (1985)

11 L.Verlet, Phys.Rev. **159**, 98 (1967)

12 J.R.Ray and A.Rahman, J.Chem.Phys. **80**, 4423 (1984)

13 H.C.Andersen, J.Chem.Phys. **72**, 2384 (1980)

14 For a review of this method see G.Ciccotti and J.P.Ryckaert, Computer Phys.Reports, **4**, 347 (1986)

15 G.Jacucci and A. Rahman, Il Nuovo Cimento **4D**, 357 (1984)

16 P.W.Anderson, La Recherche 11, 98 (1980)

17 A.Rahman and F.H. Stillinger, Phys. Rev.A **10** 368 (1974)

18 S.Alexander and McTague Phys.Rev.Lett. **41**,702 (1978)

19 C.S.Hsu and A.Rahman J.Chem.Phys. **70**,5234 (1979)

20 J.S. Tse and M.L.Klein Phys.Rev.Lett. **55**, 1672 (1987)
21 S. Nosé and M.L.Klein J.Chem.Phys. **78**, 6928 (1983)
22 M.Meyer and G.Ciccotti Molecular Physics, **56**, 1235 (1985)
M.Meyer, in Ref. 2 p. 477
23 M.Meyer, C.Marhic and G.Ciccotti Molecular Physics, **58**, 723 (1986)
24 F.Ercolessi, M.Parrinello and E.Tosatti, Surface Sci. 177, 314 (1986)
25 For a review see M.Ronchetti, Phil.Mag. **56**, 237 (1987) and the book series *Aperiodicity and Order*, M.V.Jarič ed., Academic Press 1989, 1990
26 F.Lançon, L.Billard and P.Chaudhari, Europhys.Lett. **2**, 625 (1986)
F.Lançon and L.Billard, J.Phys.France **49**, 249 (1988)
27 C.H.Bennett, J.Comp.Phys. **22**, 245 (1976)
G.M.Torrie, J.P.Valleau, J.Comp.Phys. **23**, 187 (1977)
28 L.van Hove, Phys.Rev. **95**, 249 (1954)
29 N.E.Cusack, *The Physics of Structurally Disordered Matter: An Introduction*, Adam Hilger (Bristol 1987)
30 D.Levesque, L.Verlet and J.Kürkijarvi, Phys.Rev.A **7**, 1690 (1973)
31 J.P.Hansen and M.L.Klein, J.Phys.Lett. 35, L-29 (1974)
32 S.K.Schiferl and D.C.Wallace, Phys.Rev.B 31 (1985) 7662
33 M.D.Kluge, J.R.Ray and Rahman A., J.Chem.Phys 85 (1986) 4028
34 M.D.Kluge, J.R.Ray, Phys.Rev.B 37 (1988) 4132
35 J.R.Ray, Computer Physics Reports, 8, 109 (1988)
36 C.H.Bennett, in *Diffusion in Solids: Recent Developments*, A.S.Novick and J.J.Burton eds., Chap.2, Academic Press (New York 1975)
C.H. Bennett, Colloq.Metall., **19**, 65 (1976)
C.H. Bennett, in *Algorithms for Chemical Computations*, R.E. Christoffersen ed., Chap.4, Americal Physical Soc. (Washington DC 1977)
37 G.Jacucci, in *Nontraditional Methods in Diffusion*, Murch, Birnbaum and Cost eds. The Metallurgical Society of AIME, (Warrendale PA-USA 1984) p.259
G.Jacucci, in *Diffusion in Crystalline Solids*, Murch and Nowick eds., Academic Press (London 1984) p.429

38 M.J. Gillan J.Phys.C Solid State Phys. **19**, 3391(1986) and **19**, 3517 (1986)

39 M.J. Gillan, Physica **131 B,** 157 (1985)

40 M.Dixon and M.J.Gillan, J.Phys.C.Solid State Phys. **11**, L165 (1978)

41 A.B.Walker, M.Dixon and M.J.Gillan, J.Phys.C Solid State Phys. **15** 4061(1982)

42 C.T.Chudley and R.J.Elliot Proc.Phys.Soc. **77,** 353 (1961)

43 R.E.Allen, F.W.DeWette and A.Rahman, Phys.Rev. **179**, 887 (9179)

44 R.Car and M.Parrinello, Phys.Rev.Lett. **55**, 2471 (1985), and in *Proceedings of the 18th International Conference on Physics of Semiconductors*, O.Engström ed., Vol.**2**, 1165, World Scientific 1987

45 D.Hohl, R.O.Jones R.Car and M.Parrinello Chem.Phys.Lett. **139**, 540 (1987)

D.Hohl, R.O.Jones,R.Car and M.Parrinello, J.Chem.Phys. **89** 6823 (1988)

P.Ballone, W.Andreoni,R.Car and M.Parrinello,Phys. Rev.Lett. **60** 271 (1988)

P.Ballone, W.Andreoni,R.Car and M.Parrinello, Europhys. Lett. **8**,73 (1989)

W.Andreoni, G.Pastore, R.Car, M.Parrinello,and P.Giannozzi, in *Band Structure Engineering in Semiconductor Microstructures*, R.A.Abram and M.Jaros eds, Plenum, (1989)

46 F.Buda, G.L.Chiarotti, R.Car and M.Parrinello, Phys. Rev. Lett. **63**,294 (1989)

47 R.Car and M.Parrinello, Phys. Rev. Lett. **60** 204 (1988)

G. Galli, R.M.Martin, R.Car and M.Parrinello, Phys. Rev. Lett. **62**,555 (1989)

48 W.A.Lester, B.L.Hammond, "Quantum Monte Carlo for the electronic structure of atoms and molecules", to appear in Ann.Rev.Phys. Chem.

49 M.H.Kalos, M.A.Lee P.A.Whitlock, Phys.Rev.B. 24, 115 (1981)

50 For a review see B.Berne and D.Thirumalai Ann.Rev.Phys.Chem. **37**, 401 (1986)

51 M.J.Gillan, Phil.Mag. A, 1988, 257 (1988)
52 F.F.Abraham and J.Q.Broughton, Phys.Rev.Lett. **59**,64 (1987)
53 D.C.Rapaport, Comp.Phys.Reports, **9**, 1, (1988)
D.L.Greenwell, R.K.Kalia, J.C.Patterson and P.D.Vashishta, in *Scientific Applications of the Connection Machine*, H.D.Simon ed., World Scientific 1989
54 F.F.Abraham, W.E.Rudge., D.J.Auerbach and S.W.Koch, Phys.Rev. Lett. 52, 445 (1984)
55 J.H.Harding, S.Cirolini and G.Jacucci, Rapporto Tecnico del Progetto Finalizzato del C.N.R. "Calcolo Vettoriale e Parallelo in Meccanica Statistica" (Roma, 1990)
56 G.Jacucci and C.Uhrik, J.Phys.Chem., **91**, 4981 (1987)

Reprinted Articles

Interaction potentials and their effect on crystal nucleation and symmetry[a)]

C. S. Hsu and Aneesur Rahman

Argonne National Laboratory, Argonne, Illinois 60439
(Received 27 April 1979; accepted 21 August 1979)

Molecular dynamics technique has been used to study the effect of the interaction potential on crystal nucleation and the symmetry of the nucleated phase. Four systems, namely rubidium, Lennard-Jones, rubidium-truncated, and Lennard-Jones-truncated, have been studied each at reduced density 0.95. Two types of calculations were performed. Firstly, starting from a liquid state, each system was quenched rapidly to a reduced temperature of ~0.1. The nucleation process for these systems was monitored by studying the time dependence of temperature and the pair correlation function, and the resulting crystalline structure analyzed using among other properties the Voronoi polyhedra. Only in the case of rubidium was a b.c.c. structure nucleated. In the other three cases we obtained f.c.c. ordering. Secondly, we have studied the effect of changing the interaction potential in a system which has already achieved an ordered state under the action of some other potential. After establishing a b.c.c. structure in a rubidium system, the change in the symmetry of the system was studied when the pair potential was modified to one of the other three forms. The results from both types of calculations are consistent: the rubidium potential leads to a b.c.c. structure while the other three potentials give an f.c.c. structure. Metastable disordered structures were not obtained in any of the calculations. However, the time elapse between the moment when the system is quick-quenched and the moment when nucleation occurs appears to depend upon the potential of interaction.

I. INTRODUCTION

In a molecular dynamics study of a system of 500 particles interacting via a pairwise potential appropriate for an alkali metal, namely rubidium, we have previously[1] reported the homogeneous nucleation of an ordered stable bcc phase from a liquidlike state of aggregation. In that study, one of the motivations was to gather evidence regarding the arguments presented recently by Alexander and McTague[2] which suggest that the Landau theory of second order phase transitions can be used to predict the nucleation of a body-centered cubic phase as a matter of general principle. The potential used in Ref. 1 was constructed for rubidium metal using the theory of electron screening effects in simple alkali metals by Price *et al.*[3]; with this potential the calculation[1] gives a bcc structure on nucleation. In the case of a Lennard-Jones system, Raveché and Streett[4] obtained an fcc structure (under high temperature, high density conditions) whereas Mandell *et al.*[5] obtained a bcc structure with the same potential. More recently molecular dynamics calculations have been carried out by Stillinger and Weber[6] on the Gaussian core model near its solid–fluid transition region and the results indicate that the stable phase has a bcc structure. Tanemura *et al.*,[7] whose polyhedral analysis has been used here and in Ref. 1, studied homogeneous nucleation in a system of soft (r^{-12}) spheres and observed the occurrence of both bcc and fcc structures.

It is apparent that further progress toward clarifying the above mentioned arguments of Alexander and McTague[2] requires firstly a resolution of the contradictory conclusions between Refs. 4 and 5 and secondly more work on systems interacting via other kinds of potentials. With this end in view, we have completed in addition to the work reported in Ref. 1, a study of three systems, which are named Lennard-Jones, rubidium-truncated and Lennard-Jones-truncated, depending on the potential function used. We find that starting from liquidlike states, all three systems nucleate into face-centered cubic phases. In order to further clarify the effect of the interaction potential on the crystalline structure, we made the following three calculations. After a bcc structure is well established for the rubidium system, the potential is modified to one of the three above mentioned forms and the crystalline structure is monitored. In all three cases, the structure changes to an fcc form. In one case, the process has been checked for reversibility, in the sense that the structure reverts back to a bcc form on switching back the potential to its original form: namely, the rubidium potential.

In Sec. II, the molecular dynamics method will be outlined and the quantities monitored during the calculations will be defined. In Sec. III we describe the preparation of the "quick quenched" states and the results of the time evolution thereafter. A calculation on the Lennard-Jones system at a temperature and density close to the values used by Mandell *et al.*[5] will be presented in Sec. IV. In Sec. V we present the calculations related with the effect of modifying the potential after the structure is well established for the rubidium system. The paper will be concluded with a discussion in Sec. VI. The method of eliminating the effect of thermal motion on shapes of Voronoi polyhedra is appended.

II. MOLECULAR DYNAMICS OUTLINE

The pair potentials $V(r)$ used in our calculation are shown in Fig. 1. For the rubidium system,[8a] the depth ϵ of the potential has a value 555.89×10^{-16} erg and the distance σ, where $V(r)$ crosses the axis for the first time is 4.4048 Å. For the Lennard-Jones system, the particles interact through the pair potential: $V(r) = 4\epsilon[(\sigma/r)^{12} - (\sigma/r)^6]$. In this paper the potentials referred to as truncated potentials are defined in each

[a)]Work supported by the U. S. Department of Energy.

 0021-9606/79/244974-13$01.00

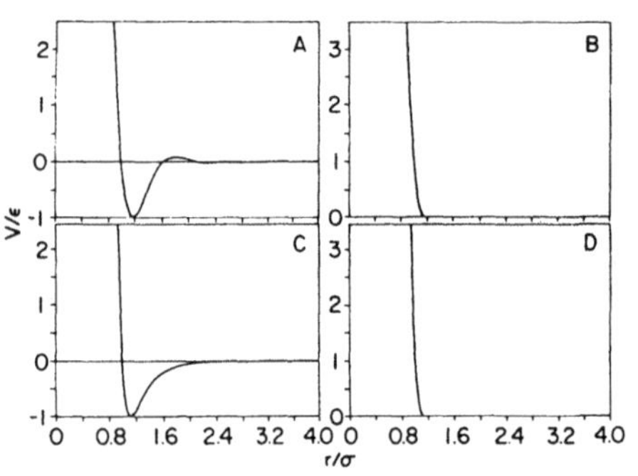

FIG. 1. Pair potentials used in the calculation. (a) rubidium potential; (b) rubidium-truncated potential; (c) Lennard-Jones potential; (d) Lennard-Jones-truncated potential. For rubidium unit of length $\sigma = 4.4048$ Å and of energy $\epsilon = 555.89\times10^{-16}$ erg.

case by elevating the corresponding potential by amount ϵ and truncating it at the minimum. We shall use the following abbreviations hereafter; RBF for rubidium, RBT for rubidium-truncated, LJF for Lennard-Jones, and LJT for Lennard-Jones-truncated. For the units of energy and length, we shall use ϵ and σ respectively. The unit of mass will be $M = 141.92\times10^{-24}$ g, the atomic mass of rubidium. Using these units, all physical variables will be expressed in their reduced, dimensionless form and will be explicitly clarified only when reduced values are not used. The unit of time τ is defined as $(M\sigma^2/\epsilon)^{1/2}$. For rubidium the value of τ is 2.2257 ps. The molecular dynamics calculations were made on 500 particle systems confined to a cubic cell and subjected to periodic boundary conditions. The reduced density was fixed at $\rho = 0.95$ for all systems. The range of interactions in both RBF and LJF systems was chosen at 2.3. The equations of motion were then integrated using time step Δt such that the total energy for each system is virtually a constant over several tens of thousands of Δt. The actual value of Δt will be given below. $T(t)$ will be used to specify the temperature at time t.

In addition to the temperature, the square of the displacement vector $\mathbf{R}(t) - \mathbf{R}(0)$ in the 1500 dimensional configuration space of the system from the point at which the calculation was started (with due account being taken of the periodic boundary conditions) was also monitored during each calculation and will be denoted by $R^2(t)$. The average pair correlation function for a certain span of time in each run was monitored to locate both the change of structure and the structural identification. This way of monitoring the calculation is for the purpose of studying the changes in the structure of the system as time proceeds. In addition, the structure factor analysis and the Voronoi polyhedra[7] construction were also performed. As mentioned in Ref. 1, Voronoi polyhedra have a very well defined signature for a body-centered cubic structure even under thermal agitation but this is not so for a face-centered cubic system. A detailed description of the problems arising out of thermal agitation has been given by Tanemura *et al.*[7] Hence to make the polyhedral construction useful in the present context, the construction has to be suitably modified; a method of removing the effect of thermal motion on the shapes of Voronoi polyhedra will be presented in the Appendix; this is only slightly different from the method used in Ref. 7.

III. QUENCHED STATES AND THEIR NUCLEATION

Throughout this part of the calculations, $\Delta t = 0.0075$ was used for RBF and RBT systems, and $\Delta t = 0.01$ was used for LJF and LJT systems.

A. Preparation of the quick quenched states

We are interested in comparing properties such as crystalline structure, etc., of different systems studied by molecular dynamics; with this end in view, in all cases the calculations were started in the same fashion: we started with one single state at $\rho = 0.9042$ and $T = 1.02$ with interaction RBT. For a 5040 Δt run with $\Delta t = 0.0075$, the behavior of $R^2(t)/N$, N being the number of particles, gives a self-diffusion constant $D \approx 0.075$. By comparing this value with the diffusion constant 0.028[8b] of RBF system at the same density but lower temperature (0.79), we know that this state is a highly mobile liquidlike state. The end configuration of this run is used to carry out the following calculations.

In a molecular dynamics calculation, the density of the system can be modified if necessary by scaling the cell size, i.e., all the position coordinates; the temperature (i.e., kinetic energy) can be modified by suitably scaling the velocities of the particles. After preparing an RBT system at $\rho = 0.9042$ and $T = 1.02$ as mentioned above, the end configuration was used to start a separate run for each of the three potentials,

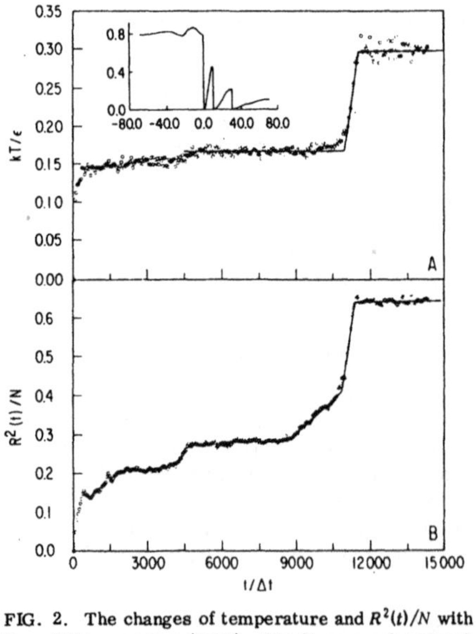

FIG. 2. The changes of temperature and $R^2(t)/N$ with time for the rubidium system (RBF). The lines are for visual guidance. Δt is the molecular dynamics integration step. This figure is taken from Ref. 1.

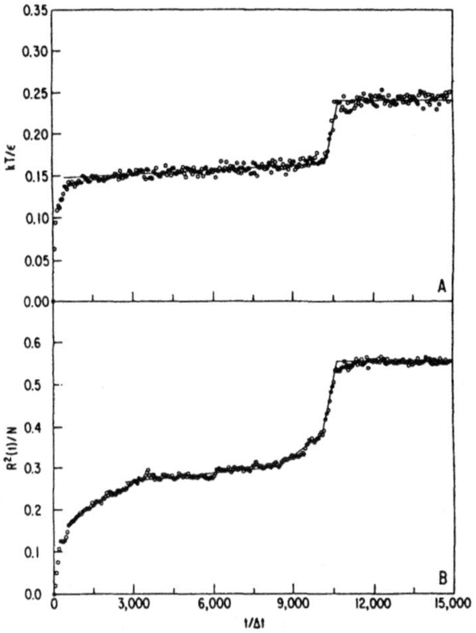

FIG. 3. The changes of temperature and $R^2(t)/N$ with time for a Lennard-Jones system (LJF_1) prepared by quenching quickly from temperature 1.0 to 0.1. The lines are for visual guidance. The length of this run is 22 970Δt. Here we only show the changes up to 15 000Δt by which time the nucleation has already taken place.

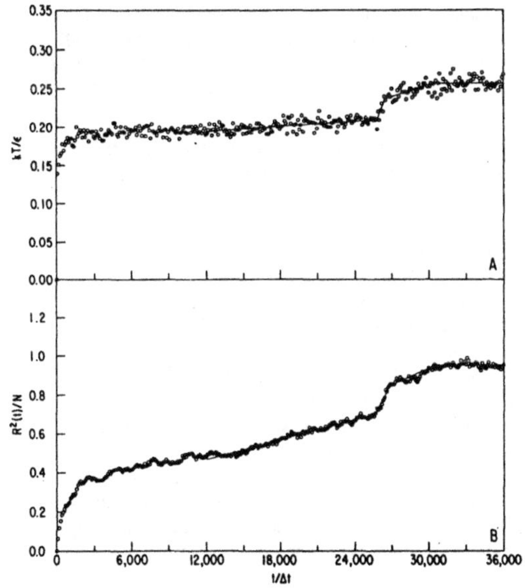

FIG. 4. The changes of temperature and $R^2(t)/N$ with time for a rubidium-truncated system (RBT) prepared by quenching quickly from temperature 1.0 to 0.1. The lines are for visual guidance.

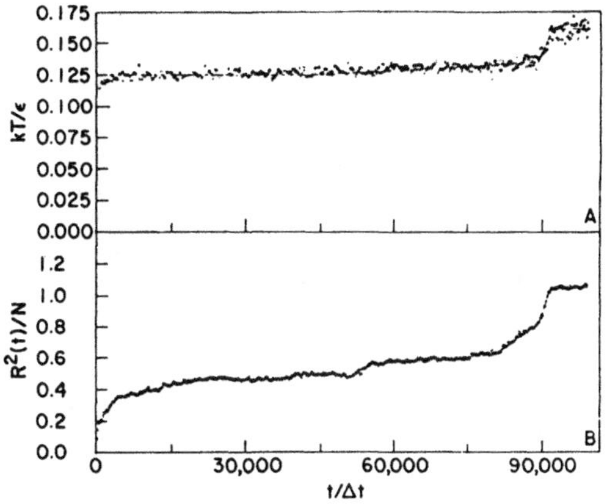

FIG. 5. Same as Fig. 4 but for a Lennard-Jones-truncated system (LJT).

namely LJF_1, RBT, and LJT. (In view of the calculation to be presented in Sec. IV below, we shall refer to the LJF system being discussed here as LJF_1.) In each case the density was increased to $\rho = 0.95$ at the start and the temperature was maintained at $T = 1.0$ for about 200Δt. Denoting the time at the end of this preparatory period as $t = 0$, each of the three systems was quick quenched as follows.

At $t = 0$ all velocities are put to zero and the system is allowed to run for 10Δt. Then all velocities are again put to zero followed by a run of another 20Δt. At this point, all velocities are once more reduced to zero. The system is then allowed to proceed without any further disturbance, i.e., with constant total energy.

B. Nucleation

In Figs. 2, 3, 4, and 5, we present the time variation of temperature $T(t)$ and displacement square $R^2(t)/N$ for RBF (taken from Fig. 3 of Ref. 1), LJF_1, RBT, and LJT, respectively. We witness that in each case the system has undergone a transition to some ordered state around a certain time t_c for each system. The values of t_c are listed in Table I. We notice that (i) in the full potential systems, namely RBF (Fig. 2) and LJF_1 (Fig. 3), the changes of $R^2(t)/N$ around times t_c are quite dramatic, while in the truncated potential systems, namely RBT (Fig. 4) and LJT (Fig. 5), the values of $R^2(t)/N$ increase gradually and the jumps

TABLE I. The estimated time t_c of relaxation before nucleation in systems mentioned in Sec. III B. The value for RBF is taken from Ref. 1.

	RBF	LJF_1	RBT	LJT
Δt	0.0075	0.01	0.0075	0.01
$t_c/\Delta t$	~11 000	~10 500	~26 500	~91 000
t_c	~80	~105	~200	~910

around times t_c are very mild; (ii) if we view the interval between the time immediately after the quick quenching and the time t_c as a relaxation process, Figs. 2, 3, 4, and 5 indicate that the relaxation rate is faster (i.e., t_c is smaller) in the full potential systems than it is in the corresponding truncated potential systems; (iii) in addition, the LJ systems, namely LJF_1 and LJT, have slower rates of relaxation than the corresponding RB systems, namely RBF and RBT respectively; (iv) we note that LJT stands apart from the other three in its relaxation: it takes more than 4 times longer than RBT and about 10 times longer than RBF or LJF_1.

It is however clear that quantitative estimates of the above mentioned differences in relaxation times can be obtained only if nucleation events of this kind are observed by repeating such calculations several times.

As regards changes in the pair correlation function, we describe here the LJF_1 system which gives a typical picture. In Fig. 6, we present the average $g(r)$ over three blocks of time: two of the blocks cover periods of time before the moment of nucleation and one after nucleation. Figure 6(a) gives the structure in the early stages of relaxation which lasts for a fairly long time (up to about $8000\Delta t$ for LJF_1, see Fig. 3). The structure reminds us of the structure of an amorphous Lennard-Jones state[9] with a double peak at distances $r/d \approx \sqrt{3}$ and $\sqrt{4}$ where d is the distance of the first maximum. The more interesting region in this figure is the part around the first minimum which is clearly asymmetric and gives indication of an incipient structure. Figure 6(b) gives the structure before but not too long before the occurrence of nucleation. Here, it is clear that

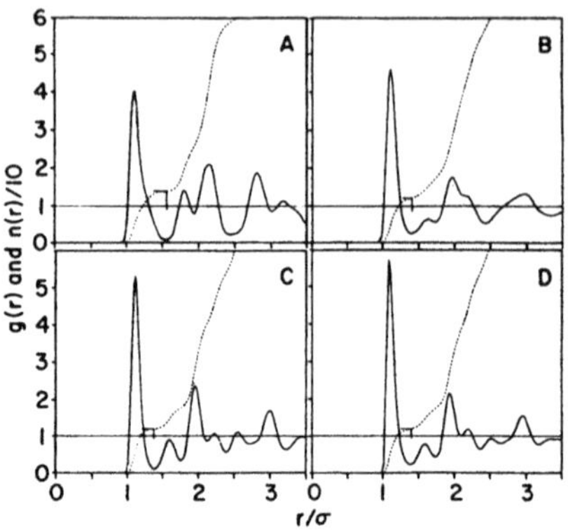

FIG. 7. The pair correlation (solid curve) and the coordination number (broken curve) for each of the four systems under investigation after an ordered stable phase has emerged from a quick quenched liquidlike state of aggregation. (a) for the rubidium system (RBF) for time ranging from 12 000 to 14 000 Δt (from Fig. 4(d), Ref. 1); (b) for the rubidium-truncated system (RBT) for time ranging from 30 000 to 32 000 Δt (see Fig. 4); (c) for the Lennard-Jones system (LJF_1) for time ranging from 12 000 to 14 000 Δt (see Fig. 3); (d) for the Lennard-Jones-truncated system (LJT) for time ranging from 97 000 to 99 000 Δt (see Fig. 5). The markers point to the first minimum in $g(r)$ and the corresponding value of $n(r)$.

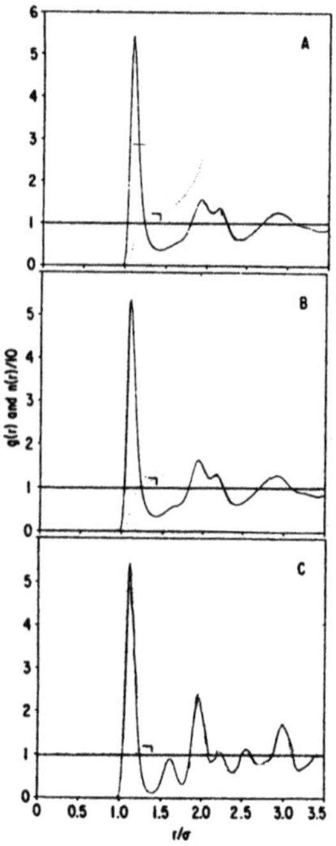

FIG. 6. The pair correlation for the Lennard-Jones system (LJF_1) for various blocks of time is shown as a solid curve. The blocks of time (see Fig. 3) for (a), (b), (c) are: 2000 to $4000\Delta t$, 8000 to 10 000 Δt, and 12 000 to 14 000 Δt respectively The broken curve shows the coordination number $n(r)$; the markers on this curve point to the first minimum in $g(r)$ and the corresponding value of $n(r)$.

there is a shoulder developing in the region immediately beyond the first minimum. The center of this shoulder does appear at a distance corresponding to the next nearest neighbor distance in an fcc configuration. After nucleation, we obtain the structure as shown in Fig. 6(c). For RBT and LJT, similar results were obtained. In Fig. 7, we display the pair correlation functions for all four systems, namely RBF, LJF_1, RBT, and LJT, after nucleation. We comment on these figures in Sec. III C below.

Following the method used by Mandell *et al.*,[5] the structure factor analysis was also performed. The structure factor $S(\mathbf{k};t)$ is defined as

$$S(\mathbf{k};t) = N^{-1}\left|\sum_{j=1}^{N} \exp i\mathbf{k}\cdot\mathbf{r}_j(t)\right|^2 .$$

The allowed values of $\mathbf{k}$ lie on the mesh $(2\pi/L)(\alpha,\beta,\gamma)$ where L is the length of the molecular dynamics cell, α, β, γ being integers. Given a configuration we monitor all integer triplets such that $65 \le (\alpha^2+\beta^2+\gamma^2) \le 100$. Among all the $\mathbf{k}$ thus generated, we pick out the seven vectors which give the highest seven values of $S(\mathbf{k};t)$. Results for LJF_1, RBT, and LJT around the moment of nucleation are presented in Table II. (Similar results for RBF are displayed in Table I of Ref. 1.) The dramatic increase in the values of several $S(\mathbf{k};t)$ around t_c clearly indicates the occurrence of nucleation.

C. Symmetry

In Table III we show the shell distances and the corresponding coordination numbers for perfect fcc and bcc structures. In Fig. 7, we have presented the pair cor-

TABLE II. The seven vectors which give the highest values of $S(\mathbf{k};t)$ as a function of t. $\mathbf{k}_i$ denotes the integer triplet (α, β, γ) and S_i denotes $S(\mathbf{k}_i;t)$, LJF_1, RBT, LJT denote the systems mentioned in Sec. III and LJF_2 the one in Sec. IV. The dotted line indicates the time of nucleation.

	$t/\Delta t$	$\mathbf{k}_1$	S_1	$\mathbf{k}_2$	S_2	$\mathbf{k}_3$	S_3	$\mathbf{k}_4$	S_4	$\mathbf{k}_5$	S_5	$\mathbf{k}_6$	S_6	$\mathbf{k}_7$	S_7
LJF_1	8000	0 8 4	64	4 −5 6	53	−1 8 3	42	−4 0 8	42	−5 0 7	38	−4 −8 4	38	5 −6 4	34
	9000	0 8 4	96	−4 0 8	50	−8 5 2	49	8 3 2	44	4 −5 6	43	−5 0 7	35	−4 −8 4	35
	10000	0 8 4	121	8 3 2	95	−4 −8 4	68	−4 0 8	66	5 −5 6	59	−8 5 2	55	−5 0 7	45
	11000	8 3 2	345	0 8 4	336	−8 5 2	256	−4 −8 4	203	−4 0 8	148	4 −6 5	124	3 2 9	121
	12000	0 8 4	374	8 3 2	354	−8 5 2	279	4 −6 5	191	3 2 9	139	−4 −8 4	121	−5 −1 7	117
	13000	0 8 4	400	8 3 2	350	−8 5 2	281	4 −6 5	187	−4 −8 4	151	3 2 9	136	−4 0 8	111
	22970	0 8 4	391	8 3 2	365	−8 5 2	293	4 −6 5	202	3 2 9	141	−4 −8 4	120	−5 −1 7	107
RBT	15000	8 0 4	37	1 4 8	35	−4 −5 6	32	5 −4 6	30	−4 −4 7	27	9 1 0	26	−1 −4 8	25
	20000	1 4 8	81	7 4 4	67	8 0 4	64	4 8 2	35	−4 −4 7	35	0 4 8	31	4 8 1	28
	25000	−7 4 4	136	1 4 8	136	8 0 4	114	3 −8 3	50	−4 −4 7	49	5 8 1	39	4 8 2	33
	26000	1 4 8	250	−7 4 4	209	8 0 4	191	5 8 1	66	0 9 1	66	−4 −4 7	61	−7 −5 3	60
	27000	−7 4 4	377	8 0 4	343	1 4 8	337	−7 −5 3	253	0 9 1	220	1 −5 7	152	−1 −4 8	94
	30000	0 9 1	337	−7 4 4	273	−7 −5 3	204	1 4 8	156	8 1 4	110	1 −5 7	108	−7 3 4	78
	33000	0 9 1	336	−7 4 4	257	−7 −5 3	195	8 1 4	116	1 4 8	101	1 5 8	95	−7 3 4	81
	36000	0 9 1	334	−7 4 4	242	−7 −5 3	189	8 1 4	131	1 4 8	122	1 −5 7	104	−7 3 4	91
LJT	82500	1 9 1	46	4 −6 5	34	−5 −5 5	33	−4 4 7	30	−1 8 3	27	−7 −4 4	24	8 −3 2	24
	85000	4 −6 5	72	−5 4 6	47	1 9 2	44	−5 −6 4	42	−4 4 7	41	−9 1 0	40	1 9 1	33
	87500	1 9 2	92	−5 −6 4	79	4 −6 5	71	−5 4 6	70	−4 4 7	61	−5 −5 5	58	−6 −5 4	37
	90000	1 9 2	194	−5 4 6	190	−6 −5 4	150	5 3 7	75	4 −7 4	72	4 −6 5	64	−5 −5 5	62
	92500	−6 −5 4	380	4 −7 4	252	6 3 6	232	1 9 2	157	−5 4 6	142	−4 5 6	111	5 2 6	37
	95000	−6 −5 4	404	4 −7 4	223	6 3 6	189	−5 4 6	169	1 9 2	139	−4 5 6	106	7 4 5	40
	99450	−6 −5 4	407	1 9 2	113	−5 4 6	110	4 −7 4	107	6 3 6	103	0 8 3	89	−6 3 7	79
LJF_2	10000	−8 −3 2	43	−2 −4 8	41	−4 4 7	40	0 −9 1	39	4 7 5	38	−4 5 6	28	1 −9 2	25
	11000	−4 4 7	80	0 −9 1	58	−8 −3 1	43	4 7 5	43	6 −2 6	40	−8 −3 2	34	−2 −4 8	31
	12000	0 −9 1	209	−2 −4 8	99	−8 −4 2	94	−8 5 1	64	−2 5 7	58	−4 4 7	57	−8 −3 2	41
	13000	0 −9 1	310	−2 5 7	166	−2 −4 8	162	6 −2 6	111	−8 −3 2	104	6 −1 6	73	−2 −5 8	70
	23000	0 −9 1	323	−2 5 7	213	−2 −4 8	167	−8 −2 3	97	6 −2 6	85	−2 4 7	80	−2 −5 8	70
	29700	0 −9 1	302	−2 5 7	196	−2 −4 8	150	−2 4 7	95	−2 −5 8	89	−8 −2 3	80	6 6 4	71

relation functions $g(r)$ and the coordination numbers $n(r)$ for RBT, LJF_1, and LJT together with those for RBF from Ref. 1. By comparing the positions of successive minima and their corresponding coordination numbers with those in Table III, we notice that all but the RBF system nucleate into fcc structures. From the structures of $g(r)$ in Fig. 7, we notice that the crystalline structure of RBT is the least ordered one. This can also be seen from the behavior of the structure factor $S(\mathbf{k};t)$. Table II shows that the largest $S(\mathbf{k};t)$ for RBT after nucleation has the value around 330 whereas for LJF_1 and LJT the value is around 400. We recall that the largest $S(\mathbf{k};t)$ for RBF after nucleation has the value around 420 (Table I of Ref. 1).

We have also constructed Voronoi polyhedra[1,7] to study the crystalline structures. In the bcc lattice the Wigner–Seitz cell (we use the nomenclature Voronoi polyhedron in the general case) is defined by the first 14 neighbors of every particle and the cell is made out of eight hexagons and six squares. As in Ref. 7, we denote this kind of cell by the signature (0608). Here each vertex is defined by *three* planes and hence the signature is very stable under thermal agitation. This is not so for an fcc lattice where the Wigner–Seitz cell of 12 rhombic faces with signature (01200) is formed by 14 vertices, six of them being common to *four* planes. A tiny thermal motion will destroy all of these six vertices and make every one of the six into a set of vertices each of which is defined by three planes. Therefore, we have to modify the way of constructing the polyhedra to enable us to study fcc structures in the presence of thermal effects. The details are given in the Appendix, but the idea is simply to allow for thermal fluctuations in assigning a signature to each polyhedron. Using this technique, the signatures of Voronoi polyhedra in the LJF_1 system were determined; the time variation of the relative population of various signatures is shown in Figs. 8(a)–8(c). Since after nucleation (Fig. 8(c)) the dominant signature is (01200), we conclude that the structural order in the LJF_1 system is fcc. Similar results were obtained for both RBT and LJT systems.

D. The nature of close packing after nucleation

In the previous sections we have consistently used the term fcc when referring to the close packed structures obtained with LJF_1, RBT, and LJT. The overlap of the various shells seen in Fig. 7 indicates

TABLE III. Shell distances and coordination numbers in fcc and bcc lattices.

Structure	i	$\mathbf{r}_i$	r_i^2/r_1^2	n_i	$\sum_{j=1}^{i} n_j$
fcc	1	110	1	12	12
	2	200	2	6	18
	3	211	3	24	42
	4	220	4	12	54
bcc	1	111	1	8	8
	2	200	4/3	6	14
	3	220	8/3	12	26
	4	311	11/3	24	50

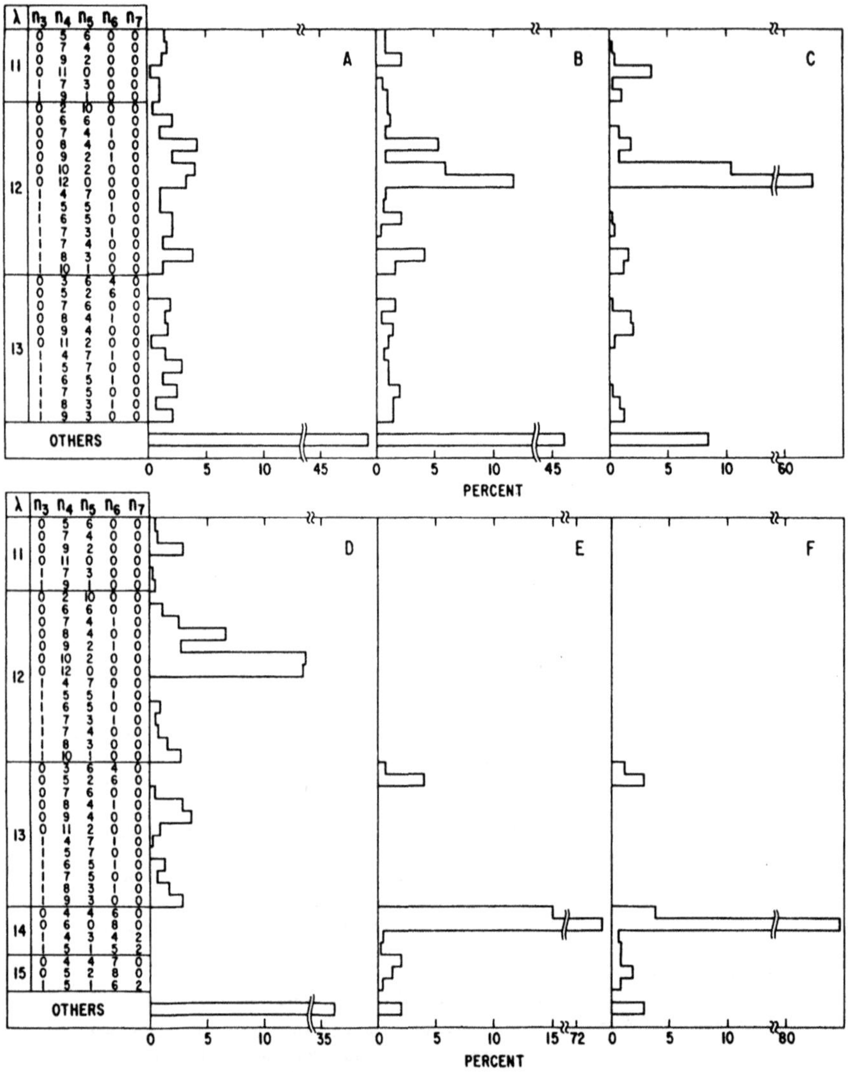

FIG. 8. Histograms of the distribution of signatures of Voronoi polyhedra for various systems. (a)–(c) for the Lennard-Jones system (LJF_1) of Sec. III at 8000 (before nucleation), 10 000 and 12 000 Δt (after nucleation), respectively; (d) for the Lennard-Jones system (LJF_2) of Sec. IV at 29 700 Δt. (e) for the rubidium system (RBF) of Sec. V A at t_1 (see Fig. 11); (f) for the rubidium system reported in Ref. 1 at 11 450 Δt. There is no overlap between the histograms (b), (c), (d) (pertaining to an fcc packing) and (e), (f) (pertaining to a bcc packing).

the presence of imperfections in the ordered structures. A visual impression of the structures can however be obtained as follows. Let **k** be the direction in which $S(\mathbf{k};t)$ takes on its maximum value (Table II) at the end of the run (e.g., vector (0, 8, 4) for LJF_1 at 22970 Δt). We have sliced the system of 500 particles with periodic boundary conditions into plates perpendicular to the direction $\mathbf{k}_1$ and plotted the planar coordinates of the particles perpendicular to $\mathbf{k}_1$ in each of these plates. These plots produce rather well defined networks formed out of equilateral triangles. Superposing these plots one can hope to see the presence of *abc* or *ab* type repetitions among successive slices perpendicular to $\mathbf{k}_1$.

It has been found that in all these plots apart from obvious point defects the *abc* type repetition is the most dominant. A typical plot is shown in Fig. 9(a). This is for the LJF_1 system of Sec. III B and clearly indicates that the close packed structure is fcc rather than hcp.

It is relevant to mention here that this manner of plotting the coordinates brings out the nature of the structure rather clearly. For comparison, we show

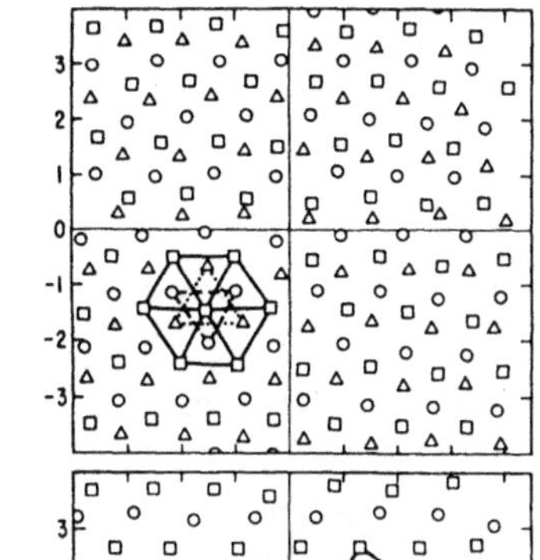

FIG. 9. The superposition of layers of particles viewed along the direction of the k vector which gives the largest $S(\mathbf{k};t)$ in Table II. Different symbols indicate the particles belonging to different layers, the order normal to the layers being □, ○, Δ. Connecting lines between symbols are for visual guidance. (a) for the Lennard-Jones system (LJF_1) of Sec. III at 22 970 Δt; (b) for the rubidium system reported in Ref. 1 at 11 450 Δt; (c) for the Lennard-Jones system (LJF_2) of Sec. IV at 29 700 Δt. The superposition of the triangles and the squares in the bottom right portion of Fig. 9(c) shows a region with an *ab* type repetition.

in Fig. 9(b) a plot for the RBF system (which was reported in detail in Ref. 1). The direction of k is obviously a 110 direction and when viewed along this direction each slice of a bcc structure shows up as a network of face centered rectangles, the positions of the particles in successive slices being related to each other through a simple translation. [Fig. 9(c) is relevant to Sec. IV below.]

E. Formation and growth of the nucleus

In Sec. III B and in Table I we have indicated the time t_c at which nucleation occurs in the four molecular dynamics systems being discussed (the nucleation event is clearly seen in Figs. 2–5). In Ref. 1 we have already reported on the formation and catastrophic growth of the nucleus in the RBF system. Since that system gave rise to a bcc structure the presence and the growth of the nucleus had to be monitored in terms of the presence of clusters of polyhedra with (0608) signature. In the LJF_1, RBT, and LJT systems obviously this had to be done with polyhedra having (01200) signature. (See Sec. III C above.)

Using the method which was explained in detail in Ref. 1, the growth of the nucleus was monitored for LJF_1 and was found to indicate a critical nucleus of about 50 particles and a growth at a velocity 0.33 (= the rate of advance of the surface of the growing nucleus). In the case of the two truncated systems the growth could not be monitored with the same degree of clarity. A rough estimate however indicated $v=0.16$ for RBT and $v=0.06$ for LJT.

IV. NUCLEATION IN A LENNARD-JONES SYSTEM AT A HIGHER TEMPERATURE

As mentioned in the previous section a Lennard-Jones system (referred to as LJF_1) with a density of 0.95 and quenched quickly from temperature 1.0 to 0.1 nucleated into an fcc structure. In view of the conclusions of Mandell *et al.*[5] that their calculations gave a bcc structure, we have performed another calculation on a Lennard-Jones system (referred to as LJF_2 to distinguish it from the one referred to as LJF_1 in Sec. III) at $\rho=0.94$ and $T=0.5$, these values of temperature and density being close to those used by Mandell *et al.*[5]

We prepared this system by allowing it to run under controlled temperature conditions at $T=0.52$, $\rho=0.94$ for $1250\,\Delta t$, with $\Delta t=0.0075$, starting from the high temperature ($T=1.02$) lower density ($\rho=0.9042$) state as stated in Sec. III A. After this initial period the system was allowed to run for $2600\,\Delta t$ without any constraints; the average temperature during these 2600 steps of integration was 0.5. We shall refer to the end of this stage as $t=0$.

On allowing the system to run for $29\,700\,\Delta t$ beyond $t=0$, (the now familiar) changes were observed in $T(t)$ and $R^2(t)/N$ which clearly indicate a transition to an ordered state of aggregation. The release of the latent heat made the temperature of the system rise to a value 0.6 and the pair correlation at the end of the 29 700 step run (for a region of time extending from $27\,700\,\Delta t$ to $29\,700\,\Delta t$) is shown in Fig. 10. The positions of the peaks and the value of the coordination number point to an fcc structure.

The analysis of the Voronoi polyhedra is shown in Fig. 8(d). This confirms our conclusion that the structure does not have a bcc packing but rather an fcc one. The behavior of $S(\mathbf{k};t)$ for this system is shown in Table II. From the set of vectors, for which $S(\mathbf{k};t)$ (at the end of the $29700\,\Delta t$ run) shows the largest values, the structure in reciprocal space is not clear. The result of projecting the coordinates in the manner already explained in Sec. III D is shown in Fig. 9(c). It appears

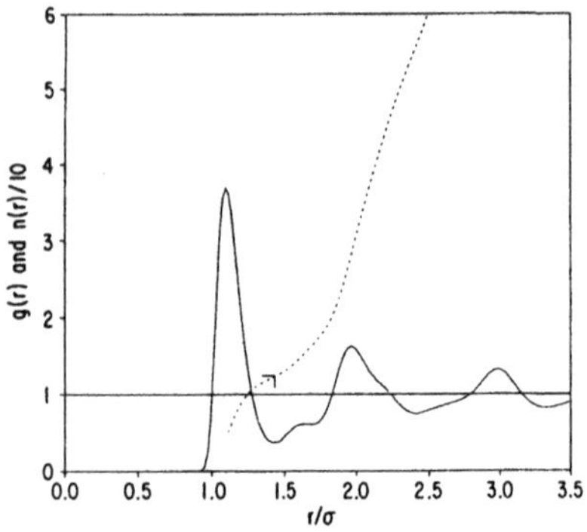

FIG. 10. The pair correlation (solid curve) and the coordination number (broken curve) for the Lennard-Jones system (LJF_2) of Sec. IV for time ranging from 27 700 to 29 700 Δt. The markers point to the first minimum in $g(r)$ and the corresponding value of $n(r)$.

from this figure, presumably because of the higher temperature, that there are large distortions in the close packing and both *abc* and *ab* type repetitions occur in the structure. On the basis of the present results we therefore conclude that a Lennard-Jones system orders itself into a close packed structure and not a bcc one. The implications of the difference between the results of Mandel *et al.*,[5] on the one hand, and of Ravesché and Streett[4] and the ones presented here, on the other, are mentioned in Sec. VIC below.

V. EFFECT OF MODIFYING THE POTENTIAL AFTER NUCLEATION OF A bcc PHASE

As mentioned in the introduction, this part of the calculation was centered on the effect of modifying the potential after a well established bcc structure for the rubidium system (RBF) has been obtained. Throughout these calculations, the time step Δt was chosen to be 0.0075 for all systems.

A. Preparation of the bcc crystal

In contrast to the quick quenching process as described in Sec. III A, here we prepare the system in a different fashion.

Starting from a liquid state of RBF system of 500 particles at $T=0.79$ and $\rho=0.90$, we first control the system to keep it in a liquid state at $T=0.60$ and $\rho=0.95$ by decreasing the temperature and increasing the density. This state is very much liquidlike as seen from the shape of $g(r)$ and of the linear increase with time of $R^2(t)/N$. A run of 3700 Δt in this state produced a near straight line behavior of $R^2(t)/N$ which gives a self-diffusion constant $D\approx 0.009$. The system is then cooled down to $T=0.10$ in four steps. Firstly, the temperature is maintained at $T=0.40$ for 630 Δt, then the system is allowed to run for 4020 Δt without any constraint. Secondly, the temperature is maintained at $T=0.30$ for 530 Δt, then the dynamics is allowed to take its course for 3870 Δt; during this time, the system underwent a transition to an ordered state and the latent heat released made the temperature of the system rise to 0.35. In the third step, the temperature is maintained at $T=0.20$ for 550 Δt and the system is allowed to run for 4460 Δt without being disturbed. Finally in the fourth step, the system is maintained at $T=0.10$ for 550 Δt and the time is reset to $t=0$. From then on, no further temperature control is employed. The temperature $T(t)$ and $R^2(t)/N$ are always monitored to check the progress of the relaxation of the system. The pair correlation function $g(r)$ is used to study the structure of the system. For a run of $t_1=3300\,\Delta t$ after the final temperature control is lifted, we notice that the system appears to be in equilibrium with $T(t)$ fluctuating around 0.10 and $R^2(t)/N$ never exceeding 0.033. The $g(r)$ shows a well developed bcc structure. The time variation of temperature $T(t)$ and the $g(r)$ in this region, namely $0<t<t_1$, are displayed in Figs. 11 (leftmost part) and 12(a), respectively.

Before proceeding with the declared theme of this section it is useful to digress briefly as follows. In Ref. 1, a rubidium system (i.e., RBF in the present nomenclature) having the same density (0.95) but prepared differently (by quenching quickly from temperature 1.0 to 0.1) also nucleated into a bcc structure. During the relatively gentle cooling process described in the previous paragraph we obtained a transition to an ordered state at a higher temperature (between 0.3 and 0.4); as described above the already ordered system was cooled to 0.1 in two further stages of cooling. The distribution of the signatures of Voronoi polyhedra at time t_1 in the rubidium system being reported in this paper and at $t=11\,450\,\Delta t$ in the system reported in Ref. 1 is shown in Figs. 8(e) and 8(f) respectively. It is evident that both distributions point to a bcc type packing. This strongly suggests that the ordering in a rubidium system will consistently turn out to be a bcc type structure irrespective of the mode of cooling the system.

B. Symmetry conversion

The bcc structure at time t_1 mentioned above is used to study the potential dependence of the crystalline structure. For this purpose, at t_1 the potential is given the distance dependence of LJF and the dynamics is allowed to take its course until time $t_2=7640\,\Delta t$ (see Fig. 11). The temperatures $T(t)$ and $R^2(t)/N$ are monitored to ob-

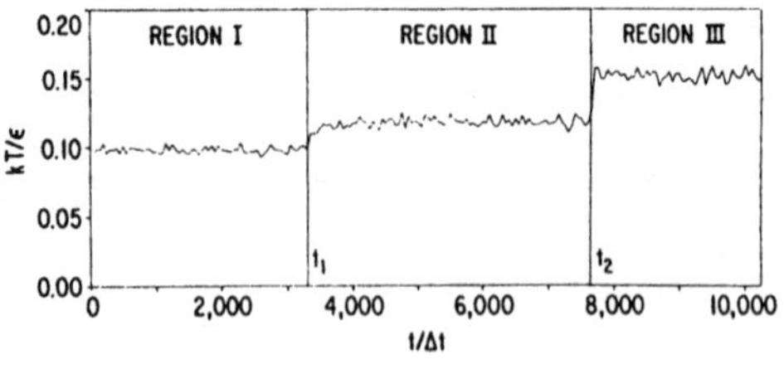

FIG. 11. The change of temperature with time for the molecular dynamics run involving potential changes discussed in Sec. V. For region I and III, the pair potential is the rubidium potential, whereas in region II, it is Lennard-Jones.

serve if new stable crystalline structures appear or not. The $g(r)$ is used to identify the crystal symmetry. At time t_2, the potential is modified back to its original form, namely RBF and the dynamics allowed to take its course.

In Fig. 11, we present the temperature $T(t)$ as a function of time starting from $t=0$. Here we separate the time span into three regions: denoted by I for time interval $0<t<t_1$, II for $t_1<t<t_2$, and III for $t_2<t$. In other words, in regions I and III, the system is an RBF system, whereas in region II it is LJF. The corresponding $g(r)$ and $n(r)$ for each region after the system reaches equilibrium, i.e., in the latter part of each region, are shown in Fig. 12. Clearly, this figure indicates the following: for RBF in region I, the structure is bcc, for LJF in region II, the structure of the system changes into an fcc phase, and for RBF in region III, the structure reverts back to bcc.

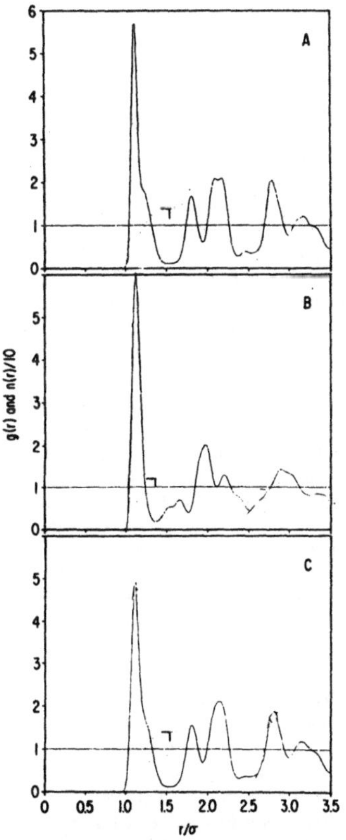

FIG. 12. The pair correlation (solid curve) and the coordination number (broken curve) of the calculation discussed in Sec. V for various blocks of time. The blocks of time for (a), (b), (c) are (see Fig. 11) 2010 to 3010 Δt in region I, 6300 to 7300 Δt region II, and 9140 to 10 140 Δt region III, respectively. The markers point to the first minimum in $g(r)$ and the corresponding value of $n(r)$. The structures in (a), (c) are clearly different from the one shown in (b).

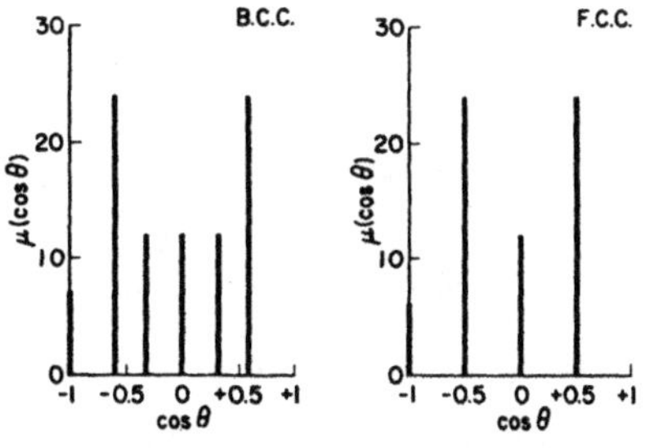

FIG. 13. The number of times angle θ occurs between bonds at a particle for ideal bcc and fcc structures counting the first 14 neighbors in the bcc case and the first 12 in the fcc case.

Following the method formulated in Ref. 1, we have also analyzed three particle correlations to clarify the symmetry of the system in different time regions. Let P be one of the N particles and let P' and P'' be two of its neighbors such that both are within a specified distance R_c from P irrespective of their mutual distance. We have monitored the distribution of the cosine of the angle $P'PP''$. For ideal bcc and fcc structures going up to R_c such that one includes the first 14 neighbors in the former case and the first 12 in the latter, we get the δ-function distribution shown in Fig. 13.

In Fig. 14, the distribution of angles is shown for

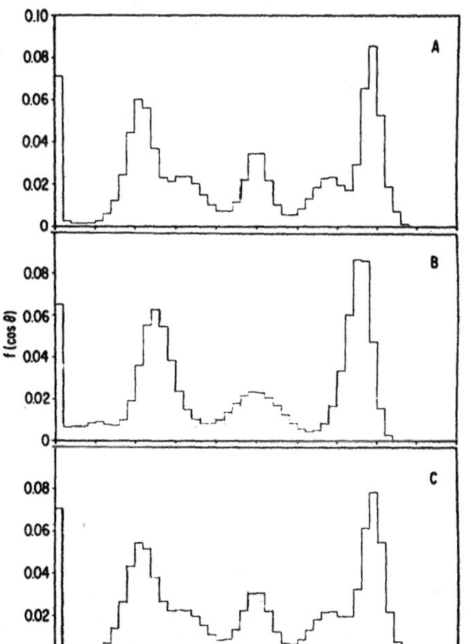

FIG. 14. Distributions of angles between bonds of the system discussed in Sec. V for the blocks of time mentioned in caption of Fig. 12. The interval $(-1, +1)$ for $\cos\theta$ is divided into 50 equal parts in the above histogram.

regions I, II, and III after the system has reached equilibrium in that region of time. In each case R_c is chosen to be the distance where $g(r)$ reaches its first minimum. The average number of neighbors up to R_c are seen in Fig. 12. From the locations of the side peaks and from the existence (or otherwise) of double peaks, we can identify the crystalline structure in each region. The conclusion is consistent with that drawn from $g(r)$ analysis: bcc structure for RBF in region I and III, and fcc structure for LJF in region II.

We have also checked whether the states in region I and III belong to the same equation of state for a crystallized RBF system. By reducing the total energy of the end configuration of region III to that of region I, we find the average temperature also reduced to that of region I. In other words, the two states in region I and III have the same mutual dependence of total energy and temperature.

Two more calculations of a similar kind as described above were made by modifying the potential at time t_1 to RBT and LJT. In Fig. 15, we show the pair correlation functions in these calculations for the time region corresponding to II during which the system is RBT or LJT. The structure in each case is clearly fcc.

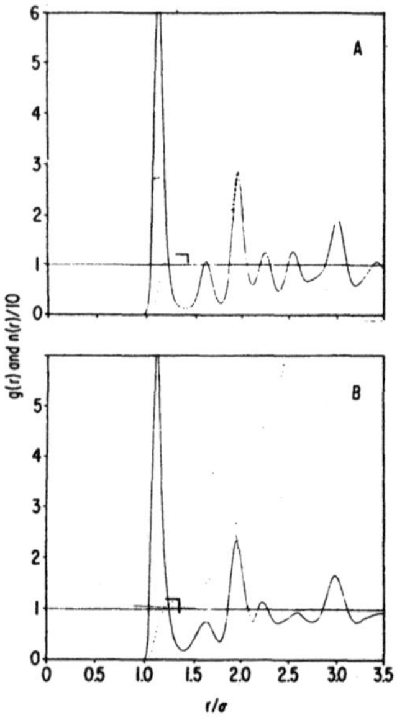

FIG. 15. The pair correlation (solid curve) and the coordination number (broken curve) with markers pointing to the minimum in $g(r)$ and the corresponding value of $n(r)$ for systems discussed in Sec. V. (a) for the system with the rubidium-truncated potential, the potential modification having been made on the rubidium system at the end of region I (in Fig. 11) and averaging done over 5800 to 6800 Δt; (b) for the system with the Lennard-Jones-truncated potential, the potential modification having been made on the rubidium system at the end of region I (Fig. 11) and averaging done over 7800 to 8300 Δt. The shapes of $g(r)$ indicate an fcc packing.

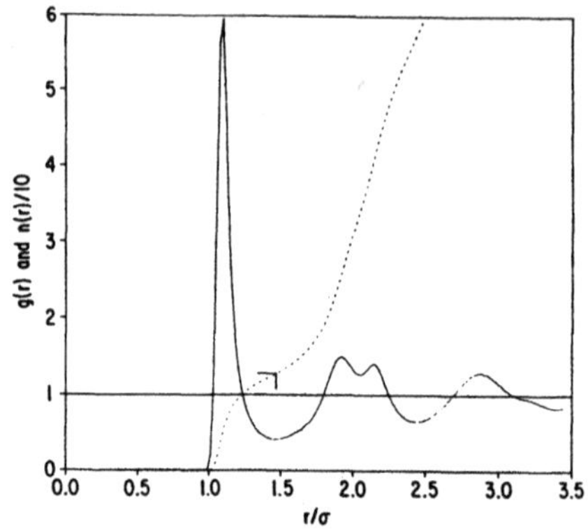

FIG. 16. The pair correlation (solid curve) and the coordination number (broken curve) for the Lennard-Jones-truncated system (LJT) of Sec. III for time ranging from 25 000 to 35 000 Δt (see Fig. 5). The markers point to the first minimum in $g(r)$ and the corresponding value of $n(r)$. Figure 5 shows that this seemingly good "glassy state" will eventually nucleate.

VI. CONCLUSIONS

(1) As has been remarked in Ref. 1 as well, in all the calculations we have made, in spite of the rapid rate of cooling, the calculations failed to give a metastable disordered structure. We note that in Figs. 2–5 there are fairly long-lived states which appear as plateaus in the increase of $R^2(t)/N$ with time and in these regions the pair correlation has the typical "double second peak" structure of the random packing of hard spheres[10]; Stillinger and Weber[11] recently have shown that this double second peak structure *also* is a function of the interaction potential! In Fig. 16, we show the pair correlation for the LJT system of Sec. III over time interval 25 000–35 000 Δt (refer to Fig. 5). It is conceivable that a molecular dynamics (or Monte Carlo) calculation can be mistakenly considered as having given a metastable disordered system if the calculation is terminated during one of these long-lived plateau regions in the behavior of $R^2(t)/N$. In their recent studies, Wendt and Abraham[12] and Stillinger and Weber[11] have dealt with properties of computer generated disordered metastable states in a systematic manner, bringing to light certain structural properties which are common to Lennard-Jones systems and to Gaussian core systems. In the light of our failure to generate amorphous metastable states with four different potentials we believe that it is necessary, by means of systematic computer simulation (and hopefully by means of fundamental statistical mechanics as well!) to study the lifetime problems of metastability in this context. This will involve a study of the problem as a function of (i) pair interaction, (ii) method of cooling, (iii) method of compressing, and (iv) the number of particles used for the calculation.

(2) The dependence on the interaction potential of the symmetry of the nucleated phase has been shown to exist in two different ways; the universality[2] of the bcc structure as the preferred phase irrespective of the interac-

tion potential has not been borne out by our calculations. Even purely repulsive potentials do not seem to form a class by themselves in this respect: LJT and RBT both gives fcc structures while the Gaussian core model[6] gives a bcc structure. With r^{-12} soft spheres[7] both types of structures have been reported.

(3) We recal that Tanemura *et al.*,[7] using $1/r^{12}$ repulsive potentials obtained both fcc and bcc structures and labeled the latter as "incompletely relaxed." Mandel *et al.*,[5] using full Lennard-Jones 6–12 potentials obtained a bcc structure while the Monte Carlo calculations of Raveché and Streett[4] resulted in an fcc arrangement. The three very different types of MD calculations as reported in this paper led to the nucleation of fcc structures with the L-J 6–12 potential. Thus it appears that the problem of homogeneous nucleation by computer calculations needs to be investigated from a variety of angles, namely (i) pair interaction, (ii) method of cooling, (iii) method of compressing and (iv) the system size.

(4) We recall the unambiguous manner in which the bcc structure for the RBF system reported in Ref. 1 was indicated in reciprocal space by the values of $S(\mathbf{k};t)$ for twelve $\mathbf{k}$ vectors forming the basis vectors of an fcc lattice. Firstly the values of S for all 12 vectors were of order 420, dropping to a value of order 10 for vectors not belonging to this set. Table II in the present paper shows that even for a relatively well ordered low temperature fcc system (see Fig. 7(c) for LJF_1) the structure in reciprocal space does not show itself up with a useful degree of clarity.

(5) As seen in Fig. 7, the ordered structures have pair correlations which are relatively more refined in the systems will full potentials than in those with truncated potentials. At the same time Figs. 2–5 indicate a more sudden change to an ordered state for the full potentials than for the truncated ones. These characteristics of the nucleation events appear to go hand in hand with the fact that, as against the systems with truncated potentials, for the full potentials the growth of the nucleus can be observed with great clarity. (See Sec. III E.)

APPENDIX

Construction of Voronoi polyhedra

The first geometrical analysis of crystallization utilizing the Voronoi polyhedra construction was performed by M. Tanemura *et al.*[7] For a given configuration of N particles, we can construct a Voronoi polyhedron π_i for each particle i. The boundaries of π_i consist of faces formed by certain of the $N-1$ perpendicular bisector planes of the line segments $\mathbf{r}_{ij}$ ($\equiv \mathbf{r}_j - \mathbf{r}_i$) with $j \neq i$. This is, in fact, a generalized version of a Wigner–Seitz cell. In the manner of Ref. 7, we use as signature a set of four or more integers $(n_3 n_4 n_5 \ldots)$ for a λ-faced polyhedron where n_l ($l \geq 3$) is the number of l-sided faces and $\lambda = \sum_l n_l$. In the bcc lattice, the Wigner–Seitz cells bear the signature (0608) with each vertex being defined by *three* planes (a property of the vertices of Voronoi polyhedra but not, in general, of the vertices of precisely geometric Wigner–Seitz cells). For an fcc lattice, the Wigner–Seitz cell of 12 rhombic faces bears the signature (01200) and is formed by 14 vertices, six of them being common to *four* planes. Obviously, each one of these "four-plane" vertices will split into standard vertices (i.e., those formed by the intersection of three planes) if an arbitrarily small amount of thermal motion is allowed. In order to include the effect of thermal fluctuations, we have made the following modification in constructing the Voronoi polyhedra. After constructing each polyhedron in the usual way, we check the lengths of the edges of the polyhedron, i.e., the distances between the vertices which form each of the edges. If any distance is smaller than a preset value ζ, the two vertices involved are counted as one and the line is eliminated. (Graphically, one example will be the case in which 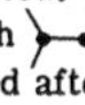 gets modified to .) The signature is assigned after the polyhedron is modified in this way.

In order to show that this kind of construction is meaningful, we have tested it on an fcc and a bcc lattice with thermal fluctuations imposed on the lattice. By using a random number generator a "fluctuation" is created, its magnitude being measured by the root mean square displacement η of the points from their regular lattice sites. The results are discussed below and shown in Tables IV(a) and IV(b).

As regards the behavior of the bcc lattice, we notice from Table IV(a) that the usual Voronoi polyhedra signature is very stable against thermal motion. Obviously the modified construction described above changes the signature if we allow a value of ζ which is too large. But for both η and ζ up to ~10% of the nearest neighbor distance d in the lattice, the modification makes no difference. *This is what is meant by stating that the Voronoi polyhedra signature for a* bcc *structure is "stable" in spite of thermal motion.*

For the fcc lattice we can see from Table IV(b), that the modification definitely helps to clear up the doubt about using Voronoi polyhedra to identify the symmetry of the structure. In this case, even a small thermal fluctuation will lead to confusion about what the Voronoi polyhedron signature indicates. Using the modified construction the signature is clarified. For example, for a fluctuation corresponding to $\eta/d \simeq 0.1$ (see Table IV(b)), $\zeta/\eta = 1$, as shown in the table, already reveals the presence of the characteristic signature F_0 (1%) of an fcc arrangement. By increasing ζ/η to 3 the fcc symmetry is clearly revealed (94% of F_0). Of course, there is an upper limit to the value of ζ/η (the length of "small" edges) for each η/d (rms displacement away from the regular sites). From the last three lines on the right side of Table IV(b), it is quite clear how in the presence of a large thermal disturbance η the character of the fcc structure is first revealed and then made ambiguous again if the value of ζ/η is too large.

The results shown in Tables IV(a) and IV(b) simply give a working quantitative basis for fairly obvious geometrical facts concerning thermally agitated lattices. The power of the method in unraveling structures generated during molecular dynamics (or Monte Carlo) calculations is evident.

TABLE IV(a): Distribution of signatures of Voronoi polyhedra in a bcc crystal with thermal fluctuation η. Nearest neighbor distance $d=\sqrt{3}$. The perfect bcc signature is $B_0(0608)$. Frequencies of occurrence of B_0, $B_1(0446)$, $B_2(0626)$ are shown. All others together are "misc." Edges of length less than ζ are discarded and the polyhedron reconstructed. $\zeta=0$ corresponds to no modification. As ζ increases bcc character of distribution gets distorted. Hence for identification purposes, *a* bcc *structure* with thermal fluctuations ***needs no modification*** of the polyhedra.

	Percentages without modification					Percentages with modification			
η/d	B_0	B_1	B_2	misc.	ζ/η	B_0	B_1	B_2	misc.
0.0577	100	0	0	0	0	100	0	0	0
					1	100	0	0	0
					2	100	0	0	0
					3	100	0	0	0
0.1155	100	0	0	0	0	100	0	0	0
					1/2	100	0	0	0
					1	95	0	5	0
					3/2	79	0	15	6
0.1732	82	10	0	8	0	82	10	0	8
					1/3	66	6	12	16
					2/3	45	3	18	34
					1	27	0	16	57

(b). Distribution of signatures of Voronoi polyhedra in an fcc crystal with thermal fluctuation η. Nearest neighbor distance $d=\sqrt{2}$. The perfect fcc signature is $F_0(01200)$. Frequencies of occurrence of F_0, $F_1(01020)$, $F_2(0840)$, $O_1(0364)$, $O_2(0366)$ are shown. All others together are "misc." Edges of less than ζ are discarded and the polyhedron reconstructed. Note that with $\zeta=0$, the signautre F_0 *cannot* appear. ***With a suitable $\zeta\neq 0$ it becomes the dominant signature.*** For the largest values of η and ζ (last line on right side of table) the construction loses its significance.

	Percentages without modification					Percentages with modification			
η/d	F_0	O_1	O_2	misc.	ζ/η	F_0	F_1	F_2	misc.
0.0354	0	25	10	65	0	0	0	0	100
					2	56	35	4	5
					3	99	1	0	0
0.0707	0	22	11	67	0	0	0	0	100
					1	2	7	12	79
					2	61	31	4	4
					3	99	1	0	0
0.1061	0	19	10	71	0	0	0	0	100
					1	1	7	12	80
					2	63	28	4	5
					3	94	1	0	5
0.1414	0	16	10	74	0	0	0	0	100
					1	1	7	12	80
					2	56	20	3	21
					3	5	0	0	95

In the light of these exploratory calculations with perfect lattice structures, we have used the following procedure for understanding the structure of molecular dynamics systems after the occurrence of a nucleation event. The usual Voronoi polyhedra construction (i.e., $\zeta=0$) is performed to check whether the predominant signature indicates a bcc structure or not. If it does not indicate a bcc structure, we use the modified construction (i.e., a nonzero ζ).

For the LJF_1 system of Sec. III, we concluded in this manner that it did not form a bcc structure on nucleation

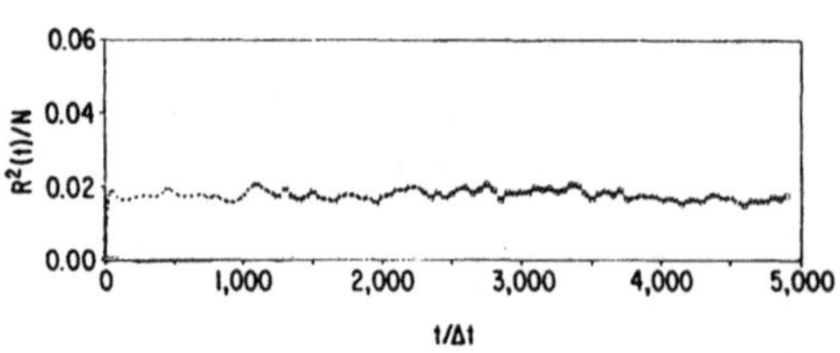

FIG. 17. $R^2(t)/N$ of the crystallized Lennard-Jones system (LJF_1) of Sec. III with the origin of time being chosen at 18 000 Δt, beyond the region of time displayed in Fig. 3.

and hence a value for ζ had to be investigated. In order to get an estimate for ζ, we proceeded as follows: In Fig. 17, we show $R^2(t)/N$ as a function of time with the origin of time being chosen at $\bar{t}=18\,000\,\Delta t$, i.e., after crystallization (far beyond the region shown in Fig. 3). From the vertical deviation (i.e., the extent of the thermal Debye–Waller cloud), we can estimate that the value of $N^{-1}\sum_{i=1}^{N}[\mathbf{r}_i(t-\bar{t})-\mathbf{r}_i(\bar{t})]^2$ will be around 0.02 where $\mathbf{r}_i(t)$ denotes the location of the ith particle at time t; this is approximately equal to twice the mean square deviation from the mean positions of the particles and therefore we infer from Fig. 17 that $\eta^2 \sim 0.01$. Since the nearest neighbor distance d is about 1.12 (see Fig. 6(c)), we obtain $\eta/d \sim 0.09$. From Table IV(b), we infer that the value of ζ/η should be about 3 in order to reveal the fcc structure. By using this value for ζ, we get the results shown in Figs. 8(a)–8(c).

For the high temperature LJF system of Sec. IV (referred to as LJF_2) the same kind of analysis of the mean square displacement gives $\eta/d \sim 0.15$. From Table IV(b), we obtain the value of $\zeta/\eta \sim 2$. By using this value for ζ, we get the result shown in Fig. 8(d).

Thus both the Lennard-Jones systems reported in this paper, when analyzed through Voronoi polyhedra, indicate that on nucleation they go into a close-packed structure.

[1] C. S. Hsu and A. Rahman, J. Chem. Phys. **70**, 5234 (1979).

[2] S. Alexander and J. McTague, Phys. Rev. Lett. **41**, 702 (1978).

[3] D. L. Price, Phys. Rev. A **4**, 358 (1971); D. L. Price, K. S. Singwi, and M. P. Tosi, Phys. Rev. B **2**, 2983 (1970).

[4] H. J. Raveché and W. B. Streett, J. Res. Natl. Bur. Stand. Sect. **80**, A 59 (1976).

[5] M. J. Mandell, J. P. McTague, and A. Rahman, J. Chem. Phys. **64**, 3699 (1976); **66**, 3070 (1977).

[6] F. H. Stillinger and T. A. Weber, J. Chem. Phys. **68**, 3837 (1978).

[7] M. Tanemura, Y. Hiwatari, H. Matsuda, T. Ogawa, N. Ogita, and A. Ueda, Prog. Theor. Phys. **58**, 1079 (1977).

[8] (a) A. Rahman, Phys. Rev. Lett. **32**, 52 (1974); Phys. Rev. A **9**, 1667 (1974). (b) This is from unpublished analysis of the work referred in 8(a).

[9] A. Rahman, M. J. Mandell, and J. P. McTague, J. Chem. Phys. **64**, 1564 (1976).

[10] J. L. Finney, Proc. R. Soc. London, Ser. A **319**, 479 (1970); **319**, 495 (1970); G. S. Cargill, J. Appl. Phys. **41**, 12 (1970); **41**, 2248 (1970); C. H. Bennett, J. Appl. Phys. **43**, 2727 (1972).

[11] F. H. Stillinger and T. A. Weber (preprint).

[12] H. R. Wendt and F. F. Abraham, Phys. Rev. Lett. **41**, 1244 (1978).

Crystal Structure and Pair Potentials: A Molecular-Dynamics Study

M. Parrinello[a] and A. Rahman
Argonne National Laboratory, Argonne, Illinois 60439
(Received 31 July 1980)

With use of a Lagrangian which allows for the variation of the shape and size of the periodically repeating molecular-dynamics cell, it is shown that different pair potentials lead to different crystal structures.

PACS numbers: 61.20.Ja, 05.70.Fh, 61.50.Cj, 64.70.Dv

Recent molecular-dynamics (MD) calculations[1] on homogeneous nucleation of a crystal out of a supercooled liquid phase have shown that the structure of the nucleated phase does depend on the pair potential. These calculations are time consuming because of the long-lived, glassy, metastable states which the system has to inhabit before nucleating. Here we present a very direct and relatively short calculation which relates the crystal structure to the pair potential in a simple manner.

Andersen[2] has shown how MD calculations can be modified to study systems under constant pressure by introducing the volume of the system as an additional dynamical variable. In this paper we show how a generalization of this idea leads to a powerful method for the study of crystal structures and their relation to pair potentials. We have performed MD calculations with a time-dependent metric tensor which allows the volume *and* the shape of the MD cell to vary with time.

Let the edges of the MD cell be $\vec{a}$, $\vec{b}$, and $\vec{c}$ (in a space-fixed coordinate system), and let them be time dependent. Periodically repeating MD cells will obviously fill up all space. Let $\underline{h}$ be the matrix formed by $\{\vec{a},\vec{b},\vec{c}\}$; $\Omega = \det \underline{h} \equiv \underline{a} \cdot b \times c$

will then be the volume of the MD cell containing, say N particles. The position of particle i will be $\vec{r}_i = \xi_i\vec{a} + \eta_i\vec{b} + \zeta_i\vec{c} = \underline{h}\vec{s}_i$, where $\vec{s}_i$ has components (ξ_i, η_i, ζ_i) each going from 0 to 1. Obviously $r_i^2 = \vec{s}_i'\underline{G}\vec{s}_i$, where $\underline{G} = \underline{h}'\underline{h}$, the transpose being denoted by a prime. Using a dot to denote time derivatives, we write the Lagrangian

$$L = \tfrac{1}{2}\sum m_i \dot{\vec{s}}_i'\underline{G}\dot{\vec{s}}_i - \sum_i \sum_{j>i} \varphi(r_{ij}) + \tfrac{1}{2}W\,\mathrm{Tr}(\dot{\underline{h}}'\dot{\underline{h}}) - p_{\mathrm{ext}}\Omega. \quad (1)$$

Obviously $r_{ij}^2 = (\vec{s}_i - \vec{s}_j)'\underline{G}(\vec{s}_i - \vec{s}_j)$; p_{ext} denotes the externally applied hydrostatic pressure; $\varphi(r)$ is the pair potential; the kinetic term associated with the time variation of $\underline{h}$ has a constant of proportionality W which has the dimension of mass.

With use of $\chi(r)$ to denote $-d\varphi/rdr$, the Lagrangian equations of motion are easily written down:

$$\ddot{\vec{s}}_i = m_i^{-1}\sum_{j\neq i}\chi(r_{ij})(\vec{s}_i - \vec{s}_j) - \underline{G}^{-1}\dot{\underline{G}}\dot{\vec{s}}_i, \quad i,j = 1,2,\ldots,N, \quad (2)$$

$$\ddot{\underline{h}} = W^{-1}(\underline{\pi} - p_{\mathrm{ext}})\underline{\sigma}. \quad (3)$$

The matrix $\underline{\sigma}$ has elements $\sigma_{ij} \equiv \delta\Omega/\delta h_{ij}$; the matrix $\underline{\pi}$ is given in dyadic tensor notation by

$$\Omega\overleftrightarrow{\pi} = \sum_i m_i\vec{v}_i\vec{v}_i + \sum_i\sum_{j>i}\chi(r_{ij})(\vec{r}_i - \vec{r}_j)(\vec{r}_i - \vec{r}_j), \quad (4)$$

the vector $\vec{v}_i$ being $\underline{h}\dot{\vec{s}}_i$. Equations (2) and (3) govern the dynamics of a system of N particles in a periodically repeating MD cell which changes with time in shape and volume. Equation (3) is the expected relation between the variation of $\underline{h}$, the microscopic stress tensor $\underline{\pi}$, and the external pressure; $\underline{\pi}$ and p_{ext} act across the various areas given by the components of $\vec{b}\times\vec{c}$, $\vec{c}\times\vec{a}$, and $\vec{a}\times\vec{b}$, which make up $\underline{\sigma}$. In Eq. (3) we clearly see the possibility of a generalization to a nondiagonal external stress tensor.

In the special case of Andersen[2] $\underline{h} = \mathrm{diag}(\Omega^{1/3}, \ldots, \Omega^{1/3})$ and $\underline{G}^{-1}\dot{\underline{G}} = 2\dot{\Omega}/3\Omega$, but his equation for $\ddot{\Omega}$ cannot be obtained from Eq. (3). However, as in Andersen,[2] it is easy to show that L of Eq. (1) generates an isoenthalpic, isobaric ensemble, apart from a small correction arising from the term in W.

We have used Eqs. (2) and (3) to investigate the Lennard-Jones 6-12 potential V_{LJ} and a pair potential[3] suitable for rubidium metal V_{Rb}.[4] Units of length, mass, and energy are chosen[1] in the standard fashion, so that all quantities carrying an asterisk are the so-called reduced variables.

The MD calculation is started with a 500-particle system forming an fcc structure in a cubic cell of length $l^* = 8.046$ appropriate to the number density $\rho^* = 0.96$. Thus at the start $\underline{h} = \mathrm{diag}(l^*, \ldots, l^*)$. A small random displacement of each particle from its lattice site provides the initial conditions for the ensuing dynamics; the MD time step was taken to be $\Delta t^* = 0.005$. The structure was monitored through the pair correlation function $g(r)$.

The summary of one of several calculations is given in Fig. 1 and described in the following. In this calculation we used $W^* = 20$.[5] At $t = 0$ V_{Rb} was taken as the pair potential and p_{ext} was taken as 4.0.[6] Using well-known MD techniques, the temperature of the system was set at a value $T^* = 0.15$ for a duration of 140 MD steps. Since the starting structure was fcc the pair correlation during this time showed clear and sharply defined shells at distances relevant to an fcc structure, namely 1, $\sqrt{2}$, $\sqrt{3}$, $\sqrt{4}$, etc. [see Fig. 1, $g(r)$ at $t^* = 0$]. However, at step 400 ($t^* = 2$) it already became evident that the MD cell was undergoing a secular modification toward a rectangular parallelepiped with two edges $\vec{a}, \vec{b}$ of about the same length and a shorter third edge $\vec{c}$ [see Fig. 1, where $(a+b)/2c$ is seen to be starting to increase with time almost as soon as the calculation starts]. The nondiagonal elements of $\underline{h}$ showed small fluctuations around zero. The volume, apart from small fluctuations, showed no secular change. After 2000 MD steps ($t^* = 10$) the structure of the system had changed from fcc to bcc [in Fig. 1 see $g(r)$ at $t^* = 14$]. The bcc structure so obtained showed no sign of any secular change for another 1500 time steps. (It is to be noted that a body-centered tetragonal lattice with edges 1, 1, and $\sqrt{2}$ is an fcc structure and, inversely, a tetragonal face-centered structure with edges $\sqrt{2}$, $\sqrt{2}$, and 1 is a bcc structure. An hcp structure also can be generated out of a body-centered tetragonal structure by appropriate changes of lengths, angles and the position of the body center.)

At this time, $t^* = 17.5$, the pair potential was changed from V_{Rb} to V_{LJ} and simultaneously p_{ext} was put to zero (which is the value appropriate for this potential at $\rho^* \sim 1.0$). At the moment of the change of potential the stable body-centered structure occupied an MD cell with sides 9.00, 9.02, and 6.41, and angles within 1° of being right angles. Immediately after the change, the cell

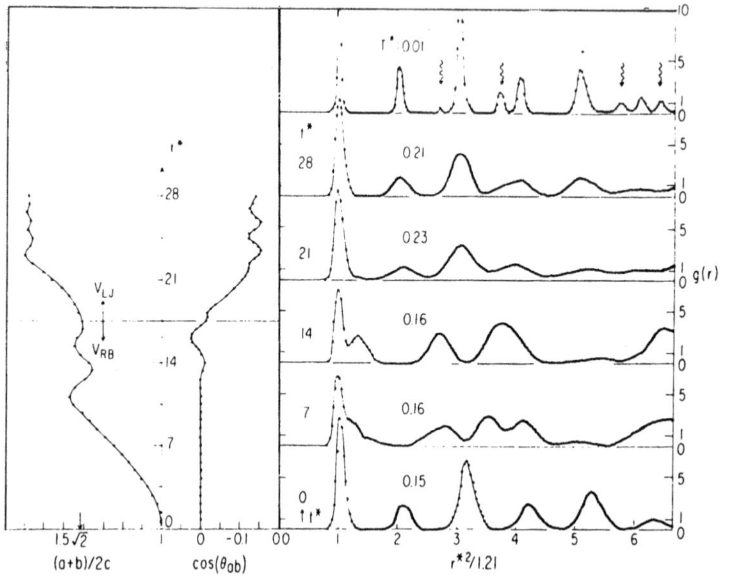

FIG. 1. The first graph on the left-hand side shows the MD cell edges $\{\vec{a}, \vec{b}, \vec{c}\}$ as a function of time t^*. The ratio $(a+b)/2c$ has been plotted; it is unity at $t^*=0$ When the potential is V_{Rb} and the MD cell is cubic; it tends to $\sqrt{2}$ with the passage of time. When V_{Rb} is changed to V_{LJ} at $t^*=17.5$, further changes occur in $\{\vec{a}, \vec{b}, \vec{c}\}$ accompanied by a change of the angle between $\vec{a}$ and $\vec{b}$. The cosine of this angle is shown as a function of time in the second graph from the left. The various times at which the $g(r)$ was monitored are indicated on the series of graphs on the right; each $g(r)$ is an average over 140 time steps; the average temperature during these time steps is also shown. The final state when quenched reveals, in the topmost figure, subsidiary peaks (wiggly arrows) due to stacking faults mentioned in the text. Note that g is plotted as a function of r^2. The ratio of the squares of shell distances is 1:2:3:4:5:6, etc., in an fcc lattice and 1:4/3:8/3:11/3:4:16/3:19/3:20/3, etc., in a bcc lattice.

and the structure started to deform. In about 1000 more time steps (i.e., at $t^*=25$) a new shape of the MD cell and a new $g(r)$ were established. The cell parameters became 9.54, 9.23, and 5.59 with an angle of 98° between the first two, the third being essentially perpendicular to them. The system acquired a density $\rho^*=1.03$. In Fig. 1 is shown the $g(r)$ at $t^*=28$. The first two peaks in $g(r)$ correspond to a close-packed structure. When the final configuration was quenched to $T^*=0.01$ with a short 100 step run, it revealed low-intensity peaks in $g(r)$ which would be absent in a perfect fcc stacking [see last $g(r)$ in Fig. 1]. A visual examination of the stacking along the direction of close packing revealed the order *ABABCBACBA* and hence the stacking faults. In Fig. 1 the history of the run is depicted in a way as to reveal the changes in the MD cell parameters and in $g(r)$ with the passage of time.

In recent years increasing attention has been given to the problem of predicting the crystalline phase of a system[7] with known particle interactions. For most ionic materials, a fairly coherent idea already exists concerning the lattice structure on the one hand and ionic sizes and charges on the other.[8] However, for monatomic systems and short-range, pairwise, central forces, which lattice is favored by the system at a certain density and temperature is in practice not an easy question to answer, except at temperatures low enough to allow a harmonic approximation to be valid.

We have shown here that the Lagrangian of Eq. (1) is a powerful tool for studying phase transitions in solids, especially in relation to the form of particle interactions. Monte Carlo studies of (N, p, T) ensembles generated by the two potential terms in Eq. (1) with $N+3$ vectors $\vec{a}$, $\vec{b}$, $\vec{c}$, and $\vec{s}_i$, $i=1,\ldots,N$, will be very suitable[5] for the study of such transitions as a function of T. In our preliminary MD studies, we found that $T^*=0.05$ was too low to trigger the changes depicted in Fig. 1 on the time scale of our calculation.

In addition to several calculations of the type reported here, many calculations on 432 particles were also made. In this case, starting with V_{LJ} and a bcc structure, one obtains a close-packed structure with stacking faults. Repeated heating and cooling of the faulty structure finally gives a perfect fcc ordering. This implies that our dynamical equations do allow the system to monitor in configuration space the subtle local minima which correspond to stacking faults in a close-packed system.

The exploratory calculations we have reported here are an example of the way our dynamical equations make it possible to relate particle interaction to particle arrangements in ordered structures. Many other applications seem possible. Generalizing from uniform p_{ext} in Eq. (3) to a general external stress tensor, the recent work of Milstein and Farber[9] on the fcc-bcc transition under (100) tensile loading can be investigated as a function of temperature and of the characteristics of the pair potential. We are at present investigating this problem.

Finally, the low-temperature phase transitions in light alkali metals can also be investigated, with the dependence of the pair potential on the density of the system taken into account.[3]

[a]On leave of absence from the Istituto di Fisica Teorica, Miramare, Trieste, Italy.

[1]C. S. Hsu and A. Rahman, J. Chem. Phys. 71, 4974 (1979).

[2]H. C. Andersen, J. Chem. Phys. 72, 2384 (1980).

[3]D. L. Price, Phys. Rev. A 4, 358 (1971); D. L. Price, K. S. Singwi, and M. P. Tosi, Phys. Rev. B 2, 2983 (1970).

[4]J. Copley and M. Rowe, Phys. Rev. Lett. 32, 49 (1974); A. Rahman, Phys. Rev. Lett. 32, 52 (1974), have shown that the V_{Rb} of Ref. 3 gives a good model of rubidium.

[5]The structural properties generated by the Lagrangian in Eq. (1) are independent of W, as they are of m_i the particle masses, in the classical systems under consideration. We have found in our MD studies that a higher value of W simply slows down the rate at which the effects described here occur.

[6]D. L. Price, Phys. Rev. A 4, 358 (1971), has shown that in alkali metals the effective ion pair potentials alone give a large positive pressure which is canceled by the metallic electron contribution.

[7]S. Alexander and J. McTague, Phys. Rev. Lett. 41, 702 (1978); J. Friedel, J. Phys. (Paris) Lett. 35, L59 (1974); T. V. Ramakrishnan and M. Yussouff, Phys. Rev. B 19, 2775 (1979).

[8]M. P. Tosi, in *Solid State Physics*, edited by H. Ehrenreich, F. Seitz, and D. Turnbull (Academic, New York, 1964), Vol. 16, p. 1.

[9]F. Milstein and B. Farber, Phys. Rev. Lett. 44, 277 (1980).

Polymorphic transitions in single crystals: A new molecular dynamics method

M. Parrinello
University of Trieste, Trieste, Italy

A. Rahman
Argonne National Laboratory, Argonne, Illinois 60439

(Received 1 July 1981; accepted for publication 14 August 1981)

A new Lagrangian formulation is introduced; it can be used to make molecular dynamics (MD) calculations on systems under the most general, externally applied, conditions of stress. In this formulation the MD cell shape and size can change according to dynamical equations given by this Lagrangian. This new MD technique is well suited to the study of structural transformations in solids under external stress and at finite temperature. As an example of the use of this technique we show how a single crystal of Ni behaves under uniform uniaxial compressive and tensile loads. This work confirms some of the results of static (i.e., zero temperature) calculations reported in the literature. We also show that some results regarding the stress-strain relation obtained by static calculations are invalid at finite temperature. We find that, under compressive loading, our model of Ni shows a bifurcation in its stress-strain relation; this bifurcation provides a link in configuration space between cubic and hexagonal close packing. It is suggested that such a transformation could perhaps be observed experimentally under extreme conditions of shock.

PACS numbers: 64.70.Kb, 61.50.Ks

I. INTRODUCTION

The behavior of solids under the combined effects of external stress and of temperature has considerable practical relevance. Yet even in the idealized case of a perfect crystal, a detailed microscopic picture of such effects is still lacking. Most of the theoretical studies have been confined to conditions at zero temperature; in addition a perfect and prefixed crystalline arrangement of the atoms has been assumed. These two assumptions may lead to useful insights for relatively small values of the stress and temperature. However, it is obviously desirable to be able to study the behavior of solids at normal temperatures and high levels of external stress. In particular at high values of the stress spontaneous defect generation and/or crystal structure transformation become possible; this makes the assumption of a perfect, even if elastically distorted, crystalline arrangement untenable. Furthermore, these processes are sensitive to temperature variations as well.

The experience of the past two decades has shown that molecular dynamics (MD) calculations can provide a valuable tool to investigate nonharmonic effects in solids.[1] The authors have recently developed[2] a new MD method which allows a crystalline system to modify its structure if the temperature and external stress conditions make such a modification favorable. This method is therefore pertinent to the discussion of stress and temperature effects just mentioned above.

It is our purpose to present work based on this new MD method; the model system which we have studied has been described in a series of papers by Milstein and collaborators.[3-5] It is a system of classical particles interacting via a pairwise additive potential of the Morse type. The parameters of the potential have been adjusted to reproduce the elastic constants and the lattice parameter of Ni.[3] We have used this model of Ni for our study. The significance of the results for laboratory experiments will be dealt with in the last section.

Two reasons have prompted our interest in this model. First, Milstein and collaborators[3-5] have made extensive calculations of stress-strain relations for this model of Ni. Their calculations are a convenient check for our method of calculation at least for those values of temperature and stress where their theoretical approach is expected to be valid.

Second, in a recent paper an interesting possibility has been suggested by Milstein and Farber.[5] They have considered an fcc crystal of Ni under a uniform tensile [100] load. As the load is increased the crystal stretches in the direction of the load and contracts in the lateral directions. Beyond a certain value of the load and hence of the extension in the [100] direction, a new path in the stress-strain relation becomes possible along which the tetragonal symmetry is broken. Along this new branch, with *decreasing* load, the system starts to *expand* in one of the two lateral directions and eventually its extension in that lateral direction "catches up" with the [100] extension, the system ending up, at zero load, again in a tetragonal face-centered structure.

We show here that at finite temperature the above-mentioned "bifurcation" does not occur. Instead very close to the bifurcation point the system actually fails. Moreover the zero load tetragonal face-centered states (except the fcc one) mentioned in Ref. 5 as possible stable states spontaneously evolve into an hcp structure even at very low temperature.

A new result we have found is that, for this model of Ni, upon uniform uniaxial compression, the fcc structure transforms into an hcp arrangement.

II. MOLECULAR DYNAMICS AT CONSTANT EXTERNAL STRESS

Molecular dynamics (MD) methods have been extensively used in the past to study a variety of physical systems; for a detailed description the reader is referred to the book by Hansen and McDonald.[6] However, for the sake of clarity we shall briefly review the main features of MD.

It is a method for studying classical statistical mechanics of well-defined systems through a numerical solution of Newton's equations. A set of N classical particles have coordinates $\mathbf{r}_i$, velocities $\dot{\mathbf{r}}_i$ and masses $m_i, i = 1,...,N$. The particles interact through a potential $V_N(\mathbf{r}_1,...,\mathbf{r}_N)$ which, in most investigations is taken to be:

$$V_N = \frac{1}{2}\sum_i\sum_j \phi(r_{ij}), \tag{2.1}$$

where $r_{ij} = |\mathbf{r}_{ij}| = |\mathbf{r}_i - \mathbf{r}_j|$. More general forms of V_N can also be used, perhaps at the cost of more computational labor. Newton's equations are then

$$m_i\ddot{\mathbf{r}}_i = \sum_{j\neq i}\frac{1}{r_{ij}}\frac{d\phi}{dr_{ij}}\mathbf{r}_{ij}, \quad i = 1,...,N, \tag{2.2}$$

and are solved numerically. As the system evolves in time it eventually reaches equilibrium conditions in its dynamical and structural properties; the statistical averages of interest are calculated from $\mathbf{r}_i(t)$ and $\dot{\mathbf{r}}_i(t), i = 1,...,N$, as temporal averages over the trajectory of the system in its phase space. For practical reasons N is restricted to at most a few thousand and for small systems surface effects are obviously very important. However, to simulate a bulk system the common practice is to use periodic boundary conditions. These are obtained by periodically repeating a unit cell of volume Ω containing the N particles by suitable translations. Periodic boundary conditions obviously give a system in which the N particles are always contained in a cell of volume Ω, and without loss of generality every particle can be thought of as being at the "center." In other words, the summation over j in Eq. (2.2) extends over the infinite system generated by the periodic boundary conditions.

As a consequence of V_N being a function of $\mathbf{r}_i$ only [Eq. (2.1)], the solution of Eq. (2.2) conserves the total energy E of the system; thus the statistical ensemble generated in a conventional MD calculation is a (Ω,E,N) ensemble or a microcanonical ensemble. We shall use (.....) to indicate quantities whose constancy characterizes a given statistical ensemble.

The restriction that the MD cell be kept constant in volume and in shape severely restricts the applicability of the method to problems involving crystal structure transformations; in such transformations changes in the shape of the cell most obviously play an essential role. (For example, in a plane four points at the vertices of a square together with one at the center can obviously be used to generate a square lattice of points; this can become a lattice of equilateral triangles only if the square is allowed to become a $1{:}\sqrt{3}$ rectangle.)

In order to overcome this difficulty we[2] have modified a method due to Andersen[7] so as to allow for changes in volume *and* shape of the MD cell containing a system of particles under constant external hydrostatic pressure. Note that in the method of Andersen[7] only changes in the volume of the MD cell were possible but not in its shape. Thus crystal structure transformations are inhibited in Andersen's method because of the suppression of the essential fluctuations, namely those in the shape of the MD cell.

This extra degree of flexibility was introduced into the MD method as follows[2]: As before the system consists of N particles in a cell that is periodically repeated to fill all space. However, the cell can have arbitrary shape and volume being completely described by three vectors $\mathbf{a}$, $\mathbf{b}$, and $\mathbf{c}$ that span the edges of the MD cell. The vectors $\mathbf{a}$, $\mathbf{b}$, and $\mathbf{c}$ can have different lengths and arbitrary mutual orientations. An alternative description is obtained by arranging the vectors as $\{\mathbf{a}, \mathbf{b}, \mathbf{c}\}$ to form a 3×3 matrix $\mathbf{h}$ whose columns are, in order, the components of $\mathbf{a}$, $\mathbf{b}$, and $\mathbf{c}$. The volume is given by

$$\Omega = \|\mathbf{h}\| = \mathbf{a}\cdot(\mathbf{b}\wedge\mathbf{c}); \tag{2.3}$$

$\mathbf{a}$, $\mathbf{b}$, and $\mathbf{c}$, in that order, are assumed to be a right-handed triad.

The position $\mathbf{r}_i$ of a particle i can be written in terms of $\mathbf{h}$ and of a column vector $\mathbf{s}_i$, with components ξ_i, η_i, and ζ_i, as

$$\mathbf{r}_i = \mathbf{h}\mathbf{s}_i = \xi_i\mathbf{a} + \eta_i\mathbf{b} + \zeta_i\mathbf{c}. \tag{2.4}$$

Obviously $0<\xi_i,\eta_i,\zeta_i<1$ is the range of variation of the numbers $\xi_i,\eta_i,\zeta_i, i = 1,...,N$. The images of $\mathbf{s}_i$ are at $\mathbf{s}_i + (\lambda,\mu,\nu)$ where λ, μ, and ν are integers from $-\infty$ to $+\infty$.

Let a prime ′ denote a transpose of a vector or a tensor in the usual way. Then the square of the distance between i and j is given by

$$r_{ij}^2 = (\mathbf{s}_i - \mathbf{s}_j)'\mathbf{G}(\mathbf{s}_i - \mathbf{s}_j), \tag{2.5}$$

where the metric tensor $\mathbf{G}$ is

$$\mathbf{G} = \mathbf{h}'\mathbf{h}. \tag{2.6}$$

To complete the notation used here we note finally that the reciprocal space is spanned by the vectors

$$\frac{2\pi}{\Omega}\{\mathbf{b}\wedge\mathbf{c},\mathbf{c}\wedge\mathbf{a},\mathbf{a}\wedge\mathbf{b}\} \equiv \frac{2\pi}{\Omega}\boldsymbol{\sigma}. \tag{2.7}$$

The matrix $\boldsymbol{\sigma} \equiv \Omega\mathbf{h}'^{-1}$, carries information concerning the size and orientation of the MD cell.

A. The case when only hydrostatic pressure is applied

In Ref. (2) variability in the shape and size of the MD cell was obtained as follows: the usual set of $3N$ dynamical variables, that describe the positions of the N particles, was augmented by the nine components of $\mathbf{h}$. The time evolution of the $3N + 9$ variables was then obtained from the Lagrangian

$$\mathcal{L} = 1/2\sum_{i=1}^{N} m_i\dot{\mathbf{s}}_i'\mathbf{G}\dot{\mathbf{s}}_i - \sum_{i=1}^{N}\sum_{j>i}^{N}\phi(r_{ij}) + 1/2W\,\mathrm{Tr}\,\dot{\mathbf{h}}'\dot{\mathbf{h}} - p\Omega, \tag{2.8}$$

where p is the hydrostatic pressure that we intended to impose on the system. We shall comment later on W, which has dimensions of mass. Whether such a Lagrangian is derivable from first principles is a question for further study; its validity can be judged, as of now, by the equations of motion and the statistical ensembles that it generates. From Eq. (2.8) the equations of motion are easily found. We get

$$\ddot{\mathbf{s}}_i = -\sum_{j\neq i} m_i^{-1}(\phi'/r_{ij})(\mathbf{s}_i - \mathbf{s}_j) - \mathbf{G}^{-1}\dot{\mathbf{G}}\dot{\mathbf{s}}_i, i = 1,...,N, \tag{2.9}$$

$$W\ddot{\mathbf{h}} = (\boldsymbol{\pi} - p)\boldsymbol{\sigma}, \tag{2.10}$$

where, using the usual dyadic notation, and writing $\mathbf{v}_i = \mathbf{h}\dot{\mathbf{s}}_i$,

$$\Omega\pi = \sum_i m_i \mathbf{v}_i \mathbf{v}_i - \sum_i \sum_{j>i} (\phi'/r_{ij}) \mathbf{r}_{ij} \mathbf{r}_{ij}. \tag{2.11}$$

When $\mathbf{h}$ = constant, i.e., when the MD cell is time independent, $\dot{\mathbf{G}} = 0$, and due to Eq. (2.4), Eq. (2.9) becomes identical to Eq. (2.2). Of course the pressure in the system cannot be controlled; its value can be obtained from 1/3 of the trace of the average of π in the usual way.

The equations derived by Hoover *et al.*[8] have a close affinity with Eq. (2.9) above. Their two first-order equations of motion are equivalent to Eq. (2.9) on identifying $\dot{\mathbf{h}}\mathbf{h}^{-1}$ with their strain rate tensor transpose. Thus $\dot{\mathbf{h}}$ = (strain-rate tensor)$'\mathbf{h}$ and hence, as desired by them, $\mathbf{h}$ is driven by the strain-rate tensor. Their equations are thus suitable for the study of externally driven nonequilibrium phenomena.

Equation (2.10), however, allows the system to be driven by the dynamic imbalance between the externally applied stress and the internally generated stress tensor [the more general case is given in Eq. (2.25) below]; thus Eq. (2.10) allows one to study nonequilibrium phenomena driven by the above mentioned imbalance. In a state of equilibrium, making the external stress have an oscillatory time dependance will also allow one to study frequency dependent response of the system to external stimuli of various kinds. From Eq. (2.10) it also follows that the mass W determines the relaxation time for recovery from an imbalance between the external pressure and the internal stress. As discussed by Andersen[7], an appropriate choice for the value of W can make this relaxation time of the same order of magnitude as that of the relaxation of a small portion of a much larger sample. His suggestion is for a choice of W such that the above-mentioned relaxation time is of the same order of magnitude as the time L/c, where L is the MD cell size and c is the sound velocity. This obviously eliminates the arbitrariness in the choice of W and makes the calculation more realistic. However, if one is interested only in static averages, W can be chosen on the basis of computational convenience. In fact, in classical statistical mechanics, the equilibrium properties of a system are independent of the masses of its constituent parts.

From Eq. (2.8) one can construct the corresponding Hamiltonian following the usual rules of mechanics. Since the system is not subject to time dependent external forces this is a constant of motion. We get

$$\mathscr{H} = \sum_i 1/2 m_i \mathbf{v}_i^2 + \sum_i \sum_{j>i} \phi(r_{ij}) + 1/2 W \,\mathrm{Tr}\, \dot{\mathbf{h}}'\dot{\mathbf{h}} + p\Omega. \tag{2.12}$$

In equilibrium, at temperature T, $9/2 k_B T$ is contributed by the term with W and $3N/2\, k_B T$ by the other kinetic terms. Therefore to an accuracy of $3{:}N$ one finds that the constant of motion $\mathscr{H}$ is nothing but the enthalpy

$$H = E + p\Omega, \tag{2.13}$$

where

$$E = \sum_i 1/2 m_i \mathbf{v}_i^2 + \sum_i \sum_{j>i} \phi(r_{ij}). \tag{2.14}$$

Hence the Lagrangian in Eq. (2.8) generates a (p, H, N) ensemble.[7]

B. The case when a general stress is applied

The above formulation of MD, briefly presented before,[2] lends itself quite naturally to the introduction of *nonisotropic* external stress. This is not difficult to realize since in the classical theory of elasticity the notion of strain is intimately connected to the variations in the metric tensor and, as has surely been noticed, $\mathbf{G}$ is a natural constituent of our MD scheme.

In order to make the above remark practicable, we need to introduce, as is usually done in elasticity theory,[9] a reference state. Using the notions already introduced, this reference state of the system can be defined by its matrix $\mathbf{h}_0$ and volume $\Omega_0 = \|\mathbf{h}_0\|$. In this reference state a point in space given by the coordinate vector $\mathbf{s}$ is at the position

$$\mathbf{r}_0 = \mathbf{h}_0 \mathbf{s}. \tag{2.15}$$

A homogeneous distortion of the system changes $\mathbf{h}_0$ to $\mathbf{h}$, moving $\mathbf{r}_0$ to $\mathbf{r}$ where

$$\mathbf{r} = \mathbf{h}\mathbf{s} = \mathbf{h}\mathbf{h}_0^{-1}\mathbf{r}_0, \tag{2.16}$$

giving the displacement $\mathbf{u}$ due to the distortion:

$$\mathbf{u} = \mathbf{r} - \mathbf{r}_0 = (\mathbf{h}\mathbf{h}_0^{-1} - 1)\mathbf{r}_0. \tag{2.17}$$

Following Landau and Lifshitz[9] to define the strain tensor ϵ, and using x_μ to denote the components of $\mathbf{r}_0$,

$$\epsilon_{\lambda\mu} = 1/2\left(\frac{\partial u_\lambda}{\partial x_\mu} + \frac{\partial u_\mu}{\partial x_\lambda} + \sum \frac{\partial u_\nu}{\partial x_\mu}\frac{\partial u_\nu}{\partial x_\lambda}\right), \tag{2.18}$$

we find, using Eq. (2.6) which defines $\mathbf{G}$ and Eq. (2.17) which defines $\mathbf{u}$, that

$$\epsilon = 1/2(\mathbf{h}_0'^{-1}\mathbf{G}\mathbf{h}_0^{-1} - 1) \tag{2.19}$$

(We give in an appendix the connection between this formal definition of ϵ in Eq. (2.19) and that given in elementary text books.)

Having identified the strain ϵ, an expression for the elastic energy, V_{el}, can now be written. If $\mathbf{S}$ is the external stress[10] and p the hydrostatic pressure,

$$V_{el} = p(\Omega - \Omega_0) + \Omega_0 \,\mathrm{Tr}\,(\mathbf{S} - p)\epsilon. \tag{2.20}$$

In the limit of small strain,

$$\mathrm{Tr}\epsilon \simeq \Delta\Omega/\Omega_0 = (\Omega - \Omega_0)/\Omega_0. \tag{2.21}$$

Hence, when Eq. (2.21) is a valid approximation, we get the more familiar expression,

$$V_{el} \simeq \Omega_0 \,\mathrm{Tr}\, \mathbf{S}\epsilon. \tag{2.22}$$

Otherwise, i.e. when Eq. (2.21) is not valid, we need Eq. (2.20) to get the correct description of the effects of hydrostatic pressure.

To generalize the Lagrangian of Eq. (2.8) we need to substitute V_{el} of Eq. (2.20) in place of $p\Omega$ in Eq. (2.8) for $\mathscr{L}$. This gives us the new Lagrangian $\mathscr{L}_s$,

$$\mathscr{L}_s = \mathscr{L} - 1/2\, \mathrm{Tr}\, \mathbf{\Sigma}\mathbf{G}, \tag{2.23}$$

where the symmetric tensor $\mathbf{\Sigma}$ is related to the stress $\mathbf{S}$:

$$\mathbf{\Sigma} = \mathbf{h}_0^{-1}(\mathbf{S} - p)\mathbf{h}_0'^{-1}\Omega_0. \tag{2.24}$$

In deriving Eq. (2.23) we have dropped inconsequential constant terms in the energy viz. $p\Omega_0$ and Ω_0 Tr($\mathbf{S} - p$). We have also used the identity Tr(**AB**) = Tr(**BA**).

Using Eq. (2.23) to write the Lagrangian equations of motion we get Eq. (2.9) as before but Eq. (2.10) is now replaced by

$$W\mathbf{h} = (\pi - p)\sigma - \mathbf{h}\Sigma. \qquad (2.25)$$

It is easy to see that, analogous to Eq. (2.13), the Lagrangian $\mathscr{L}_s$ gives rise to a ($\mathbf{S}$,H_s,N) ensemble where the generalized enthalpy is

$$H_s = E + V_{el}, \qquad (2.26)$$

where E is given by Eq. (2.14) and V_{el} by Eq. (2.20).

The equations of motion Eq. (2.25) imply that a state of equilibrium will necessarily give zero for the average value of the right side of Eq. (2.25). This makes, using the definition of Σ, and writing σ_0 for the equivalent of Eq. (2.7),

$$\langle(\pi - p)\sigma\rangle = \langle\mathbf{h}\rangle\mathbf{h}_0^{-1}(\mathbf{S} - p)\sigma_0. \qquad (2.27)$$

This suggests, as is otherwise obvious intuitively, that for a system in equilibrium, to relate the constant matrix Σ [which controls the trajectories via Eq. (2.25)] to the external stress matrix $\mathbf{S}$ through Eq. (2.24), the most reasonable choice for the reference state appears to be $\mathbf{h}_0 = \langle\mathbf{h}\rangle$.

III. SUMMARY OF PREVIOUS STATIC CALCULATIONS

Before discussing our MD calculations in the next section and to put them in proper context, it is appropriate to recapitulate some of the results obtained by Milstein and collaborators.[3-5] In their model for Ni they assume a pairwise additive potential for the system, the pair potential being

$$\phi(r) = D\{\exp[-2\alpha(r - r_0)] - 2\exp[-\alpha(r - r_0)]\}. \qquad (3.1)$$

The constants D, α, and r_0 are fixed from a fit to the elastic constants c_{11} and c_{12} and to the lattice constant a_0 of fcc nickel.[3] Using this potential, all the properties are calculated at zero temperature, starting with a perfect fcc arrangement. The details of the procedure to calculate the stress-strain relation are given in Ref. 5. We only recall that in these calculations the lengths a_1, a_2, and a_3 of the initially cubic cell are allowed to adjust to changing stress conditions, but the angles between the cell edges are constrained to remain right angles. Moreover, for each set of a_1, a_2, and a_3 values, the atoms are given appropriately modified lattice positions but of course without any thermal disorder. The results that are obtained by Milstein and Farber[5] are as follows, when the system is subjected to a homogeneous [100] load.

Along a "primary path" in the stress-strain relation (as indicated in Fig. 1) one has $a_1 \neq a_2 = a_3$ and along this path three structures can be identified at zero load (Fig. 1). One of the three obviously is the original undistorted fcc state for which $a_1 = a_2 = a_3 = a_0$. The second, $B^{(1)}$, is a bcc structure obtained in compression (i.e. $a_1 < a_0$) at $a_1 \simeq 0.80\, a_0$, $a_2 = a_3 = \sqrt{2}a_1$. The third, $T^{(1)}$, is a tetragonal state, also obtained in compression with $a_1 \simeq 0.75 a_0$, $a_2 = a_3 \simeq 1.59 a_1$. The fcc state is at the absolute minimum of energy, $B^{(1)}$ at a local maximum and $T^{(1)}$ at a local minimum.[5] From a consideration of the Born stability criteria one concludes[4] that $B^{(1)}$

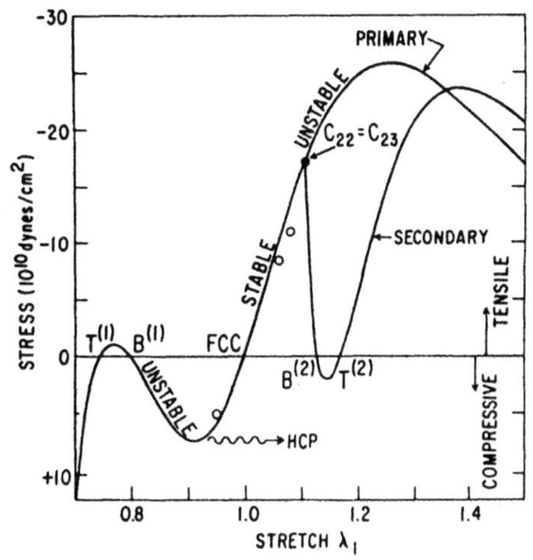

FIG. 1. Stress-strain relation under uniaxial load. O denote our results, lines are the static calculations of Milstein and Farber.[5] We found: (i) System failure at point (marked $c_{22} = c_{23}$) of intersection of the primary and secondary paths of Ref. 5; (ii) B and T tetragonal states on the zero load line spontaneously evolve into hcp structures; (iii) Under extreme compressive loading (marked by ⁓→) the system changes to an hcp structure.

and $T^{(1)}$ should be unstable states.

Along the same "primary path" but when a_1 is increased under the action of a tensile load the system is predicted[4] to fail at a value of stress $= 16\times 10^{10}$ dyne/cm^2 because at that point the Born stability condition $c_{22} - c_{23} > 0$ is violated. At this point $a_1 = 1.107\, a_0$.

However, it was discovered by Milstein and Farber[5] that at the point where $c_{22} = c_{23}$, a "secondary path" branches out of the primary path of extension. Along the secondary path, with decreasing load, the tetragonal symmetry is broken, i.e. $a_2 \neq a_3$ along this secondary path. This point, at which $c_{22} = c_{23}$, is hence a bifurcation point. Along the secondary path the zero load condition is encountered at two points where tetragonal symmetry is reestablished. One, a bcc state, $B^{(2)}$ at $a_1 = a_2 = \sqrt{2}a_3 = 1.1293\, a_0$ and the other a tetragonal state, $T^{(2)}$, with $a_1 = a_2 = 1.5701\, a_3 = 1.1696\, a_0$; $B^{(2)}$ is at a local energy maximum and $T^{(2)}$ at a local minimum.

The existence of this secondary path was envisaged in Ref. 5 as a possible mechanism for an fcc to bcc transition under conditions of a strictly uniaxial [100] tensile load.

IV. MOLECULAR DYNAMICS RESULTS

An MD calculation on Ni using the Lagrangian $\mathscr{L}_s$ of Eq. (2.23) makes it possible to check the validity of the results summarized above. This is because in MD with the Lagrangian $\mathscr{L}_s$ the restrictions of zero temperature and preassigned crystalline arrangement can both be removed.

As is customary we shall use reduced (i.e. dimensionless) quantities to specify various physical parameters. These reduced quantities will be denoted by an asterisk.

We shall use $D = 0.35059\times 10^{-12}$ erg to be the unit of energy; $L = r_0/2^{1/6} = 2.2518$ Å that of length; m the mass of the Ni atom, that of mass. Then $\tau = (mL^2/D)^{1/2} = 0.375 \times 10^{-12}$ sec will be the unit of time. D and r_0 are the quantities occurring in the potential[3] Eq. (3.1). All calculations were made with $p^* = 0$, $W^* = 20$. (This choice of W^* is suggested by our previous experience[2] in using the Lagrangian $\mathcal{L}$). Since we were interested in static averages no attempt was made to calibrate W^* so as to obtain a realistic value for the relaxation times related with the behavior of Eq. (2.25). As discussed in Sec. 2 the choice of W^* only affects the calculation of dynamical correlations.

Throughout, we have monitored the structure of the system by calculating the pair correlation function $g(r)$ as was done previously.[2] In certain situations it was found useful to plot out all the particle coordinates in suitable two-dimensional "slices" of the three-dimensional system so as to get a visual impression of the structure.

Since the various MD calculations fall into distinct categories, these will now be identified through subsections and suitable subheadings.

A. Preparation of a system under conditions of zero stress

The genesis of this and all subsequent calculations was a 500-particle system of Ni atoms on a perfect fcc lattice. The initial value of $\mathbf{h}^*$ was $h^*_{ij} = 1^*_0\delta_{ij}$, $1^*_0 = 7.8244$. Hence the MD cell at the start was a perfect cube. A small random displacement of the particles from the lattice sites and zero velocities provided the initial conditions for the ensuing dynamics. The equations of motion, Eqs. (2.9) and (2.25), were solved, with $\Delta t^* = 0.01$, using the predictor-corrector algorithm.[11] For this calculation Σ was put equal to zero in Eq. (2.25). Initially the temperature of the system was controlled with the standard procedures of MD. After a long period of "aging" to allow for the establishment of equilibrium, an MD run was made in which the temperature fluctuated around the mean value $T^* = 0.14$ (i.e.350° K). The MD cell remained a cube to high accuracy (i.e., the nondiagonal elements of $\mathbf{h}$ fluctuated around essentially zero values). The mean value of the three cell edges was the same, being $1^*_{0.14} = 7.88 \pm 0.02$. The pair correlation showed an unmodified fcc structure.

B Compressive uniaxial loading

A configuration (i.e., the values and the derivatives of all the dynamical variables) of the equilibrium run just described was used as the initial condition for a calculation in which a uniaxial [100] compressive load was applied. In other words Σ^*_{11} was nonzero positive, all other Σ^*_{ij} being 0. Σ^*_{11} was raised to a value $\Sigma^*_{11} = 15$ using two short intermediate runs at $\Sigma^*_{11} = 4$ and 8. Under the action of such a load the matrix $\mathbf{h}$ starts to change in a very well defined manner, i.e., the MD cell starts to distort away from its initial cubic shape. As expected there is a contraction in the [100] direction and an expansion in [010] and [001] directions while preserving the tetragonal symmetry to high accuracy. At $\Sigma^*_{11} = 15$ a long run of 5000 Δt^* was made. All averages were calculated with the last 2290 Δt^* of this run. We found

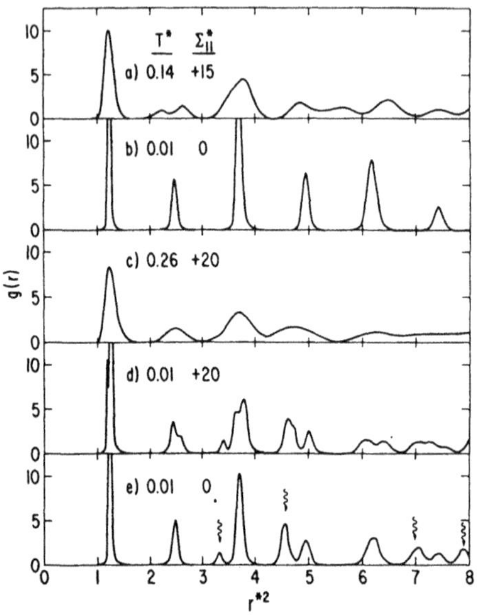

FIG. 2. Plot of pair correlation $g(r)$ to show structural transformation produced during compressive [100] loading. Note that abscissa denotes r^{*2} (not r^*). If r^*_i is the distance of the ith shell, r^{*2}_i/r^{*2}_1 have the values 1, 2, 3, 4, 5, etc. for fcc and 1, 2, 8/3, 3, 11/3, 4, 5, etc. for hcp ordering. (a) Shows fc tetragonal structure at $\Sigma^*_{11} = 15$ (compressive [100] load) and $T^* = 0.14$. Note the splitting of the second peak corresponding to a "stretch" $\lambda_1 = 0.95$ (Sec. B). (b) System reverts back to fcc structure on unloading. Removal of thermal effects by quenching[12] to $T^* = 0.01$ shows perfect fcc shell structure (Sec. B). (c) At $\Sigma^*_{11} = 20$ large changes occur in the MD cell parameters (shown in Fig. 3). $g(r)$ shows some differences at large r, compare with (a) above (Sec. C). (d) At $\Sigma^*_{11} = 20$, after completion of structural changes, quenching reveals shells of hcp ordering in addition to peak splitting as in (a) above (Sec. C). (e) Unloading the system shown in (d) to zero load, letting it equilibrate at $T^* = 0.21$ then quenching the system leads to unambiguous shell structure of hcp ordering (Sec. C).

$\langle h^*_{11}\rangle = 7.45 \pm 0.05$, $\langle h^*_{22}\rangle = \langle h^*_{33}\rangle = 8.08 \pm 0.05$, giving a "stretch" $\lambda_1 = \langle h^*_{11}\rangle/1^*_{0.14} = 0.95$. The nondiagonal $\langle h^*_{ij}\rangle$ were zero to within ± 0.05. Using $\langle \mathbf{h}^*\rangle$ for $\mathbf{h}^*_0$ one gets, from Eq. (2.24), $S_{11} = 5.25 \times 10^{10}$dynes/cm^2 for Ni, as seen in Fig. 1. This is in good agreement with the stress-strain relation given by Milstein and Farber.[5] (We note here that the nondiagonal elements of $\langle \mathbf{h}^*\rangle$ were so small as to be negligible; their inclusion or otherwise has no effect on the value quoted above for S_{11}).

The pair correlation of the system in equilibrium at $\Sigma^*_{11} = 15$, $T^* = 0.14$, is shown in Fig. 2(a). Note that $g(r)$, the pair correlation, is monitored as a function of $\mathbf{r}^2$; the second peak in Fig. 2(a) is split because the six original next nearest equidistant neighbors of the fcc structure break up into sets of two and four neighbors at slightly different distances. This splitting is not visible in the first peak because of thermal motion. The behavior of $\langle \mathbf{h}^*\rangle$ already made it clear that we had obtained a face-centered tetragonal structure. The $g(r)$ simply was a confirmation.

The fact that this equilibrium state under a uniaxial compressive load is reversibly connected to the original fcc state was easy to demonstrate. Using the end of the above $\Sigma^*_{11} = 15$, $T^* = 0.14$ equilibrium run as the initial condition, the load was reduced to $\Sigma^*_{11} = 0$ in several short runs with successively smaller values of Σ^*_{11}, the temperature in the final $\Sigma^*_{11} = 0$ run being $T^* = 0.14$. Finally, on reaching a no load condition, the system was quenched to a very low temperature to observe the pair correlation with the effects of thermal motion removed.[12] This is shown in Fig. 2(b) leaving no doubt that one has regained the original perfect fcc structure. Of course, all other indicators, namely the components of $\mathbf{h}^*$, pointed to the same conclusion.

C. Structure transformation under further compression

After completing the $\Sigma^*_{11} = 15$ study the load was raised to $\Sigma^*_{11} = 20$ and the dynamics was allowed to take its course according to the dictates of the equations of motion [Eqs. (2.9) and (2.25)].

The behavior of the system was remarkably different as might have been guessed by the title given to this subsection. Figure 3 shows the details of the changes with the passage of time. The MD cell, i.e., the $\mathbf{h}$ matrix, undergoes large and swift changes which cannot possibly be described as elastic deformations. In fact, as Fig. 3 shows, when equilibrium was reached the average values of the components of $\mathbf{h}^*$ were $\langle h^*_{11} \rangle = 5.54 \pm 0.03$, $\langle h^*_{22} \rangle = 9.76 \pm 0.06$,

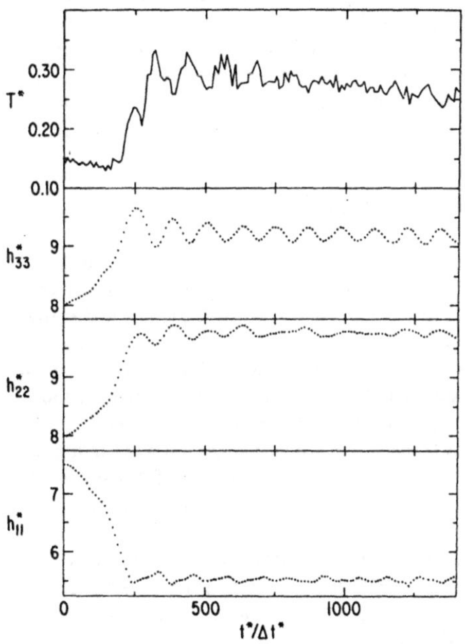

FIG. 3. Behavior of the MD cell parameters and the temperature of the system as it evolves in time when the compressive [100] load is increased from $\Sigma^*_{11} = +15$ to $+20$ (the latter value shown by ⤳ in Fig. 1). After a rapid change h^*_{ii} settle down to values at which an hcp structure can be accommodated in the MD cell (Fig. 2(c) shows the $g(r)$ at the end of the above time elapse). The rise in temperature is due to the release of elastic energy as the transformation occurs. See Sec. C for details.

$\langle h^*_{33} \rangle = 9.18 \pm 0.09$. The average of the nondiagonal elements was essentially zero. We note here that the tetragonal symmetry of the initially stressed state is destroyed as a result of this transformation; one gets instead an orthorhombic system, still under [100] compression. (See below for the description of this orthorhombic system under zero load.)

As seen in Fig. 3, contemporaneously with the rapid changes in $\mathbf{h}^*$, the temperature T^* increased from ~ 0.15 to ~ 0.30, finally settling down to ~ 0.25 or perhaps somewhat less. This is obviously a manifestation of an abrupt release of elastic energy in the relatively short time interval of $\sim 100 \Delta t^*$ (or ~ 0.4 ps). Of course we recall that the $\Sigma^*_{11} = 15$ to 20 change was made in one Δt^*.

The pair correlation at the end of the $\Sigma^*_{11} = 20$ run is shown in Fig. 2(c). In spite of the large deformations in $\mathbf{h}$ mentioned above, the $g(r)$ in Fig. 2(c) is very similar to the one in Fig. 2(a) and does not show clear evidence of a new arrangement of particles. However, the quenching technique does indicate that a new structure has been formed. The $g(r)$ after quenching is shown in Fig. 2(d). This figure displays not only the shell structure of a hexagonal close packed system but it also shows that under this loaded state otherwise single shells show up as split into two. Visual examination of particle positions suitably displayed showed that stacking faults were present in the hcp arrangement. It is interesting to note here that the direction of the compressive stress is normal to the c axis of the new, close-packed structure.

Having obtained the above-described transformation under a compressive load we reduced the load from $\Sigma^*_{11} = 20$ to $\Sigma^*_{11} = 0$ using several intermediate steps. The temperature in the final $\Sigma^*_{11} = 0$ run was $T^* = 0.21$. There was no structural change evident during this process. To reveal the structure clearly at the end of this process (of reducing the load from a high value to zero) we used the usual quench technique. The $g(r)$ is shown in Fig. 2(e). The shell

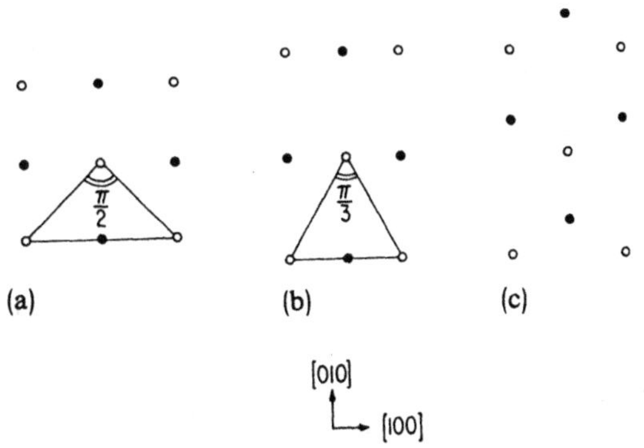

FIG. 4. Two planes of an fcc structure perpendicular to [001] are shown by O and ● respectively. (a)-(b) shows how the face-centered square structure changes to a triangular lattice on suitable compression in the [100] direction. (b)-(c) shows the necessary translation of the ● planes to achieve hcp ordering. At the same time spacing between O and ● planes has to correspond to the "c/a" value of an hcp arrangement, namely $\sqrt{8/3}$.

structure in Fig. 2(e) shows unambiguously an hcp arrangement. We recall that on reducing the load from $\Sigma^*_{11} = +15$ to $\Sigma^*_{11} = 0$ the system had gone back to its original fcc state (Sec. B 1).

The structure of the system which when quenched gave the $g(r)$ shown in Fig. 2(e) is of considerable interest. The distortions in the original cubic MD cell which allow an hcp structure to be accommodated were as follows: We found $\langle h^*_{22}\rangle/\langle h^*_{11}\rangle = \sqrt{3} \pm 0.01$, $\langle h^*_{33}\rangle/\langle h^*_{11}\rangle = \sqrt{8/3} \pm 0.01$, and an almost precise ($\langle h^*_{ij}\rangle \simeq 0.01, i \neq j$) rectangular parallelopiped. The ratio $\sqrt{3}$ is necessary to transform the [001] square-centered plains of the fcc structure into hexagonal plains [see Figs. 4(a)–4(b)]. The ratio $\sqrt{8/3}$ is the "c/a" value of the hcp structure. We note also that these distortions alone do not bring an fcc into an hcp structure; they have to be accompanied by the slip of alternate [001] planes as illustrated in Figs. 4(b)–4(c).

Given the highly coordinate nature of this transformation it is not surprising that in a rapid decompression defects such as stacking faults are produced.

D. System under tensile uniaxial loading

A study similar to the one just described was made by applying a tensile load; starting from the same configuration as was used to initiate the compressive runs, we raised the tensile load from 0 to $\Sigma^*_{11} = -20$ through a series of small intermediate steps at $\Sigma^*_{11} = -4, -8, -12, -16$. During these runs the MD cell remained tetragonal, elongated in the [100] direction and shortened equally in the [010] and [001] directions. At $\Sigma^*_{11} = -20$ the system was perfectly stable for the duration of a 4610 Δt^* MD run. During this run we obtained $T^* = 0.13$, $\langle h^*_{11}\rangle = 8.35 \pm 0.01$, $\langle h^*_{22}\rangle = \langle h^*_{33}\rangle = 7.75 \pm 0.02$, $\langle h^*_{ij}\rangle \simeq 0$ for $i \neq j$. The stress S evaluated from Eq. (2.24), using $\mathbf{h}_0^* = \langle \mathbf{h}^* \rangle$, gave $S_{11} = -8.6 \times 10^{10}$ dyne/cm^2 for Ni, the strain, expressed as a stretch, being $\lambda_1 = \langle h^*_{11}\rangle / l^*_{0.14} = 1.06$. These values are in complete accord with the results of Milstein and Farber[5] as seen in Fig. 1.

Increasing the tensile load to $\Sigma^*_{11} = -25$ gave a system like the one just mentioned. The final values were $T^* = 0.13$, $\langle h^*_{11}\rangle = 8.49 \pm 0.01$, $\langle h^*_{22}\rangle = \langle h^*_{33}\rangle = 7.72 \pm 0.04$, $\langle h^*_{ij}\rangle \simeq 0$ for $i \neq j$. This gave $\lambda_1 = 1.08$ and $S_{11} = -10.9 \times 10^{10}$ dyne/cm^2 again in accord with Ref. 5 as shown in Fig. 1. Note the proximity to the $c_{22} = c_{23}$ bifurcation point of Milstein and Farber.[5]

E. System failure under tensile uniaxial loading

On going from $\Sigma^*_{11} = -25$ to -30 the system "failed" as is described below. Figure 5 shows the values of h^*_{ii} as a function of time. The arrow indicates a region of time before which it has been found that it is possible to recuperate by rapidly reducing the load to more normal values. Beyond this time recovery may be difficult or impossible to achieve. This will be discussed in the following subsection.

As is clear from Fig. 5 the system does show a nontetragonal behavior at this increased load. This is in accord with the prediction of Ref. 5 regarding tetragonality in this region of strain, i.e., near the bifurcation point, made on the basis of static calculations. However, the fact of system failure really substantiates the conjectural remark of Milstein and Farber[5] that the very presence of the bifurcation point may lead to failure.

F. Recovery from failure

The configurations 100Δt^* before and after the time indicated by the arrow in Fig. 5 were used as the initial conditions in a series of calculations at tensile loads smaller than the $\Sigma^*_{11} = -30$ value which leads to failure.[13] This was done in an attempt to capture, but without success, a stable point on the secondary stress-strain path of Milstein and Farber[5] i.e., a system without tetragonal symmetry.

In the first attempt, with a configuration occurring earlier than the position of the arrow in Fig. 5, Σ^*_{11} was brought to zero in a series of intermediate runs of about 300 Δt^* each (i.e., with slow unloading), with Σ^*_{11} reduced by five units every time. The final $\Sigma^*_{11} = 0$ state was found to be a system in which a cubic MD cell was reestablished; the structure had fallen back on to the undistorted fcc state of the primary path (see Fig. 1).

In the second attempt, with a configuration occurring later than the position of the arrow in Fig. 5, the length of the intermediate runs was 100 Δt^* (i.e., rapid unloading). Already at $\Sigma^*_{11} = -25$, i.e., during the very first intermediate run, the differences between h^*_{22} and h^*_{33} tended to increase monotonically. This tendency continued during the $\Sigma^*_{11} = -20$ run and the system failed.

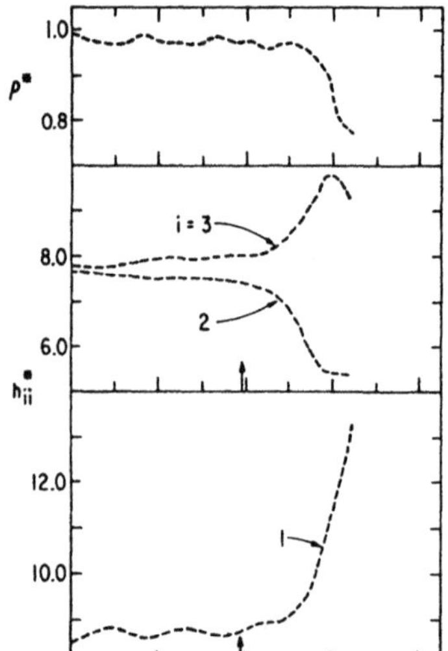

FIG. 5. At a tensile load of $\Sigma^*_{11} = -30$ the system fails after about 400 Δt. Initial breaking of tetragonal symmetry is clear ($h^*_{22} \neq h^*_{33}$). Arrow shows the critical moment after which even rapid unloading cannot prevent failure. Before the critical moment even slow unloading leads to the recovery of the original fcc state (see Fig. 1).

We conclude that along the secondary path the system is unstable at high values of stress. As is shown in the next subsection *this is so even at zero load.*

G. Stable structures at zero load

As indicated in Sec. 3, the $B^{(2)}$ and $T^{(2)}$ states of Ref. 5 occur on the secondary path and at zero load. A calculation completely analogous to the one we reported[2] on a Lennard-Jones system was made on the Ni system now under consideration. Starting at $B^{(2)}$ i.e., with $h^*_{11} = h^*_{22} = \sqrt{2}\ h^*_{33} = 8.7826$, $h^*_{ij} = 0$ for $i \neq j, T^* = 0.11$ and $\Sigma^* = 0$, an MD calculation was initiated with the equations of motion given by $\mathscr{L}_s$ (which is the same as $\mathscr{L}$, because $\Sigma = 0$) of Eq. (2.23). The system spontaneously started to produce distortions in the MD cell and after only 890 Δt^* the angles between the otherwise orthogonal axes became $\sim 89°$, $89°$, and $102°$ with an uncertainty of $\pm 3°$ in each case), the lengths of the edges of the MD cell became $l^*_1 \simeq l^*_2 = 8.88 \pm 0.12$, $l^*_3 = 6.42 \pm 0.08$, while the temperature of the system rose to $T^* = 0.20$. These changes were accompanied by changes in the pair correlation function $g(r)$. As usual we quenched the system and this revealed a $g(r)$ corresponding to hcp ordering with stacking faults.

Instead of starting at the $B^{(2)}$ point, we started a calculation like the one described above from a configuration near the $T^{(2)}$ point on the secondary path of Ref. 5. This also transformed very fast to a system with hcp ordering with stacking faults. Note that the $T^{(2)}$ point is very little different from the $T^{(1)}$ state of the primary path. Since $T^{(1)}$ was found by Milstein[4] to be unstable it is not surprising that the same happens for $T^{(2)}$. Same remark applies a fortiori to the zero load bcc state since obviously it is the same state whether it is on one path or the other.

We did start a calculation from the $T^{(1)}$ state as the initial condition and, quite as we expected, the system evolved and gave an hcp state. In this case, the final state had no stacking faults.

We thus conclude that at zero load and finite (and rather low!) temperatures for Ni only the fcc structure is stable among the ones considered by Milstein and Farber.[5] However, as our calculations have clearly shown, a new, locally stable state is possible and this is an hcp structure.

The relative stability of the fcc and hcp structures is of course a difficult question to answer. An answer to this question requires a detailed and accurate computation of the free energy difference. From our calculations we can only infer that the local free energy minima in configuration space seem to be locally very stable at the low temperatures we have investigated.

V. CONCLUDING REMARKS

In this paper we have illustrated some of the possibilities opened up by our new MD method for the investigation of the elastic behavior of solids. Our calculations have reproduced results obtained by others using static methods.

In Fig. 1 we have displayed how the stress-strain relation in the system is in good agreement with the results of Milstein and Farber.[5] We can go even further to state that the departure of our results away from their static results is itself in the right direction since the system will be "softer" in the presence of thermal agitation; Fig. 1 shows this indeed to be so. However, because of its greater flexibility, the new method has allowed us to predict new results. A clear example of this is a genuine bifurcation point: the *fcc→hcp transition under compression*. In principle such a transition could have been predicted by static calculations[5] of Milstein and Farber if they had searched for *all* possibilities instead of restricting themselves to the case of face-centered tetragonal states. In an MD calculation the system follows a dynamical trajectory determined by well defined dynamical equations and is free to assume the crystal structure most suited to the interaction potential and the ambient conditions of temperature and stress. Being a fully dynamical method it even allows the study of the kinetics of temperature and stress induced transformations.

A few words of caution are in order especially if one wants to make comparisons with laboratory experiments. One limitation comes from the small system size and periodic boundary conditions used for the dynamical simulation. This probably reduces but certainly does not eliminate the occurrence of extended defects such as grain boundaries or dislocations; these defects play an important role in the plastic behavior of solids. This limitation is, however, not intrinsic and can be much reduced with the use of much larger MD systems. The possibility of creating extended defects probably accelerates the breakdown of the system under applied stress. Thus the tensile strength we have obtained ($S_{11} \sim 11 \times 10^{10}$ dyne/cm^2) is probably only an upper bound to the true value.

The second weakness of model calculations is the use of potentials like the one in Eq. (3.1). This potential is purely empirical without much justification from a microscopic point of view. Moreover, the normal state properties of crystalline systems are determined by values of the potential and its derivatives only at distances where the various neighbors are situated. The detail of the short-range repulsion probably plays a crucial role in the phenomena discussed in this paper and is not so well determined when one uses normal state properties of the crystal to determine the parameters of the empirical potential. Especially in Ni a proper account of the role of d-electrons may call for the use of many body forces in a more realistic calculation. This again is not a intrinsic problem with simulation methods.

Some of the conclusions reached in this paper are expected to be consequences only of the symmetry; as such they have a greater range of validity than the potential used in arriving at those conclusions. For instance it is possible that the fcc-hcp transition in a [100] compression is observable. Since the many-body terms in the "true" potential function are expected to be of short range their presence or absence cannot play a determining role in this particular structural transformation.

From the point of view of structural transformation in monotonic systems it will be most useful to have a systematic study of model systems with pair interactions of the type $1/r^m - 1/r^n$ when these systems are put under various forms of external stress. Such calculations and also the ones we have presented are an exact consequence (apart from problems related with the numerical solutions of differential equations) of the potential function used. As such they serve the function of data for testing approximate theoretical models of the phenomena under investigation.

There is the intriguing possibility that the fcc-hcp transition might actually occur in Ni single crystals under extreme conditions of shock.

ACKNOWLEDGMENT

This work was supported by the U.S. Department of Energy.

APPENDIX A

The connection between the formal definition of the strains, Eq. (2.19), and that given in elementary textbooks[14] is as follows. Consider a reference state in which the MD cell is a rectangular parallelopiped with edges parallel to the Cartesian reference frame. In this case $\mathbf{h}_0$ is diagonal, the diagonal elements being the lengths of the three edges. In a distorted state the MD cell is given by $\mathbf{h} = \mathbf{h}_0 + \Delta\mathbf{h}$ say. Up to linear terms in $\Delta\mathbf{h}$, we find from Eq. (2.6),

$$\mathbf{G} = \Delta\mathbf{h}'\mathbf{h}_0 + \mathbf{h}_0'\Delta\mathbf{h} + \mathbf{h}_0'\mathbf{h}_0. \tag{A.1}$$

Using this in Eq. (2.19), due to the diagonal form of $\mathbf{h}_0$ we get

$$\epsilon_{ii} = (h_{ii} - h_{0,ii})/h_{0,ii}, \tag{A.2}$$

and for $i \neq j$

$$\epsilon_{ij} = 1/2(h_{ij}/h_{0,jj} + h_{ji}/h_{0,ii}). \tag{A.3}$$

Equation (A2) connects the diagonal terms of the strain to the length variation of the edges of the cell while Eq. (A3) gives the expected relation between the off-diagonal terms and the changes in the angles between the edges.

Up to order Δh, the volume change is given, as in Eq. (2.21), by $\Delta\Omega = \Omega_0 \operatorname{Tr} \epsilon$.

[1]E. R. Cowley, G. Jacucci, M. L. Klein, and I. R. McDonald, Phys. Rev. B **14**, 1758 (1976).
[2]M. Parrinello and A. Rahman, Phys. Rev. Lett. **45**, 1196 (1980).
[3]F. Milstein, J. Appl. Phys. **44**, 3825 (1973).
[4]F. Milstein, J. Appl. Phys. **44**, 3833 (1973).
[5]F. Milstein and B. Farber, Phys. Rev. Lett. **44**, 277 (1980).
[6]J. P. Hansen and I. R. McDonald, *Theory of Simple Liquids* (Academic, London, 1976).
[7]H. C. Andersen, J. Chem. Phys. **72**, 2384 (1980).
[8]W. G. Hoover, D. J. Evans, R. B. Hickman, A. J. Ladd, W. T. Ashurst, and B. Moran, Phys. Rev. A **22**, 1690 (1980).
[9]L. D. Landau and E. M. Lifshitz, *Theory of Elasticity* (Pergamon, Oxford, 1959).
[10]Note that in the definition of **S** through Eq. (2.20) we have adopted a certain sign convention: $\mathbf{S} = +p$ corresponds to a system under hydrostatic pressure p.
[11]A. Rahman, in NATO Advanced Study Institute Series: *Correlation Functions and Quasiparticle Interactions in Condensed Matter*, edited by J. W. Halley (Plenum, New York, 1978).
[12]This procedure, used systematically in Ref. 2, is extremely useful for the purpose of resolving peaks in $g(r)$ which otherwise would merge in neighboring peaks especially if the corresponding coordination numbers are very different. For example, in an hcp structure the third and the fouth shells are very close to each other with a population, respectively, of 2 and 18 neighbors.
[13]In an MD calculation the "failure" of a system is manifested by a monotonic change in the h_{ii} and in an obvious disruption of the shell-like structure of $g(r)$. A monotonic drop in the value of the density is simply a consequence of the behavior of the h_{ii}.
[14]C. Kittel, *Introduction to Solid State Physics* (Wiley, New York, 1968).

Structural Transformations in Solid Nitrogen at High Pressure

Shuichi Nosé and Michael L. Klein
Chemistry Division, National Research Council of Canada, Ottawa, Ontario K1A 0R6, Canada
(Received 24 January 1983)

The Andersen-Parrinello-Rahman constant-pressure molecular-dynamics technique has been generalized to study molecular crystals. The method, which appears to have wide applicability, is used to study structural transformations in solid nitrogen at high pressures.

PACS numbers: 64.70.Kb, 64.30.+t

The constant-pressure computer-simulation technique proposed by Andersen[1] for the study of liquids and elegantly generalized by Parrinello and Rahman,[2] in order to discuss the relationship of crystal structure to pair potentials in atomic systems, has been further extended to handle the case of molecular solids and liquids. As a pedagogical example we discuss structural transformations that we observe upon isobarically quenching a model solid-nitrogen crystal initially held at room temperature and high pressure. In particular, at a pressure of 70 kbar and $T = 140$ K we observe a spontaneous shear distortion of the nitrogen crystal that transforms the structure from cubic to trigonal. This distortion is accompanied by preferential alignment of certain of the molecules with the trigonal axis. The results of our calculations will likely have important implications for the interpretation of high-pressure experiments that can now be carried out with diamond anvil cells.[3,4]

It has been known for some time that depending upon the conditions of temperature and pressure solid nitrogen can exist in different crystal structures.[5] The relative stability of the phases at lower pressures depends upon fine details of the intermolecular potentials such as the balance between the short-range forces that are responsible for the shape of the molecule and the longer-ranged electrostatic interactions that occur between the quadrupole moments.[6-8] Recent investigations, using diamond anvil cells to generate the required high pressure, have revealed that at room temperature solid nitrogen exists in a cubic plastic crystal phase, with eight molecules per unit cell. This *Pm3n* crystal structure is akin to the famous *A*15 structure of high-temperature superconductors.[4] In the case of solid nitrogen there are chains of disklike, orientationally disordered molecules along the cube directions (D_{2d} sites) together with an interpenetrated bcc lattice of more spherically disordered molecules (T_d sites). Previous computer-simulation studies of this structure[9] have confirmed that the disorder is dynamic and not static and this in turn is consistent with the intramolecular Raman spectrum which consists of two distinct lines.[3]

At room temperature solid nitrogen is therefore isomorphous with the high-temperature low-pressure structure of solid oxygen.[10] The analogy is even more striking when one notes that the ratio of intermolecular separation to bond length is virtually identical in these two solids. Since it is known that in solid oxygen the electrostatic quadrupole-quadrupole interaction is unimportant it is perhaps reasonable to conclude that at the pressures generated in the diamond anvil cell, the same situation pertains in solid nitrogen. Accordingly, one should then be able to model the intermolecular interactions by simple atom-atom potentials and this what we have done here.

When the *Pm3n* phase of solid oxygen is cooled it transforms into a rhombohedral structure, $R\bar{3}m$, appropriate to the close packing of rodlike objects, namely two-dimensional rafts of close-packed molecules stacked in the fcc sequence *ABCABC*..., with all the molecular axes parallel to the stacking direction.[11] It is natural to ask whether or not the isobaric cooling of solid nitrogen from room temperature also generates the $R\bar{3}m$ structure found in solid oxygen and indeed recent static energy calculations have given support to this possibility.[12] However, unpublished experiments carried out at Los Alamos National Laboratory by the group of Mills and Schiferl show no evidence for such a transition.[13] In order to cast some light on this somewhat puzzling observation we have carried out a series of constant-pressure molecular-dynamics (MD) calculations for solid nitrogen at high pressure. Since in this method of calculation[2] the MD cell can change both its size and shape in response to the net imbalance between the thermally generated internal stresses and the externally applied pressure P_{ex}, it offers the possibility of studying isobaric structural transformations at finite temperatures.

Accordingly, we have in mind the study of the structure and dynamics of a system of rigid dumbbell molecules interacting through atom-atom potentials whose parameters are chosen to approximate solid nitrogen. The equations of motion for the system follow from the Lagrangian formulation of Parrinello and Rahman (PR) suitably augmented to allow for rotational motion. In particular, if $\underline{h}$ is the matrix formed by three time-dependent vectors $\vec{a}$, $\vec{b}$, and $\vec{c}$ that specify the periodically replicated MD cell, then its volume $V=\det\underline{h}\equiv\vec{a}\cdot\vec{b}\times\vec{c}$. The position of molecule i in the cell is given by $\vec{r}_i=\underline{h}\,\vec{s}_i$ where the components of $\vec{s}$ range from 0 to 1. It then follows that $r_i^2=\vec{s}_i{}'\underline{G}\,\vec{s}_i$, where $\underline{G}=\underline{h}'\underline{h}$ and the prime denotes a transposed quantity. We write the augmented PR Lagrangian as

$$L=\tfrac{1}{2}\sum_i m_i\dot{\vec{s}}_i{}'\underline{G}\dot{\vec{s}}_i-\sum_i\sum_{j>i}\varphi(r_{ij},\vec{\alpha}_i,\vec{\alpha}_j)$$
$$+\tfrac{1}{2}W\mathrm{Tr}(\dot{\underline{h}}'\dot{\underline{h}})+\tfrac{1}{2}\sum\vec{\omega}_i\underline{I}_i\vec{\omega}_i-P_{ex}V,$$

where $\vec{\omega}_i$ and $\underline{I}_i$ are the angular velocity and inertia tensor of molecule i whose mass is m_i. The potential energy, which depends upon the orientations $\vec{\alpha}$, is assumed to be the sum of interactions between site k on molecule i and site l on molecule j, separated by the distance $R_{ij}{}^{kl}$, thus

$$\varphi(r_{ij},\vec{\alpha}_i,\vec{\alpha}_j)\equiv\varphi_{ij}=\sum_{kl}\varphi_{ij}{}^{kl}.$$

The equations of motion for the scaled center-of-mass coordinates are then formally identical to those of PR, namely

$$\ddot{\vec{s}}_i=\underline{h}^{-1}\vec{f}_i/m_i-\underline{G}^{-1}\dot{\underline{G}}\dot{\vec{s}}_i.$$

Here $\vec{f}_i$ is the net force acting on molecule i, which in turn is the sum of all the forces acting on its individual interaction sites, k:

$$\vec{f}_i=\sum_k\vec{f}_i{}^k=\sum_{j\neq i}\sum_{kl}-(\partial\varphi_{ij}{}^{kl}/\partial\vec{R}_{ij}{}^{kl}).$$

The equations of motion for the angular coordinates are identical to those in the usual constant-volume MD method and follow by relating the time rate of change of the angular momentum $\vec{M}_i=\underline{I}_i\vec{\omega}_i$ to the torque; thus we have

$$\dot{\vec{M}}_i=\sum_k\vec{d}_i{}^k\times\vec{f}_i{}^k,$$

where $\vec{d}_i{}^k$ is the position vector of site k relative to the center of mass. In our application we have rewritten the above equations to describe the motion of the intramolecular bond vector.

Finally, the equations governing the dynamics of the unit cell are also similar to those of PR, namely

$$W\ddot{\underline{h}}=(\underline{\pi}-P_{ex})\underline{\sigma},$$

where $\underline{\pi}$ is a stress tensor and the matrix $\underline{\sigma}\equiv V\underline{h}'^{-1}$.

We give only brief technical details of the calculations because the method and more extensive applications will be presented elsewhere. The centers of mass of 64 rigid nitrogen molecules, interacting through a simple Lennard-Jones atom-atom potential $\varphi(R)=4\epsilon[(\sigma/R)^{12}-(\sigma/R)^6]$ with $\epsilon=37.3$ K and $\sigma=3.31$ Å and fixed half bond length $d=0.1646\sigma$, were initially arranged in the observed high-pressure room-temperature $Pm3n$ structure with appropriate, randomly chosen, orientations.[9] A MD calculation was carried out at constant volume with use of a time step of 0.005 ps and the equations of motion were integrated with a fifth-order algorithm for the translational coordinates, $\vec{s}_i$, and a fourth-order one for angular coordinates, $\vec{\alpha}_i$. This initial calculation confirmed the results of an earlier study of this crystal by conventional MD methods.[9] The condition of constant volume was then replaced by the constraint of constant pressure, which in our case was fixed by $P_{ex}\sim70$ kbar, and the calculation continued. Since the MD cell is now free to undergo volume and shear fluctuations we were able to examine whether or not the initial $Pm3n$ configuration is at least locally stable for our adopted potential.

The left-most panel in Fig. 1 shows the evolution of the room-temperature high-pressure MD run for 2500 time steps. Also shown is the evolution of the lengths of the MD cell vectors and the angles between them. Although there are considerable fluctuations in both of these quantities, which are related to the elastic compliances of the crystal,[14] there is no evidence for any deviation from a cubic unit cell. More detailed investigation of the behavior of individual molecules confirmed the $Pm3n$ structure. In particular, the disklike and spherical disordered molecules retained their identities.

The next panels in Fig. 1 show the effect of sequentially quenching the room-temperature solid-nitrogen structure in steps of 50 K down to 100 K while maintaining $P_{ex}=70$ kbar. The volume is seen to decrease systematically as the temperature is lowered and the three MD cell vectors remain equal in length, within the statistical un-

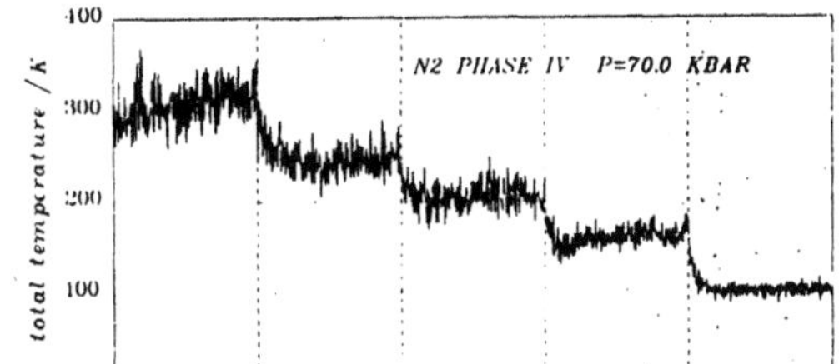

FIG. 1. Time evolution of quantities characterizing the basic molecular-dynamics cell used for solid nitrogen at 70 kbar. The left-most panel of the figure refers to $T = 300$ K and displays results for 2500 time steps, the first 500 of which are used to scale the temperature of the system. Subsequent panels contain simillar information but for temperatures sequentially lowered by about 50 K. The last panel, which refers to $T = 100$ K, clearly reveals the trigonal distortion that takes the crystal from $Pm3n$ to $R3c$ (see text).

certainties. However, below 150 K the MD cell angles are no longer 90°; the system has undergone a spontaneous shear distortion of 3°. An inspection of the structure so generated reveals that the spherically disordered molecules are now all aligned along the trigonal direction, the crystal having $R3c$ symmetry, but not the $R\bar{3}m$ structure of solid O_2. At the pressure set in the MD calculation (70 kbar) the experimental molar volume is about 1 cm^3 mol^{-1} less than that found in the simulation.[4] This discrepancy could probably be rectified by examining a more realistic intermolecular potential.[11] However, the aim of this study was not simply to obtain a quantitative modeling of solid nitrogen; it was intended also to provide a demonstration of the elegance and utility of constant-pressure MD calculations. Additional calculations using a potential model that included a quadrupole-quadrupole interaction yield essentially the same result, confirming our earlier discussion. Careful heating and cooling of the crystal reveals little hysteresis and locates the shearing transition at 140 K. These studies also show that this transition is proceeded, at about 230 K, by the alignment of the disklike molecules parallel to a cube axis. Both transformations involve small volume changes, certainly less than 0.1 cm^3 mol^{-1}.

In summary, we have reported computer-simulation calculations on the effect of isobaric cooling of a nitrogen crystal. At 70 kbar we observe cubic ($Pm3n$, $z = 8$) → cubic($I2_13$, $z = 64$) → trigonal ($R3c$, $z = 64$) phase transformations at around $T = 230$ K and $T = 140$ K, respectively. Our prediction of a trigonal distortion at low temperatures agrees with two recent energy minimization calculations.[15,16] It will be of considerable interest to see if our calculation is consistent with structural and spectroscopic measurements on high-pressure nitrogen crystals.

We thank Aneesur Rahman and Roger Impey for important contributions to our understanding of constant-pressure MD calculations. We also thank Bob Mills, D. Schiferl, and Richard LeSar for discussing their unpublished work on solid nitrogen at high pressures. Koji Kobashi, Dick Etters, Chandran Chandrasekharan, Sam Trevino, Don Tsai, and Ian McDonald have, at various times, freely discussed with us their own related, but different, calculations on nitrogen crystals. One of us (S.N.) is the recipient of a fellowship from the Natural Sciences and Engineering Research Council of Canada.

[1]H. C. Andersen, J. Chem. Phys. 72, 2384 (1980).
[2]M. Parrinello and A. Rahman, Phys. Rev. Lett. 45, 1196 (1980).
[3]R. LeSar, S. A. Ekberg, L. H. Jones, R. L. Mills, L. A. Schwalbe, and D. Schiferl, Solid State Commun. 32, 131 (1979).
[4]D. T. Cromer, R. L. Mills, D. Schiferl, and L. A. Schwalbe, Acta Crystallogr. Sec. B 37, 8 (1981).
[5]T. A. Scott, Phys. Rep. 27C, 89 (1976).

[6]J. C. Raich and R. L. Mills, J. Chem. Phys. 55, 1811 (1971).
[7]J. C. Raich and N. S. Gillis, J. Chem. Phys. 66, 846 (1977).
[8]T. Luty, A. van der Avoid, and R. M. Berns, J. Chem. Phys. 73, 5305 (1980).
[9]M. L. Klein, D. Lévesque, and J. J. Weis, Can. J. Phys. 59, 530 (1981).
[10]T. H. Jordan, W. E. Streib, H. W. Smith, and W. N. Lipscomb, Acta Crystallogr. 17, 777 (1964).
[11]E. M. Hörl, Acta Crystallogr. 15, 845 (1962), and Acta Crystallogr. Sec. B 25, 2515 (1969).
[12]K. Kobashi, A. A. Helmy, R. D. Etters, and I. L. Spain, Phys. Rev. B 26, 5996 (1982).
[13]We thank Dr. R. L. Mills and Dr. D. Schiferl of Los Alamos National Laboratory for allowing us to quote this information.
[14]M. Parrinello and A. Rahman, J. Chem. Phys. 76, 2662 (1982).
[15]V. Chandrasekharan, R. D. Etters, and K. Kobashi, Phys. Rev. B (to be published).
[16]R. LeSar, unpublished calculations.

New High-Pressure Phase of Solid ^{4}He Is bcc

Dominique Lévesque and Jean-Jacques Weis

Laboratoire de Physique Théorique et Hautes Energies, Université Paris-Sud, F-91405 Orsay, France

and

Michael L. Klein

Chemistry Division, National Research Council of Canada, Ottawa, Ontario K1A 0R6, Canada

(Received 6 June 1983)

The effect of isobarically heating solid ^{4}He at high density is investigated with use of constant-pressure molecular-dynamics calculations and a realistic interatomic pair potential. At $P \sim 16$ GPa the observed sequence of stable phases with increasing temperature is fcc → bcc → liquid. The presence of a new, thermally stabilized, bcc phase for high-density solid ^{4}He is thus confirmed.

PACS numbers: 67.80.Gb, 62.50.+p, 64.70.Dv, 64.70.Kb

The behavior of solid helium at high densities continues to attract attention.[1-6] The recent observation of a cusp point on the melting curve of solid ^{4}He around room temperature suggested a triple point and hence the presence of a new high-pressure solid phase.[6] While no direct evidence was obtained for a solid → solid phase transition, the characteristics of melting changed above the cusp point; the enhanced premelting and smaller volume change were interpreted as evidence for the presence of a bcc solid.[6]

The interatomic pair potential for He is well established.[7] Moreover, computer simulation techniques have now advanced to the point that it is possible to explore directly the relationship between an interatomic potential and its preferred crystal structure at finite temperature.[8,9] Hence, the question of a possible solid → solid phase transition occurring in solid ^{4}He at high pressures and the nature of the phases involved are amenable to direct investigation. Accordingly, we have carried out a series of isobaric molecular-dynamics simulations.[8] Under an external pressure corresponding to about 16 GPa, we observed the following sequence of stable phases with increasing temperature: fcc → bcc → liquid. The presence of a stable bcc phase preceding the melting of high-density ^{4}He thus confirms the speculations based upon the anomalous behavior of the melting curve.[6] If corresponding-states arguments are invoked, other insulating crystals will likely also exhibit a bcc phase when subjected to analogous conditions of temperature and pressure.

The constant-pressure molecular-dynamics (MD) technique that we employed is documented in the literature[8] and so we omit most of the details. Essentially, the equations of motion are integrated by standard techniques for a system of 432 ^{4}He atoms initially arranged on a body-centered tetragonal lattice with periodic boundary conditions. If the lengths of the basis vectors of the MD cell a, b, and c are in the ratios $1{:}1{:}\sqrt{2}$ the system is equivalent to an fcc lattice, whereas if the ratios of $a{:}b{:}c$ are 1:1:1 and the angles (a,b), (b,c), (a,c) are still 90°, the system is bcc. The new MD method allows the system to change both its volume and shape in response to any instantaneous imbalance between an externally applied pressure and spontaneously generated thermal stresses. By this technique several solid → solid phase transitions have been successfully investigated.[8,9] In particular, a previous study of the relative stability of fcc and bcc lattices employing the Lennard-Jones 12-6 potential established that at low temperature and pressures the close-packed structure is preferred.[8] Although still widely used as an effective pair potential for the rare gases, the 12-6 potential is unfortunately a poor approximation to their true potentials.[10]

Accordingly, in our simulations that are designed specifically to model He, we have employed a realistic pair potential.[7] Before describing our results we mention two questions, not specifically addressed here, that perhaps will deserve further study: quantum effects and three-body forces. Our defense for their omission from our calculations is based largely on pragmatic and intuitive arguments. First, it is known that quantum effects have only a modest influence on the location of the He melting line around room temperature.[4] Second, if solid He can be approximated by an oscillator model obeying the Grüneisen equation of state, then the leading quantum contribution to the pressure is given by $\Delta P_Q = (3\gamma RT/20V)(\theta/T)^2$. With use of extrapolations of lower-pressure data[3] to estimate the Grüneisen parameter γ and the Debye temperature θ, we obtain $\Delta P_Q V/RT \sim 1$ which is small compared with the pressure of interest to us here ($PV/RT \sim 20$).

Finally, we note that of all the rare gases, three-body forces are the smallest in helium[10]; we estimate $\Delta P_3 V/RT \sim 0.4$. Moreover, we are interested in a possible fcc→bcc phase transition and since it is known that at constant density three-body forces are essentially identical in these two structures[11] they are not likely to greatly influence such a transition. We now describe our results.

All of the calculations were carried out under a nominal pressure of 15.2 GPa but when due allowance is made for the contribution from ΔP_Q and ΔP_3 the effective external pressure is approximately 16 GPa. Figure 1(a) shows how at 310 K an initial ideal bcc structure, whose MD cell started with $a=b=c=14.44$ Å and $(a,b)=(b,c)=(a,c)=90°$, evolved spontaneously to a fcc-like structure whose MD cell had $\vec{a}=(12.52$ Å, -1.28 Å, $0)$, $\vec{b}=(1.40$ Å, 13.39 Å, 1.00 Å$)$, $\vec{c}=(0.03$ Å, -0.18 Å, 17.53 Å$)$, and $(a,b)=90°$, $(b,c)=89.2°$, $(a,c)=89.6°$. The transformation was completed over a period of 1500 time steps of 1.23 fs and the resulting fcc structure was stable when followed for a further 6500 steps. We conclude therefore that under the conditions $T=310$ K, $P=16$ PGa, solid ^{4}He is fcc. However, at higher temperatures, for example at 360 K [see Fig. 1(b)], the reverse behavior was exhibited, namely starting from an fcc MD cell with $a=b=12.85$ Å, $c=18.17$ Å, $(a,b)=(b,c)=(a,c)=90°$, the system evolved spontaneously to become predominantly bcc with an MD cell $\vec{a}=(12.04$ Å, -3.50 Å, 0.56 Å$)$, $\vec{b}=(-0.66$ Å, 12.52 Å, -0.10 Å$)$, $\vec{c}=(0.36$ Å, -0.97 Å, 20.35 Å$)$, $(a,b)=109°$, $(b,c)=95°$, and $(a,c)=87°$. Between these two temperatures the system is observed to oscillate from fcc to bcc as a function of time but with a definite preference for bcc as the temperature increases. We have attempted to classify the character of these "mixed" crystals by monitoring the distribution of neighbors around a given atom and also the distribution of angles specified by the interatomic bond vectors. In this way individual atoms are classified as either bcc-like or fcc-like and we have an efficient means to follow the time evolution of the crystal structure. Typical plots are shown inset in Fig. 2(a), which also shows the variation of the enthalpy H as a function of

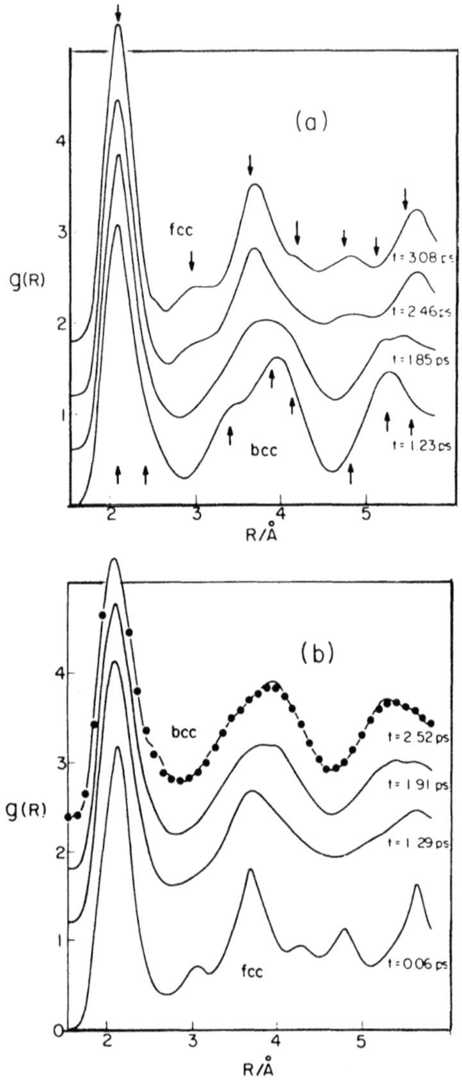

FIG. 1. (a) Time evolution of the pair distribution function, $g(r)$, for solid ^{4}He at $P=16$ GPa and $T=310$ K starting from a bcc lattice at $t=1.23$ ps and finishing at $t=3.08$ ps with an fcc lattice. The arrows indicate peak positions in the ideal bcc and fcc lattices. (b) Time evolution of $g(r)$ at $P=16$ GPa and $T=360$ K starting from an fcc lattice at $t=0.06$ ps and finishing with a bcc lattice at $t=2.52$ ps; the dots are results for $t=6.47$ ps.

temperature for the three phases. Figure 2(b) gives an analogous plot for the density variations.

The arrow in Fig. 2(a) indicates the melting temperature predicted for the Aziz ^{4}He potential[7]

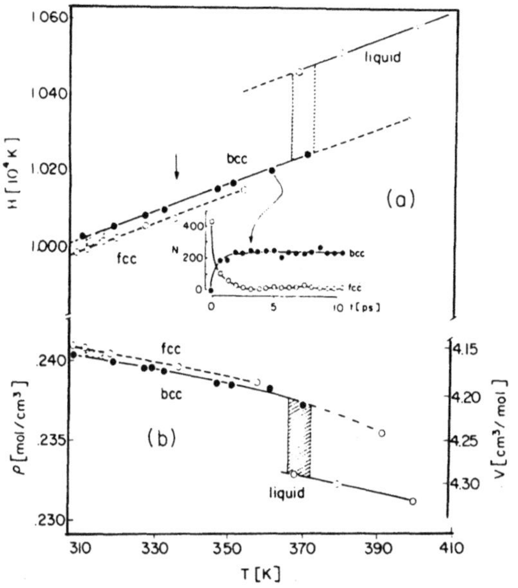

FIG. 2. (a) Temperature dependence of the enthalpy H for three phases of ^4He. The inset indicates the evolution of local structure (circles, fcc; dots, bcc). The arrow indicates the melting temperature of the fcc phase calculated in Ref. 6 (see text) while the shaded regions indicate our estimated transitions. (b) Temperature dependence of the density ρ for solid ^4He at $P = 16$ GPa.

based upon a perturbation theory that used as a reference system Monte Carlo results for an inverse twelfth-power potential.[4] In our constant-pressure molecular-dynamics calculations, the system is still solid under these conditions, and hence our results point to shortcomings in the perturbation theory, under these extreme conditions. We note from Fig. 2(b) that the volume change on melting is about 2.1% at $T \sim 370$ K. The experimental estimate[6] is certainly greater than 1%.

If we use corresponding-states scaling, an fcc-bcc-liquid triple point would occur in Xe at about $P = 1.2$ Mbar and $T = 10^4$ K. Unfortunately, under these conditions Xe may well no longer be an insulator[12-16] and in any case it may already be bcc for other reasons.[17] The rare gases Ne and Ar would appear to be more serious candidates to exhibit this phenomenon since their metallization occurs under more extreme conditions. However, the most promising candidate for study may well be solid hydrogen, which at room temperature and high pressure appears to be hcp.[18-20] We estimate a possible hcp-bcc-liquid triple point at $P = 26.5$ GPa and $T = 950$ K. In studies of solid H_2 the requisite pressure (density) has already been far exceeded, but studies have yet to be reported above room temperature.

In summary we have observed that at $P \sim 16$ GPa, solid fcc ^4He transforms to bcc before melting. Further work is needed to establish the range of existence of this new phase and to explore more deeply the reasons for its stability. These questions will be taken up elsewhere.

One of us (M.L.K.) thanks the Centre National de la Recherche Scientifique–National Research Council of Canada exchange agreement for sponsoring a visit to Orsay during which this work was completed, and Professor J. M. Besson for his comments. The Laboratoire de Physique Théorique et Hautes Energies is a Laboratoire associé au Centre National de Recherche Scientifique.

[1]J. M. Besson and J. P. Pinceaux, Science 206, 1073 (1979).

[2]J. P. Franck and W. B. Daniels, Phys. Rev. Lett. 44, 259 (1980).

[3]B. L. Mills, D. H. Liebenberg, and J. C. Bronson, Phys. Rev. B 21, 5137 (1980).

[4]P. Loubeyre and J. P. Hansen, Phys. Lett. 80A, 181 (1980).

[5]D. A. Young, A. K. McMahan, and M. Ross, Phys. Rev. B 24, 5119 (1981).

[6]P. Loubeyre, J. M. Besson, J. P. Pinceaux, and J. P. Hansen, Phys. Rev. Lett. 49, 1172 (1982); J. M. Besson, private communication.

[7]R. Aziz, V. P. S. Nain, J. S. Carley, W. L. Taylor, and G. T. McConville, J. Chem. Phys. 70, 4430 (1979).

[8]M. Parrinello and A. Rahman, Phys. Rev. Lett. 45, 1196 (1980), and 50, 1073 (1983), and J. Appl. Phys. 52, 7182 (1981).

[9]S. Nosé and M. L. Klein, Phys. Rev. Lett. 50, 1207 (1983).

[10]J. A. Barker, in *Rare Gas Solids*, edited by M. L. Klein and J. A. Venables (Academic, London, 1976).

[11]R. D. Murphy and J. A. Barker, Phys. Rev. A 3, 1037 (1971).

[12]K. S. Chan, T. L. Huang, T. A. Grybowski, T. J. Whetton, and A. L. Ruoff, Phys. Rev. B 26, 7116 (1982); D. A. Nelson and A. L. Ruoff, Phys. Rev. Lett. 42, 383 (1979).

[13]M. Ross and A. K. McMahan, Phys. Rev. B 21, 1658 (1980).

[14]K. Asaumi, T. Mori, and Y. Kondo, Phys. Rev. Lett. 49, 837 (1982).

[15]I. Makarenko, G. Weill, J. P. Itie, and J. M. Besson, Phys. Rev. B 26, 7113 (1982); K. Syassen, Phys. Rev. B 25, 6548 (1982); D. Schiferl, R. L. Mills, and L. E.

Trimmer, Solid State Commun. 46, 783 (1983).
[16]H. Niki, H. Nagara, H. Miyagi, and T. Nakamura, Phys. Lett. 79A, 428 (1980); J. Hama and S. Matsui, Solid State Commun. 37, 889 (1981).
[17]A. K. Ray, S. B. Trickey, and A. B. Kunz, Solid State Commun. 41, 351 (1982).
[18]I. F. Silvera and R. J. Wijngaarden, Phys. Rev. Lett. 47, 39 (1981).
[19]H. Shimizu, E. M. Brody, H. K. Mao, and P. M. Bell, Phys. Rev. Lett. 47, 128 (1981).
[20]J. van Straaten, R. J. Wijngaarden, and I. F. Silvera, Phys. Rev. Lett. 48, 97 (1982).

IL NUOVO CIMENTO Vol. 4 D, N. 4 Ottobre 1984

Relative Stability of f.c.c. and b.c.c. Structures for Model Systems at High Temperatures (*) (**).

A. Rahman

Materials Science and Technology Division, Argonne National Laboratory Argonne, Ill. 60439, U.S.A.

G. Jacucci

Dipartimento di Fisica dell'Università di Trento - 38050 Povo, Italia

(ricevuto il 20 Marzo 1984)

Summary. — Free-energy, entropy and volume differences between face-centered and body-centered cubic structures have been evaluated for model rare gas and alkali metal crystals by using the method of overlapping distributions. Stable phases are predicted in agreement with the behaviour of real materials in the regions of validity of classical mechanics and in agreement with the results of previous dynamical-simulation studies of crystal nucleation from the melt and of polymorphic transformations. The existence of a stable b.c.c. phase at high pressure and temperatures is predicted in this way for Lennard-Jones solids, while no high-pressure f.c.c. phase is expected for model Rb and Cs systems. We also show the possibility of making calculations of free-energy barriers to displacive crystalline transformations along a prescribed trajectory in configuration space.

PACS. 64.70. – Phase equilibria, phase transitions, and critical points of specific substances.

(*) To speed up publication, the authors of this paper have agreed to not receive the proofs for correction.

(**) Work supported by the U.S. Department of Energy, Istituto per la Ricerca Scientifica e Tecnologica, Trento, and Gruppo Nazionale di Struttura della Materia del C.N.R., Italy.

1. – Introduction.

To understand how different lattice structures arise in nature, one needs to postulate the interaction potential between various particles and then to use it to evaluate the relevant thermodynamic potential which ensures the stability of one structure relative to another. For complex materials the first already is a formidable problem. For certain classes of monatomic materials (rare-gas solids, alkali metals) the problem of the interaction potential is understood well enough so as to warrant a serious attack on the problem of evaluating thermodynamic-potential differences.

However, attempts have been made in the literature to solve the problem of the relative stability of lattice structures without detailed knowledge of the interactions. ZENER (1), in particular, suggested that failure of certain shear modes could be responsible for low-temperature structural transformations in the light alkali metals, whereas FRIEDEL attempted to show (2) that the b.c.c. phase has favorable higher entropy than the close-packed phase, this arising from first-neighbor interactions and the topology of the BCC structure. ALEXANDER and MCTAGUE went even further (3) and used the Landau theory of melting transitions (4) to show the « universal » preference of b.c.c. structures in all systems.

In recent years molecular dynamics and Monte Carlo calculations (5) have made it possible to go beyond the realm of conjecture and unrealistic general arguments. In a study of homogeneous nucleation in monatomic supercooled fluids it has been found that the structure of the nucleated phase is sensitive to the pair potential (6). Specifically, a potential V_1 (say, a Lennard-Jones pair interaction) gave rise to structure S_1 (a close packed structure) by homogeneous nucleation, while V_2 (an alkali metal) gave S_2 (a b.c.c. structure). Already in these studies it was shown that a system with potential function V_1 nucleating into structure S_1 would spontaneously change its structure to S_2 when the potential function was switched to V_2.

(1) C. ZENER: in *Phase Stability in Metals and Alloys*, edited by P. S. RUDMAN, J. STRINGER and R. I. JAFFEE (McGraw-Hill, New York, N.Y., 1967), p. 25.
(2) J. FRIEDEL: *J. Phys. (Paris) Lett.*, **35**, 159 (1974).
(3) S. ALEXANDER and J. MCTAGUE: *Phys. Rev. Lett.*, **41**, 702 (1978).
(4) L. D. LANDAU: *Phys. Z. Sowjetunion*, **11**, 26, 545 (1937).
(5) See, for example, *a*) *Modern Theoretical Chemistry*, Vol. **5**, edited by B. J. BERNE (Plenum Press, New York, N.Y., 1977) and *b*) W. W. WOOD and J. J. ERPENBECK: *Annu. Rev. Phys. Chem.*, **27**, 319 (1976).
(6) C. S. HSU and A. RAHMAN: *a*) *J. Chem. Phys.*, **70**, 5234 (1979); *b*) **71**, 4974 (1979).

Even more recently [7] it has become possible to study, by molecular-dynamics techniques, the behavior of perfect crystalline solids at constant external stress in such a way as to allow the solid the freedom to adopt a different structure if the ambient temperature and pressure conditions make a change favorable.

This new method of studying polymorphic transitions has also been applied successfully to the study of such transitions in the relatively more complicated case of binary ionic systems (KCl, AgI) [8].

Thus it has now become necessary to attempt to calculate, by fundamental statistical mechanics, the free-energy differences between various crystalline structures, given the potential of interaction. The methodology for pursuing such a program now exists and various facets of this methodology will be dealt with in an appropriate section below.

The purpose of the present paper is to report the results of some initial calculations on two potential functions and two structures. For a Lennard-Jones system we have evaluated F_F-F_B, where F is the Helmholtz free energy and subscripts F and B stand for f.c.c. and b.c.c., respectively.

Another class of systems for which a quantitatively accurate theoretical Hamiltonian is available in the literature is that of alkali metals. The perturbation theory provides an effective ionic-pair interaction between the alkali ions embedded in a sea of conduction electrons. Because of the quantitative reliability of these predictions, we have considered it profitable to undertake a study of crystalline alkali metals from the point of view of structural and free-energy differences.

Lennard-Jones systems and alkali metals have an added attraction for a study of f.c.c., b.c.c. structures and of free-energy differences between the structures because in the laboratory rare-gas solids (for which LJ is a good model system) crystallize as a close-packed structure and alkali metals as a b.c.c. structure.

From the point of view of fundamental theory, recent developments [9] have shown how to calculate the change in thermodynamic potentials as systems freeze from liquid to solid; the same theory also shows how to calculate the free-energy difference between various structures in terms of correlation functions in the liquid state out of which the structures are nucleated. This is another reason why a serious attempt at calculating free-energy differences between various structures and for various model systems is called for.

---

[7] M. PARRINELLO and A. RAHMAN: *Phys. Rev. Lett.*, **45**, 1196 (1980); *J. Appl. Phys.*, **52**, 7182 (1981).

[8] *a*) M. PARRINELLO and A. RAHMAN: *J. Phys. (Paris)*, **42**, C6-511 (1981); *b*) M. PARRINELLO, A. RAHMAN and P. VASHISHTA: *Phys. Rev. Lett.*, **50**, 1073 (1983).

[9] T. V. RAMAKRISHNAN and M. YUSSOUFF: *Phys. Rev. B*, **19**, 2775 (1979).

2. – Free-energy calculation.

In classical statistical mechanics the central problem is the evaluation of the configurational integral Q that appears in the partition function Z:

$$Q = \int_{\Omega} \exp[-\beta V(\{\boldsymbol{r}_i\})]\, d\boldsymbol{r}_1 \dots d\boldsymbol{r}_i \dots d\boldsymbol{r}_N ,$$

$$Z = \exp[-\beta F] = (2\pi m/\beta h^2)^{3N/2} Q .$$

F is the Helmholtz free energy, β the inverse temperature, h Planck's constant and Ω the volume of the N-particle system. The problem we wish to solve is that of the free-energy difference of two systems with potential functions V_1 and V_2 spanning the same configuration space $\{\boldsymbol{r}_i\} \equiv R$. In order to calculate this free-energy difference, we need

$$Q_1/Q_2 = \exp[-\beta(F_1 - F_2)] = \int_{\Omega} \exp[-\beta V_1(R)]\, d\tau \Big/ \int_{\Omega} \exp[-\beta V_2(R)]\, d\tau .$$

The integration extends over the same volume in the configuration space R. Simple algebraic manipulation gives

$$(1) \qquad \begin{cases} Q_1 = \int \exp[-\beta(V_1 - V_2)] \exp[-\beta V_2]\, d\tau , \\ Q_1/Q_2 = \langle \exp[-\beta(V_1 - V_2)] \rangle_2 , \end{cases}$$

where the symbol $\langle\ \rangle_2$ has an obvious meaning. This « one-sided » expression for Q_1/Q_2 and similarly a mirror image expression with 1, 2 interchanged is a practical way of evaluating $\Delta F = F_1 - F_2$ by sampling configurations from one of the two canonical ensembles.

A « two-sided », more symmetric, formulation is obtained as follows by constructing normalized distribution functions $h_1(\Delta)$ and $h_2(\Delta)$ in the canonical ensembles generated by V_1 and V_2, respectively. Thus, by definition,

$$h_1(\Delta) = \int \delta(V_1 - V_2 - \Delta) \exp[-\beta V_1]\, d\tau / Q_1 = \langle \delta(V_1 - V_2 - \Delta) \rangle_1$$

$$h_2(\Delta) = \int \delta(V_1 - V_2 - \Delta) \exp[-\beta V_2]\, d\tau / Q_2 = \langle \delta(V_1 - V_2 - \Delta) \rangle_2 .$$

But, by the same manipulation as before,

$$h_1(\Delta) = \int \delta(V_1 - V_2 - \Delta) \exp[-\beta(V_1 - V_2)] \exp[-BV_2]\, d\tau / Q_1 =$$

$$= \exp[-\beta\Delta] \int \delta(V_1 - V_2 - \Delta) \exp[-\beta V_2]\, d\tau / Q_1 = \exp[-\beta\Delta]\, h_2(\Delta) Q_2/Q_1 .$$

Hence we have

$$Q_1/Q_2 = \exp[-\beta\Delta]\, h_2(\Delta)/h_1(\Delta)\,. \tag{2}$$

The left-hand side being independent of Δ, the right side gives a practical method of evaluating the fraction Q_1/Q_2. The functions $h_i(\Delta)$ can be estimated in a Monte Carlo sampling by compiling a histogram $h_i^*(\Delta)$ of the frequency of occurrence of configurations with $V_1 - V_2$ between $\Delta - \delta/2$ and $\Delta + \delta/2$.

The introduction of these energy distribution overlap methods for the calculation of free-energy differences is due to VALLEAU and collaborators ([10]) and to BENNETT ([11]), building upon earlier work of McDonald and Singer ([12]). An alternative equation for the two-sided evaluation was proposed by BENNETT ([11]); this was denoted by him as the acceptance ratio method. This and other methods, including eq. (2), are discussed and compared in an illuminating paper ([11]) by BENNETT, in which a complete treatment of the relative statistical error can also be found. Bennett's equation can be derived from the following identity:

$$\int_\Omega w(R)\exp[-\beta(V_1+V_2]\,d\tau = \langle w(R)\exp[-\beta V_2]\rangle_1 Q_1 = \langle w(R)\exp[-\beta V_1]\rangle_2 Q_2\,,$$

or

$$\frac{Q_2}{Q_1} = \frac{\langle w(R)\exp[-\beta V_2]\rangle_1}{\langle w(R)\exp[-\beta V_1]\rangle_2}\,, \tag{3}$$

where $w(R)$ is an arbitrary function of the configuration; the function w was introduced by BENNETT in order to optimize the estimation of the free-energy difference $(1/\beta)\ln(Q_1/Q_2)$. According to him ([11]) statistical arguments can be used to show that the choice

$$w(R) = \text{const } x\left(\frac{Q_2}{n_2}\exp[-\beta V_1] + \frac{Q_1}{n_1}\exp[-\beta V_2]\right)^{-1}, \tag{4}$$

where n_1, n_2 are the numbers of statistically independent configurations from each Monte Carlo sample, minimizes the expectation value of $(\Delta F_{est} - \Delta F)^2$. Substituting eq. (4) in eq. (3) gives

$$\frac{Q_2}{Q_1} = \frac{\langle f(\beta(V_1-V_2)+c)\rangle_1}{\langle f(\beta(V_1-V_2)-c)\rangle_2}\exp[+c]\,, \tag{5}$$

([10]) J. P. VALLEAU and G. M. TORRIE: *Modern Theoretical Chemistry*, Vol. **5**, edited by B. J. BERNE (Plenum Press, New York, N.Y., 1977) and references therein.
([11]) C. H. BENNETT: *J. Comput. Phys.*, **22**, 245 (1976).
([12]) I. R. MCDONALD and K. SINGER: *Discuss. Faraday Soc.*, **43**, 40 (1967); *J. Chem. Phys.*, **47**, 4766 (1967); **50**, 2308 (1969).

where $c = \ln(Q_2 n_1/Q_1 n_2)$ and $f(x) = 1/(1+\exp[x])$ is the Fermi function. Equation (5) is true for any value of the shift constant c, but the particular value specified minimizes the expected square error [11]. The magnitude σ^2 of this minimum square error can be conveniently expressed in terms of n_1, n_2 and of the normalized variances of f in ensembles 1 and 2 [11]:

$$\sigma^2 = \frac{1}{n_1}\frac{\langle f^2\rangle_1 - \langle f\rangle_1^2}{\langle f\rangle_1^2} + \frac{1}{n_2}\frac{\langle f^2\rangle_2 - \langle f\rangle_2^2}{\langle f\rangle_2^2}. \tag{6}$$

The use of Monte Carlo sampling to evaluate free-energy differences with eq. (2) or eq. (5) rests on, there being sufficient overlap between h_1^* and h_2^*. This is clearly seen in eq. (2), since, if there is no overlap, the ratio h_1^*/h_2^* cannot be estimated. The use of eq. (1) is even more critical, in that it requires essentially complete overlap of one of the two histograms by the other. We first note that, from eq. (1) and the definition of $h_1(\Delta)$ and $h_2(\Delta)$,

$$\int h_1(\Delta)\mathrm{d}\Delta = 1 = \frac{Q_2}{Q_1}\langle\exp[-\beta(V_1-V_2)]\rangle_2 = \frac{Q_2}{Q_1}\int_{-\infty}^{+\infty} h_2(\Delta)\exp[-\beta\Delta]\mathrm{d}\Delta\,.$$

In regions of Δ where $h_2(\Delta)$ is small, the histogram $h_2^*(\Delta)$ will be zero. If $h_1 = (Q_2/Q_1)h_2\exp[-\beta\Delta]$ is significant in these regions, the use of h_2^* will lead to very large errors since Q_1/Q_2 will be determined from the area in those regions. No such problem exists for the two-sided methods eqs. (2) and (5) if partial overlap of h_1^* and h_2^* exists. The statistical error, however, will depend strongly on the extent of the overlap [11].

Where no overlap occurs, bridging distributions may be employed. These correspond to potential-energy functions intermediate between V_1 and V_2. When necessary, several such intermediate ensembles may be employed. This technique called «multistage sampling» rests on repeated applications of eqs. (2) or (5). Alternatively, if the gap between the two histograms h_i^* is not much larger than their widths, a graphical method, proposed by BENNETT [11], may be employed to extract the value of Q_2/Q_1. One gets from eq. (2)

$$\beta(F_2-F_1) + \tfrac{1}{2}\beta\Delta + \log h_1(\Delta) = \log h_2(\Delta) - \tfrac{1}{2}\beta\Delta\,. \tag{7}$$

We assumed that no values of Δ exist for which h_1^* and h_2^* are both nonzero. Hence eq. (7) is not directly useful. However, a plot of the quantities $\frac{1}{2}\beta\Delta + \log h_1(\Delta)$ and $\log h_2(\Delta) - \frac{1}{2}\beta\Delta$ against Δ produces two curves whose extrapolations into the gap are separated by the constant vertical displacement $\beta(F_2-F_1)$. The free-energy difference can then be read easily from the graph (see fig. 1).

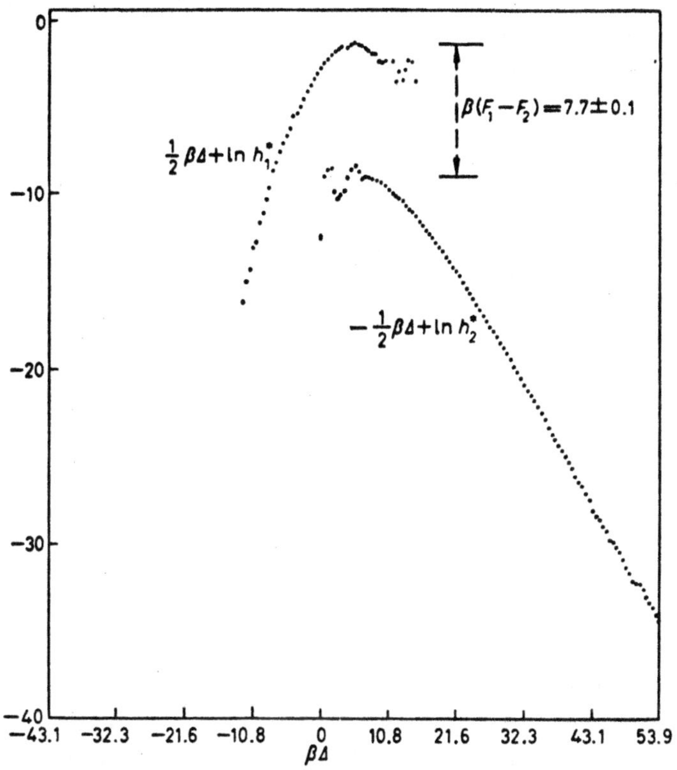

Fig. 1. – Use of the graphical method (ref. [1]) for the evaluation of free-energy differences in the case of little or no overlap between $h_1^*(\Delta)$ and $h_2^*(\Delta)$ (see the text). Data are taken from the comparison of an LJ crystal of f.c.c. structure to the corresponding quasi-harmonic model. From the above graph we get $\beta\,\Delta F$ per particle $= 0.143 \pm$ $\pm\, 0.002$ at $T = 0.5$ and $\varrho = 1.0$.

3. – Strategy and methodology of the present work.

This section is divided in three subsections on the following topics:

a) crystal free energy and the f.c.c.-b.c.c. transformation,

b) alternative thermodynamical paths for multistage sampling,

c) choice of sampling algorithm.

The reader interested in technical procedures for the evaluation of ΔF and of its statistical error should refer to the preceding paper in this review (called I henceforth) and to ref. [11] for material not included in sect. **2**. Here we only note that the results presented in sect. **5** to **7** below are obtained with Bennett's method [11], eq. (5) of sect. **2**. The error quoted will be the expected error or square root of the variance given by eq. (6) of sect. **2**. The numbers n_1 and n_2 representing the number of statistically independent configurations from each Monte Carlo sample are obtained by making use of the correlation length τ in the way described in I. Results have also been obtained by using the simple two-sided overlap method in the form used by VALLEAU and and collaborators [10]. In that case the quoted error is the root mean square deviation of the r.h.s. of eq. (2) of sect. **2** in the domain of values of the variable Δ where overlap of the histograms $h_1^*(\Delta)$ and $h_2^*(\Delta)$ occurs.

a) Crystal free energy and the f.c.c.-b.c.c. *transformation.* The standard way of evaluationg the free energy of crystals is from a normal-mode expansion of the potential energy [13]. In the quasi-harmonic approximation knowledge of the normal-mode frequencies ν_j as functions of the density is all the information needed, besides the value of the potential energy of the geometrically perfect lattice. At high temperatures the partition function Z of a three-dimensional harmonic crystal containing N particles has the limiting form

$$Z_N = \exp[-\beta V_0] \prod_{j=1}^{3N-3} (k_B T/h\nu_j) Z_{c.m.},$$

where V_0 is the potential energy of the static lattice and $Z_{c.m.}$ is the center-of-mass contribution. Quite a bit is known about harmonic crystals, particularly with nearest-neighbor (n-n) interaction. For example, the entropy for small crystals has been investigated [14,15], and the excess (with respect to the Einstein model) Born-von Kármán harmonic entropies for the two close-packed three-dimensional periodic (n-n) crystals have been obtained as functions of N by HOOVER [15]:

$$(S^e/Nk_B)_{f.c.c.} = 0.24680 - (\ln N)/N, \qquad (S^e/Nk_B)_{HCP} = 0.24541 - (\ln N)/N.$$

Furthermore, for f.c.c. solids, thermodynamic properties including pressure, Gruneisen γ and elastic constants have been calculated with lattice dynamics for both the Lennard-Jones 6-12 and the exponential six pair potentials [16]; the interaction potential is extended to convergence, but the harmonic approximation is made for the forces. The application of this general « lattice dynamics » method to the f.c.c.-b.c.c. polymorphic transformation has been carried out by HOOVER *et al.* [17] for inverse power potentials $\varphi(r) = \varepsilon(\sigma/r)^n$.

---

(13) G. JACUCCI and N. QUIRKE: *Free energy calculations for crystals*, in *Computer Simulation in the Physics and Chemistry of Solids*, in *Lect. Notes in Phys.* (Springer Verlag, Berlin, 1982).
(14) A. BEYERLEIN and Z. W. SALSBURG: *J. Chem. Phys.*, **47**, 3763 (1967).
(15) W. G. HOOVER: *J. Chem. Phys.*, **49**, 1981 (1968).
(16) A. C. HOLT, W. G. HOOVER, S. G. GRAY and D. R. SHORTLE: *Physica*, **49**, 61 (1970).
(17) W. G. HOOVER, D. A. YOUNG and R. GROVER: *J. Chem. Phys.*, **56**, 2207 (1972).

Lattice dynamics can be applied only to stable structures, having real ν_j. The close-packed f.c.c. structure satisfies the mechanical-stability requirement in the quasi-harmonic approximation for many systems: n-n interaction, the inverse-power and the Lennard-Jones potentials and for hard spheres. The b.c.c. crystal is unstable to shear for n-n, hard spheres, and for the inverse power potential for $n > 7$ ([17]). We have verified that it is also unstable for the Lennard-Jones (12-6) potential summed to convergence, in the quasi-harmonic approximation presently discussed.

The result of the inverse-power potential investigation has been that, for $n < 7$, the system undergoes a polymorphic transformation from the close-packed structure at low temperature to the body-centered structure before melting ([17]). It has been concluded that the looser packing of the body-centered arrangement induces more low-frequency shear modes. Their effect can be seen in the larger values of the excess entropy:

$$\Delta S/Nk_{\mathrm{B}} = 1/(N-1) \sum_{j=1}^{3N-3} \ln\,(\nu_{\mathrm{E}}/\nu_j)$$

and of the mean square displacement relative to the Einstein model: $\langle r^2\rangle/\langle r_{\mathrm{E}}^2\rangle = 1/(3N-3) \sum_{j=1}^{3N-3} (\nu_{\mathrm{E}}/\nu_j)^2$. Taking into account the different Helmholtz free energy of the Einstein model for the two structures, the polymorphic transition can be located in thermodynamic space. The interest of this investigation partly resides in the fact that some metallic crystals show this type of polymorphic transformation and that the repulsive interaction in these systems is certainly much softer than that in the (12-6) LJ.

Analysis of experimental data, for example on alkali halides ([18]) and on argon ([19]) crystals, has shown, however, that the anharmonic contributions to the free energy of the lattice cannot be neglected. More frequently these days the study of disorder and of anharmonicity in condensed systems is made by using computational rather than analytical tools. For the f.c.c. Lennard-Jones solid, for example, the comparison of lattice dynamics approximations with Monte Carlo calculations ([16]) has clearly demonstrated the contribution to thermodynamic properties of terms in the energy beyond quadratic in the particle displacements. Methods described in sect. **2**, in particular, have been used in a «direct Monte Carlo calculation of anharmonic free energies» by Pollock ([20]); this is an application of the one-sided method of eq. (1) above; the system studied is a periodically repeating LJ, nearest-neighbor, f.c.c. crystal containing 32 particles. The values of the free energy obtained by this method

---

([18]) M. P. Tosi and F. Fumi: *Phys. Rev.*, **131**, 1485 (1963).
([19]) J. Kuebler and M. Tosi: *Phys. Rev. A*, **137**, 1617 (1965).
([20]) E. L. Pollock: *J. Phys. C*, **9**, 1129 (1976).

are in better argeement with experimental rare-gas crystal data than the previous estimates from conventional anharmonic theory. We note in passing that from fig. 1 of Pollock ([20]) we read $\Delta F/NkT = 0.12$ at $T = 0.5$ and $\varrho = 1.0$. From our figure 1 we can see that two-sided overlap evaluation of $\Delta F/NkT$ gives 0.143 ± 0.002.

More recently, KRATKY ([21]) has investigated the relative stability of f.c.c. and HCP hard-sphere crystals using a method for the direct evaluation of the partition function by an importance sampling technique and taking account of the special features of the hard-sphere potential. He finds the f.c.c. structures favored by an entropy difference of $0.02k_B$.

To our knowledge no machine calculations exist for the free-energy difference between various polymorphs of a crystal with continuous pair interaction.

b) Possible thermodynamic paths in multistage of the f.c.c.-b.c.c. *transformation.* A simple geometrical transformation relates the f.c.c. and b.c.c. lattices. By using the cubic directions in a b.c.c. lattice for reference, a deformation which stretches one cubic direction by a factor $2^{\frac{1}{3}}$ and contracts the other two, each by a factor $2^{-\frac{1}{6}}$, produces a body-centered tetragonal lattice where the body center is at the center of a square of side $2^{\frac{1}{2}}$. It is easy to see that the resulting structure is f.c.c. and that the density is unchanged in this transformation. This rather special feature of the transformation will be used in devising various paths for the multistage sampling method.

The methodology described in sect. **2** requires the use of two Hamiltonians defined in the same configuration space. In the f.c.c.-b.c.c. transformation we shall choose two identical Hamiltonians which contain the same number of particles interacting with the same pair potential; but the two configurations upon which the Hamiltonians act will be related by a one-to-one correspondence via the space transformation mentioned above. In other words we wish to evaluate ΔF between two regions in configuration space using the same Hamiltonian. The geometric deformation is used to relate the configurations in the two regions in a one-to-one correspondence. An alternative, but operationally equivalent description would be to interpret the calculation as one for two different Hamiltonians H_1, H_2 having the same configuration space; for example, with the usual b.c.c. structure the pair interaction will be $\varphi(r^2 = x^2 + y^2 + z^2)$ for H_1, and for H_2 it will be anisotropic and given by $\varphi(s)$ with $s^2 = 2^{-\frac{1}{3}}(x^2 + y^2) + 2^{\frac{2}{3}}z^2$.

The free energy F_0 of a subdomain of volume Ω_0 of configuration space is determined by the configurational integral Q_0, the integration being restricted to Ω_0. If the domain Ω_0 corresponds to a well-defined pocket of the potential-energy surface V and is well separated from other such pockets by a ridge of

([21]) K. W. KRATKY: *Chem. Phys.*, **57**, 167 (1981).

high values of V, then Ω_c is a metastable region. Trajectories given by the solution of the equations of motion will be trapped in the pocket for a long time. While the value of F pertaining to the unrestricted domain Ω is the relevant quantity in the thermodynamic limit and will be dominated by the truly stable configurations, the values of F_0 obtained from different regions of metastability will describe the relative stability of such metastable regions. The escape time from the metastable regions will depend on certain properties of V: the height of the ridges, the existence of low-lying saddle points along paths joining neighboring metastable regions, and of course of the temperature.

In the case of a LJ crystal a low-energy barrier path is available [7] for a b.c.c. to f.c.c. transformation to take place. A way of constraining the system to remain in the b.c.c. structure is to use fixed periodic boundary conditions to inhibit the model crystallite from taking the path of the transformation irrespective of the existence of a potential-energy barrier. Similarly, imposing the value of the ratio between the sides of the rectangular parallelepiped through (periodic) boundary conditions is a way of constraining the system in order to construct intermediate ensembles for multistage sampling. The values of the constrained free energy F_ξ for different values of the « reaction co-ordinate » ξ locating the position of the system along the transformation path will show whether a barrier exists and will yield an estimate of its numerical value. These matters will be discussed further in sect. 7.

Let us describe three possible « routings » for the multistage sampling:

i) A simple linear combination of the two Hamiltonians, namely $H(\lambda) = \lambda H_1 + (1 - \lambda) H_2$; varying λ gradually switches on one Hamiltonian and switches off the other.

ii) Whenever the harmonic model H_h in both structures is stable, a good choice is to evaluate F(f.c.c.)-F_h(f.c.c.) and F(b.c.c.)-F_h(b.c.c.) separately. The advantage is that anharmonicity usually gives a relatively small correction; hence the harmonic Hamiltonian H_h may be expected to retain much of the many-body correlation effects. As a consequence the overlap of energy distributions can be expected to be larger than with other routings.

iii) Whenever a « physical » path exists between the two structures as, for example, the one found by PARRINELLO and RAHMAN [8a] in KCl which goes from rock-salt to the « CsCl » polymorph, it can be employed in the way alluded to above.

In the present work method i) has been mainly used. We anticipated that iii) will be discussed in more detail in sect. 7.

c) About sampling algorithms. The distributions $h(\Delta)$ introduced in sect. 2 can be evaluated by using various algorithms to sample configuration space. In principle, $h(\Delta)$ is defined in the canonical ensemble. Not all algorithms apply to the canonical ensemble: the usual MD, for instance, samples the microcanonical ensemble. It can be argued, however, that the different distributions of say the fluctuations of the potential energy per particle in the two ensembles will not reflect itself in the distribution $h(\Delta)$ of Δ. We have investigated this point looking for differences in the estimation of $h(\Delta)$ in the two ensembles without finding any within statistical error. In the following we shall present data obtained in the canonical as well as microcanonical ensembles. No difference is found in the evaluation of $F_{\text{f.c.c.}}$-$F_{\text{b.c.c.}}$.

A canonical version of MD is, of course, presently available ([22]). In paper I we have compared the efficiency of MC and two MD algorithms for configuration space sampling and for gathering quantitative information on $h(\Delta)$ in particular. While this efficiency was found to be not too different for the two algorithms, one particular feature stood out in favor of MD and MDM (by using the terminology in I) in particular, namely the variance of the mean of the potential energy in MDM runs for crystals was shown to be much smaller than in MDM or MC. This was interpreted to be a consequence of the oscillatory behavior of the energy fluctuations of the system in the microcanonical ensemble. It was shown further that MDC calculations can be treated in a way as to correct for the uncertainty in the reading of the mean potential energy associated with the slow random fluctuations of the internal energy of the system resulting from the time integral of the heat flux to and from the temperature bath.

In view of the role played by the internal energy in determining entropies from free-energy estimates and given the fact that the magnitude of the expected error of potential-energy averages in standard MC or MD simulations is quite substantial and will be found to be one order of magnitude larger than that of free-energy differences, it is essential to make the optimal choice (MDM) for the algorithm to be used in the estimation of potential-energy averages.

The insight presented in paper I is largely a result of the present study rather than pre-existing wisdom; hence the choice of algorithms we have made for this work is not necessarily one we would repeat. However, although the data have been gathered with various algorithms, later larger N runs were MD calculations.

4. – Model systems and procedures.

We have used the methods of sect. **2** and **3** to investigate the Lennard-Jones 6-12 potential and also the pair potentials suitable for rubidium and cesium metals. The alkali metal effective pair interactions were constructed

([22]) H. C. ANDERSEN: *J. Chem. Phys.*, **72**, 2384 (1980).

by PRICE *et al.* [23] using the theory of electron screening of the bare ion-ion Coulomb potential.

We note that there are other effective pair interactions available in the literature. The DRT (Dagens, Rasolt, Taylor [24]) potential using the Geldart-Taylor [25] dielectric function has been used successfully for studying a variety of properties [26].

Eventually, for ΔF calculations of the kind reported here, it would be very satisfactory to compare the results based on different assumptions in the theory of metals. Our present work was based on effective alkali pair potentials of Price *et al.* [23] which have already been used in the investigation of homogeneous liquid-to-solid nucleation [6], on the one hand, and of polymorphic structural transformations [7], on the other.

At this point we digress somewhat towards the thermodynamics of density, volume and free-energy changes in the vicinity of equilibrium conditions. The motivation of this digression is easily seen in the fact that the pseudopotential approach to the cohesion in metals, up to second order in the perturbation caused by the pseudopotential, gives the energy V of the metal system as

$$V = V_1(\Omega) = \sum_{l}^{N} \sum_{j>l}^{N} V_{\mathrm{EPP}}(|\boldsymbol{r}_l - \boldsymbol{r}_j|; \Omega),$$

N being the number of ions in volume Ω. EPP stands for effective pair potential. The ionic positions are at $\boldsymbol{r}_l$, $l = 1, \dots, N$. Note that V_{EPP} depends on Ω as well as on the interparticle distance $|\boldsymbol{r}_{lj}|$. The least satisfactory part of the theory of V given above is $V_1(\Omega)$. Hence, in a polymorphic transition between two structures, if there is a density change, then the potential difference has to be evaluated by using the change in $V_1(\Omega)$.

To deal with this problem one can take advantage of the common practice in the calculations of defect properties which is to work at constant volume [26]. The desired constant-pressure transformation characteristics are then deduced with the use of the thermodynamic-equilibrium condition.

It is easily shown [26,27] that the Gibbs free-energy difference for variations at constant pressure are equal to the Helmholtz free-energy differences at constant volume. Similarly volume changes $\Delta\Omega$ occurring when a polymorphic transition takes place at constant pressure are related to Δp when they occur

---

[23] D. L. PRICE: *Phys. Rev. A*, **4**, 358 (1971); D. L. PRICE, K. S. SINGWI and M. P. TOSI: *Phys. Rev. B*, **2**, 2983 (1970).
[24] M. RASOLT and R. TAYLOR: *Phys. Rev. B*, **11**, 2717 (1975); L. DAGENS, M. RASOLT and R. TAYLOR: *Phys. Rev. B*, **11**, 2726 (1975).
[25] D. J. W. GELDART and R. TAYLOR: *Can. J. Phys.*, **48**, 169 (1970).
[26] G. JACUCCI and R. TAYLOR: *J. Phys. F*, **9**, 1489 (1979) and references therein.
[27] C. P. FLYNN: *Point Defects and Diffusion* (Clarendon, Oxford, 1972).

at constant volume. The above-mentioned relations are derived from first-order truncations of suitable Taylor expansions and hence are valid to first order in relative changes like $\Delta\Omega/\Omega$ etc.

Let us derive the $\Delta\Omega$, Δp relation with the f.c.c.-b.c.c. transformation in mind. $\Delta\Omega = \Omega_F - \Omega_B$ and $\Delta p = p_F - p_B$ are changes in volume and pressure accompanying the f.c.c.-b.c.c. transformation at constant p and Ω, respectively. If we set up the system F (*i.e.* f.c.c.) in such a way that on transformation to B (*i.e.* b.c.c.) the latter attains the same state point p_B, Ω_B in both constant-volume and constant-pressure transformations, F would have started at p_F, Ω_B in one case and p_B, Ω_F in the other. Thus, for F, the volume change $\Delta\Omega = \Omega_F - \Omega_B$ is attained by a pressure change $-\Delta p = p_B - p_F$. Hence $\Delta\Omega = \Omega\chi_T^F\Delta p$ where χ_T^F is the compressibility of F, by using a symmetric argument $\Delta\Omega = \Omega\chi_T^B\Delta p$, all to first order in small quantities. Neglecting $(\chi_T^F - \chi_T^B)/\chi_T$, we write $\Delta\Omega = \Omega\chi_T\Delta p$. Since the compressibility cannot be evaluated without knowning $V_1(\Omega)$, we shall use the experimental values for the alkali metal in question. All the above transformations are at constant temperature T.

In the case of volume-independent interaction, LJ systems for instance, constant-pressure calculations can now be made without special difficulty (7). In the calculations reported in this paper, however, the LJ system has been treated in the same way as the alkalis, namely at constant volume.

We note here, as a small digression, that, at constant volume Ω, calculations for an alkali require but one table of values of $V_{EPP}(r; \Omega)$ as a function of r; at constant pressure, however, one needs several such tables to cover the range of Ω values that the system fluctuates through at the ambient pressure and temperature, in addition to the functions $V_1(\Omega)$. Such a numerical scheme is actually being used for constant-pressure quench studies of crystal nucleation from the melt (28).

The next item to mention in this section of procedures is that of the truncation of the potential. The usual procedure is to truncate the potential and to introduce corrections for the error thus committed. The larger the system and the longer the cut-off, lesser is the error. However, a central feature of the ΔF calculational methods is the width of the energy distributions and various overlaps between such distributions. The larger the system, the narrower are the distributions and greater is the difficulty of obtaining practically reasonable overlapping histograms. Thus a compromise has to be made to allow for a balance between conflicting requirements.

Liquid-state calculations are not encumbered by this question of balance. Firstly, the pair correlations approach their asumptotic value of unity at rather short distances allowing accurate evaluation of the interactions beyond the cut-off; secondly, the entropy per particle of the liquid-state system has been found to be insensitive to system size; in fact, 32-particle systems yielded the

(28) Y. LIMOGE and A. RAHMAN: private communication.

same free energies as 108-particle systems after correcting for cut-off effects; this was found to be so for monatonic LJ systems ([10]) as well as for diatomic systems ([29]), a molecule being two LJ centers rigidly fused together.

In the work being reported here, on LJ and alkali crystal systems, we have found that below a certain size the usual procedure to correct for neighbors beyond the cut-off led to poor results; extensive calculations with 54- and 128-particle systems brought us to this conclusion. However, for 144 and 320 particle systems (the latter requiring heavy multistage calculation) we obtained systematically consistent results. Our conclusion is that for the reliability of free-energy difference data a rule of thumb is $N \geqslant 144$. We consistently chose a cut-off such that for the f.c.c. and b.c.c. structure the number of interacting neighbors came as close as possible. This number was 54 for f.c.c., 58 for b.c.c. in a 144-particle system and 140, 136, respectively, for the 320 system. Note specially (in fig. 1 of ([6a])) that the alkali metal potential we are using can, in practice, be truncated in the usual way because of the rapid decrease in the amplitude of Friedel oscillations.

Some of the results for the 128-particle system will also be discussed below to display the small difference with the results for a 144-particle system.

As is commonly done in setting up dynamical matrices for small unit cell crystal vibrations, one can use a small-sized system and the neighbors in the adjoining cells to take all particles into account which fall within the cut-off radius around a given particle; this may be a way of dealing with the conflicting requirements of the long cut-off and broad energy distributions. This is still to be systematically tested. Our observation that the entropy per particle is not insensitive to the cut-off for $N < 144$ leads us to doubt the possibility of getting accurate results from systems much smaller than this.

In concluding this section on procedures we shall state a variety of details which need to be put down. Let ε and σ denote units of energy and length. For LJ systems these parameters appear in their usual way in the analytical form of the potential. For alkali metals we have chosen for all our calculation $\varepsilon = 555.86 \cdot 10^{-16}$ erg and $\sigma = 4.4048$ Å; all quantities will be expressed in their dimensionless reduced form by using the above ε, σ, besides k_B. Extensive quantitites will be given per particle, throughout.

As for the state parameters, we studied the LJ system most thoroughly at number density $\varrho = 1.0054$ and temperature $T = 0.500$, *i.e.* some $\frac{2}{3}$ of the way to the triple point. We also studied this system at $\varrho = 0.8378$, $T = 0.500$ and at $\varrho = 1.0054$, $T = 1.02$.

Two alkali metals, Rb, Cs were studied at STP and under pressure at room temperature. Rb was also investigated at $\varrho = 1.564$ and at low T, namely 0.1.

These data made it possible to assess the variation of ΔF with ϱ and T for these model systems.

([29]) N. QUIRKE and G. JACUCCI: *Mol. Phys.*, **45**, 823 (1982).

5. – Results for the Lennard-Jones potential.

a) Evaluation of ΔF and ΔS. In this subsection we present numerical estimates for ΔF between b.c.c.-f.c.c. structures of the Lennard-Jones crystal. All data refer to the thermodynamic state defined by $T = 0.500$ and $\varrho = 1.0054$ (in the usual LJ units).

The procedure used was the multistage overlapping distribution method applied to the linear combination Hamiltonian $H(\lambda) = \lambda H_B + (1 - \lambda) H_F$ described in sect. **3**.

A 320-particle system interacting up to a distance $R_c = 3.20$ was studied by using MDC and 11 intermediate stages for evaluating ΔF. After equilibration each ensemble was sampled by using a 2000 step run. The results are given in table I, together with analogous results corresponding to a cut-off distance $R_c' = 2.35$.

As seen from the last line in table I, the result for $F_F - F_B$ was $-0.1757 \pm \pm 0.0015$ and -0.1044 ± 0.0011 for the cut-off distances R_c and R_c', respectively. We can check whether this discrepancy can be accounted for by the difference in cut-off by making use of the values of the potential energy at $T = 0$; by using a superscript to denote zero temperature, $V_F^0 - V_B^0$ is found to be $-8.1629 + + 7.8810 = -0.2819$ and $-7.7793 + 7.5680 = -0.2113$ for R_c and R_c', respectively. Thus $-0.2819 + 0.2113 = -0.0706$ satisfactorily accounts for the discrepancy $-0.1757 + 0.1044 = -0.0713$ in ΔF values. This suggests that the change in cut-off does not affect the calculated value of ΔS.

In fact, the potential-energy difference $V_F - V_B$ at temperature $T = 0.500$

TABLE I. – *Values of $\Delta F = F_{\text{f.c.c.}} - F_{\text{b.c.c.}}$. U_λ is $\lambda U_{\text{f.c.c.}} + (1-\lambda) U_{\text{b.c.c.}}$. The system is an LJ comprising 320 particles. R_c is the potential cut-off distance.*

λ		$R_c = 2.35$		$R_c = 3.20$	
		eq. (2)	eq. (5)	eq. (2)	eq. (5)
b.c.c.					
0	0.03	0.0177 ± 0.0010	0.0176 ± 0.0003	0.0165 ± 0.0015	0.0165 ± 0.0005
0.03	0.07	0.0160 ± 0.0011	0.0160 ± 0.0003	0.0140 ± 0.0012	0.0140 ± 0.0004
0.07	0.125	0.0132 ± 0.0009	0.0135 ± 0.0003	0.0095 ± 0.0010	0.0096 ± 0.0004
0.125	0.1875	0.0085 ± 0.0008	0.0084 ± 0.0003	0.0037 ± 0.0010	0.0035 ± 0.0003
0.1875	0.25	0.0025 ± 0.0015	0.0032 ± 0.0002	0.0018 ± 0.0008	-0.0022 ± 0.0003
0.25	0.375	-0.0044 ± 0.0009	-0.0043 ± 0.0004	-0.0144 ± 0.0015	-0.0156 ± 0.0006
0.376	0.5	-0.0142 ± 0.0013	-0.0146 ± 0.0004	-0.0233 ± 0.0015	-0.0242 ± 0.0005
0.5	0.625	-0.0238 ± 0.0006	-0.0237 ± 0.0003	-0.0317 ± 0.0009	-0.0323 ± 0.0005
0.625	0.75	-0.0311 ± 0.0012	-0.0311 ± 0.0003	-0.0386 ± 0.0015	-0.0399 ± 0.0004
0.75	0.875	-0.0392 ± 0.0010	-0.0393 ± 0.0004	-0.0486 ± 0.0006	-0.0476 ± 0.0005
0.875	1.0	-0.0503 ± 0.0011	-0.0501 ± 0.0004	-0.0575 ± 0.0007	-0.0575 ± 0.0006
	f.c.c.				
	ΔF:	-0.1051 ± 0.0035	-0.1044 ± 0.0011	-0.1704 ± 0.0041	-0.1757 ± 0.0015

is found to be -0.2429 ± 0.0036 and -0.1703 ± 0.0032 for R_c and R'_c, respectively, and the difference between these is -0.0726 ± 0.0048.

It is thus apparent that the calculated ΔF can be corrected for the finite cut-off by making appropriate lattice sums from the cut-off to infinity and also that the entropy difference $S_F - S_B = -0.1344 \pm 0.0078$ (for R_c) and $-0.1318 \pm \pm 0.0068$ (for R'_c) should be in satisfactory agreement with each other.

In addition to the 320-particle system, a 144-particle system also has been studied but by MC using the cut-off $R'_c = 2.35$ and 5 stages. Several runs of 5000 macrosteps have been carried out, thus obtaining the following results, uncorrected for contributions beyond cut-off: $\Delta F = -0.0940 \pm 0.0036$ and $\Delta V = -0.1640 \pm 0.0060$, yielding $\Delta S = -0.140 \pm 0.014$, again in satisfactory agreement with the corresponding MDC results with the same cut-off R'_c.

In summary, the entropy difference is found to be insensitive to the cut-off and to the number of particles in the system for the reported values of R_c and N. The best estimate is obtained by averaging the above MDC results: $S_F - S_B = -0.1331 \pm 0.0049$.

The free-energy difference, corrected for the long-range contribution in the way anticipated above, is $F_F - F_B = -0.1566 \pm 0.0015$ and $-0.1559 \pm \pm 0.0011$ from calculations employing R_c and R'_c, respectively.

b) ΔF at neighboring state points. The value of $\Delta F \equiv F_F - F_B$ decides the relative stability of the two structures. Additional information as well as a check of the thermodynamic consistency of the results can be obtained by evaluations of ΔF at different state points. We have carried out investigations along these lines with a 128-particle system. The size of this system (and the cut-off used) has turned out to be somewhat smaller than needed to provided values of ΔS independent of N and R_c, within statistical error. The results to be discussed below, although internally consistent, deserve, therefore, a word of caution, if extrapolated to predict the behavior of infinite systems.

i) The derivative of ΔF with respect to the volume Ω is related to the pressure change Δp observed in a structural transformation at constant Ω. To first order in small quantities, this Δp in turn is related to the volume change $\Delta\Omega$ the system would undergo if the transformation happens at constant pressure p. Now $p = -(\partial F/\partial\Omega)_T$ gives $\Delta p = -(\partial\Delta F/\partial\Omega)_T$ and, as already shown, $\Delta\Omega = \Omega\chi_T\Delta p$. We hence get the following value for the volume change connected with the transformation at $T = 0.5$, by measuring the pressure (using the virial theorem), and hence Δp:

$$\Delta\Omega/\Omega = (\Omega_F - \Omega_B)/\Omega = (-0.43 \pm 0.08)\%,$$

i.e. the relative volume change is smaller than one percent and is negative.

ii) Taking into account the eventual curvature c of ΔF as a function of T, one can expect the following relation to hold between quantities relative to temperatures T_1 and T_2:

$$\Delta F_2 = \Delta F_1 - \Delta S_1(T_2 - T_1) + c(T_2 - T_1)^2 ,$$
$$\Delta F_1 = \Delta F_2 - \Delta S_2(T_1 - T_2) + c(T_1 - T_2)^2 ,$$

i.e.

$$\Delta F_2 = \Delta F_1 - \tfrac{1}{2}(\Delta S_1 + \Delta S_2)(T_2 - T_1) . \tag{8}$$

To check whether the calculated values of ΔF obey eq. (8), runs were carried out at $T_1 = 0.50$ and $T_2 = 0.85$, at a density $\varrho = 1.00$. The value of $\Delta F_2 = -0.030 \pm 0.006$ is in good accord with the right-hand side of eq. (8), which is found to be -0.027.

In summary, this additional investigation has provided a check that ΔF increases with increasing temperature, in accord with the negative sign, and the value, of ΔS; and it also increases with increasing volume, in accord with the negative sign, and the value, of $\Delta\Omega$.

It would have been more satisfactory to obtain data on the variation of ΔF with Ω and T by using the large system. This was not done in view of the heavy computation that would have been required. However, we believe that the thermodynamic consistency of the results has been satisfactorily demonstrated.

The signs of ΔS and $\Delta\Omega$ show that the relative stability of the f.c.c. structure (*i.e.* the value of $(\Delta G)p$) increases upon increasing the pressure, but decreases upon increasing the temperature. However, a linear extrapolation from $T = 0.5$, where $\Delta F = -0.156 \pm 0.001$ and $\Delta S = -0.133 \pm 0.005$, by using $\partial \Delta F/\partial T = -\Delta S$, shows that there will be no f.c.c. $\rightarrow$ b.c.c. transition in this system before melting the crystal at normal pressure.

The location of the f.c.c. $\leftrightarrow$ b.c.c. transition line in the p-T diagram can be derived approximately from the values of ΔS and $\Delta\Omega$ by expanding [26,27]

$$(\Delta G(p, T))_p \equiv G_F(p, T) - G_B(p, T) \simeq (\Delta F(p, T))_V$$

about (p_0, T_0) to first order in $\delta p = p - p_0$ and $\delta T = T - T_0$:

$$(\Delta G(p, T))_p = (\Delta G(p_0, T_0))_p - (\Delta S)_p \, \delta T + (\Delta\Omega)_p \, \delta p .$$

Approximating $(\Delta S)_p \simeq (\Delta S)_V + (\alpha/\chi_T)(\Delta\Omega)_p$, where α is the coefficient of thermal expansion and $\alpha/\chi_T \sim 9$ for argon [30], we get

$$(\Delta G(p, T))_p = \Delta F - 1.3 \, \Delta S \, \delta T + \Delta\Omega \, \delta p ,$$

[30] *Argon, Helium and the Rare Gas*, edited by G. A. Cook (Interscience, New York, N.Y., 1961).

in terms of our quantitites ΔF, ΔS and $\Delta\Omega$ estimated at (p_0, T_0). The polymorphic transition line is identified by $(\Delta G(\tilde{p}, \tilde{T}))_p = 0$:

$$\tilde{T} - T_0 = \frac{\Delta F}{1.3\Delta S} + \frac{\Delta\Omega}{1.3\Delta S}(\tilde{p} - p_0) .$$

Inserting numerical values in this relation we estimate that the f.c.c.-b.c.c. transition line should cross the experimental melting line of argon under pressure ([30]) at about 230 K and 0.8 GPa, the stable b.c.c. phase region being located above this point towards higher pressures and temperatures. As a consequence, we expect that at pressures of ten to hundred (in reduced units ε/σ^3) and up there should exist a temperature region where the Lennard-Jones solid prefers to take up the BCC structure.

6. – Results for alkali metals.

a) *Rubidium at* STP. The model system consisted of 144 particles interacting with the pair potential described in sect. **4**. With a cut-off $R_c = 2.435$ each particle has ~ 54 neighbors in the f.c.c. structure and ~ 58 in the b.c.c. one, *i.e.* it includes the first six shells in both structures, as in the MC calculation performed with 144 LJ particles. Two independently equilibrated MC runs 5000 macrosteps long have been carried out at $\varrho = 0.92237$ and $T = 0.728$, corresponding to standard temperature and pressure (STP), thus obtaining the following results:

$$F_F - F_B \quad \begin{matrix} 1) & 0.1050 \pm 0.0028 \\ 2) & 0.1076 \pm 0.0025 \end{matrix} \Bigg\} = 0.1063 \pm 0.0019 ,$$

$$V_F - V_B \quad \begin{matrix} 1) & -4.552 + 4.639 = 0.087 \pm 0.016 \\ 2) & -4.553 + 4.662 = 0.109 \pm 0.015 \end{matrix} \Bigg\} = 0.098 \pm 0.011 ,$$

$$T(S_F - S_B) \qquad = 0.008 \pm 0.011 .$$

The values of the correlation length were about 50 macrosteps for the Fermi function and 100 for V. The multistage calculations consisted of 5 ensembles, corresponding to $\lambda \equiv (0.0, 0.25, 0.50, 0.75, 1.0)$. The free-energy difference ΔF is very precisely determined and it is positive as expected. The potential-energy difference ΔV coincides with this value within statistical error, indicating that the entropy difference ΔS for the two lattices is very small: ΔV is only about 2 % of the energy per particle.

b) $\Delta\Omega$ *and thermodynamic checks for* Rb. Two analogous MC runs were carried out at $T = 0.728$ and $\varrho = 1.564$. The density dependence of the effective pair interaction in alkali metals mentioned in sect. **4** makes it imperative to use the potential relevant to this higher density. The results of the MC runs are

$$F_F - F_B \qquad \begin{matrix} 1) & & 0.1861 \pm 0.0028 \\ 2) & & 0.1918 \pm 0.0025 \end{matrix} \Bigg\} = 0.1890 \pm 0.0019\,,$$

$$V_F - V_B \qquad \begin{matrix} 1) & 3.3779 - 3.1570 = 0.2209 \pm 0.0058 \\ 2) & 3.3599 - 3.1523 = 0.2076 \pm 0.0141 \end{matrix} \Bigg\} = 0.2143 \pm 0.0076\,,$$

$$T(S_F - S_B) \qquad = 0.0253 \pm 0.0078\,.$$

Number of particle, cut-off and length of the MC runs were as mentioned above. Values of the correlation lengths were very similar. The result shows that increasing the density favors still more the b.c.c. structure in terms of both potential and free energy, although now we have $S_F > S_B$ rather unexpectedly.

This result can be combined with the previous one to estimate the value of the volume change connected with the transformation. From the observations made above we can already predict that $\Omega_F - \Omega_B$ will be positive. We shall make use of the atomic volume Ω at melting $92.60 \cdot 10^{-24}\,\text{cm}^3$ and we shall approximate the isothermal bulk modulus B_T by the value of the adiabatic one $B_S = 2.14 \cdot 10^{10}\,\text{dyn/cm}^3$ at a density $\varrho = 1.530$, *i.e.* close to melting. Again we have

$$\Delta\Omega/\Omega = -(B_T)^{-1}\left(\frac{\partial \Delta F}{\partial \Omega}\right)_T = (0.56 \pm 0.02)\,\%\,.$$

One other temperature was investigated at $\varrho = 1.564$, namely $T = 0.100$. Nine ensembles were used in the multistage sampling scheme to get sufficient overlap. The values of τ were about half of the previous ones. The other conditions were the same. Here are the results:

$$F_F - F_B \qquad \begin{matrix} 1) & & 0.2020 \pm 0.0006 \\ 2) & & 0.2008 \pm 0.0007 \end{matrix} \Bigg\} = 0.2014 \pm 0.0005\,,$$

$$V_F - V_B \qquad \begin{matrix} 1) & 2.3814 - 2.1803 = 0.2011 \pm 0.0021 \\ 2) & 2.3836 - 2.1798 = 0.2038 \pm 0.0019 \end{matrix} \Bigg\} = 0.2025 \pm 0.0014\,,$$

$$T(S_F - S_B) \qquad = 0.0011 \pm 0.0015\,.$$

The entropy difference is again very small. We can perform a thermodynamic check to verify that the value of $T\Delta S$ larger than the statistical error found at $T = 0.728$ is indeed meaningful. $T_1 = 0.1$, $T_2 = 0.728$:

$$\Delta F_2 = \Delta F_1 - \tfrac{1}{2}(\Delta S_1 + \Delta S_2)(T_2 - T_1) = 0.1840 \pm 0.0056\,,$$

which compares favorably with the value found at T_2: $\Delta F_2 = 0.1890 \pm 0.0019$. The problem of locating the f.c.c.-b.c.c. transition line in the p-T diagram has been dealt with in the previous section. In the scheme presented there, which uses linear extrapolation as the central approximation, the value of ΔS appears in the denominator in determining the value of the transition temperature. The small value of ΔS in rubidium leads us to the conclusion that the transition cannot occur before melting.

c) *Comparison* Rb-Cs. Using a somewhat smaller system, containing 128 particles with an interaction potential cut-off $R_c = 2.304$, we have carried out a parallel calculation of ΔF for Rb and Cs. All runs were MC calculations 2000 macrosteps long. Each comprised 5 overlaps, corresponding to $\lambda = (0.0, 0.25, 0.50, 0.75, 0.90, 1.00)$. The results are shown in table II.

TABLE II. – *Calculations of* ΔF *for* Rb *and* Cs *at standard temperature and pressure* (STP) *and under compression.* The asterisks * denote mean values over the three runs. The energy unit $\varepsilon = 555.86 \cdot 10^{-16}$ erg is the well depth of the Rb pair potential at STP.

Metal	ϱ	$(F_F - F_B)/\varepsilon$	$(U_F - U_B)/\varepsilon$	$T(S_F - S_B)/\varepsilon$
Rb	ϱ_{STP}	0.103 ± 0.003 *	0.126 ± 0.008 *	0.023 ± 0.009 *
Cs	ϱ_{STP}	0.101 ± 0.005	0.111 ± 0.018	0.010 ± 0.019
Rb	$1.25 \times \varrho_{STP}$	0.158 ± 0.006	0.204 ± 0.017	0.046 ± 0.108
Rb	$1.70 \times \varrho_{STP}$	0.216 ± 0.008	0.310 ± 0.010	0.094 ± 0.013
Cs	$1.72 \times \varrho_{STP}$	0.212 ± 0.005	0.306 ± 0.019	0.094 ± 0.020

The results for Rb are somewhat different from the ones obtained with 144 particles and a longer cut-off. The more extensive study performed with the Lennard-Jones system showed that convergence of the entropy difference was only reached with the 144-particle system, where the cut-off used consistently included 54 particles, up to the fourth shell of neighbors, in the f.c.c. structure and 58 particles, up to the fifth shell of neighbors, in the b.c.c. one. Accordingly, we shall consider the results obtained with the 128-particle systems precise but inaccurate for the description of the given potential model and we report them primarily for pointing out the similarity for the alkali metals. Nonetheless, it should be noted that the new results for Rb are in semi-quantitative accord with the one given in subsections *a*) and *b*). ΔF and ΔV are about equal at STP, their value being about 0.1, indicating stability of the b.c.c. structure dictated by potential energy. $T\Delta S$ is essentially zero at STP, but becomes positive under pressure. On the other hand, compressing the crystal increases ΔV so much that ΔF increases as well, indicating a positive value for the volume change $\Delta\Omega$ for the transformation: the value obtained for $\Delta\Omega/\Omega$ is $(+0.77 \pm 0.06)\%$ for Rb, not too different from the 144-particle value, and $+0.74 \pm 0.05$ for Cs, by using a value of $1.63 \cdot 10^{11}$ dyn/cm^2 for B_T of Cs.

The striking feature exhibited by this data is the coincidence of the values of the various energies expressed in the *same units* for the two metals at STP or under pressure (while the values of the density ϱ_{STP} differ by 26%). With the potential model used, the properties of the heavier alkali metals, for what concerns the stability of crystal lattices, may be the same. In particular, we see no evidence of the experimentally observed transition to f.c.c. in Cs at high compression.

7. – Free-energy barriers.

The method of overlapping distributions for the calculation of free-energy differences relies, in principle, on just one requirement, namely that the potential functions to be investigated, V_1, V_2 say, span the same configuration space. $V_\lambda = \lambda V_1 + (1 - \lambda) V_2$ obviously satisfies this criterion if V_1 and V_2 do and, therefore, is a suitable potential function for a multistage determination of ΔF. The weight parameter λ has no obvious physical meaning.

There is a clear gain in using V_λ as written above, *i.e.* as a linear combination of V_1 and V_2. The gain is that, as intermediate stages are introduced, the histogram already calculated can be systematically re-used. Suppose we start with V_1 and V_2 and get the histograms for $h_1(\Delta)$ and $h_2(\Delta)$ finding that there is no overlap. We would then decide to use V_λ with $\lambda = \frac{1}{2}$. One of the two distributions between V_1, V_2 would be $h_1[(1 - \lambda)\Delta]$. Obviously the histogram for this h_1 is obtained from the previous histogram for $h_1(\Delta)$ by relabelling the axis with factor $(1 - \lambda)$; similarly for the stage V_λ, V_2. Hence the gain mentioned above.

However, in physical terms the function V_λ has no particular significance. We know from laboratory and from computational experience that crystalline structures do «travel» in their configuration space from one point to another along paths which have a particularly simple description. For such displacive transformations, it would be particularly desirable to be able to calculate the free-energy barrier.

We have developed a method of calculating ΔF between the end points of such paths with the aim not only of evaluating ΔF by the method of overlapping distributions, but also of evaluating the free-energy barrier that needs to be overcome in a crystalline-structure transformation.

The example we present below is only partially satisfactory from the above point of view. It is simply an application of the methodology to a b.c.c.-f.c.c. transformation in an LJ crystal and not an example of a free-energy barrier along the chosen path because along this path there is no barrier.

Let $l_1 = l_2 = l_3 = a$ denote the cubic-cell edges of a b.c.c. lattice. The tetragonal body-centered lattice with $l_1 = a\xi$, $l_2 = l_3 = a/\sqrt{\xi}$, $1 < \xi < 2^{\frac{1}{3}}$ is b.c.c. at $\xi = 1$ and f.c.c. at $\xi = 2^{\frac{1}{3}}$ for which $l_1/l_2 = 2^{\frac{1}{2}}$. The volume per particle is $2/a^3$ irrespective of the value of ξ.

From the point of view of the determination of the reaction path and the reaction co-ordinate, a much more complicated case is that of rock-salt $\leftrightarrow$ CsCl structural change for many alkali halides ([8a]). We shall not go into details here except to say that perhaps a barrier does exist in this case.

It is clear that the tetragonal structures whose potentials can be symbolized by V_ξ span a specific f.c.c. $\leftrightarrow$ b.c.c. path. The multistage overlapping method can obviously be used with V_ξ instead of V_λ except that the « gain » mentioned above for V_λ is not available while using V_ξ, simply because $V_\xi - V_{\xi'} \neq (\xi - \xi')(V_1 - V_2)$.

Using a LJ system of 320 particles, we calculated the ΔF with seven equally spaced intermediate values of ξ with the following results:

$$T = 0.7\,, \qquad \varrho = 1.0054 \text{ LJ system}\,,$$

$$F_F - F_B = \Delta F = -\,0.088 \pm 0.003\,,$$

$$\Delta V = -\,0.184 \pm 0.007\,, \qquad \Delta S = -\,0.137 \pm 0.008\,.$$

Let us check these results with those of subsect. **5***a*).

Using the identity $(\partial/\partial\beta)\beta F = V$, we can construct finite differences arising out of the two temperatures 0.5 and 0.7. This gives, respectively,

$$\frac{-\,0.088 \pm 0.003}{0.7} + \frac{0.104 \pm 0.001}{0.5}$$

and

$$\tfrac{1}{2}(-\,0.184 \pm 0.007 - 0.170 \pm 0.003)\cdot(1/0.7 - 1/0.5)\,,$$

i.e. 0.082 and 0.101 with an expected error of about 0.01.

8. – Comments and conclusions.

Free-energy, entropy and volume differences between face-centered and body-centered cubic structures have been evaluated for model crystals. Because these differences only amount to about one percent of the respective quantities per particle in the two phases, high-precision difference methods based on ensemble sampling are employed. Statistical errors of one percent or less are obtained in most cases.

The relative stability of phases of model rare gas and alkali metal crystal is assumed within classical mechanics, but without approximations like harmonicity or the like. Results are found to be in agreement with the behaviour of real materials as well as of model systems studied in dynamical simulations of crystal nucleation ([6]) and of polymorphic transformations ([7]). A linear extrapolation to high temperature and pressure (which neglects differences in second-order derivatives of the Gibbs free energy G with respect to p and T for the two structures) of the dependence of ΔG on p and T estimated locally permits one to draw the line of the polymorphic transformation in the p-T diagram for the f.c.c.-b.c.c. transition. The existence of a stable b.c.c. phase at high pressure and temperature is predicted in this way for Lennard-Jones solids. In contrast, no high-pressure f.c.c. phase is found for model systems of Rb and Cs.

In sect. **5** we showed that for the LJ system at $T = 0.5$ and $\varrho = 1.0$, $S_{\text{f.c.c.}} < S_{\text{b.c.c.}}$. Thus the relative stability of the f.c.c. lattice is due to the potential energy making the dominant contribution to the stability of f.c.c. But no general statement can be made about the sign of the difference $S_{\text{f.c.c.}} - S_{\text{b.c.c.}}$. We see from sect. **6** that for an alkali metal at high density $S_{\text{f.c.c.}} > S_{\text{b.c.c.}}$ but this fails to make f.c.c. relatively more stable. Thus arguments based on the « looseness » of the b.c.c. structure as being the cause of stability are not quite valid.

Moreover, in the cases considered here, ΔV and ΔF are of the same sign, *i.e.* the stability is dictated by the potential energy and not by entropy.

The alkali potential functions used here fail to predict the experimentally observed transition to f.c.c. in Cs at high compression. This is presumably due to a failure of the pseudopotential second-order perturbation theory to take the changes of electronic structure properly into account. Recently MD calculations have been reported by showing the existence of such a transformation for a model system describing Li ([31]) akin to the ones employed in the present study of the heavier alkali metals. Therefore, we conclude that our calculations would have given a different result if applied to the model for Li.

The methods outlined and used in this paper on a few model systems show the possibility of making precise calculations of free-energy barriers to displacive crystalline transformations along a prescribed trajectory in configuration space. Therefore, relevant information on transition probabilities can also be obtained in addition to assessing the magnitude of the relative stability (*i.e.* ΔG) of different crystal structures. When the barrier is absent or low enough, dynamical-simulation studies are very useful in detecting the sign of ΔC, in locating the transition line (*i.e.* $\Delta G = 0$) and in showing the route spontaneously taken by the system in undergoing the transformation. If the barrier is too high, however, dynamical simulation studies may observe the system being trapped in a metastable state during the whole time of the simulation. As an extreme example of this circumstance, we note that spontaneous polymorphic transformations of the crystalline structures have been observed in molecular dynamic studies in which nontraditional periodic boundary conditions were used; these allow the system the additional freedom to change the shape of the elementary box. This should be contrasted to the use of the usual rigid boundary conditions made in this work. Free-energy

---

([31]) R. G. MUNRO and R. D. MOUNTAIN: *Phys. Rev. B*, to be published.

difference calculations here have been performed in such a way that the system is constrained to remain in a well-defined region in configuration space.

An interesting result arising from the nature of the method of overlapping distributions is that it is fallacious to think of the future simply in terms of studying « bigger » systems even if bigger computers are available. Increase in system size makes the calculations difficult, in principle, by making the distributions narrow and thus preventing their overlap.

* * *

We have profited from many useful discussions with M. PARRINELLO in the early stages of this work. GJ is grateful for the repeated warm hospitality received in the Materials Science and Technology Division at Argonne National Laboratory.

● RIASSUNTO

La differenza di energia libera, entropia e volume tra strutture cubiche a facce centrate e a corpo centrato sono valutate, per cristalli modello di gas rari e metalli alcalini, con il metodo delle distribuzioni sovrapposte. Nella regione di validità della meccanica classica si predice la stabilità delle fasi osservate per le sostanze reali anche in accordo con i risultati di precedenti studi di simulazione della nucleazione del cristallo dalla fase liquida e di trasformazioni polimorfe. Inoltre si predice in questo modo l'esistenza di una fase stabile di struttura cubica a corpo centrato per i solidi di Lennard-Jones, mentre non ci si aspetta una fase di struttura cubica a facce centrate per i metalli alcalini sotto pressione. Si mostra inoltre la possibilità di effettuare il calcolo della barriera di energia libera che impedisce le trasformazioni cristalline con spostamento lungo una traiettoria prefissata nello spazio delle configurazioni.

Относительная устойчивость FCC и BCC структур для модельных систем при высоких температурах.

Резюме (*). — Оцениваются различия свободных энергий, энтропий и объемов между гранецентрированными и объемоцентрированными кубическми структурами для модельных редких газов и щелочно-галоидных кристаллов, используя метод перекрывающихся распределений. Предсказываются устойчивые фазы в соответствии с поведением реальных материалов в областях справедливости классической механики и в согласии с результатами предыдущих динамических рассмотрений зарождения кристаллов из расплава и полиморфных образований. Предсказывается существование устойчивой BCC фазы при высоких давлениях и температурах для твердых тел Леннарда-Джонса, тогда как не ожидается устойчивой FCC фазы при высоком давлении для модельных Rb и Cs систем. Мы также показываем возможность проведения вычислений барьеров свободной энергии для кристаллических превращений вдоль предсказанной траектории в конфигурационном пространстве.

(*) *Переведено редакцией.*

Strain Accumulation in Quasicrystalline Solids

Franco Nori, Marco Ronchetti,[a] and Veit Elser[b]
Institute for Theoretical Physics, University of California, Santa Barbara, California 93106
(Received 24 March 1988)

We study the relaxation of 2D quasicrystalline *elastic networks* when their constituent bonds are perturbed homogeneously. Whereas ideal, quasiperiodic networks are stable against such perturbations, we find significant accumulations of strain in a class of disordered networks generated by a growth process. The grown networks are characterized by root mean square phason fluctuations which grow linearly with system size. The strain accumulation we observe in these networks also grows linearly with system size. Finally, we find a dependence of strain accumulation on cooling rate.

PACS numbers: 62.30.+d, 61.70.−r

There is a widely held belief that in certain solids (e.g., glasses) there are large accumulations of strain even though external stresses are absent.[1] In crystalline solids the strain fields of isolated defects are well understood but fail to show any accumulation. In amorphous solids, where the phenomena is believed to exist, identification of an appropriate strain-free reference structure has always been a major problem.[2] Quasicrystalline solids[3] represent an interesting intermediate case in that the number of local structural elements is finite and yet these compose a structure with (possibly) positive configurational entropy. In this Letter, we exhibit a two-dimensional (2D) quasicrystalline model, a "decagon aggregate," where the phenomenon of strain accumulation is well defined. Numerical studies of our model show that the accumulation of strain is correlated with the behavior of the phason field.

The building blocks of our model are decagons packed edge to edge (see Fig. 1). In any aggregate of edge-sharing decagons, any two decagons can always be related by a pure translation. Moreover, such a translation can always be expressed as an integral linear combination of four basis vectors $\mathbf{e}_i^{\parallel}$ $(i=1,\dots,4)$. The latter fact is equivalent to the statement that the possible decagon centers may be obtained by projecting a suitable 4D lattice. Consequently, each decagon is associated with a pair of two-component vectors: $\mathbf{x}^{\parallel}$, the location of its center; and $\mathbf{x}^{\perp}$, its "phason" coordinates. The pair $(\mathbf{x}^{\parallel},\mathbf{x}^{\perp})$ comprise the 4D lattice point. Details of the projection technique are given at length elsewhere.[4] For our purposes, it is sufficient to note that if the separation (in the physical plane) of two decagons is given by $\Delta\mathbf{x}^{\parallel}=\sum_{i=1}^{4} n_i\mathbf{e}_i^{\parallel}$, then $\Delta\mathbf{x}^{\perp}=\sum_{i=1}^{4} n_i\mathbf{e}_i^{\perp}$ gives the separation of their phason coordinates. The geometry of the $\mathbf{e}_i^{\perp}$ vectors is shown in the inset of Fig. 1.

The geometrical structure of edge-sharing decagons is our strain-free reference solid. We constrain the local structure by the requirement that the next-nearest-neighbor (nnn) decagon separation is $\tau=(1+\sqrt{5})/2$ times the nearest-neighbor (nn) or edge-sharing separation (see Fig. 1). If we depict only the pattern of nn bonds in the decagon aggregate, a network such as shown in Fig. 2 results. The connectivity of the network is significantly increased by including nnn bonds. Our model uses both types of bonds to stabilize the network mechanically. Specifically, we impose Lennard-Jones potentials between pairs of decagons joined by a bond. The Lennard-Jones scale parameter is given by σ_1 for nn

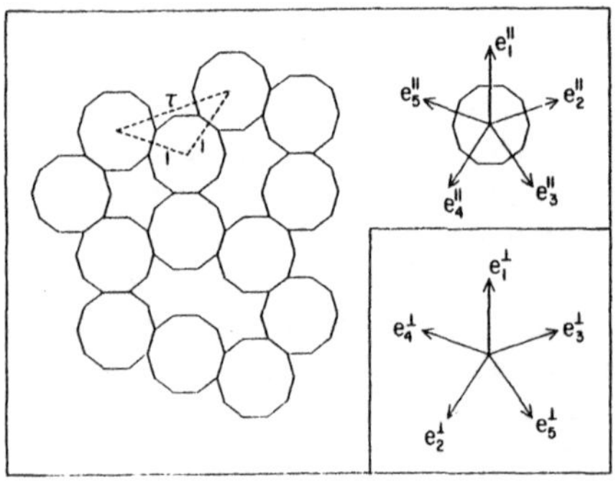

FIG. 1. Example of a decagon packing and projected lattice generators $\mathbf{e}_i^{\parallel}$. Inset: projected lattice generators $\mathbf{e}_i^{\perp}$.

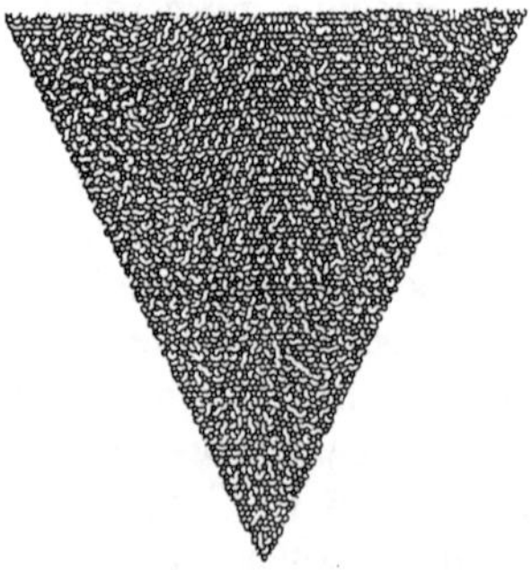

FIG. 2. Example of a grown decagon packing showing the network formed by nn bonds.

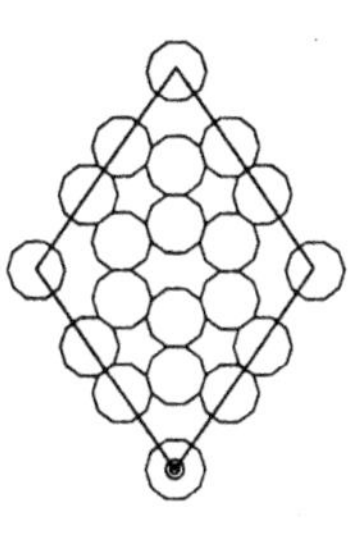

FIG. 3. Decoration of the 2D Penrose rhombi with decagons.

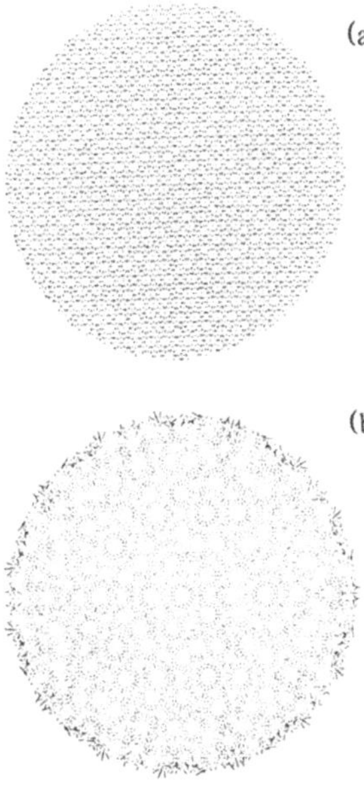

FIG. 4. (a) Phason field of the quasiperiodic packing; (b) fiftyfold magnification of the displacement field of the same packing.

bonds and σ_τ for the nnn bonds. Both Lennard-Jones depths are the same. The original edge-sharing structure, or reference solid, is stabilized by the choice $\sigma_\tau/\sigma_1 = \tau$. We are interested in the formation of strains (deformation of the reference solid) when σ_τ/σ_1 differs *infinitesimally* from τ. In this limit (*elastic regime*) the actual form of the potential is irrelevant since the harmonic behavior dominates the dynamics.

The constraints on the local geometry of our decagon packing still permit a large number of possible strain-free reference solids. We have studied two kinds of packings that can easily be distinguished by the behavior of the phason coordinate. The first is a quasiperiodic packing constructed by our decorating the 2D Penrose tiling of rhombi with decagons as shown in Fig. 3. The variation of $\mathbf{x}^\perp$ from decagon to decagon is shown in Fig. 4(a) by means of a vector proportional to $\mathbf{x}^\perp$, based at $\mathbf{x}^\parallel$ for each decagon with 4D coordinates $(\mathbf{x}^\parallel,\mathbf{x}^\perp)$. Although there are rapid local variations in $\mathbf{x}^\perp$, there are no systematic changes on long length scales. In fact, a necessary condition for quasiperiodic long-range order (of the reference solid) is simply that the $\mathbf{x}^\perp$ differences are bounded.

The second kind of decagon packing we have studied was generated by a growth algorithm. Our algorithm[5] is an extension of earlier aggregation models[6] which avoids the formation of 1D defects, or "tears," in the connectivity of the bond network. The growth geometry is a triangle with one edge moving at constant velocity v away from the opposite vertex. Using coordinates $\mathbf{x}^\parallel = (x,y)$, where y represents the growth direction, the interior of the triangle is given by $2|x| < y$, $0 < y < y_0$. The growth nucleus is a single decagon placed at (0,0). High connectivity of the network is achieved by Metropolis annealing with a linear temperature field $T(y) = h(y - y_0)$. The growth velocity is established by the motion of the zero-temperature isotherm: $\dot{y}_0 = v$. Each decagon-decagon bond, both nn with length 1 and nnn with length τ, is assigned a cohesive energy of -1. Growth and annealing occur in the region $T > 0$ $(y > y_0)$. A single growth-thermalization process consists of the following two operations applied to one of the N decagons, say D_1, chosen at random from the region $y > y_0$. (1) A bond emanating from D_1 is chosen at random and if a decagon D' placed at the other end of the bond satisfies two properties it is added to the structure: (i) It is simultaneously bonded to at least one other decagon, say D_2, and (ii) the distance between D' and other decagons to which it is not bonded is greater than τ. (2) D_1 may be removed according to the Metropolis criterion: a random number r, uniform in (0,1) is chosen and if $r < \exp(-n_{\mathrm{bond}}/T)$, then D_1 is removed. Here n_{bond} is the loss of cohesive energy given by the number of bonds removed when D_1 is removed and $T(y)$ is the local temperature. At the completion of each growth-thermalization process the zero-temperature isotherm is advanced according to $y_0 \to y_0 + v/N$.

We find that the parameter values $h = 0.3$ and $v < 0.001$ produce satisfactory networks without tears at the length scales considered here. Figure 5(a) shows the phason field of a circular region excised from the center of the triangular aggregate shown in Fig. 2. A striking feature of Fig. 5(a) is the long-wavelength variation in $\mathbf{x}^\perp$. This is especially remarkable in view of the uniformity of the growth process. The same feature has been observed in analogous simulations of a 3D icosahedral model.[7]

The rigid geometries of our two reference solids, the

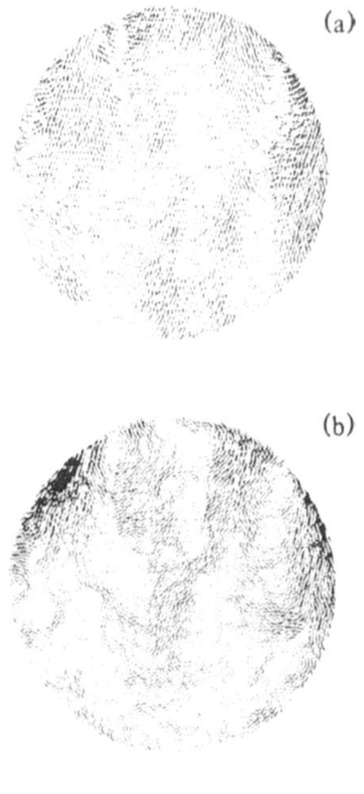

FIG. 5. (a) Phason field of a grown decagon packing; (b) tenfold magnification of the displacement field of the same packing.

quasiperiodic and grown decagon packings, correspond to a very nongeneric case of interparticle potentials. There are no symmetry principles that require that the ratio of the bond lengths is precisely τ or that the angles between bonds are precise multiples of 36°. Indeed, by imposing pair potentials with $\sigma_\tau/\sigma_1=\tau+\delta$, $\delta\neq 0$ these properties disappear. We have generated relaxed decagon packings for $\delta=0.1$ using a molecular-dynamics algorithm. This value of δ is small enough that the response (i.e., displacement field) scales linearly with δ. In the relaxed packings, not all the lengths of nn and nnn bonds are exactly at the minimum of their respective pairwise potentials.

A trivial consequence of modifying the potential is a uniform strain, e.g., isotropic contraction or expansion, of the unrelaxed packing. The uniform component of the strain was eliminated with the method of least squares. Let $\mathbf{x}_i$ ($\mathbf{x}_i''$) denote the unrelaxed (relaxed) positions of the ith decagon and $\mathbf{x}_i'$ the linear change applied to the unrelaxed position, i.e., $\mathbf{x}_i'=\mathbf{x}_i+\mathbf{A}\mathbf{x}_i+\mathbf{b}$. The uniform strain matrix $\mathbf{A}$ and translation vector $\mathbf{b}$ are determined by our minimizing the expression $\Delta^2=N^{-1}\sum_{i=1}^{N}(\mathbf{x}_i''-\mathbf{x}_i')^2$. Δ gives the root mean square (rms) displacement, i.e., the "random" strain when the trivial effects given by a linear transformation (rotation, expansion, translation, shear, etc.) have been eliminated.

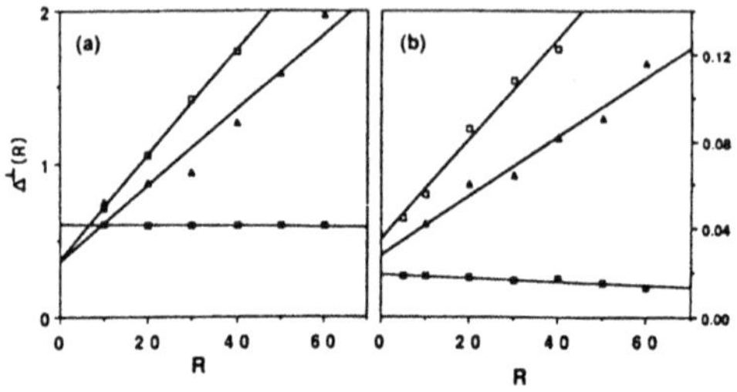

FIG. 6. Root mean square (a) phason and (b) ordinary displacements for the three decagon packings: quasiperiodic (■); grown, with $h=0.3$, $v=0.0015$ (▲); grown, with $h=0.3$, $v=0.0005$ (□). The data are plotted for circular packings with radius R.

The random-strain field of the quasiperiodic packing is shown in Fig. 4(b). Although the strain field has destroyed the bond-length–angle relationships of the rigid geometry, it is clear that the relaxed structure is still quasiperiodic. To see this we note that the strain field itself is quasiperiodic. This is evident from Fig. 4(b) and is easily explained since (i) each decagon displacement is determined by its environment (in the reference solid) and (ii) the set of similar environments forms a quasiperiodic pattern. Figure 4(b) also suggests that the decagon displacements (after subtraction of the uniform component) are bounded. We believe this is generally true for quasiperiodic structures, provided the forces are short ranged and the perturbation of the potential is sufficiently small.

The behavior of the strain fields of the grown decagon packings is quite different, as Fig. 5(b) shows. Again, we have subtracted the uniform component of the strain so that only the random component, associated with inhomogeneities, remains. There is clearly an accumulation of strain in that the large displacements away from the reference solid involve the coherent motion of many decagons. The growth of the rms displacement, $\Delta(R)$, with the radius R of the packing is shown in Fig. 6(b). In the quasiperiodic packing, ΔR quickly saturates as a function of R whereas each of the grown packings we have studied show a linear rise in $\Delta(R)$. A comparison of Figs. 5(a) and 5(b) suggests that the inhomogeneity responsible for the accumulation of strain in the grown packings is the long-wavelength variation of the phason coordinate. It is interesting that a plot of $\Delta^\perp(R)$, the rms phason displacement (with linear component subtracted) also shows a linear rise [Fig. 6(a)].

Several diffusion experiments[8,9] have established a linear growth of peak width with phason momentum $\mathbf{G}^\perp$ in quasicrystals. The apparent linear growth of $\Delta^\perp(R)$ with R in Fig. 6(a), seen also in the 3D icosahedral model,[7] is consistent with this behavior. Specifically, if $\mathbf{x}_i^\parallel$ is the position of the ith decagon in the reference solid and

$\delta\mathbf{x}_i$ its displacement after relaxation, then the scattering phase angle is given by[4]

$$\mathbf{G}^{\parallel}\cdot(\mathbf{x}_i^{\parallel}+\delta\mathbf{x}_i) = -\mathbf{G}^{\perp}\cdot\mathbf{x}_i^{\perp}+\mathbf{G}^{\parallel}\cdot\delta\mathbf{x}_i \ (\mathrm{mod}\, 2\pi) . \quad (1)$$

Fluctuations in both $\mathbf{x}_i^{\perp}$ and $\delta\mathbf{x}_i$ lead to peak broadening. In particular, for peaks with $|\mathbf{G}^{\perp}| \gg |\mathbf{G}^{\parallel}|$ one considers a coherence radius R_c defined by $|\mathbf{G}^{\perp}|\Delta^{\perp}(R_c)\sim\pi$. From the behavior $\Delta^{\perp}(R)\sim aR$ shown in Fig. 6(a) one then obtains a peak broadening $\delta G\sim\pi/R_c\sim a|\mathbf{G}^{\perp}|$.

Systematic departures from linear $|\mathbf{G}^{\perp}|$ peak broadening have been noted for diffraction peaks in the opposite limit: $|\mathbf{G}^{\parallel}| \gg |\mathbf{G}^{\perp}|$. The interpolating form $\delta G^2 = |a\mathbf{G}^{\perp}|^2+|b\mathbf{G}^{\parallel}|^2$ has been fitted to experimental data with some success[8,9] and Horn *et al.*[8] have argued that a $\mathbf{G}^{\parallel}$ term is a consequence of dislocations. The same argument given above, but applied to the rms fluctuations in $\delta\mathbf{x}_i$, leads to peak broadening of the form $\delta G\sim b|\mathbf{G}^{\parallel}|$, where now b comes from the behavior $\Delta(R)\sim bR$ shown in Fig. 6(b). Thus, our model, which is *free of dislocations*, reproduces the main features of peak broadening in quasicrystals. Experimentally, the ratio a/b is large.[9] Our results, where b depends linearly on δ, give a similarly large ratio, suggesting that the analog of δ in real quasicrystals is also small. Finally, we find (e.g., see Fig. 6) a dependence of strain accumulation on cooling rate, in that slow cooling induces larger strain. This surprising result is consistent with recent experiments in Ga-Mg-Zn.[10]

We acknowledge useful conversations with C. Henley. We are indebted to Eduardo Fradkin and the computing center of the Materials Research Laboratory of the University of Illinois for their invaluable assistance. This work has been supported in part by the NSF through Grants No. MRL-DMR-86-12860, No. PHY82-17853, supplemented by funds from NASA, and by DOE through Grant No. DE84-ER-45108.

(a)Permanent address: Dipartimento di Fisica, Universitá degli Studi di Trento, 38050 Povo (TN), Italy.

(b)Permanent address: AT&T Laboratories, Murray Hill, NJ 07974.

1J. C. Phillips, Phys. Today **35**, No. 2, 27 (1982).

2D. R. Nelson, Phys. Rev. B **28**, 5515 (1983); P. J. Steinhardt, D. R. Nelson, and M. Ronchetti, Phys. Rev. Lett. **47**, 1297 (1981), and Phys. Rev. B **28**, 784 (1983).

3Our definition of the word "quasicrystal" corresponds to the experimental one: A solid having a sharp but noncrystallographic diffraction pattern. In addition to quasiperiodic structures this definition includes random aggregation models such as the icosahedral glass.

4An excellent survey of the literature is given by C. L. Henley, Comments Condens. Mater. Phys. **13**, 59 (1987).

5V. Elsin, in *Proceedings of the Fifteenth International Colloquium on Group Theoretical Methods in Physics*, edited by R. Gilmore and D. H. Feng (World Scientific, Singapore, 1987), Vol. 1, p. 162.

6D. Schechtman and I. Blech, Metall. Trans. A **16**, 1005 (1985); P. W. Stephens and A. I. Goldman, Phys. Rev. Lett. **56**, 1168 (1986), and **57**, 2331 (1986).

7V. Elser, in *Proceedings of the NATO Advanced Research Workshop on New Theoretical Concepts in Physical Chemistry*, edited by A. Amann, L. S. Cederbaum, and W. Gans (Reidel, Dordrecht, 1988).

8P. M. Horn *et al.*, Phys. Rev. Lett. **57**, 1444 (1986); P. A. Heiney *et al.*, Science **238**, 660 (1987).

9D. Gratias *et al.*, to be published.

10W. Ohashi and F. Spaepen, Nature (London) **330**, 555 (1987); H. S. Chen *et al.*, Phys. Rev. B **38**, 1658 (1988).

Au (100) Surface Reconstruction

F. Ercolessi[a] and E. Tosatti[a]
IBM Zurich Research Laboratory, 8803 Rüschlikon, Switzerland

and

M. Parrinello
Dipartimento di Fisica Teorica, University of Trieste, I-34014 Trieste, Italy
(Received 19 May 1986)

We study the structure of the reconstructed Au (100) surface, using a phenomenological Hamiltonian, including a many-body force term (the "glue"), carefully optimized to account for a vast variety of properties of solid and liquid gold. The optimal atomic configuration of (100) slabs is obtained by a molecular-dynamics strategy. We find that the glue term drives the reconstruction into a denser, quasitriangular surface layer. By variation of cell size and atom number, the lowest-energy configuration is found to be roughly (1×5), and more precisely (34×5), close to (26×48) suggested by experiment.

PACS numbers: 68.35.Bs, 61.50.Lt

Among the noble metals, gold stands out for its remarkable surface properties. All its low-index surfaces reconstruct in such a way as to give rise to close-packed layers.[1] This behavior has been particularly well characterized for Au(100). LEED,[1] He-scattering,[2] and scanning-tunneling-microscope (STM)[3] studies indicate a rather complicated reconstruction pattern, whose main character, however, is simple: The topmost layer has switched from a square arrangement to a slightly rotated and somewhat contracted triangular packing. While plausible but speculative arguments have been advanced to justify such behavior, there exists as yet no detailed theoretical description of the phenomenon. At the *ab initio* microscopic level, with present-day possibilities, the task is certainly a very difficult one. Heine and Marks[4] have described a possible microscopic mechanism at the origin of this contraction, without attempting a quantitative modeling. Even a phenomenological sort of theory, for example, of the effective-Hamiltonian type proposed for W(100),[5] could be of use; but in the case of Au an extension to include many-body forces is clearly inescapable.

The missing ingredient, which two-body forces cannot simulate, is of course electronic cohesion. The important element in a gain in *d*-band and *s*-band energy in noble metals is a good atomic coordination: All atoms should be surrounded as much as possible by other atoms. Hence, a phenomenological scheme better suited for Au (and similar metals) should contain many-body forces, capable of describing and of rewarding atomic coordination. While qualitatively similar ideas as well as several calculational schemes based on this concept have recently been described,[6–8] we have independently developed our own approach,[9] which is based like that of Ref. 6 on a two-body force plus a many-body force henceforth called the "glue." The purpose of this Letter is to show that by use of a carefully constructed glue Hamiltonian, the occurrence of elaborate surface reconstructions can be naturally understood, together and on the same footing as many properties of bulk Au.

The total potential energy of the system has the same form as Ref. 6:

$$V = \frac{1}{2} \sum_{\substack{ij \\ j \neq i}} \phi(|\mathbf{r}_i - \mathbf{r}_j|) + \sum_i U(n_i),$$

where $n_i = \sum_{j \neq i} \rho(|\mathbf{r}_i - \mathbf{r}_j|)$. Here, ϕ is a two-body potential, ρ is a "density function" associated with each atom, and finally U is the glue, which provides good energetics only for a properly coordinated atom. In a stable $T = 0$ bulk, all $n_i = n_0$ which we arbitrarily choose to be $n_0 = 12$. Thus, n_i can be thought of as an effective coordination number. This model is very well suited for computer simulation, since the forces depend only upon the distances between pairs of particles.

Although further theoretical justifications of this scheme can be found in the so-called "embedded-atom method,[6] we do not believe that a truly first-principles derivation can be given. Rather, we have empirically constructed suitable functions $\phi(r)$, $\rho(r)$, $U(n)$ which account for all known lattice propreties of Au. The details of this exercise are too long to be presented here.[10] Our choice, shown in Fig. 1, fits exactly the $T = 0$ lattice parameter, the cohesive energy, the surface tension, the bulk modulus, and the X-point transverse phonon frequency. It also reproduces fairly well the vacancy formation energy, the Cauchy pressure, the melting temperature and latent heat, and the thermal expansion coefficient. The thermal properties, as well as the stability of the fcc structure up to the melting point, depend upon details of ϕ, ρ, U in a

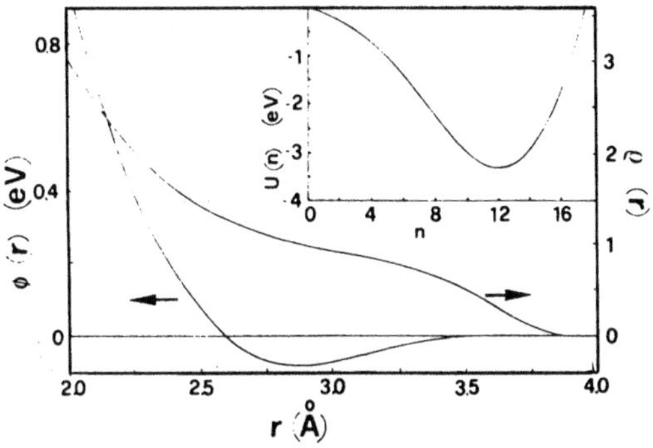

FIG. 1. The three functions characterizing the model: the two-body potential $\phi(r)$, the "atomic density" function $\rho(r)$, and (inset) the glue function $U(n)$.

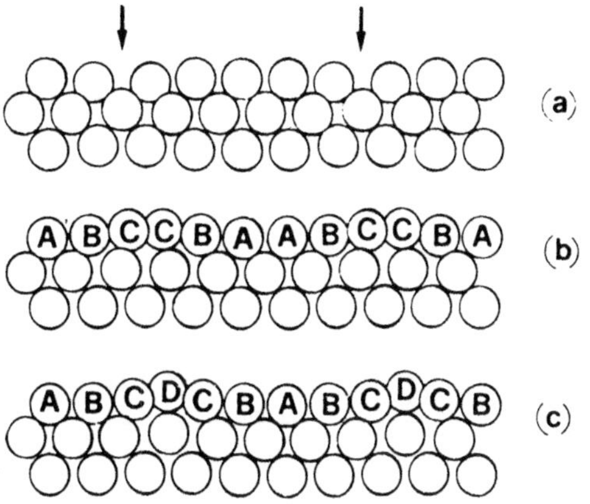

FIG. 2. Side views of minimum-energy configurations of (100) slabs after thermal equilibration and subsequent annealing. Two molecular-dynamics cells are shown for clarity. (a) The starting configuration was a perfect (100) slab. The atoms have shrunk into five-row wide stripes, leaving gaps (indicated by the arrows) in between. The surface energy is 109.6 meV/Å^2. (b) Five additional adatoms are present in the starting configuration. They are absorbed, giving rise to a 20% denser quasitriangular reconstructed first layer with an *ABCCBA* stacking. The surface energy is 102.3 meV/Å^2. (c) Same as (b), but the registry is different and the stacking is *ABCDCB*. The surface energy is 102.6 meV/Å^2. All atomic positions shown to scale (not schematic), but atom radii are arbitrary. Vertical (z) direction [100], horizontal (y) direction [01$\bar{1}$].

complex way, and were studied by molecular dynamics (MD). Both surface and vacancy-formation energies were measured on equilibrated MD samples to take relaxations into account. The experimental surface energy was fitted to that of a relaxed but not reconstructed (111) face.

We have studied the Au (100) surface with this glue Hamiltonian. This is done by MD studies of slabs with in-plane periodic boundary conditions and initially $5\times5=25$ atoms per (100) plane. The area of the slab and its square shape are kept rigid to prevent transformation into a (111) slab. The area is readjusted at each temperature to match the bulk thermal expansion calculated independently. We found that a number of layers $L=14$ is sufficient to decouple the two surfaces. We first of all studied the energetics, and used MD mainly as a tool to generate the minimal-energy configuration. Our typical procedure consists of warming the slab up to about $T_{\mathrm{melting}}/2$ and, after equilibration, of gradually cooling back to $T=0$. The total length of the cycle is of the order of 10 000 MD steps (1 step $\cong 7\times10^{-15}$ sec). This method does not guarantee attainment of the absolute minimum. However, it is always possible to improve one's confidence in a given configuration by trying different annealing schedules, and by starting from different initial conditions.

This procedure was first applied to the clean, unreconstructed (100) faces. Figure 2(a) shows the appearance of the first atomic layer after annealing. Note that the surface atoms have shrunken together, leading to formation of close-packed stripes (five atomic rows each) separated by a gap. This gap in turn can be seen as leading to the formation of two monatomic steps, here still very near to one another. The second layer has remained a basically perfect (100) plane. This is a clear indication that our (100) surface wants to reconstruct into a denser layer, even within the constraint of our small 5×5 cell. To pursue this idea further, we made a series of runs where a number n of extra adatoms was added on top of the first layer; n was varied throughout the range from $n=1$ to $n=25$. For n small, the extra atoms were absorbed into the first layer giving rise, after annealing, to a denser packing. At the same time, we found a decrease of surface energy, defined as $\sigma(n)=[E(N)-N\epsilon_c]/2A$, where $N=25L+2n$ (L is the number of layers in the slab) is the total number of particles in the sample, $E(N)$ the total energy of the slab, ϵ_c the cohesive energy, and A the surface area, and the factor 2 accounts for two surfaces. A minimum of $\sigma(n)$ was obtained for $n=5$, as shown by Fig. 3. The corresponding first-layer arrangement of Fig. 2(b) is a good candidate for explaining the Au (100) reconstruction. The second and deeper layers retain a strained (100) reconstruction, in agreement with the experimental findings.[11] The amplitudes of the corrugation predicted for the first four layers are $\xi_1=0.47$ Å, $\xi_2=0.21$ Å, $\xi_3=0.13$ Å, and $\xi_4=0.08$ Å. The relaxations of the distances between average layer positions are $\Delta d_{12}=+3.6\%$, $\Delta d_{23}=+2.2\%$, and $\Delta d_{34}=-0.2\%$. The increase of d_{12} and d_{23} is due to excessive coordination in the second layer, caused by the first-layer

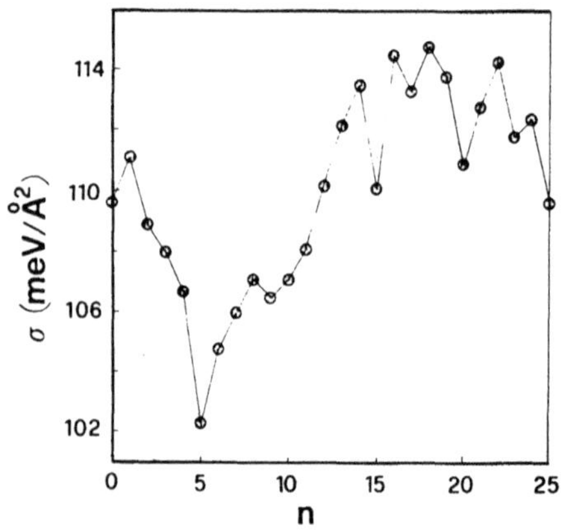

FIG. 3. Surface energy of the final configurations as a function of the number of adatoms present at the start in a 5×5 (100) slab. The minimum at $n=5$ corresponds to the configuration in Fig. 2(b).

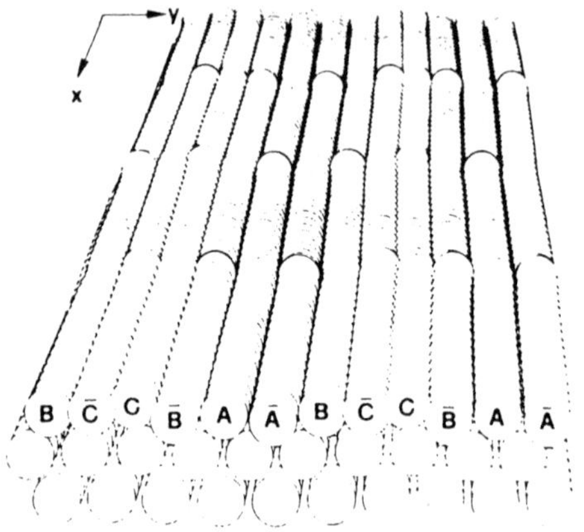

FIG. 4. Perspective view of the (34×5) reconstructed cell. The surface energy is 102.2 meV/Å^2.

reconstruction. The stacking of the rows is *ABCCBA*, as in Fig. 2(b). We also found a local energy minimum at another stacking *ABCDCB*, shown in Fig. 2(c). In this arrangement, the surface energy is slightly higher and the corrugation larger ($\xi_1=0.74$ Å). In all cases, the strain is not uniformly distributed; the surface density is higher in a hilltop row, and lower in a valley row, where the atoms are not far from their ideal hollow-site positions over the square substrate.

Of course, the role played by the cell size in these calculations is not a minor one. The $n=5$ "best" configuration has a (1×5) reconstructed structure which fits very well in our 5×5 cell. On the other hand, the use of different cells may reveal the existence of reconstructed surfaces with a still lower surface energy. Following this idea, we performed calculations of the same kind, with cells suited to the following reconstruction patterns: (1×7), (1×12), (1×8), and (1×3), respectively, of the kind 8 onto 7, 14 onto 12, 10 onto 8, 4 onto 3 (in increasing order of surface density). All these surfaces reconstructed into a denser overlayer, and the surface energies were (in the order) 103.7, 103.2, 103.3, 108.6 meV/Å^2. Since they are all higher than the (1×5) surface energy (102.3 meV/Å^2 for the *ABCCBA* registry, or 102.6 meV/Å^2 for the *ABCDCB* registry), the (1×5) is the preferred pattern.

So far, we have considered only the reconstruction along the fivefold y $[01\bar{1}]$ direction, which gives rise to the basic (1×5) pattern. In this geometry, the quasitriangular overlayer is contracted on the average by 3.77% in the y direction, while it is in registry with the underlying lattice in the x [011] direction. On the other hand, the unit cell has been regarded as a (20×5) for a long time,[1] indicating a 4.76% average contraction along x. More recent STM measurements[3] suggest a more accurate (26×48) structure, where the contractions are 3.83% along x and 4.42% along y. We searched for the optimal contraction in the x direction by studying the surface energies of reconstructed $(M\times5)$ twelve-layer slabs. The details of this important refinement are somewhat elaborate and will appear elsewhere.[10] The lowest surface energy is found on the $\bar{A}B\bar{C}C\bar{B}A$ (34×5) cell (the soliton structure now makes A and $\bar{A}$, B and $\bar{B}$, C and $\bar{C}$ inequivalent). It is about 0.1% less than the *ABCCBA* (1×5) surface energy, and only 0.003% less than (26×5), which is very close to (26×48), i.e., the minimum at $M=34$ is extremely flat.

Figure 4 shows a perspective view of the first three layers of this slab. The contraction along x induced by the density increase is not uniform but localized in soliton-type regions, stacked to give rise to a slightly distorted centered rectangular superlattice. In correspondence with the soliton centers, the corrugation in the y direction has a double-maximum-double-minimum structure and is large (0.8 Å). In the middle of the regions between the solitons, the corrugation and the overall structure are very similar to the *ABCCBA* (1×5) cell. Here, the y corrugation agrees well with the He-scattering[2] and STM[3] estimates. The solitons slightly compress the surface atoms in these middle regions. The resulting glue-energy gain more than compensates the formation energy of the solitons. Also, the multilayer relaxations, generally close to those of the (1×5) above, become soliton modulated. Our structure does indeed show the alternation of single-maximum–single-minimum smooth areas ("smooth ribbons") and double-maximum–double-minimum soliton areas ("rough ribbons") seen by STM.[3] While differing in several ways from the hard-sphere model used in Ref. 3, the present structure seems generally more plausible, and no less compatible with the data. A thorough, detailed discussion will be given elsewhere.[10]

In summary, we have presented the first detailed theoretical study of the Au (100) surface reconstruction, realized with the same many-body forces which work for bulk Au. We find a triangular overlayer whose main features seem to agree well with known experimental facts. Further studies, such as temperature behavior, other Au surfaces and defects, small clusters, and extension to other metals, are presently being pursued.

We are grateful to M. Garofalo and H. Rohrer for useful discussions. Two of us (F.E. and E.T.) wish to thank T. Schneider and A. R. Williams for hospitality and assistance at the IBM Zurich Research Laboratory.

[a]Permanent address: International School for Advanced Studies, Strada Costiera 11, I-34014 Trieste, Italy.

[1]M. A. Van Hove, R. J. Koestner, P. C. Stair, J. P. Bibérian, L. Kesmodel, I. Bartoš, and G. A. Somorjai, Surf. Sci. **103**, 189, 218 (1981), and references therein.

[2]K. H. Rieder, T. Engel, R. H. Swendsen, and M. Manninen, Surf. Sci. **127**, 223 (1983).

[3]G. K. Binnig, H. Rohrer, Ch. Gerber, and E. Stoll, Surf. Sci. **144**, 321 (1984); H. Rohrer, private communication.

[4]V. Heine and L. D. Marks, Surf. Sci. **165**, 65 (1986).

[5]A. Fasolino, G. Santoro, and E. Tosatti, Phys. Rev. Lett. **44**, 1684 (1980).

[6]M. S. Daw and M. I. Baskes, Phys. Rev. B **29**, 6443 (1984).

[7]M. W. Finnis and J. E. Sinclair, Philos. Mag. A **50**, 45 (1984).

[8]D. Tománek and K. H. Bennemann, Surf. Sci. **163**, 503 (1985).

[9]F. Ercolessi, Ph.D. thesis, University of Trieste, 1983 (unpublished); M. Garofalo, thesis, International School for Advanced Studies, Trieste, 1984 (unpublished).

[10]F. Ercolessi, M. Parinello, and E. Tosatti, to be published.

[11]D. M. Zehner *et al.*, J. Vac. Sci. Technol. **12**, 454 (1975).

Strain fluctuations and elastic constants[a)]

M. Parrinello

University of Trieste, Italy

A. Rahman

Argonne National Laboratory, Argonne, Illionis 60439

(Received 20 August 1981; accepted 23 November 1981)

It is shown that the elastic strain fluctuations are a direct measure of elastic compliances in a general anisotropic medium; depending on the ensemble in which the fluctuation is measured either the isothermal or the adiabatic compliances are obtained. These fluctuations can now be calculated in a constant enthalpy and pressure, and hence, constant entropy, ensemble due to recent developments in the molecular dynamics techniques. A calculation for a Ni single crystal under uniform uniaxial 100 tensile or compressive load is presented as an illustration of the relationships derived between various strain fluctuations and the elastic modulii. The Born stability criteria and the behavior of strain fluctuations are shown to be related.

I. INTRODUCTION

It was in a paper by Lebowitz, Percus, and Verlet[1] that a study was made of the ensemble dependence of fluctuations; in the same paper it was shown that fluctuations of the kinetic energy in a molecular dynamics calculation (which generates members of a microcanonical ensemble of states) is related to the constant volume heat capacity. It has now become standard practice in molecular dynamics and Monte Carlo calculations to use fluctuations of various phase space functions to determine the thermodynamic properties of the system.

Recently, we have[2] presented a Lagrangian formulation for molecular dynamics calculations in which the ensemble of states corresponds to constant stress **S**, enthalpy H (heat function), and number of particles N. A special case of this is where the constant stress is a hydrostatic pressure, thus giving a (p, H, N) ensemble. This was presented in another paper.[3]

In the following paper we show how the fluctuations of strain in an $(\mathbf{S}, H, N)$ ensemble are related to adiabatic (constant entropy S) compliances and hence to the elastic modulii of the system. In the case of the (p, H, N) ensemble the relevant compliance is the adiabatic compressibility.

II. THE (p,H,N) ENSEMBLE

There is no difference in principle between the treatment of the (p, H, N) and the $(\mathbf{S}, H, N)$ ensemble for the purpose in hand. Hence, first because one is more familiar with the former and second, because of the simplicity in writing equations involving variables without indices, we shall first deal briefly with the (p, H, N) ensemble.

The thermodynamic relations are stated in the Appendix for completeness. As shown in Landau and Lifshitz,[4] in a (p, β, N) ensemble the volume fluctuations are

$$\langle(\Delta V)^2\rangle_{p,\beta,N} = (V/\beta)\chi_\beta , \quad (1)$$

where χ_β, the isothermal compressibility $= -(1/V)(\partial V/\partial p)T$; using Eqs. (A1) and (A3) and combining with Eq. (1)

$$\langle(\Delta V)^2\rangle_{p,\beta,N} + (\partial\beta/\partial H)_p(\partial V/\partial\beta)_p^2 = (V/\beta)\chi_s . \quad (2)$$

From Lebowitz *et al.*[1] [their Eq. (2.12)], we see that the left-hand side of the above equation is precisely $\langle(\Delta V)^2\rangle_{p,H,N}$. Hence (as has already been shown by Haile and Graben[5])

$$\langle(\Delta V)^2\rangle_{p,H,N} = (V/\beta)\chi_s . \quad (3)$$

Thus, volume fluctuations in a system with constant pressure, constant enthalpy, and constant number of particles are just a measure of the adiabatic compressibility. (See the last remark in the Appendix.)

III. THE $(\mathbf{S},H,N)$ ENSEMBLE

In dealing with anisotropic media under the influence of a general stress tensor the elastic energy, V_{el}, is usually written as

$$V_{el} = \Omega_0 \,\mathrm{Tr}(\mathbf{S}\epsilon) , \quad (4)$$

where Ω_0 is the unstrained volume and ϵ the strain tensor, assumed small enough for the above expression to be a valid approximation. (The sign convention is usually taken to be $\mathbf{S} = -p$ when $\mathbf{S}$ is isotropic.)

With the above V_{el}, the expression for the probability of a fluctuation takes the form[4]

$$w \propto \exp[-\beta/2\{\Delta T\Delta S + \Omega_0 \,\mathrm{Tr}(\Delta\mathbf{S}\Delta\epsilon)\}] . \quad (5)$$

But

$$\Delta S = C_\epsilon \Delta T/T - \Omega_0 \,\mathrm{Tr}[(\partial\mathbf{S}/\partial T)\Delta\epsilon] ,$$

where C_ϵ is the heat capacity at constant strain. Hence

$$w \propto \exp[-\beta/2\{(C_\epsilon/T)(\Delta T)^2 + \Omega_0(\partial S_{ij}/\partial\epsilon_{kl})_T\Delta\epsilon_{ij}\Delta\epsilon_{kl}\}] , \quad (6)$$

where summation over repeated indices is understood. From this it follows that in a $(\mathbf{S}, \beta, N)$ ensemble

$$\langle\Delta\epsilon_{ij}\Delta\epsilon_{kl}\rangle_{\mathbf{S},\beta,N} = (kT/\Omega_0)(\partial\epsilon_{ij}/\partial S_{kl})_T . \quad (7)$$

The equation corresponding to Eq. (2.12) of Ref. (1) is now

[a)]Work supported by the U. S. Department of Energy.

 0021-9606/82/052662-05$02.10

$$\langle\Delta\epsilon_{ij}\Delta\epsilon_{kl}\rangle_{\mathbf{S},H,N}=\langle\Delta\epsilon_{ij}\Delta\epsilon_{kl}\rangle_{\mathbf{S},\beta,N}+(\partial\beta/\partial H)_{\mathbf{S}}(\partial\epsilon_{ij}/\partial\beta)_{\mathbf{S}}(\partial\epsilon_{kl}/\partial\beta)_{\mathbf{S}}\,. \quad (8)$$

Analogous to Eq. (A3) we have $(\partial\beta/\partial H)_{\mathbf{S}}=-k\beta^2/C_{\mathbf{S}}$, where $C_{\mathbf{S}}$ is the constant stress heat capacity; the equation analogous to Eq. (A1) is

$$(\partial\epsilon_{ij}/\partial S_{kl})_{\beta}=(\partial\epsilon_{ij}/\partial S_{kl})_{S}+\Omega_0(k\beta^3/C_{\mathbf{S}})(\partial\epsilon_{ij}/\partial\beta)_{\mathbf{S}}(\partial\epsilon_{kl}/\partial\beta)_{\mathbf{S}}\,. \quad (9)$$

Using Eq. (7) and Eq. (9) we get from Eq. (8)

$$\langle\Delta\epsilon_{ij}\Delta\epsilon_{kl}\rangle_{\mathbf{S},H,N}=(kT/\Omega_0)(\partial\epsilon_{ij}/\partial S_{kl})_S\,. \quad (10)$$

It is customary to use elastic moduli $C^S_{kl,ij}\equiv(\partial S_{kl}/\partial\epsilon_{ij})_S$ rather than the compliances. Let us denote by $\mathbf{C}^S$ the 9×9 matrix of the adiabatic elastic moduli so that the modulus $C^S_{ij,kl}$ is the (ij,kl) element of $\mathbf{C}^S$. Then we can go from the moduli to the compliances by a matrix inversion, to get

$$\langle\Delta\epsilon_{ij}\Delta\epsilon_{kl}\rangle_{\mathbf{S},H,N}=(kT/\Omega_0)(\mathbf{C}^S)^{-1}_{ij,kl}\,. \quad (11)$$

IV. BORN STABILITY CRITERIA

We have shown in Sec. III that at constant stress, constant enthalpy, and constant number, (in other words at constant entropy) the strain–strain correlation function is given by the adiabatic elastic compliance. The matrix of the adiabatic elastic moduli $C^S_{ij,kl}=(\partial S_{ij}/\partial\epsilon_{kl})_S$, was denoted by $\mathbf{C}^S$. In other words, $\mathbf{S}=\mathbf{C}^S\boldsymbol{\epsilon}$ and hence $\boldsymbol{\epsilon}=(\mathbf{C}^S)^{-1}\mathbf{S}$; the compliance $(\partial\epsilon_{ij}/\partial S_{kl})_S=(\mathbf{C}^S)^{-1}_{ij,kl}$ [Eq. (11)]. This element of $(\mathbf{C}^S)^{-1}$ is the algebraic complement of the element $C^S_{kl,ij}$ of $\mathbf{C}^S$ divided by the determinant of $\mathbf{C}^S$. [Similar statements of course can be made for the (S,β,N) ensemble as well except that instead of the superscript S for constant entropy one will show T to indicate isothermal elastic moduli.]

The stability criteria of Born state that a *necessary* condition for crystal stability is that the quadratic form $\boldsymbol{\epsilon}\mathbf{C}\boldsymbol{\epsilon}\equiv C_{ij,kl}\epsilon_{ij}\epsilon_{kl}$ be positive definite.[8] Instability can occur with the vanishing of a principal minor of the determinant of C. This implies that a divergence in some of the correlations $\langle\Delta\epsilon_{ij}\Delta\epsilon_{kl}\rangle$ will occur when the Born stability criteria are violated.

V. MOLECULAR DYNAMICS AT CONSTANT APPLIED STRESS

The three stages prior to the formulation of new molecular dynamics equations of relevance here were the following.

(i) The traditional case of generating a microcanonical ensemble of states using the Lagrangian

$$\mathcal{L}_1=\frac{1}{2}\sum m_i\mathbf{v}_i^2-V(\mathbf{r}_1,\ldots,\mathbf{r}_N)\,. \quad (12)$$

Periodic boundary conditions are applied, most often, in the form of a repeating cubic cell of volume L^3. The point to note is that L is a constant and can be used as the unit of length.

(ii) The introduction of a time dependent volume $\Omega(t)$ by Andersen[7] who used

$$\mathcal{L}_2=\Omega^{2/3}\frac{1}{2}\sum_i m_i\dot{\mathbf{s}}_i^2-V+\tfrac{1}{2}C\dot{\Omega}^2-p\Omega\,, \quad (13)$$

where $\mathbf{s}_i=\mathbf{r}_i/\Omega^{1/3}$. Periodic boundary conditions of the usual kind give a pulsating cubic box which changes in time according to a Lagrangian equation of motion.[7] The role of the constant C in Eq. (13) is discussed by Andersen.[7]

(iii) The introduction of a time dependent shape by Parrinello and Rahman[3] who used vectors $\mathbf{a}(t)$, $\mathbf{b}(t)$, and $\mathbf{c}(t)$ to define the molecular dynamics cell and used (a prime indicating the transpose)

$$\mathcal{L}_3=\frac{1}{2}\sum m_i\dot{\mathbf{s}}_i'\mathbf{G}\dot{\mathbf{s}}_i-V+\tfrac{1}{2}W\,\mathrm{Tr}\,\dot{\mathbf{h}}'\dot{\mathbf{h}}-p\Omega\,, \quad (14)$$

where $\mathbf{r}_i=\mathbf{h}\mathbf{s}_i$, $\mathbf{h}(t)=\{\mathbf{a},\mathbf{b},\mathbf{c}\}$, $\mathbf{G}=\mathbf{h}'\mathbf{h}$, and $\Omega=\|\mathbf{h}\|$ and W is a mass associated with the coordinates $h_{\lambda\mu}$. Periodic boundary conditions of the usual kind give a pulsating molecular dynamics cell of arbitrary shape which changes according to Lagrangian equations of motion.[3]

The final step in this development is the introduction of an anisotropic stress tensor $\mathbf{S}$ in place of p in $\mathcal{L}_3$.

This was done by the authors[2] in the following manner. The elements that are used in constructing $\mathcal{L}_3$ naturally lend themselves to the introduction of the notion of strain. The concept of strain and that of the metric tensor are intimately connected and, as surely has been noticed, $\mathbf{G}$ the metric tensor is an integral part of the Lagrangian $\mathcal{L}_3$.

In describing a strained state of a system one needs a so-called reference state; for this we shall use the matrix $\mathbf{h}_0$ and the corresponding volume $\Omega_0=\|\mathbf{h}_0\|$. The matrix $\mathbf{h}_0$ can be used to set up a mapping between space points $\mathbf{r}$ and a dimensionless vector $\boldsymbol{\xi}$, i.e., $\mathbf{r}=\mathbf{h}_0\boldsymbol{\xi}$. A homogeneous distortion changes $\mathbf{h}_0$ to $\mathbf{h}$, moving $\mathbf{r}$ to d, where $\mathbf{d}=\mathbf{h}\boldsymbol{\xi}=\mathbf{h}\mathbf{h}_0^{-1}\mathbf{r}$. Hence the displacement $\mathbf{u}$ (at $\mathbf{r}$) due to the distortion is $\mathbf{d}-\mathbf{r}$ or

$$\mathbf{u}=(\mathbf{h}\mathbf{h}_0^{-1}-1)\mathbf{r}\,. \quad (15)$$

The strain tensor $\boldsymbol{\epsilon}$ is defined as[8]

$$\epsilon_{\lambda\mu}=\frac{1}{2}\left(\frac{\partial u_\lambda}{\partial r_\mu}+\frac{\partial u_\mu}{\partial r_\lambda}+\sum_\nu\frac{\partial u_\nu}{\partial r_\mu}\frac{\partial u_\nu}{\partial r_\lambda}\right)\,. \quad (16)$$

Hence, we find

$$\boldsymbol{\epsilon}=\tfrac{1}{2}(\mathbf{h}_0'^{-1}\mathbf{G}\mathbf{h}_0^{-1}-1)\,. \quad (17)$$

If $\mathbf{S}$ denotes the external stress the elastic contribution to the energy will be, from Eq. (4),

$$V_{el}=\tfrac{1}{2}\,\Omega_0\,\mathrm{Tr}\,\mathbf{S}(\mathbf{h}_0'^{-1}\mathbf{G}\mathbf{h}_0^{-1}-1)\,. \quad (18)$$

The new Lagrangian $\mathcal{L}_s$, which takes account of the anisotropic strain will then be, on leaving out inconsequential constant energy terms,

$$\mathcal{L}_s=\frac{1}{2}\sum m_i\dot{\mathbf{s}}_i\mathbf{G}\dot{\mathbf{s}}_i-V+\tfrac{1}{2}W\,\mathrm{Tr}\,\dot{\mathbf{h}}'\dot{\mathbf{h}}+\tfrac{1}{2}\mathrm{Tr}\sum\mathbf{G}\,, \quad (19)$$

where

$$\Omega_0^{-1}\sum=\mathbf{h}_0^{-1}\mathbf{S}\mathbf{h}_0'^{-1}\,. \quad (20)$$

The equations of motion arising out of $\mathcal{L}_s$ are simple to write down and have been given elsewhere. The point of interest here is that the above Lagrangian gives a constant of motion $\mathcal{H}$ which is

$$\mathcal{H}=\sum_i \tfrac{1}{2}m_i\mathbf{v}_i^2+V+\tfrac{1}{2}W\,\mathrm{Tr}\,\dot{\mathbf{h}}'\dot{\mathbf{h}}-\Omega_0\,\mathrm{Tr}\,\mathbf{S}\boldsymbol{\epsilon}\ , \tag{21}$$

with $\mathbf{v}_i$ indicating $\mathbf{h}\dot{\mathbf{s}}_i$.

In equilibrium, at temperature T, the term in W contributes $(9/2)k_BT$ while the term with m_i's contributes $(3N/2)k_BT$. Hence to an accuracy of $3:N$, the enthalpy

$$H=E-V_{el}\ , \tag{22}$$

$$E=\sum_i \tfrac{1}{2}m_i\mathbf{v}_i^2+V\ , \tag{23}$$

is a constant of the motion.

The formal development of Sec. III and the fact that under $\mathcal{L}_s$ one generates an $(\mathbf{S},H,N)$ ensemble allows us to conclude that in a molecular dynamics calculation using $\mathcal{L}_s$ the elastic constants of the system can be determined from Eq. (10) of Sec. III.

VI. SPECIAL CASE OF TETRAGONAL SYMMETRY

As an application of the above general development we shall consider a system having tetragonal symmetry with [100] as the direction of tetragonal symmetry. We can write the following set of nonredundant equations [using for notation $D=C_{11,11}(C_{22,22}+C_{22,33})-2C_{11,22}^2$ and $\epsilon^{\pm}=(\epsilon_{22}\pm\epsilon_{33})/\sqrt{2}$],

$$\langle\Delta\epsilon_{11}\Delta\epsilon_{11}\rangle=(kT/\Omega_0)(C_{22,22}+C_{22,33})/D\ , \tag{24.1}$$

$$\langle\Delta\epsilon_{11}\Delta\epsilon^{+}\rangle=(kt/\Omega_0)\sqrt{2}(-C_{11,22})/D\ , \tag{24.2}$$

$$\langle\Delta\epsilon^{+}\Delta\epsilon^{+}\rangle=(kT/\Omega_0)\,C_{11,11}/D\ , \tag{24.3}$$

$$\langle\Delta\epsilon^{-}\Delta\epsilon^{-}\rangle=(kt/\Omega_0)/(C_{22,22}-C_{22,33})\ , \tag{24.4}$$

$$\langle\Delta\epsilon_{23}\Delta\epsilon_{23}\rangle=(kT/\Omega_0)/C_{23,23}\ , \tag{24.5}$$

$$\langle\Delta\epsilon_{12}\Delta\epsilon_{12}\rangle=(kT/\Omega_0)/C_{12,12}\ . \tag{24.6}$$

The Born conditions for this case are

$$C_{11,11}>0;\quad C_{11,11}C_{22,22}-C_{11,22}^2>0\ ;$$

$$(C_{22,22}-C_{22,33})D>0;\quad C_{23,23}>0;\quad C_{12,12}>0\ .$$

These are equivalent to

$$C_{11,11}>0\ ;\quad C_{22,22}-C_{22,33}>0;\quad D>0;\quad C_{23,23}>0;\quad C_{12,12}>0\ .$$

Thus the four denominators in Eqs. (24.1)–(24.6) are the quantities whose vanishing defines the boundary of the elastic stability region

$$C_{11,11}(C_{22,22}+C_{22,33})=2C_{11,22}^2\ , \tag{25.1}$$

$$C_{22,22}=C_{22,33}\ , \tag{25.2}$$

$$C_{23,23}=0\ , \tag{25.3}$$

$$C_{12,12}=0\ . \tag{25.4}$$

The condition $C_{11,11}=0$ need not be included in the list of conditions that define the boundary of the region of stability since for a sufficiently small and positive $C_{11,11}$ the condition in Eq. (25.1) is already satisfied.

VII. CONCLUDING REMARKS AND AN ILLUSTRATION

The general result given in Eq. (11) and hence also the special case of tetragonal symmetry shown in Eq. (24), implies that certain strain fluctuations will be enhanced as a result of a reduction in the value of one or more of the principal minors of the matrix of elastic modulii. In this context of enhanced strain fluctuations we recall the Lindemann criterion for mechanical instability in crystals and assert that mechanical failure will occur when these fluctuations are so large that the underlying atomic displacements become a sizable fraction of the atomic spacing.

However, one should not overlook the possible presence of short wavelength phonons which due to the vanishing of the vibration frequency can also lead to instability even when none of the minors of the matrix of elastic modulii is small enough to produce large strain fluctuations that are being discussed here.

The possible divergent behavior of strain fluctuations is similar to what occurs at the liquid–gas phase transition where the bulk modulus goes to zero and volume fluctuations diverge. Hence if there is an elastically driven second order transition between different polymorphic crystalline phases then there will occur divergences in the strain fluctuations for the appropriate values of temperature and stress.

From Eq. (11) we expect that one or more of the strain fluctuations will be enhanced if, under suitable conditions of temperature and external stress, the elastic constants are brought close to the Born condition $\det\mathbf{C}=0$. Light scattering experiments using a crystal at a suitable temperature and/or suitable conditions of external stress might indeed confirm this phenomenon.

As regards the special case treated in Sec. VI, there are certain aspects of Eqs. (24) and Eqs. (25) which are worth dwelling upon. The four conditions, Eq. (25.1) to Eq. (25.4), are independent, and hence when one holds the other three need not be satisfied.

When Eq. (25.1) is satisfied, i.e., when D defined in Sec. VI is zero, then Eqs. (24.1), (24.2), and (24.3) show a divergence whereas Eq. (24.4), (24.5), and (24.6) remain finite. Thus, when the condition $D=0$ is approached, ϵ_{22} and ϵ_{33} must fluctuate in phase so as to prevent $\epsilon^{-}=(\epsilon_{22}-\epsilon_{33})/\sqrt{2}$ from having large fluctuations. On the other hand, when the condition $C_{22,22}-C_{22,33}=0$ is approached we must have ϵ_{22} and ϵ_{33} fluctuating out of phase.

Thus, even if the theoretical stability limits [Eqs. (25.1) to (25.4)] are not reached, and this will be so in most practical situations, we expect that the trends described above may be observable under suitable temperature and stress conditions. We have recently reported on a molecular dynamics study[2] of the effect of uniaxial stress on a single crystal of Ni. This data is being analyzed from the point of view of the behavior of strain fluctuations dealt with in the present paper. To illustrate the observation made above about the phase relation between the fluctuations of ϵ_{22} and ϵ_{33} we will

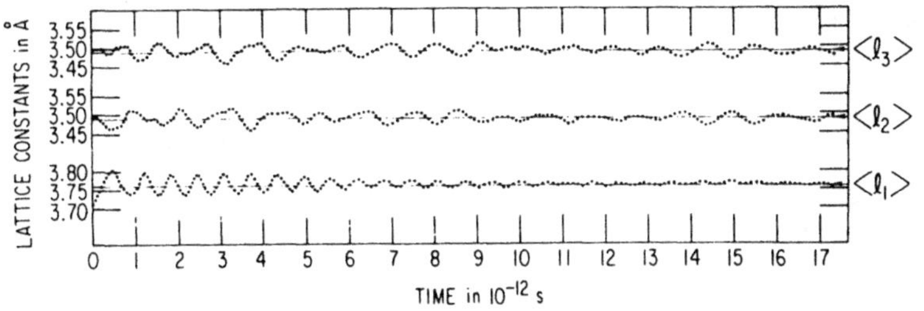

FIG. 1. Fluctuation of the three lattice constants of a tetragonal structure generated by applying a tensile load to an fcc crystal of Ni. Strain components can be calculated from the l_i shown above; $(l_i - \langle l_i \rangle)/\langle l_i \rangle$ is a good approximation (Ref. 2) for ϵ_{ii}. The condition $C_{22,22} = C_{22,33}$ is close to being satisfied at this load; see Eqs. (24.3) and (24.4) for consequences.

briefly mention one of the several calculations on Ni already reported.[2]

On applying a [100] tensile load to a single crystal of fcc Ni one obtains a face-centered tetragonal structure. Under conditions of zero load and a temperature of 356 K the model[9] of Ni we have used[2] gave a stable perfect fcc lattice in thermal motion, the lattice constant being 3.55 ± 0.09 Å. As the load was increased the lattice became tetragonal and at a tensile load of 8.6×10^{10} dyn cm^{-2} and 330 K, the two lattice constants became 3.76 ± 0.01 and 3.49 ± 0.01 Å. This state of the system was perfectly stable over a long MD calculation of 1.7×10^{-11} s. Further details about many such calculations are in Ref. 2. One more detail from Ref. 2 is relevant. According to static calculations of Milstein[10] at about 16×10^{10} dyn cm^{-2}, the Born condition [Eq. (25.2)] $C_{22,22} = C_{22,33}$ is satisfied. (Our dynamical calculation[2] showed system failure to occur already at $\sim 11 \times 10^{10}$ dyn/cm^2.)

The time behavior of the three lattice constants for the system described above, i.e., for a stable tetragonal structure at 330 K and tensile [100] load of 8.6×10^{10} dyn cm^{-2} is shown in Fig. 1. In spite of the noisy features we see from this figure that ϵ_{22} and ϵ_{33} are out of phase, in complete accord with the prediction of Eqs. (24.3) and (24.4).

On applying a compressive [100] load the static calculations of Milstein[10] show that at a load of about 7×10^{10} dyn cm^{-2} the instability criterion $D = 0$ [Eq. (25.1)] is satisfied. (Our dynamical calculations[2] showed a polymorphic transition to occur at a value slightly lower than 7×10^{10} dyn cm^{-1}.)

The time behavior of the three lattice constants for a stable tetragonal structure at 356 K and compressive [100] load of 5.25×10^{10} dyn cm^{-2} is shown in Fig. 2. In spite of the noise we can see (i) that ϵ_{22} and ϵ_{33} are in phase, (ii) that ϵ_{11} is out of phase with those two, and (iii) that the fluctuation ϵ_{11} is larger than that of the other two. Statement (i) is in accord with Eqs. (24.3) and (24.4), (ii) is a consequence of Eq. (24.2) since[11] $C_{11,22} > 0$, and (iii) is in accord with Eq. (24.1).

The time scale of the fluctuations shown in Figs. 1 and 2 is dependent on the choice of W, see Eq. (19). The statistical averages we have dealt with here do not depend on this choice. This has been discussed by Andersen,[7] Haile and Graben,[5] and Parrinello and Rahman.[2,3]

We finally note that a different method for the numerical calculation of the isothermal elastic constants of crystalline systems has been proposed by Squire *et al.*[12] An extension of their method to noncrystalline systems

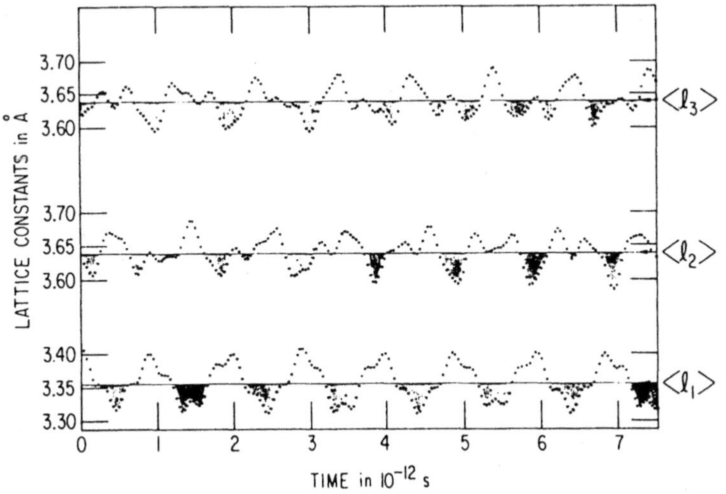

FIG. 2. System under a compressive load. The condition $D = C_{11,11}(C_{22,22} + C_{22,33}) - 2C^2_{11,22} = 0$ is close to being satisfied; see Eqs. (24.1), (24.2), (24.3), and (24.4) for consequences. The shading is to show the phase relation between $\Delta\epsilon_{11}$, $\Delta\epsilon_{22}$, and $\Delta\epsilon_{33}$.

does not seem to be straightforward. In contrast, the computational scheme described above seems to be suitable for all systems, crystalline or otherwise. This remark is of particular significance in the study of superionic conductors. These have normal solid-like elastic behavior but one or more of the constituents have large liquid-like constants of self-diffusion. However, extensive calculations are necessary before one can assess the practical merits and limitations of the method. We plan to undertake such calculations in the near future.

APPENDIX

Using p, T as variables,

$$\left(\frac{\partial V}{\partial p}\right)_S = \left(\frac{\partial V(p,T)}{\partial p}\right)_{S(p,T)} = \left(\frac{\partial V}{\partial p}\right)_T + \left(\frac{\partial V}{\partial T}\right)_p \left(\frac{\partial T}{\partial p}\right)_S$$

$$= \left(\frac{\partial V}{\partial p}\right)_T - \left(\frac{\partial V}{\partial T}\right)_p \left(\frac{\partial T}{\partial S}\right)_p \left(\frac{\partial S}{\partial p}\right)_T$$

$$= \left(\frac{\partial V}{\partial p}\right)_T - (T/C_p)\left(\frac{\partial V}{\partial T}\right)_p \left[-\left(\frac{\partial V}{\partial T}\right)_p\right] \qquad \text{(A1)}$$

or

$$\left(\frac{\partial V}{\partial p}\right)_\beta = \left(\frac{\partial V}{\partial p}\right)_S - k\beta^3/C_p\left(\frac{\partial V}{\partial \beta}\right)_p^2 . \qquad \text{(A1)}$$

We also have the enthalpy $H = U + pV$, and

$$dH = dS/k\beta + V dp , \qquad \text{(A2)}$$

$$\left(\frac{\partial \beta}{\partial H}\right)_p = -k\beta^2/C_p . \qquad \text{(A3)}$$

We note finally from Eq. (A2) that for p and H constant $dS = 0$.

[1]J. L. Lebowitz, J. K. Percus, and L. Verlet, Phys. Rev. **153**, 250 (1967).
[2]M. Parrinello and A. Rahman, J. Appl. Phys. **52**, 7182 (1981).
[3]M. Parrinello and A. Rahman, Phys. Rev. Lett. **45**, 1196 (1980).
[4]L. D. Landau and E. M. Lifshitz, *Statistical Physics* (Addison-Wesley, Reading, Mass., 1958).
[5]J. M. Haile and H. W. Graben, Mol. Phys. **40**, 1433 (1980).
[6]M. Born and K. Huang *Dynamical Theory of Crystal Lattices* (Clarendon, Oxford, 1954).
[7]H. C. Andersen, J. Chem. Phys. **72**, 2384 (1980).
[8]L. S. Landau and E. M. Lifshitz, *Theory of Elasticity* (Addison-Wesley, Reading, Mass., 1959).
[9]F. Milstein, J. Appl. Phys. **44**, 3825 (1973).
[10]F. Milstein, J. Appl. Phys. **44**, 3833 (1973).
[11]As is shown by Milstein (Ref. 10), for central pairwise interactions, the conditions $C_{23,23}$ and $C_{22,22} > 0$ are equivalent to $C_{11,22} > 0$. In the same paper Milstein has given a plot of $C_{11,22}$ against stress.
[12]D. R. Squire, A. C. Holt, and W. G. Hoover, Phys. **42**, 388 (1969).

PHYSICAL REVIEW B VOLUME 32, NUMBER 2 15 JULY 1985

Molecular dynamics calculation of elastic constants for a crystalline system in equilibrium

John R. Ray and Michael C. Moody
Kinard Laboratory of Physics, Clemson University, Clemson, South Carolina 29631

Aneesur Rahman
Argonne National Laboratory, Argonne, Illinois 60439
(Received 25 February 1985)

We have performed molecular dynamics calculations for a crystalline system in equilibrium to show that by using a fluctuation formula involving the internal stress tensor it is possible to calculate the elastic constants at the ambient temperature with ease and accuracy. The method also allows one to calculate the elastic constants when the system is subjected to an arbitrary external stress.

The molecular dynamics method developed by Parrinello and Rahman[1] furnishes a means for studying structural phase transformations in solids; such transformations are, of course, nonequilibrium processes. However, this molecular dynamics method can also be used to calculate the equilibrium properties of a system; the trajectories generated by the equations of motion belong to an ensemble with constant enthalpy H, constant thermodynamic tension t, and constant particle number N, or briefly an (H,t,N) ensemble; the equilibrium properties may therefore be calculated by using the fluctuation formulas of the (H,t,N) ensemble. The details are given in Ref. 2 where it has also been shown how the formulation becomes relevant for the theory of finite elasticity.

It was shown by Parrinello and Rahman (see Ref. 2 for details) that the (isentropic or adiabatic) compliance tensor S_{ijkl} is given in terms of the strain fluctuations by the equation

$$\delta(\epsilon_{ij}\epsilon_{kl})=\langle\epsilon_{ij}\epsilon_{kl}\rangle_{\rm av}-\langle\epsilon_{ij}\rangle_{\rm av}\langle\epsilon_{kl}\rangle_{\rm av}=\frac{k_BT}{V}S_{ijkl}\ , \quad (1)$$

where ϵ_{ij} is the strain tensor, and V the equilibrium volume of the N particle system.

However, it has been found by Sprik *et al.*[3] and by others[4] that from the point of view of convergence to statistically significant results Eq. (1) is unsatisfactory.

We present in this paper an alternative approach which shows much greater promise of producing desirable results even with molecular dynamics runs of only moderate length.

In Ref. 2 we discussed not only the (H,t,N) ensemble but also the (E,h,N) ensemble; the latter is a generalization of the familiar (E,V,N), i.e., the microcanonical, ensemble; in the (E,h,N) ensemble the 3×3 matrix h, to be defined below, is kept constant; this keeps not only the volume ($=\det h$) constant but also holds fixed the shape of the periodically repeating molecular dynamics cell containing the N particles.

In the (E,h,N) ensemble, as shown in Ref. 2, the adiabatic elastic constants C_{ijkl} can be expressed in terms of a fluctuation formula involving fluctuations in the internal stress tensor P. Several other average quantities also intervene and one needs to give some notational details.

The constant matrix h has as its columns the elements of the vectors $\mathbf{a}$, $\mathbf{b}$, and $\mathbf{c}$ which span the molecular dynamics cell; $h=(\mathbf{a},\mathbf{b},\mathbf{c})$. For simplicity we assume the potential energy of the system to be pair-wise additive; the pair potential is denoted by $u(r)$. Let χ denote $r^{-1}u'$ and f denote $r^{-2}(u''-\chi)$. The internal stress tensor is then

$$P_{ij}=\frac{1}{V}\left[\sum_a p_{ai}p_{aj}/m_a-\sum_{b>a}\chi(r_{ab})x_{abi}x_{abj}\right]\ , \quad (2)$$

p_{ai} being the momentum components, $\mathbf{x}_{ab}$ the vector joining a and b of length r_{ab}, and V the volume containing the N particles.

Then the adiabatic elastic constant C_{ijkl} is given, under conditions of zero stress, by

$$C_{ijkl}=-\frac{V_0}{k_BT}\delta(P_{ij}P_{kl})+\frac{2Nk_BT}{V_0}(\delta_{il}\delta_{jk}+\delta_{ik}\delta_{jl})+\frac{1}{V_0}\left\langle\sum_{b>a}f(r_{ab})x_{abi}x_{abj}x_{abk}x_{abl}\right\rangle_{\rm av}\ . \quad (3)$$

Equation (3) is obtained from results in Ref. 2 by using appropriate values of the averages involving particle momenta only. The more general relation which gives the elastic constants for a system under stress will be given below [see Eq. (4)]. One further remark before we present the results of our calculations.

Under an arbitrary state of stress the molecular dynamics cell will adopt a certain shape and volume (using the Parrinello-Rahman form of molecular dynamics); in addition the particles in the cell will take on positions which can be quite varied when there is more than one atom per Bravais unit cell. To make use of Eq. (4) for determining elastic constants under an arbitrary state of stress an efficient procedure is to make an (H,t,N) calculation first at the desired stress and temperature; this will furnish an average for h and for all particle positions $\mathbf{x}_a$. These are then the input values in the (E,h,N) calculation of the elastic constants at the same temperature using Eq. (4).

TABLE I. Elastic constants in units of Nk_BT/V_0 for our molecular dynamics run and Cowley's Monte Carlo run. Cowley's results are for a reduced temperature of 0.3 while the temperature of the molecular dynamics run was 0.298. For the C_{12} and C_{44} elastic constants we give only the symmetry-averaged quantity. For argon ($\epsilon_0=120$ K, $\sigma=3.4$ Å), $Nk_BT/V=11.7$ MPa. The value of PV/Nk_BT was -0.03 ± 0.03 for the molecular dynamics run while Cowley gives 0.02 ± 0.02 for his Monte Carlo run. The time step Δt was equal to 0.005 which for argon corresponds to about 10^{-14} sec. The reduced density of the system was 0.934.

Time	C_{11}	C_{22}	C_{33}	$\frac{(C_{11}+C_{22}+C_{33})}{3}$	$\frac{(C_{12}+C_{13}+C_{23})}{3}$	$\frac{(C_{44}+C_{55}+C_{66})}{3}$
				Molecular dynamics data		
5 000Δt	187.6	179.8	185.9	184.4±4.1	93.7±2.8	84.4±2.3
10 000Δt	185.3	180.9	184.1	183.4±2.3	93.7±1.6	82.4±1.1
15 000Δt	185.9	181.5	182.6	183.3±2.3	94.2±1.7	82.4±1.9
20 000Δt	185.4	181.5	182.6	183.2±2.0	94.5±1.1	82.8±2.2
25 000Δt	185.8	181.6	181.9	183.1±2.3	94.2±0.9	83.0±1.6
30 000Δt	185.4	181.8	182.2	183.1±2.0	94.7±0.9	82.7±1.4
35 000Δt	185.0	181.6	182.1	182.9±1.8	94.9±0.9	83.0±1.5
40 000Δt	185.1	182.3	182.6	183.3±1.5	94.8±1.0	82.9±1.4
				Monte Carlo data[a]		
≃25 000Δt				182.0±0.5	94.1±0.5	82.2±0.2

[a]Reference 5.

We have employed Eq. (3) in various (E,h,N) molecular dynamics runs and find that the elastic constants are determined accurately and efficiently. In Table I we show, along with other quantities, the three elastic constants C_{11}, C_{22}, C_{33} calculated using Eq. (3) for a molecular dynamics system consisting of 500 particles forming a fcc lattice and interacting via the nearest-neighbor Lennard-Jones (12,6) potential. The calculation is for zero pressure and a reduced temperature of about 0.3. Various other thermodynamic quantities were also calculated during this run and are displayed in Table II.

Cowley[5] has calculated various quantities in a Monte Carlo calculation using 108 particles interacting with the nearest-neighbor Lennard-Jones potential at a temperature of 0.3. Cowley does not give C_{11}, C_{22}, and C_{33} separately but only their average which he identifies as $C_{11}=(C_{11}+C_{22}+C_{33})/3$. Cowley's calculations correspond to roughly 25 000 molecular dynamics time steps according to the estimates made in Ref. 3.

TABLE II. Specific heats at constant volume and pressure in units of Nk_B and adiabatic and isothermal bulk moduli in units of Nk_BT/V_0 for our molecular dynamics run and Cowley's Monte Carlo run. These data correspond to the same conditions as the data in Table I.

Time	C_v	C_p	B_s	B_T
		Molecular dynamics data		
5 000Δt	2.82	3.53	123.9	99.0
10 000Δt	2.86	3.63	123.6	97.5
15 000Δt	2.83	3.55	123.9	98.7
20 000Δt	2.81	3.51	124.0	99.3
25 000Δt	2.85	3.59	123.8	98.2
30 000Δt	2.78	3.45	124.3	100.2
35 000Δt	2.80	3.49	124.1	99.6
40 000Δt	2.77	3.43	124.3	100.4
		Monte Carlo data[a]		
≃25 000Δt	2.82	3.53	123.4	98.6

[a]Reference 5.

The formula used by Cowley to calculate elastic constants may be obtained from Eq. (3) by identifying C_{ijkl} as the isothermal elastic constants and interpreting averages as canonical ensemble averages. In this way all momenta can be integrated out of the resulting equation, which is then in a form suitable for a Monte Carlo calculation. The resulting formula for the elastic constants in the canonical, or (T,h,N), ensemble was first given by Squire *et al.*[6] (see also Wallace *et al.*[7]). A direct transformation between the (E,h,N) fluctuation formula (3) and its (T,h,N) counterpart can be obtained by using generalized ensemble theory. We shall discuss this latter point in detail in a later paper.

It is clear from the results of Table I that Eq. (3) furnishes a very efficient method for calculating elastic constants in molecular dynamics. The error quoted, for the symmetry averaged C_{11} in our calculation, is found from the mean-square deviation of the three numbers which make up the symmetry averaged C_{11}. Cowley's error was determined by breaking the Monte Carlo chain into segments and calculating the quantity $(C_{11}+C_{22}+C_{33})/3$ for each segment. The relative efficiency of the molecular dynamics and Monte Carlo calculation of elastic constants needs to be investigated.

Looking in detail at our calculation shows that if one is only interested in 5% accuracy for the elastic constants then the molecular dynamics calculation need only be run for 2000 time steps. The other two elastic constants, C_{12} and C_{14}, as well as other thermodynamic variables like specific heats, showed this same rapid convergence and

TABLE III. Elastic constants for our system under tensile loading in units of Nk_BT/V_0. The stretch of the system is 5% along the [001] direction. The temperature of the system is 0.299. The only component of the stress that is nonzero is the 33 component where the 3 axis is parallel to the [001] direction. As the system is stretched its area perpendicular to the [001] direction decreases. The reduced density of the system is 0.909.

Time	C_{11}	C_{22}	C_{33}	C_{12}	C_{13}	C_{23}	C_{44}	C_{55}	C_{66}
				Molecular dynamics data					
$5000\Delta t$	180.6	181.9	107.2	123.6	59.2	60.7	49.0	48.7	101.7
$10000\Delta t$	180.6	179.4	106.2	122.9	62.5	61.9	50.7	49.5	102.1
$15000\Delta t$	179.5	178.7	104.7	121.2	61.6	60.9	51.1	50.0	104.1

precision.

The generalization of Eq. (3) to the case of a system with nonzero stress applied to it has the form[2]

$$V_0 h_{0ip}^{-1} h_{0jq}^{-1} h_{0lr}^{-1} h_{0ms}^{-1} C_{pqrs} = \frac{-4}{k_B T}\delta(M_{ij}M_{lm}) + 2Nk_B T(G_{mi}^{-1}G_{jl}^{-1} + G_{li}^{-1}G_{jm}^{-1}) + \left\langle \sum \sum_{b>a} f(r_{ab}) s_{abi} s_{abj} s_{abm} s_{abl} \right\rangle_{\text{av}}, \quad (4)$$

where $M = -(V/2)h^{-1}Ph'^{-1}$, s_{abi} is a scaled coordinate related to the real coordinate by $x_{abi} = h_{ij}s_{abj}$, G is the metric tensor $G = h'h$, h_0 is value of the matrix for the case of zero stress, and h is its value when the system is under the prescribed stress. The values of h_0 and h to be used in Eq. (4) are obtained by carrying out appropriate (H,t,N) runs as already discussed above.

In Table III we show the calculation of the elastic constants using Eq. (4); the system was stretched by about 5% along the [001] direction by the application of a suitable tension. For argon this load would be 62 MPa for nearest-neighbor interactions. The rapid convergence and accuracy of this calculation was the same as the previous case of zero stress, Table I. Notice that the elastic constants for this calculation show the necessary tetragonal symmetry. The differences between the elastic constants in Table I (zero stress) and Table III (finite stress) are related to the higher-order elastic constants of the material.[8]

The (T,h,N) ensemble fluctuation formula for elastic constants has the same form as Eq. (4) except that we must identify the elastic constants as isothermal and the averages as canonical ensemble (T,h,N) averages. On performing the momentum averages in the (T,h,N) formula corresponding to Eq. (4) one obtains a result suitable for calculating elastic constants at finite stress in a Monte Carlo calculation. The results in this paper are, apparently, the first accurate calculation of elasctic constants using molecular dynamics.

We shall publish a more detailed discussion of our calculations in a later paper.

One of us (J.R.R.) thanks H. W. Graben and M. J. Skove for helpful discussions and the Research Corporation for support of this project.

[1]M. Parrinello and A. Rahman, Phys. Rev. Lett. **45**, 1196 (1980).

[2]J. R. Ray and A. Rahman, J. Chem. Phys. **80**, 4423 (1984).

[3]M. Sprik, R. W. Impey, and M. L. Klein, Phys. Rev. B **29**, 4368 (1984).

[4]M. Parrinello and A. Rahman and independently J. R. Ray have attempted to use Eq. (1) to obtain elastic constants with unsatisfactory results.

[5]E. R. Cowley, Phys. Rev. B **28**, 3160 (1983).

[6]D. R. Squire, A. C. Holt, and W. G. Hoover, Physica **42**, 388 (1969).

[7]D. C. Wallace, S. K. Schiferl, and G. K. Straub, Phys. Rev. A **30**, 616 (1984).

[8]R. F. S. Hearmon, in *Landolt-Bornstein: Crystal and Solid State Physics*, edited by K.-H. Hellweg (Springer-Verlag, Berlin, 1979), Vol. II.

PHYSICAL REVIEW B VOLUME 13, NUMBER 2 15 JANUARY 1976

Dynamical structure factor $S(\vec{Q}, \omega)$ of rare-gas solids

Jean Pierre Hansen
Laboratoire de Physique Théorique des Liquides, Université Paris VI, 75230 Paris, France*

Michael L. Klein
Chemistry Division, National Research Council of Canada, Ottawa, Canada K1A 0R6
(Received 26 September 1973; revised manuscript received 5 August 1975)

The classical molecular-dynamics (MD) method was used to make a systematic study of the dynamical structure factor $S(\vec{Q},\omega)$ for systems of N particles disposed on an fcc lattice with periodic boundary conditions and interacting with n shells of neighbors via a Mie–Lennard-Jones (12-6) potential. Calculations were carried out for $N(n)$ = 108(3), 256(3 or 5), 864(3), and 2048(1) at reduced densities and temperatures corresponding to roughly the melting point of the model and one-half this temperature. Phonons of a particular wave vector $\vec{Q}$ were identified by the peaks in $S(\vec{Q},\omega)$. The one-phonon approximation to $S(\vec{Q},\omega)$ was investigated, as was the $\vec{Q}$ dependence of certain phonons. The temperature shifts of phonons for both constant density and pressure were studied along with their shifts with density at constant temperature. The one-phonon $S(\vec{Q},\omega)$ for certain zone-boundary phonons is compared with the predictions of a self-consistent phonon approximation that includes phonon damping. Even at half the melting temperature the latter theory is not totally adequate and possible reasons for this are discussed. Finally, as a by-product of the MD calculations, certain equilibrium properties were also obtained and compared with perturbation-theory results that use hard spheres as a reference system. Results of the latter procedure were very poor.

I. INTRODUCTION

The molecular-dynamics (MD) method, as developed for hard-sphere systems by Alder and Wainwright[1] and for continuous potentials by Rahman[2] and Verlet[3] has been extensively used to simulate the equilibrium and time-dependent properties of a wide variety of liquids and dense fluids. The related Monte Carlo method developed by Metropolis *et al.*[4] and by Wood[5] has also been widely applied to the study of equilibrium (or static) properties of liquids and of solids. In particular the elastic constants of rare-gas solids have been computed in this way[6] and compared to the predictions of self-consistent-phonon theories including anharmonic corrections.[7] However, the dynamical properties of solids can only be simulated by the molecular-dynamics method and one aim of the present work was to provide a test of the self-consistent-phonon theory dynamical properties in rare-gas solids over a wide range of temperatures. The only previous computer simulation of the lattice dynamics of rare-gas solids[8] was restricted to the computation of self-correlation functions and to an indirect determination of phonon dispersion curves from the response of a crystal to external disturbances. However, lattice vibrations are highly collective modes directly related to the spectrum of the density fluctuations which in turn are related to the differential cross section for coherent inelastically scattered thermal neutrons. This spectrum is the dynamical structure factor $S(\vec{Q}, \omega)$, the Fourier transform with respect to time of the correlation function of the density operator

$$\rho_{\vec{Q}}(t) = \sum_{i=1}^{N} e^{i\vec{Q}\cdot\vec{r}_i(t)} , \qquad (1)$$

where the sum is over the N atoms of the system, and

$$S(\vec{Q}, \omega) = \int_{-\infty}^{+\infty} e^{i\omega t} F(\vec{Q}, t)\, dt , \qquad (2)$$

$$F(\vec{Q}, t) = (1/N) \langle \rho_{\vec{Q}}(t) \rho_{-\vec{Q}}(0) \rangle , \qquad (3)$$

where the angular brackets denote statistical average; $F(\vec{Q}, t)$ is frequently referred to as the "intermediate scattering function" and can be calculated "exactly" by the MD method for systems of several hundred atoms. The first calculation of this type was made by Levesque *et al.*[9] for rare-gas liquids near the triple point. Since then the dynamical structure factor has been computed for a variety of fluid or liquid systems, including liquid metals[10] and the classical one-component plasma.[11] In the present work we apply essentially the same technique to the case of crystalline rare gases. A brief preliminary report of parts of this work was presented elsewhere.[12] Extensions to the case of solid alkali halides, solid nitrogen, and solid alkalis will be published elsewhere.[13]

The recent availability of experimental data on inelastic neutron scattering from rare-gas single crystals at high temperatures[14] renders the present work particularly timely. In fact, to allow a direct comparison between experiment and computer simulation, several calculations were made for wave vectors and under density-temperature conditions very close to some of the experimental conditions. The simulation of dynamical properties of solids at high temperatures is also relevant for quantitative comparison with the recent experimental spectra of depolarized light scattering from rare gas crystals,[15] as has been done recently for simple liquids.[16] That part of the work will be presented separately.[17] Here we restrict ourselves to a presentation and discussion of the $S(\vec{Q}, \omega)$ data.

The paper is organized as follows: Technical details of our calculations are presented in Sec. II. Section III deals with the equilibrium properties which have been obtained as a byproduct of the MD computations. The results are compared to the predictions of the thermodynamic perturbation theory.[18] The $S(\vec{Q}, \omega)$ data are presented in Sec. IV and compared with available experimental results. Section V is devoted to a confrontation of our results with the prediction of self-consistent-phonon theory. Some concluding remarks are contained in Sec. VI.

II. COMPUTATION

In the usual manner, the time evolution of a system of N atoms was determined over a certain time interval τ, by solving numerically Newton's equations of motions for the N atoms using Verlet's finite difference algorithm.[3] The usual periodic boundary conditions were assumed and the N atoms interacted through the standard Mie–Lennard-Jones two-body potential

$$v(r) = 4\epsilon[(\sigma/r)^{12} - (\sigma/r)^6] . \quad (4)$$

This potential was used, rather than more "realistic" potentials, because it yields satisfactory results for the dense rare-gas liquids and because it has been widely used in the self-consistent-phonon calculations. Moreover, the results expressed in "reduced units," i.e., with σ, ϵ, and $\tau_0 = (m\sigma^2/48\epsilon)^{1/2}$ chosen as length, energy, and time units, can be easily converted to absolute units for any one of the rare gases, by an appropriate choice of the potential parameters ϵ and σ. Our energy unit τ_0^{-1} corresponds roughly to about 18 cm^{-1}, 0.54 THz, or 2.23 meV for ^{36}Ar, if one uses $\sigma = 3.405$ Å and $\epsilon/k = 119.8$ K.

Runs were made in the reduced temperature range $0.28 \le T^* = kT/\epsilon \le 1.2$ and in the reduced density range $0.97 \le \rho^* = N\sigma^3/V \le 1.031$, for systems of 108, 256, 864, and 2048 atoms. We recall that the triple point of a Lennard-Jones system corresponds to $T^* \simeq 0.68$ and $\rho^*_{\text{solid}} \simeq 0.96$.[19] In each run the initial configuration was chosen to be a perfect fcc lattice and we used a time increment of $0.032\tau_0$ in the finite difference algorithm; this corresponds to $\sim 10^{-14}$ sec for argon. The integration of the equations of motion was carried out for up to 40 000 time steps in each run; consequently τ, the total time interval, was of the order of 4.10^{-10} sec for argon. Fourteen independent runs were made, some of them under very similar $\rho^* - T^*$ conditions, but for different N, in order to study the N dependence of various quantities.

In the various runs, each of the N atoms was made to interact with a fixed number of shells of neighbors n. The contribution of the remaining shells to the thermodynamic properties (e.g., the equation of state) was accounted for by adding static lattice sum corrections. Because the rms displacement of the atoms from their lattice sites is only a small fraction of the lattice parameter (of the order of 14% at melting[20]) this procedure introduces negligible errors for $n \ge 3$. For $N = 108$, 256, and 864, n was chosen equal to 3 or 5 (42 or 78 neighbors); however, for the largest system ($N = 2048$), n was taken equal to 1 (12 neighbors) in order to stay within reasonable limits for the computer time (which is proportional to $N \times n$).

It should be stressed that the MD calculations are purely classical and ignore all quantum effects. In the temperature range considered here, this is justified for the heavier rare gases. Their contribution to the equilibrium properties (in particular to the Helmholtz free energy) can be estimated by calculating the $\hbar^2$ term of the Wigner expansion[21]; this expansion has been shown to be strongly convergent even in the vicinity of the triple point of neon[22] and will be used in Sec. III. The most important result of the present work is the computation of the dynamical structure factor (2), as a function of ω, for several $\vec{Q}$ vectors.

The dynamical structure factor $S(\vec{Q}, \omega)$ can be computed in either of two ways. The first method consists in computing the intermediate scattering function (3) by correlating the real and imaginary parts of the density operator (1) at different times and finally taking the Fourier transform (2). Typical $F(\vec{Q}, t)$ are reproduced in Fig. 1 for two $|\vec{Q}|$ values (two different phonons) along the [100] direction. Figure 2 shows the corresponding $S(\vec{Q}, \omega)$. The difficulties of this method are at once apparent from Fig. 1: for low-$|\vec{Q}|$ phonons, the correlation function decays very slowly (long lifetime) and one has to numerically autocorrelate the density operator over a long time interval in order not to introduce significant truncation errors.

The second method has been described in Ref. 12 and consists in computing directly (in the course of the MD run) the Fourier-Laplace transform $\rho_{\vec{Q}}(\omega)$ of the density operator, using the formula

$$S(\vec{Q}, \omega) = \frac{1}{N} \int_{-\infty}^{+\infty} e^{i\omega t} \langle \rho_{\vec{Q}}(t) \rho_{-\vec{Q}}(0) \rangle \, dt$$

$$= \lim_{T \to \infty} \int_0^T e^{i\omega t} \rho_{\vec{Q}}(t) \, dt \int_0^T e^{-i\omega t'} \rho_{-\vec{Q}}(t') \, dt'$$

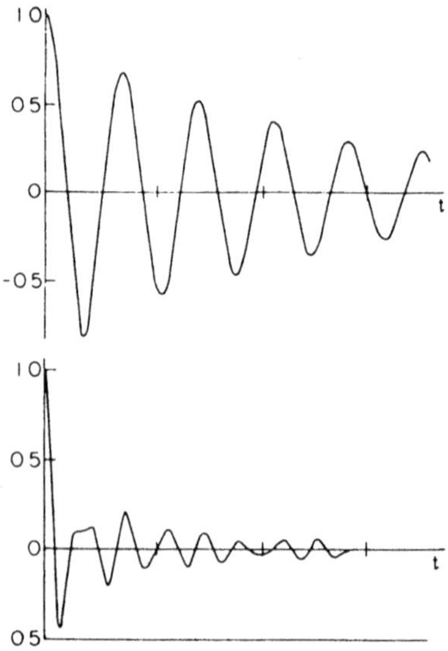

FIG. 1. "Intermediate scattering function" $F(\vec{Q}, t)$, for two $\vec{Q}$ values whose Fourier transforms are shown in Fig. 2. The increment on the time scale corresponds to $6.4\tau_0$ units.

$$S(\vec{Q}, \omega) = \frac{1}{N} \lim_{\tau\to\infty} |\rho_{\vec{Q}}(\omega)|^2 .$$

For MD calculations running over 20 000–40 000 time steps, the truncation error is negligible, but the disadvantage of this method is that the statistical "noise" is more difficult to eliminate, because it is automatically included in the Fourier-Laplace transform. In view of this statistical noise, we convoluted the raw data with a Gaussian width function which usually had a full width at half-maximum (FWHM) $= 0.1\tau_0^{-1}$. This is essentially equivalent to truncating $F(\vec{Q}, t)$ at $t \simeq 40\pi\tau_0$ (approximately 4000 time steps). Comparing the results obtained by the two methods, they appear to yield very similar $S(\vec{Q}, \omega)$. The second method is slightly more convenient and was used in most of the work.

The appearance of the scalar product $\vec{Q}\cdot\vec{r}$ in the density operator (1) allows only longitudinal response to be observable in symmetry directions, such as $[00\zeta]$; this situation is the same as in isotropic fluids. However, the transverse response can be observed for $\vec{Q}$ vectors that have nonzero reciprocal-lattice vectors $\vec{K}$, i.e., outside the first Brillouin zone.

In the classical limit, $S(\vec{Q},\omega)$ is even in ω. However, if quantum effects are duly taken into account, the well-known detailed balance condition

$$S(\vec{Q}, -\omega) = e^{-\beta\hbar\omega} S(\vec{Q}, \omega) \tag{5}$$

implies the presence of odd terms in $S(\vec{Q}, \omega)$; consequently, the intermediate scattering function (3) has an imaginary part in the quantum case (which is a direct consequence of the fact that the atom positions at different times do not commute). This imaginary part can be calculated to first order in $\hbar$ from the classical (real) intermediate scattering function.[23] However, this procedure has not been applied. This point is discussed further in Sec. V when we compare our MD calculations with self-consistent-phonon calculations.

III. EQUILIBRIUM PROPERTIES

The MD method yields equilibrium as well as dynamical properties; we first discuss our results for the equilibrium properties. We have computed the following thermodynamic quantities (a) the excess internal energy per particle

$$\frac{U}{N} = \frac{1}{N}\left\langle \sum_{i<j} v(r_{ij}) \right\rangle , \tag{6}$$

(b) the equation of state, calculated from the virial theorem

$$Z = \frac{P}{\rho kT} = 1 - \frac{1}{3NkT}\left\langle \sum_{i<j} r_{ij} \frac{dv(r_{ij})}{dr_{ij}} \right\rangle , \tag{7}$$

and (c) the specific heat at constant volume (per particle) which, in the microcanonical ensemble, can be calculated from the fluctuation of the temperature[24]

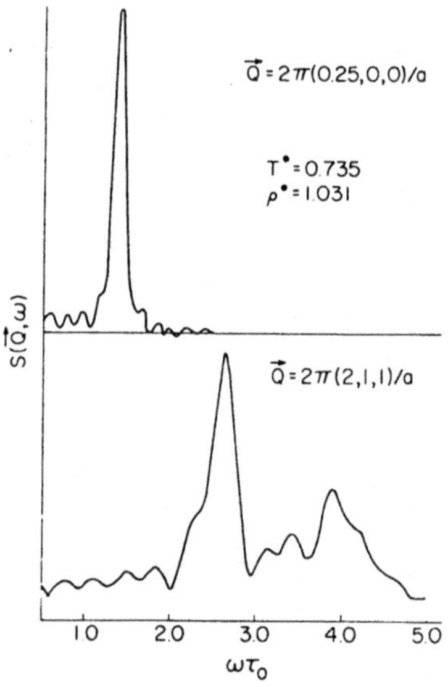

FIG. 2. Two typical $S(\vec{Q}, \omega)$ curves taken from run No. 13 of Table I. The upper curve shows the longitudinal phonon while the lower one shows both the transverse and longitudinal zone-boundary phonons in the $[\zeta 00]$ direction.

TABLE I. Summary of equilibrium properties determined in the MD runs and comparison of selected quantities with a perturbation theory that uses a hard-sphere reference system.

Run	N	n	T^*	ρ^*	$10^2 \langle u^{*2} \rangle$	Z	$Z^{\text{pert.}}$	$\frac{U}{N}$	$\frac{C_v}{Nk}$	$\frac{\langle \Delta V^* \rangle}{24}$	$\frac{F^{\text{pert.}}}{NkT}$	$\frac{F^{(1)}}{NkT}$ (Ne)
1	256	5	0.2814	1.031	0.477	−2.933	−1.386	−8.106	1.61	40.75	−22.36	4.66
2	108	3	0.3185	1.031	0.494	−1.604	8.16	−8.058	1.74	41.82	−19.11	3.67
3	256	5	0.319	1.031	0.535	−1.600	8.21	−8.055	1.78	41.73	−19.07	3.67
4	864	3	0.321	1.031	0.556	−1.595	8.47	−8.054	1.74	41.90	−18.92	3.64
5	2048	1	0.3375	1.030	0.564							
6	864	3	0.501	1.003	1.013	−0.407	−0.762	−7.710	1.95	41.12	−10.42	1.468
7	864	3	0.504	1.003	1.056	−0.354	−0.732	−7.706	1.95	41.20	−10.33	1.453
8	864	3	0.6364	0.98	1.496	0.100	0.460	−7.429	2.11	40.55	−7.20	0.897
9	864	3	0.647	0.98	1.522	0.227	0.656	−7.416	2.10	40.81	−7.01	0.872
10	2048	1	0.726	0.97	1.662							
11	864	3	0.7326	0.97	1.864	0.693	1.406	−7.255	2.34	41.32	−5.72	0.690
12	864	3	0.735	0.97		0.714	1.443	−7.252	2.28	41.37	−5.68	0.686
13	256	3	0.735	1.031	1.200	3.816	4.141	−7.505	2.33	42.21	−5.52	0.866
14	256	3	1.197	1.031	1.983	5.274	6.294	−6.916	2.87	63.26	−1.76	0.396

$$\frac{C_v}{Nk} = \frac{3}{2[1-(3/2\rho^* T^*)][\langle T^{*2}\rangle - \langle T^* \rangle^2]} \,. \qquad (8)$$

As is well known, the MD and Monte Carlo methods do not readily yield the Helmholtz free energy (or alternatively the entropy) directly, but this can be calculated by integrating numerically the equation of state[25]; in this way the Helmholtz free energy for rare-gas solids was computed for several isotherms above the triple point,[19] but because the primary aim of the present work is the study of the dynamical properties of rare gas solids we have not attempted to perform the similar computations in the ρ^* - T^* range considered here. We have however calculated the Helmholtz free energy per particle F/NkT, using thermodynamic perturbation theory based on a hard-sphere reference system.[18] This theory yields reasonable results in the triple-point region.[18] We have also calculated the equation of state using perturbation theory and compared it to our "exact" results in Table I listing the various equilibrium properties which we have obtained.

The first quantum correction to the Helmholtz free energy is given by[21,22]

$$F^{(1)}/NkT = \tfrac{1}{24}\langle \Delta V^* \rangle (\Lambda/T^*)^2 \,, \qquad (9)$$

where

$$\langle \Delta V^* \rangle = \frac{1}{N\epsilon}\left\langle \sum_{l=1}^{N} \nabla_l^2 \left(\sum_{i<j} v(r_{ij}) \right) \right\rangle$$

is the average of the Laplacian of the total potential energy expressed in reduced units, and $\Lambda^2 = \hbar^2/\sigma^2 m\epsilon$ is the dimensionless de Boer quantum parameter. For neon $\Lambda^2 \sim 0.896\times10^{-2}$ if one uses $\sigma = 2.74$ Å and $\epsilon/k = 35.6$ K and for argon, $\Lambda^2 \sim 0.874\times 10^{-3}$ if one uses $\sigma = 3.405$ Å and $\epsilon/k = 119.8$ K. Thus the first-order quantum correction is seen to be ten times smaller for argon than for neon at the same $\rho^* - T^*$ and it is even smaller for the heavier rare gases Kr and Xe. In Table I we list the values obtained for $\frac{1}{24}\langle \Delta V^* \rangle$ as well as the corresponding values of $F^{(1)}/NkT$ in the case of Ne. Comparing this to the classical free energy obtained from perturbation theory, it is seen that in the $\rho^* - T^*$ range considered in this work the quantum corrections to the free energy are rather large for neon but they are only about 2% for argon.

Finally, we have computed the mean-square displacement of an atom from its lattice site

$$\langle u^2 \rangle = \frac{1}{N}\left\langle \sum_i u_i^2 \right\rangle \,, \qquad (10)$$

where $\vec{u}_i = \vec{r}_i - \vec{R}_i$ is the displacement and $\vec{R}_i$ is the lattice site of the ith atom. The results for $\langle u^2 \rangle$ are listed in Table I along with the thermodynamic properties. The strong N dependence of $\langle u^2 \rangle$ is quite apparent by comparing the results in run Nos. 2, 3, and 4; $\langle u^2 \rangle$ increases with N, in agreement with the predictions of harmonic theory. On the other hand the N dependence of the equation of state and internal energy appears to be negligible. Note that the relative statistical error on Z, $\langle u^2 \rangle$ and U/NkT is less than 1%, but of the order of 5% for C_v/Nk. In the case of the 2048-particle system, the thermodynamic properties are not listed because, as we have explained in Sec. II, the corresponding runs were made for interactions with one shell of neighbors only.

Comparison of the "exact" equation-of-state results and the predictions of the perturbation theory clearly shows the inadequacy of a perturbation theory of the solid based on a hard-sphere reference system, especially at low temperatures. This was already emphasized in Ref. 18; the predicted free energies agree within a few percent

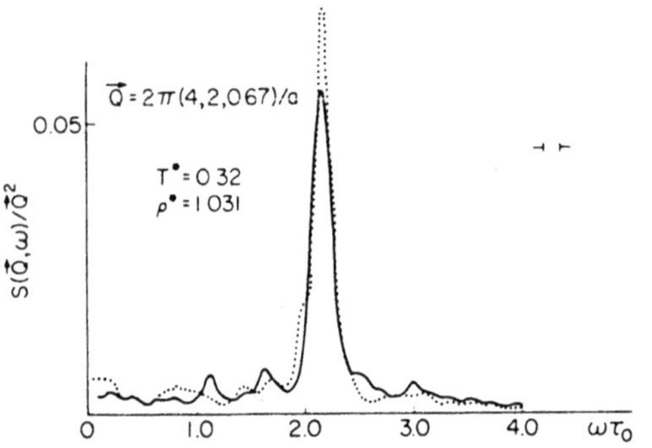

FIG. 3. Dependence of $S(\vec{Q},\omega)$ on the system size. The solid curve is taken from run No. 2 (108 particles) while the dotted curve is from run No. 4 (864 particles).

with the computer results in the triple point region, but at lower and higher temperatures the agreement becomes rapidly worse. The equation-of-state results, obtained by differentiating the free energy, are not surprisingly even poorer and the data listed in Table I clearly illustrate the complete failure of the theory at low temperatures. This is in marked contrast with the situation in the fluid, where the perturbation theory works remarkably well.[26] The fundamental reason seems to lie in the inadequacy of the completely anharmonic hard-sphere reference system. In Ref. 17 it is also shown that the hard-sphere solid is completely inadequate to describe the dynamics of rare-gas crystals, as probed by depolarized light scattering.

IV. DYNAMICAL STRUCTURE FACTOR

For each of the 14 runs listed in Table I, $S(\vec{Q},\omega)$ has been computed for selected wave vectors. Moreover, we calculated the corresponding one-phonon approximation

$$S_1(\vec{Q},\omega)=\frac{e^{-2W}}{N}\lim_{T\to\infty}\frac{1}{T}|\tilde{\rho}_{\vec{Q}}(\omega)|^2\,, \tag{11}$$

where $\tilde{\rho}_{\vec{Q}}(\omega)$ is the Fourier-Laplace transform of

$$\tilde{\rho}_{\vec{Q}}(t)=\sum_{i=1}^{N}e^{i\vec{Q}\cdot\vec{R}_i}\vec{Q}\cdot\vec{u}_i(t) \tag{12}$$

and $2W=\frac{1}{3}|\vec{Q}|^2\langle u^2\rangle$.

Except for large $|\vec{Q}|$, these results are essentially identical with the full $S(\vec{Q},\omega)$. This feature is not unexpected in view of the sum rules which must be satisfied by S and S_1,

$$\frac{1}{2\pi}\int_{-\infty}^{+\infty}\omega^2 S(\vec{Q},\omega)\,d\omega=\frac{kTQ^2}{m}\,, \tag{13}$$

$$\frac{1}{2\pi}\int_{-\infty}^{+\infty}\omega^2 S(\vec{Q},\omega)\,d\omega=e^{-2W}\frac{kTQ^2}{m}\,. \tag{14}$$

Except at the lowest $\vec{Q}$ vectors, where the phonon peaks are very sharp, our numerical data satisfy these sum rules to within 10%.

If $\vec{Q}$, ω, and $S(\vec{Q},\omega)$ are expressed in reduced units, (13) can be recast in the form

$$\frac{1}{2\pi}\int_{-\infty}^{+\infty}\omega^{*2}S^*(\vec{Q}^*,\omega^*)\,d\omega^*=\frac{Q^{*2}T^*}{48}\,. \tag{15}$$

In this form the sum rule will serve as a basis of our qualitative understanding of our results.

A. N dependence

Figure 3 shows $S(\vec{Q},\omega)$ for $\vec{Q}=(2\pi/a)(4,2,\frac{2}{3})$, $T^*=0.32$, $\rho^*=1.031$, computed with systems of $N=108$ and $N=864$ atoms. Within the resolution and statistical uncertainties of our data, the two results are identical. For all other cases where comparisons were possible, peak positions showed no significant N dependence.

B. Dispersion curves

Figures 4(a) and 4(b) show the dispersion curves for the $[00\zeta]$ longitudinal phonons for $T^*=0.34$, $\rho^*=1.030$ and $T^*=0.73$, $\rho^*=0.97$. For $\vec{Q}$ vectors close to the zone boundary, the peaks are seen to broaden substantially, even at $T^*=0.32$, in qualitative agreement with quasiharmonic perturbation theory.[27] As expected the higher T^* phonons are considerably less well defined. This explains why recent neutron scattering experiments were unable to give clear evidence of their existence in solid Kr and Ar near melting.[28,29]

C. Q dependence

In Fig. 5 we show the longitudinal phonon with $\vec{q}=\vec{Q}-\vec{K}=(2\pi/a)(\frac{1}{2},0,0)$, for two reciprocal-lattice vectors, $\vec{K}=(2\pi/a)(0,0,0)$ and $\vec{K}=(2\pi/a)(4,0,0)$ at $T^*=0.337$, $\rho^*=1.030$, and $T^*=0.73$, $\rho^*=0.97$. For easier comparison $S(\vec{Q},\omega)$ for larger $\vec{Q}$ has been divided by the ratio of the squares of the wave vectors. The ratio of the peak height, is close to the ratio of the respective Debye-Waller factors e^{-2W}, as one expects for one-phonon scattering. However, the larger $\vec{Q}$ phonons show enhanced background (multiphonon) scattering. This background contributes a fraction $(e^{2W}-1)$ of the total scattering[30] and is necessarily large when $\langle u^2\rangle$ is large and hence e^{-2W} is small. This phonon appears to have a shoulder on the left of the main peak which grows with T^*. The same phonon but with $\vec{K}=(2,0,0)\,2\pi/a$, has been studied experimentally in ^{36}Ar and exhibits a similar feature.[29] The large frequency shift of the main peak should be noted. Temperature and density shifts of the phonon peaks will now be discussed in more detail.

D. Temperature shifts at constant density

Figure 6 shows the transverse phonon at $\vec{Q}=(2\pi/a)(2,\frac{1}{2},0)$ for a fixed density $\rho^*=1.031$ and two different temperatures: $T^* = 0.735$ and $T^* = 1.20$; this phonon exhibits a sizable negative shift as T^* increases. Figure 7 shows the longitudinal phonon at $\vec{Q} = (2\pi/a)(\frac{1}{2}, 0, 0)$ under identical temperature and density conditions. This time the shift is small and positive. The apparent increase in intensity as T^* is increased is governed entirely by the sum rule equation (15).

E. Temperature shifts at constant pressure

Figure 8 shows the temperature shifts for transverse phonons propagating in $[00\zeta]$ direction for roughly the same $\rho^* - T^*$ states studied experimentally by Batchelder *et al.*[14] As expected the peaks shift to lower frequencies and broaden as T^* increases, and in addition the multiphonon background grows. This behavior is again compatible with the sum rule equation (15).

Figure 9 compares the MD percentage frequency shifts with the experimental results[14,29] and the results of a self-consistent-phonon calculation.[31] The MD results are in better agreement with the experimental data than the phonon calculations.

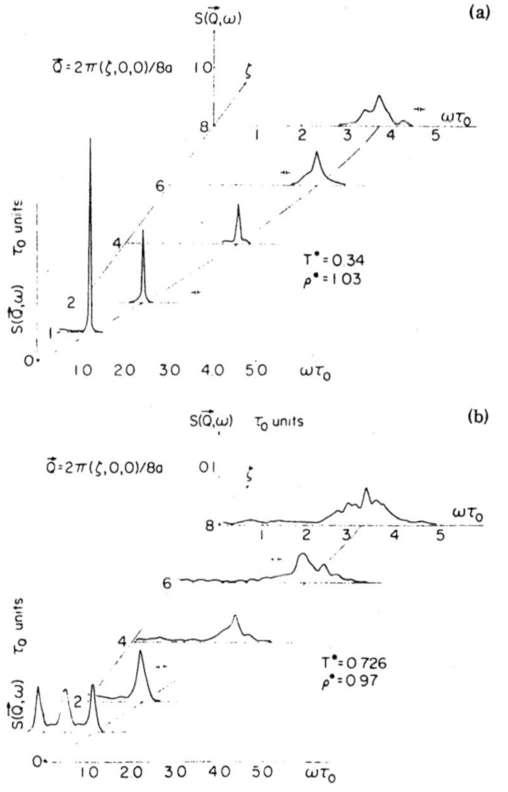

FIG. 4. (a) Dispersion curve for longitudinal phonons propagating in the $[00\zeta]$ direction taken from run No. 5. (b) Dispersion curve for longitudinal phonons propagation in the $[00\zeta]$ direction taken from run No. 10. Note the Rayleigh-Brillouin triplet for the smallest wave vector.

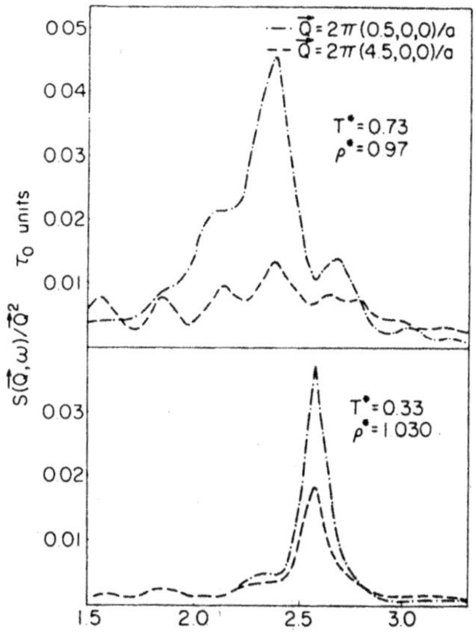

FIG. 5. $\vec{Q}$ dependence of $S(\vec{Q}, \omega)$ for the longitudinal phonon halfway to the zone boundary in the $[00\zeta]$ direction. The upper curve corresponds to roughly the triple point of the (12-6) potential while the lower curve corresponds to about one-half the melting point at zero pressure. Data taken from run Nos. 10 and 5, respectively.

F. Density shifts at constant temperature

Returning to Fig. 7 we see the strong negative shift in the longitudinal $(\frac{1}{2}, 0, 0)$ phonon peak position on changing the density from 1.031 to 0.97 at $T^* = 0.735$. The relative shift amounts to about 15% which compares well with predictions using

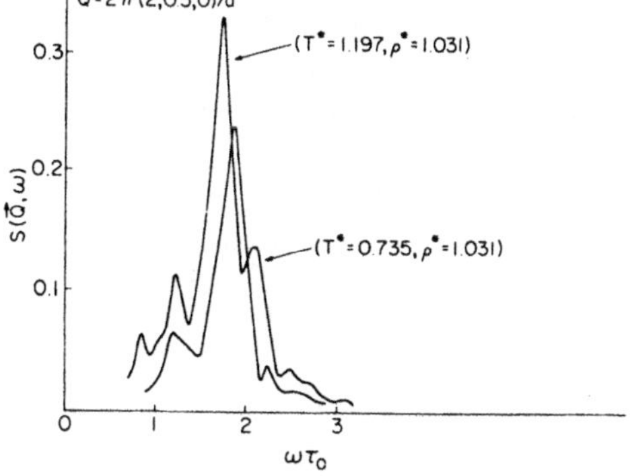

FIG. 6. Temperature dependence of $S(\vec{Q}, \omega)$ at constant density for the transverse phonon halfway to the zone boundary in the $[00\zeta]$ direction. Taken from run Nos. 13 and 14.

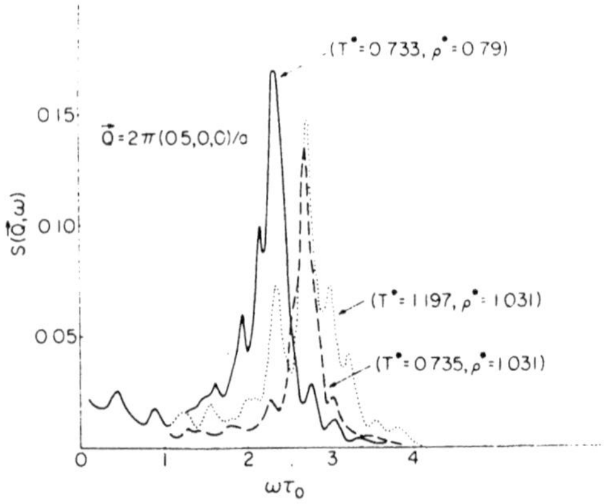

FIG. 7. Temperature dependence at constant density and density dependence at constant temperature of $S(\vec{Q}, \omega)$ for the longitudinal phonon halfway to the zone boundary in the [00ζ] direction. Data taken from run Nos. 13 (dashed) 14 (dots) and 11 (full curve).

the Grüneisen approximation. The increase in intensity accompanying the negative frequency shift is a consequence of the sum rule equation (15).

G. Rayleigh-Brillouin triplet

The lowest $\vec{Q}$ phonon studied was obtained with 2048-particle system and corresponds to $\vec{Q} = (2\pi/a)(\frac{1}{8}, 0, 0)$. This phonon is shown in Fig. 4(b) and clearly exhibits a Rayleigh-Brillouin triplet. From the width of the Rayleigh peak $\Gamma_R = DQ^2$, one can extract an estimate of the thermal diffusivity, which is related to the thermal conductivity by $D = \kappa/\rho C_p$. We find $D \cong 0.1\,\tau_0^{-1}$ which compares favorably with the value derivable from the experimental value of κ for Ar.[32] Less-well-resolved results were obtained with the 864-particle system, for which the smallest allowable $\vec{Q}$ vector is $(2\pi/a)(\frac{1}{6}, 0, 0)$. The Brillouin peak position in the latter case agrees within statistical uncertainties with the sound velocities obtained by the Monte Carlo method[6] for the same system.

V. COMPARISON WITH SELF-CONSISTENT PHONON CALCULATIONS

A. Preamble

Following the work of Maradudin and Fein[33] and Ambegaokar, Conway, and Baym[30] we can write the intermediate scattering function as a power series in $\vec{Q}$:

$$S(\vec{Q}, t) = S_0 + S_1 + S_{\text{int}} + S_2 + \cdots . \qquad (16)$$

These terms describe successively the elastic (Bragg) scattering, one-phonon inelastic scattering, the interference between the one-phonon and two-phonon scattering, and the two-phonon inelastic scattering, etc. A detailed expression for S_1 has been given earlier in Eqs. (11) and (12). For a harmonic crystal [either quasiharmonic, or self-consistent harmonic (SCHA)] the appropriate ensemble average occurring in S_1 is readily evaluated and one obtains

$$S_1^0(\vec{Q}, \omega) = \frac{\hbar}{2m} e^{-2W} \sum_{q\lambda} \Delta(\vec{Q} - \vec{q}) \frac{(\vec{Q} \cdot \vec{e}_{q\lambda})^2}{\omega_{q\lambda}} A_{q\lambda}^0(\omega) , \qquad (17)$$

$$A_{q\lambda}^0(\omega) = 2\pi[(\bar{n}_{q\lambda} + 1)\delta(\omega - \omega_{q\lambda}) + \bar{n}_{q\lambda}\delta(\omega + \omega_{q\lambda})], \qquad (18)$$

where $\omega_{q\lambda}$ and $\vec{e}_{q\lambda}$ are the harmonic phonon frequen-

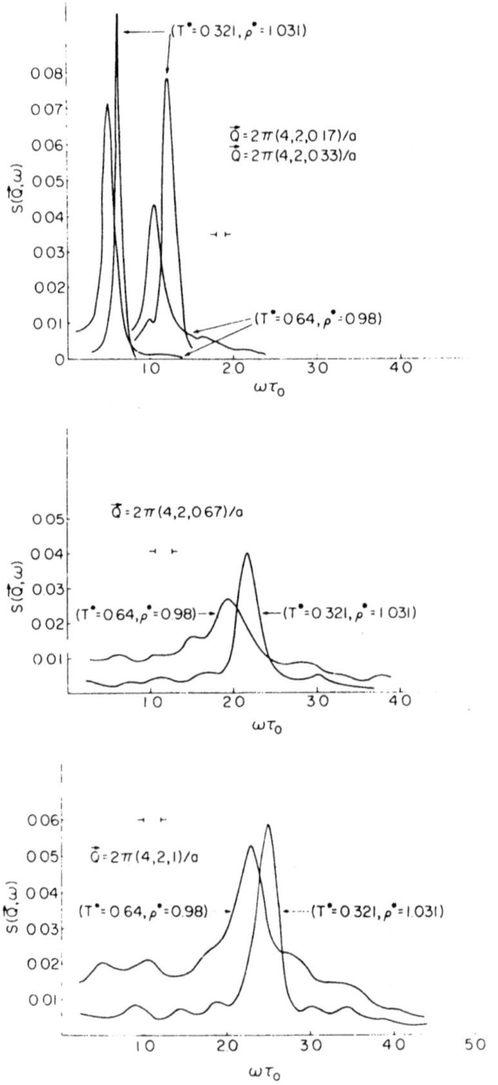

FIG. 8. Temperature dependence of $S(\vec{Q}, \omega)$ at roughly zero pressure. The FWHM of the Gaussian filter function used to smooth the raw $S(\vec{Q}, \omega)$ data is also shown. The data were taken from run Nos. 4, 8, and 9.

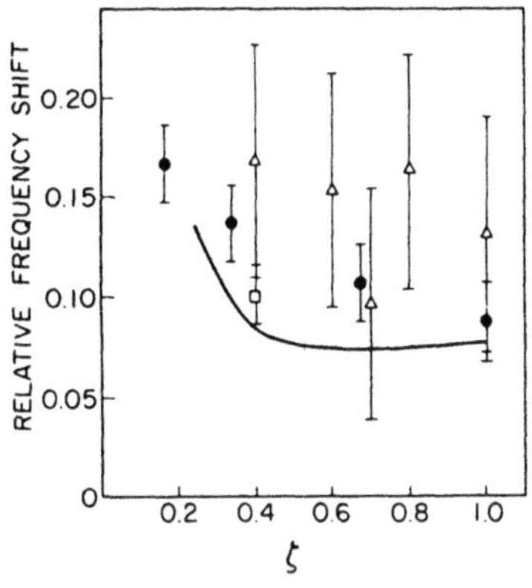

FIG. 9. Relative shift of the peaks in $S(\vec{Q}, \omega)$ taken from Fig. 8 (full circles) compared with experimental data for ^{36}Ar under similar conditions taken from Refs. 29 (open square) and 14 (open triangles). The solid line is taken from self-consistent-phonon calculation of Ref. 31.

cies and eigenvectors and $\bar{n}_{q\lambda}$ is the phonon occupation number. The delta function $\Delta(\vec{Q}-\vec{q})$ indicates that the momentum transfer must be equal to the phonon wave vector plus or minus a reciprocal-lattice vector $\vec{K}$. The superscript 0 denotes the harmonic approximation. $S_1^0(Q, \omega)$ has δ-function peaks at the appropriate phonon frequencies.

In the harmonic approximation S_{int} vanishes while S_2 gives rise to a broad (structured) background extending to twice the maximum frequency.

In an anharmonic crystal the δ-function peaks become broadened and shifted and one can show that (at least for the high-symmetry directions such as [100], [110], and [111] in the fcc lattice)

$$S_1(\vec{Q}, \omega) = \frac{\hbar}{2m} e^{-2W} \sum_{q\lambda} \Delta(\vec{Q}-\vec{q}) \frac{(\vec{Q}\cdot\vec{e}_{q\lambda})^2}{\omega_{q\lambda}} A_{q\lambda}(\omega) , \quad (19)$$

where

$$A_{q\lambda}(\omega) = (n_\omega + 1) B_{q\lambda}(\omega), \quad n_\omega = (e^{\beta\hbar\omega} - 1)^{-1} \quad (20)$$

and

$$B_{q\lambda}(\omega) = \frac{8\omega_{q\lambda}^2 \Gamma_{q\lambda}(\omega)}{[\omega^2 - \omega_{q\lambda}^2 - 2\omega_{q\lambda}\Delta_{q\lambda}(\omega)]^2 + [2\omega_{q\lambda}\Gamma_{q\lambda}(\omega)]^2} . \quad (21)$$

The rigorous derivation of this equation follows from anharmonic many-body perturbation theory[33] and will not be discussed here. We state only that Δ and Γ are the shift and width (inverse lifetime) due to phonon-phonon interactions. The lowest order contribution to Γ arises from the possibility of a phonon decaying into two other phonons. The first detailed calculations for a Mie–Lennard-Jones potential that allowed for this effect were those of Bohlin and Hogberg.[27] Applications using self-consistent-phonon theory have been reported by Koehler, Goldman *et al.*, and others.[7] In Koehler's work only the lowest-order contribution to Δ and Γ are included and in our subsequent application we will label such results by $\sum_K$. Goldman *et al.*[7] included some addition contributions to Δ and Γ and in what follows calculations based upon their approximation will be labeled $\sum_{GHK}$. For details the reader is referred to the original papers.

The interference term S_{int} which describes the coupling of the one-phonon peak to the multiphonon background was discussed by Maradudin and Ambegaokar[33] and more recently by Horner[34] and Glyde.[35] The main point here is that such terms are odd in $\vec{Q}$ and will oscillate in sign (growing in magnitude) as $\vec{Q}$ increases. Since such terms necessarily vanish at a Brillouin-zone boundary we will for convenience restrict ourselves to phonons such as these and hence not have to consider S_{int} any further. Furthermore, by evaluating S_1 in both the MD calculation and the phonon case we need not consider S_2, etc. The effect of such terms can be estimated by comparing the full MD results with the one phonon ones.

B. Self-consistent phonon calculations

We present a comparison of the phonon calculations based upon Eqs. (17) and (19) with the MD results derived from Eq. (11) in Figs. 10 and 11. Although calculations are in the one phonon approximations the phonon calculation is quantum while the MD one is classical. We have, therefore, replaced the factor $(n_\omega + 1)$ occurring in Eq. (20) of the phonon calculation by the factor $kT/\hbar\omega$ while leaving the factor $B(\omega)$ fully quantum. This procedure is essentially equivalent to applying Schofield's quantum correction[23] to the classical MD calculations in order that they at least satisfy the detailed balance condition [Eq. (5)].

Figure 10 shows the zone-boundary [100] phonons for the nearest-neighbor ($n = 1$) 2048-particle system and a (12-6) potential at about one-half the melting temperature, Fig. 10(a), and the melting point, Fig. 10(b). The calculated response using self-consistent-phonon theory depends little on whether one uses Koehler's self-energy $\sum_K$ or that of Goldman *et al.*, $\sum_{GHK}$. There is fairly good agreement with the MD data at the lower temperature except that the relative intensity of the two peaks in the longitudinal response seems to be incorrect. This is likely in part due to the fact that phonon calculation is quantum, while the MD is classical. At the higher temperature there are larger discrepancies between the two calculations. Figure 11 shows this effect more clearly for the zone-boundary [100] transverse phonon of the 864-particle system [(12-6) potential with $n = 3$]. The temperature and density correspond roughly to the melting point of the (12-6) system (actually corresponding to slightly metastable real solid argon). Here we see the rather large discrepancy between the MD predicted peak in S_1 and that obtained using self-consistent-phonon theory. The temperature is now sufficiently high that the MD results will likely be unaffected by quantum effects. Hence the discrepancy indicates a need to improve the phonon calculations, at high temperatures. To do so it will probably be necessary to explicitly include the hard-core effects as discussed by

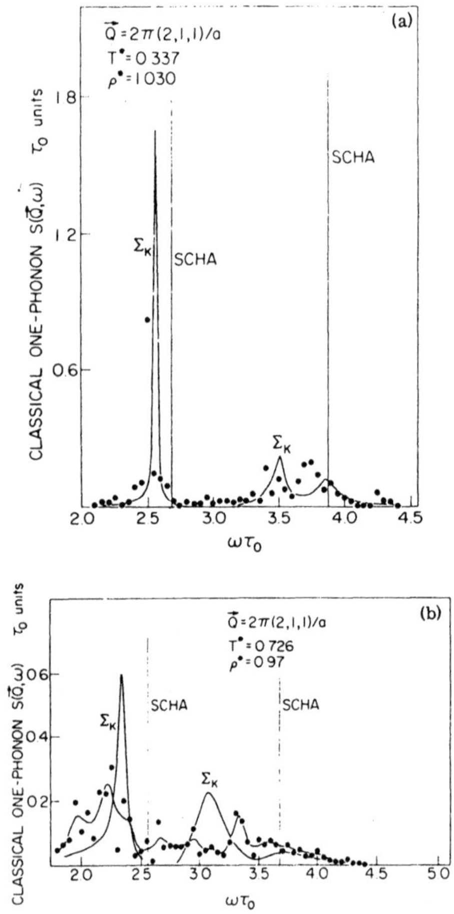

FIG. 10. (a) Detailed comparison of the one-phonon $S_1(\vec{Q}, \omega)$ with the self-consistent harmonic approximation (SCHA) and anharmonic self-consistent-phonon calculations utilizing the Σ_k self-energy (see text). The dots are the unsmoothed MD data from run No. 5. (b) Detailed comparison of the one-phonon $S_1(\vec{Q}, \omega)$ with the self-consistent-harmonic approximation (SCHA) and anharmonic self-consistent-phonon calculations utilizing the Σ_k self-energy (see text). The dots are the unsmoothed MD data from run No. 10. The continuous solid line through the dots is the smoothed MD data.

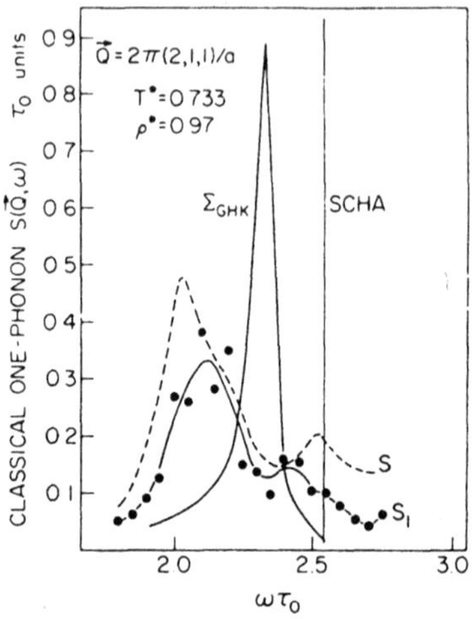

FIG. 11. Comparison of the one-phonon $S_1(\vec{Q}, \omega)$ with the self-consistent-harmonic approximation (SCHA) and anharmonic self-consistent-phonon calculations utilizing the Σ_{GHK} self-energy (see text). The dots are the MD S_1 data from run No. 11 and the solid curve is the smoothed S_1 data. The dashed curve is the smoothed full $S(\vec{Q}, \omega)$ which includes the multiphonon scattering.

Horner[36] and more recently by Kanney and Horton. ton.[37] Such a calculation although extremely interesting would go beyond the scope of this paper. Finally, Fig. 11 shows the full S as well as S_1 and hence indicates the extent to which multiphonon effects can distort the peak in $S(\vec{Q}, \omega)$. Detailed comparisons between experiment at high temperatures and theory would necessarily involve including multiphonon effects.

VI. CONCLUSION

We have demonstrated that it is feasible to calculate the dynamical structure factor $S(\vec{Q}, \omega)$ for solids using the computer simulation MD technique. Results for the (12-6) Mie–Lennard-Jones potential show many of the features recently observed in inelastic neutron scattering experiments on rare gas solids. At high temperatures, the MD results for $S_1(\vec{Q}, \omega)$, the one-phonon approximation to $S(\vec{Q}, \omega)$ have revealed the inadequacy of self-consistent-phonon theories that do not explicitly treat the hard-core problem.

ACKNOWLEDGMENTS

We both thank D. Levesque for his invaluable help in constructing our $S(\vec{Q}, \omega)$ program and J. J. Weis for valuable discussions. The self-consistent-phonon calculations were kindly provided by V. V. Goldman. This work was completed while one of us (M. L. K) was a visitor to the Laboratoire de Physique Théorique des Liquides, Université Paris VI, under the CNRS-NRC exchange agreement.

*Laboratoire associé au Centre National de la Recherche Scientifique.

[1]B. J. Alder and T. E. Wainwright, J. Chem. Phys. 31, 459 (1959).

[2]A. Rahman, Phys. Rev. 136, A405 (1964).

[3]L. Verlet, Phys. Rev. 159, 98 (1967).

[4]N. Metropolis, A. W. Rosenbluth, M. N. Rosenbluth, A. M. Teller, and E. Teller, J. Chem. Phys. 21, 1087 (1953).

[5]See, e.g., the extensive review article by W. W. Wood, in *Physics of Simple Liquids*, edited by H. N. V. Temperley, J. S. Rowlinson, and G. S. Rushboroke (North-Holland, Amsterdam, 1968).

[6]A. C. Holt, W. G. Hoover, S. G. Gray, and D. R. Shortle, Physica (Utr.) 49, 61 (1970); M. L. Klein and W. G. Hoover, Phys. Rev. B 4, 537 (1971).

[7]T. R. Koehler, Phys. Rev. Lett. 22, 777 (1969); V. V. Goldman, G. K. Horton, and M. L. Klein, Phys. Rev. Lett. 24, 1424 (1970).

[8]J. M. Dickey and A. Paskin, Phys. Rev. 188, 1407 (1969).

[9]D. Levesque, L. Verlet, and J. Kürkijarvi, Phys. Rev. A 7, 1690 (1973).

[10]A. Rahman, Phys. Rev. Lett. 32, 52 (1974).

[11]J. P. Hansen, E. L. Pollock, and I. R. McDonald, Phys. Rev. Lett. 32, 277 (1974).

[12]J. P. Hansen and M. L. Klein, J. Phys. Lett. 35, L-29 (1974).

[13]M. L. Klein, G. Jacucci, and I. R. McDonald, J. Phys. Lett. 36, L-97 (1975); J. J. Weis and M. L. Klein, J. Chem. Phys. 63, 2869 (1975).

[14]D. N. Batchelder, B. C. G. Haywood, and D. M. Saunderson, J. Phys. C 4, 910 (1971).

[15]P. A. Fleury, J. M. Worlock, and M. L. Carter, Phys. Rev. Lett. 30, 591 (1973); H. E. Jackson, D. Landheer, and B. P. Stoicheff, *ibid.* 31, 296 (1973).

[16]B. J. Alder, H. Strauss, and J. J. Weis, J. Chem. Phys. 59, 1002 (1973).

[17]B. J. Alder, J. P. Hansen, M. L. Klein, H. Strauss, and J. J. Weis (unpublished).

[18]J. J. Weis, Molec. Phys. 28, 187 (1974).

[19]J. P. Hansen and L. Verlet, Phys. Rev. 184, 151 (1969).

[20]J. P. Hansen, Phys. Rev. A 2, 221 (1970).

[21]E. Wigner, Phys. Rev. 40, 149 (1932).

[22]J. P. Hansen and J. J. Weis, Phys. Rev. 188, 314 (1969).

[23]P. Schofield, Phys. Rev. Lett. 4, 239 (1960).

[24]J. L. Lebowitz, J. K. Percus, and L. Verlet, Phys. Rev. 153, 250 (1967).

[25]W. G. Hoover and F. H. Ree, J. Chem. Phys. 47, 4873 (1967); see also E. L. Pollock (unpublished).

[26]H. C. Andersen, J. D. Weeks, and D. Chandler, Phys. Rev. A 4, 1597 (1971); L. Verlet and J. J. Weis, *ibid.* 5, 939 (1972).

[27]L. Bohlin and T. Högberg, J. Chem. Phys. Solids 29, 1805 (1968); L. Bohlin, Solid State Commun. 9, 141 (1971); 10, 1219 (1972).

[28]J. Skalyo, Y. Endoh, and G. Shirane, Phys. Rev. B 9, 1797 (1974).

[29]Y. Fujii, N. A. Lurie, R. Pynn, and G. Shirane, Phys. Rev. B 10, 3647 (1974).

[30]V. Ambegaokar, J. Conway, and G. Baym, in *Lattice Dynamics* edited by R. F. Wallis (Pergamon, New York, 1965), p. 261.

[31]G. K. Horton, V. V. Goldman, and M. L. Klein, J. Appl. Phys. 41, 5138 (1970).

[32]F. Clayton and D. N. Batchelder, J. Phys. C 6, 1213 (1973).

[33]A. A. Maradudin and A. E. Fein, Phys. Rev. 128, 2589 (1962); A. A. Maradudin and V. Ambegaokar, *ibid.* 135, A1071 (1964).

[34]H. Horner, Phys. Rev. Lett. 29, 556 (1972).

[35]H. R. Glyde, Can. J. Phys. 52, 2281 (1074).

[36]H. Horner, Solid State Commun. 9, 79 (1971); see also Chap. 8 in *Dynamical Properties of Solids*, edited by G. K. Horton and A. A. Maradudin (North-Holland, Amsterdam, 1974).

[37]L. B. Kanney and G. K. Horton, Phys. Rev. Lett. 34, 1565 (1975).

PHYSICAL REVIEW B VOLUME 14, NUMBER 4 15 AUGUST 1976

Anharmonic effects in the phonon spectra of sodium chloride

E. R. Cowley
Physics Department, Brock University, St. Catherines, Ontario L2S 3A1, Canada

G. Jacucci*
Centre Européen de Calcul Atomique et Moléculaire, Faculté des Sciences, 91405 Orsay, France

M. L. Klein†
Laboratoire de Physique Théorique des Liquides, Université Pierre et Marie Curie, 75230 Paris, France

I. R. McDonald
Chemistry Department, Royal Holloway College, Egham, Surrey TW20 0EX, England
(Received 9 February 1976)

Results are reported of a series of molecular-dynamics "experiments" on solid NaCl at temperatures of 80, 302, 954, and 1153 K. Attention is focused primarily on the computation of the dynamical structure factor and its one-phonon approximation; comparison of the two allows the isolation of contributions from multiphonon and interference terms. Anharmonic effects are analyzed in terms of perturbation theory and the theory is found to give satisfactory results for the phonon frequencies at temperatures up to 80% that of melting. Most of the calculations are carried out for a simple rigid-ion potential, but the effects of polarization and its incorporation in molecular-dynamics calculations are briefly discussed.

I. INTRODUCTION

The lattice vibrations of NaCl, the protype ionic crystal, were first analyzed in detail in the classic paper of Kellermann.[1] Kellermann's calculations were based on the assumption that the crystal is composed of rigid ions interacting through a potential in which a short-range repulsion is superimposed on the long-range Coulombic term. This simple model, even in the more elaborate version of Tosi and Fumi,[2] is unable to account quantitatively for the dynamical properties of NaCl, as revealed by neutron spectroscopy,[3] because it neglects the effects of ionic polarization. Several models exist which remedy this defect,[4-6] and one of these (the shell model of Dick and Overhauser[4]) is discussed briefly in Sec. IV F. However, the motivation for the work presented here lies in the question of the general nature of the lattice vibrations in NaCl-type crystals at high temperatures and their interpretation in terms of anharmonic perturbation theory. For that reason we have carried out most of our calculations with the Tosi-Fumi rigid-ion potential, a model which is known to give a good fit to many equilibrium crystal properties.[7]

The melting point of NaCl at atmospheric pressure is 1073 K, or approximately three times its Debye temperature. Above room temperature it may be treated as essentially a classical solid and classical theories of anharmonic lattice dynamics[8] may therefore be applied. Such methods have indeed been used successfully in the interpretation of thermodynamic[9] and optical data[10] on NaCl. However, there exist few direct data on the phonons at elevated temperatures (above two-thirds that of melting), where a breakdown of simple perturbation approaches to the problem of vibrational anharmonicity[11] can be expected. To make it possible to study anharmonic effects in the range of temperature in which experimental data are lacking, we have carried out a series of computer simulations by the method of molecular dynamics, the primary purpose being the calculation of the dynamic structure factor $S(\vec{Q}, \omega)$. The latter, as is well known, is related in a simple way to the double differential cross section for coherently scattered thermal neutrons, whereas the intensity of x-ray scattering is related to the integral of $S(\vec{Q}, \omega)$ over all ω. The basis of the molecular-dynamics method is the solution of Newton's equations of motion for a finite system of particles contained in a box with periodic boundary conditions. From the classical phase-space trajectories which this procedure yields, ensemble averages of the type encountered in statistical mechanics are readily evaluated. The outstanding virtue of the method is that it yields results which are essentially exact *for the system studied*. This means, in the present case, that the predictions of perturbation theory can be tested in an unambiguous way if carried out for the same model. We wish to emphasize at the outset that our definition of the model includes the fact that a small periodic system is used. In particular, the lattice-dynamical calculations have been made over a mesh of points in the Brillouin zone which is consistent with the size of the molecular-dynamics system.

The computer "experiments" yield with equal ease both the full multiphonon $S(\vec{Q}, \omega)$ and its one-phonon approximation $S_1(\vec{Q}, \omega)$. This makes it possible to investigate the convergence of the standard phonon expansion of $S(\vec{Q}, \omega)$. Moreover, a study of the $\vec{Q}$ dependence of various phonons can be expected to reveal the interesting interference effects of the type first identified in the beautiful work of Cowley and Buyers[12] and also discussed in an important paper by Horner.[13]

Four separate molecular-dynamics runs have been carried out. The first, for a temperature of 80 K, was designed to test the accuracy of the simulation by comparison with the results of quasiharmonic lattice dynamics for the same interionic potential. The second, at 302 K, was made in order to compare with calculations based on anharmonic perturbation theory in a range of temperature where the theory should work well. The other two "experiments," at 954 and 1153 K, were carried out specifically to probe possible inadequacies in the perturbation theory of anharmonic effects in the region where these effects are largest. Anticipating some of our conclusions, we shall see in Sec. IV E that the perturbation results are surprisingly good, but that important differences in detail remain. We shall also see that all the interesting manifestations of vibrational anharmonicity are contained in our results. The method, of course, can easily be used in the study of other alkali halides.

II. MOLECULAR-DYNAMICS CALCULATIONS

The molecular-dynamics calculations were carried out on a system of 216 particles, consisting of 108 positive (Na^+) and 108 negative (Cl^-) ions disposed on the interpenetrating fcc lattices of the NaCl crystal structure and lying within a cubic cell of length L. Periodic boundary conditions were imposed and the classical equations of motion of the ions were solved by means of a simple finite difference algorithm,[14] with a time step in the numerical integrations of 7×10^{-15} sec. The interionic pair potential used was the generalized Huggins-Mayer potential with parameters deduced from solid-state thermodynamic data by Tosi and Fumi,[2] namely

$$\phi_{ij}(r_{ij}) = z_i z_j / r_{ij} + c_{ij} b \exp[(\sigma_i + \sigma_j - r_{ij})/\rho] - C_{ij}/r_{ij}^6 - D_{ij}/r_{ij}^8 , \qquad (1)$$

where z_i and z_j are the charges on ions labeled i and j, located at $\vec{r}_i, \vec{r}_j$ and separated by a distance $r_{ij} = |\vec{r}_i - \vec{r}_j|$. The second term on the right-hand side of Eq. (1) describes the overlap repulsion between ions. In this term the factor b takes the same value for all alkali halides which crystallize in the NaCl-type structure, σ_i and σ_j are lengths ("basic radii") characteristic of the ions, ρ is a "hardness parameter" characteristic of the particular salt, and the c_{ij} are numerical constants introduced by Pauling. The last two terms in Eq. (1) represent the contributions, respectively, from dipole-dipole and dipole-quadrupole dispersion forces. As we have already remarked, the major weakness of the Tosi-Fumi model, at least insofar as the calculation of lattice vibrations is concerned, is the fact that no explicit account is taken of the effects of ionic polarization. We shall return briefly to this question in Sec. IV F. Values of the potential parameters used in the calculations are listed in Table I and details of the thermodynamic states which were studied are summarized in Table II. In Table II we also give the total number of time steps which were generated at each state point.

TABLE I. Parameters in the Tosi-Fumi potential for NaCl.

$b = 3.38\times10^{-13}$ erg
$c_{++} = 1.25$
$c_{+-} = 1$
$c_{--} = 0.75$
$\sigma_+ = 1.170$ Å
$\sigma_- = 1.585$ Å
$\rho = 0.317$ Å
$C_{++} = 1.68\times10^{-60}$ erg cm^6
$C_{+-} = 11.2\times10^{-60}$ erg cm^6
$C_{--} = 116\times10^{-60}$ erg cm^6
$D_{++} = 0.8\times10^{-76}$ erg cm^8
$D_{+-} = 13.9\times10^{-76}$ erg cm^8
$D_{--} = 233\times10^{-76}$ erg cm^8

Use of a periodic boundary condition makes it possible to employ the Ewald method for the calculation of the electrostatic energy. In doing so we have exploited an idea due to Singer (private communication) whereby the total electrostatic energy E_C is written in the exact form given by

$$E_C = \tfrac{1}{2} \sum_{n\neq0} A_n(\eta)(C_{\vec{n}}^2 + S_{\vec{n}}^2) + \sum_i z_i \sum_{j>i} z_j \frac{\operatorname{erfc}(\eta r_{ij})}{r_{ij}} - \sum_i \frac{z_i^2 \eta}{\pi^{1/2}} . \qquad (2)$$

The symbol erfc is used to denote the complementary error function, $\operatorname{erfc}(x) = 1 - \operatorname{erf}(x)$, and

$$A_n(\eta) = \frac{\exp(-\pi^2 n^2/\eta^2 L^2)}{n^2} , \qquad (3)$$

$$C_{\vec{n}} \equiv \sum_i c_{\vec{n}i} = \sum_i z_i \cos\left(2\pi \frac{\vec{n}\cdot\vec{r}_i}{L}\right), \qquad (4)$$

TABLE II. Selected equilibrium properties of NaCl.

Source	T (K)	V ($cm^3\,mol^{-1}$)	pV/NkT	$-U$ ($kJ\,mol^{-1}$)	$\langle u_+^2\rangle^{1/2}$ (Å)	$\langle u_-^2\rangle^{1/2}$ (Å)	Time steps (1000's)
MD [a]	80.3	26.92	0.08	773.5	0.126	0.113	10
Expt. [b]	298.0	27.0	0.0	764.0			
MC [c]	298.0	27.65	0.0	762.6			
MD	301.7	27.60	0.13	762.1	0.234	0.227	6
MD	953.8	29.50	0.87	727.3	0.494	0.460	10
Expt.	1073.0	30.0	0.0	719.6			
MC	1073.0	31.37	0.0	717.4			
MD	1153.0	31.37	0.21	713.2	0.611	0.577	28.8

[a] MD, results from present molecular-dynamics calculations.
[b] Expt., experimental data (Ref. 7).
[c] MC, results from Monte Carlo calculations (Ref. 7).

$$S_{\vec{n}} \equiv \sum_i s_{\vec{n}i} = \sum_i z_i \sin\left(2\pi \frac{\vec{n}\cdot\vec{r}_i}{L}\right) . \tag{5}$$

The quantity η is a disposable parameter (having dimensions length^{-1}) which governs the relative rate of convergence of the two series in Eq. (2), one of which is an expansion in reciprocal space, the other being a sum in real space. The first term on the right-hand side of Eq. (2) involves a summation over reciprocal-lattice vectors n of the simple cubic structure of cells of side L, together with a sum over particles within one such cell. The advantage of writing the reciprocal-space term in the particular form quoted here is that the summation over ions runs only over *single* ions rather that pairs, with a correspondingly large reduction in the length of the computations. The real-space summation in Eq. (2) is taken over pairs i,j within a single cell. Finally, differentiation of E_C with respect to the coordinates of ion i yields the Coulombic force $\vec{F}_{Ci}$ on ion i as

$$\vec{F}_{Ci} = -\sum_{n\neq 0} \vec{n}\left(\frac{2\pi}{L}\right) A_n(\eta)(c_{\vec{n}i}S_{\vec{n}} - s_{\vec{n}i}C_{\vec{n}}) + z_i \sum_{j\neq i} \vec{r}_{ij} z_j \left(\frac{\operatorname{erfc}(\eta r_{ij})}{r_{ij}^3} + \frac{2\eta}{\pi^{1/2}}\frac{\exp(-\eta r_{ij}^2)}{r_{ij}^2}\right) . \tag{6}$$

Equation (6) has been used in all our molecular dynamics calculations. The parameter η was taken as $5.6/L$ and the reciprocal-space term was evaluated for 309 pairs of vectors, account being taken of the fact that vectors $\vec{n}$ and $-\vec{n}$ make identical contributions both to E_C and to $\vec{F}_{Ci}$. The real-space term was truncated at $r_{ij} = \frac{1}{2}L$, the same cutoff in the potential being used for the non-Coulombic terms in (1).

Our main effort has been directed at the calculation of the dynamical structure factor $S(\vec{Q}, \omega)$, which we have computed in the manner of Levesque, Verlet, and Kürkijarvi[14] from the classical expression

$$S(\vec{Q}, \omega) = \frac{1}{N}\int_0^\infty e^{i\omega t}\langle \rho_{\vec{Q}}(t)\rho_{-\vec{Q}}(0)\rangle\, dt = \lim_{T\to\infty}\frac{1}{NT}\int_0^T e^{-i\omega t}\rho_{\vec{Q}}(t)\,dt \int_0^T e^{i\omega t'}\rho_{-\vec{Q}}(t')\,dt' = \frac{1}{N}\lim_{T\to\infty}\frac{1}{T}\,|\rho_{\vec{Q}}(\omega)|^2 , \tag{7}$$

where N is the number of ions of each type (i.e., the number of unit cells), $\hbar\vec{Q}$ is the momentum transfer, $\hbar\omega$ is the energy transfer, and $\rho_{\vec{Q}}(\omega)$ is the Fourier-Laplace transform of the particle density $\rho_{\vec{Q}}(t)$. The latter we write in terms of partial densities as

$$\rho_{\vec{Q}}(t) = \rho_{\vec{Q}}^{+}(t) + \rho_{\vec{Q}}^{-}(t) , \tag{8}$$

with

$$\rho_{\vec{Q}}^{+}(t) = \sum_{\text{cations}} e^{i\vec{Q}\cdot\vec{r}_i(t)} , \quad \rho_{\vec{Q}}^{-}(t) = \sum_{\text{anions}} e^{i\vec{Q}\cdot\vec{r}_j(t)} . \tag{9}$$

Thus the calculation of $S(\vec{Q}, \omega)$ reduces to the calculation of partial dynamical structure factors $S_{\alpha\beta}(\vec{Q}, \omega)$, i.e., the Fourier transforms of correlation functions $\langle\rho_{\vec{Q}}^{\alpha}(t)\rho_{-\vec{Q}}^{\beta}(0)\rangle$, where $\alpha, \beta = +, -$ and $S_{+-}(\vec{Q}, \omega) = S_{-+}(\vec{Q}, \omega)$. Specifically,

$$S(\vec{Q}, \omega) = S_{++}(\vec{Q}, \omega) + S_{--}(\vec{Q}, \omega) + 2S_{+-}(\vec{Q}, \omega) . \tag{10}$$

On the other hand, the cross section for the coherent inelastic scattering of neutrons, $S^n(\vec{Q}, \omega)$, is constructed by weighting the partial quantities occurring in Eq. (10) by neutron scattering lengths. The x-ray scattering intensity would be obtained by weighting with the appropriate form factors and integrating over ω. Similarly, the spectrum of charge density fluctuations, representing the optic modes of vibrations, is obtained by weighting with the charges of the ions. Thus

$$S^n(\vec{Q},\omega)=b_+^2S_{++}(\vec{Q},\omega)+b_-^2S_{--}(\vec{Q},\omega)+2b_+b_-S_{+-}(\vec{Q},\omega) \tag{11}$$

and

$$S^z(\vec{Q},\omega)=z_+^2S_{++}(\vec{Q},\omega)+z_-^2S_{--}(\vec{Q},\omega)+2z_+z_-S_{+-}(\vec{Q},\omega)\,. \tag{12}$$

The phonon frequencies measured in a neutron experiment can be identified with the peaks in $S^n(\vec{Q},\omega)$, in the calculation of which we have used the reduced scattering lengths $b_+=0.52$ and $b_-=1.47$.

During the course of the molecular-dynamics runs we also evaluated the one-phonon approximation to $S(\vec{Q},\omega)$, denoted by $S_1(\vec{Q},\omega)$, which may be computed in the same way as $S(\vec{Q},\omega)$ itself except that $\rho_{\vec{Q}}(t)$ in Eq. (7) is replaced by $\bar{\rho}_{\vec{Q}}(t)$, defined as

$$\bar{\rho}_{\vec{Q}}(t)=\bar{\rho}^+_{\vec{Q}}(t)+\bar{\rho}^-_{\vec{Q}}(t)\,, \tag{13}$$

where, for example,

$$\bar{\rho}^+_{\vec{Q}}(t)=d_+(Q)\sum_{\text{cations}} e^{i\vec{Q}\cdot\vec{R}_i}\vec{Q}\cdot\vec{u}_i(t)\,. \tag{14}$$

The quantity $d_+(Q)$ is related to the Debye-Waller factor (see Sec. III below) and

$$\vec{u}_i(t)=\vec{r}_i(t)-\vec{R}_i \tag{15}$$

is the instantaneous displacement of the ith ion from its lattice site $\vec{R}_i$. The one-phonon approximations to $S^n(\vec{Q},\omega)$ [i.e., $S_1^n(\vec{Q},\omega)$] and $S^z(\vec{Q},\omega)$ [i.e., $S_1^z(\vec{Q},\omega)$] are constructed in an analogous way, account being taken of the different weighting of the partial structure factors.

The length of the molecular dynamics cell (for $N=108$) is $L=3a$, where a is the lattice constant, i.e., twice the separation d of neighboring Na^+ and Cl^- lattice sites. From the periodic nature of the system it follows that the *independent* values of momentum transfer which we can study are limited to $\vec{Q}=(2\pi/3a)(n_1,n_2,n_3)$, where the n_i are integers. For example, in the ⟨100⟩ direction, we are limited to studying three different values of the phonon wave vector $\vec{q}$. However, we can study the same "phonon" for several different values of the momentum transfer because we can always write $\vec{q}=\vec{Q}\pm\vec{g}$, where $\vec{g}$ is any Bragg vector. It should also be noted that the appearance of terms of the form $\vec{Q}\cdot\vec{u}$ in both (9) and (14) means that we cannot study transverse phonons in the first Brillouin zone, a limitation which applies with equal force in a real neutron scattering experiment.

Finally, we note that our computations must satisfy the Placzek sum rule

$$\frac{1}{2\pi}\int_0^\infty \omega^2 S(\vec{Q},\omega)\,d\omega=\tfrac{1}{2}kTQ^2\left(\frac{1}{m_+}+\frac{1}{m_-}\right), \tag{16}$$

where m_+, m_- are the ionic masses. This provides a useful check on the accuracy of the molecular-dynamics results. In practice the rule is satisfied to within a few percent, except at the lowest temperature. There, because the peak in $S(\vec{Q},\omega)$ is very sharp, the fact that the spectrum is sampled only at discrete values of ω can cause the second moment to be considerably in error.

III. LATTICE DYNAMICS

In this section we briefly indicate the relationship of usual phonon calculations[8-10] to the computer simulation work reported here. Our starting point is an expression for the time correlation function occurring in Eq. (7), i.e.,

$$S(\vec{Q},t)\equiv\frac{1}{N}\langle\rho_{\vec{Q}}(t)\rho_{-\vec{Q}}(0)\rangle=\frac{1}{N}\sum_{i,j}e^{X_{ij}}\langle e^{x_i}e^{y_j}\rangle\,, \tag{17}$$

where $X_{ij}=i\vec{Q}\cdot(\vec{R}_i-\vec{R}_j)$, $x_i=i\vec{Q}\cdot\vec{u}_i(t)$, and $y_j=-i\vec{Q}\cdot\vec{u}_j(0)$. The labels $i\equiv l,\kappa$ and $j\equiv l,\kappa'$ are used to denote the κth (or κ'th) ion in the lth (or l'th) unit cell; thus $\vec{u}_i(t)$ denoted the displacement of the κth ion in the lth unit cell from its equilibrium position $\vec{R}_i$.

Ambegoakar, Conway, and Baym[15] have shown that from an expression of the type of (17) a Debye-Waller factor can be rigorously separated out. In the classical case we may write

$$\langle e^{x_i}e^{y_j}\rangle=d(x_i)d(y_j)[1+\langle x_iy_j\rangle+\tfrac{1}{2}(\langle x_i^2y_j\rangle-\langle x_iy_j^2\rangle)+\tfrac{1}{2}\langle x_iy_j\rangle^2+\cdots]\,. \tag{18}$$

Successive terms on the right-hand side of this equation are the correlation functions corresponding to elastic, one-phonon, lowest-order interference, and lowest-order two-phonon contributions to $S(\vec{Q},t)$. The Debye-Waller factor $d(x_i)$ is defined in terms of a cumulant expansion as

$$\ln d(x_i)=(1/2!)\langle x_i^2\rangle+(1/4!)(\langle x_i^4\rangle-3\langle x_i^2\rangle^2)+\cdots\,. \tag{19}$$

For a cubic crystal, Eq. (19) can be written in a more transparent notation as $d(x_i)\equiv d_\kappa(Q)$, with

$$d_\kappa(Q)\simeq\exp(-\tfrac{1}{6}Q^2\langle u_\kappa^2\rangle)\,. \tag{20}$$

The one-phonon approximation for $S(\vec{Q},t)$ follows immediately. On substituting (18) into (17) we find that

$$S_1(\vec{Q},t)=\frac{1}{N}\langle\bar{\rho}_{\vec{Q}}(t)\bar{\rho}_{-\vec{Q}}(0)\rangle\,, \tag{21}$$

where $\bar{\rho}_{\vec{Q}}(t)$ is the density operator defined by (13). We now introduce the quantities $\hat{\rho}^+_{\vec{Q}}(t)$ and $\hat{\rho}^-_{\vec{Q}}(t)$ where, for example,

$$\hat{\rho}_{\vec{Q}}^{+}(t)=\sum_{\text{cations}} e^{i\vec{Q}\cdot\vec{R}_i}\vec{\xi}\cdot\vec{u}_i(t)\,, \tag{22}$$

with $\vec{Q}=Q\vec{\xi}$. With the aid of (22) we can rewrite Eq. (21) in the form

$$\begin{aligned}(N/Q^2)S_1(Q,t)&=d_+^2(Q)\langle\hat{\rho}_{\vec{Q}}^{+}(t)\hat{\rho}_{-\vec{Q}}^{+}(0)\rangle\\&+d_-^2(Q)\langle\hat{\rho}_{\vec{Q}}^{-}(t)\hat{\rho}_{-\vec{Q}}^{-}(0)\rangle\\&+2d_+(Q)d_-(Q)\langle\hat{\rho}_{\vec{Q}}^{+}(t)\hat{\rho}_{-\vec{Q}}^{-}(0)\rangle\\&\equiv[Qd(Q)]^2F\,,\end{aligned} \tag{23}$$

which serves as a formal definition of the quantity $d(Q)$; F is a correlation function which is independent of Q.

To proceed further in lattice-dynamical calculations it is necessary to evaluate the correlation function $\langle\vec{u}_i(t)\cdot\vec{u}_j(0)\rangle$. This is done by introducing normal coordinates derived from the eigenfrequencies $\omega_{\vec{q}\lambda}$, i.e., the harmonic frequencies of wave vector $\vec{q}$ and polarization λ, and their associated eigenvectors $\vec{e}^{\kappa}_{\vec{q}\lambda}$. In the harmonic approximation (and the classical limit) the result is

$$S_1(\vec{Q},\omega)=\frac{kT}{\omega}\sum_{\lambda}\Delta(\vec{Q}-\vec{q})\,|F_{\vec{q}\lambda}(\vec{Q})|^2A^0_{\vec{q}}(\omega)\,, \tag{24}$$

where $\Delta(\vec{Q}-\vec{q})$ is the crystal δ function, the spectral function $A^0_{\vec{q}\lambda}(\omega)$ is given by

$$A^0_{\vec{q}\lambda}(\omega)=2\pi[\delta(\omega-\omega_{\vec{q}\lambda})-\delta(\omega+\omega_{\vec{q}\lambda})]\,, \tag{25}$$

and the one-phonon inelastic structure factor $F_{\vec{q}\lambda}(\vec{Q})$ is defined as

$$F_{\vec{q}\lambda}(\vec{Q})=\sum_{\kappa}d_\kappa(Q)\,e^{i\vec{q}\cdot\vec{R}_\kappa}(i\vec{Q}\cdot\vec{e}^{\kappa}_{\vec{q}\lambda})/(2m_\kappa\omega_{\vec{q}\lambda})^{1/2} \tag{26}$$

Thus the spectrum is given as a sum of δ-function peaks of appropriate weights.

For an anharmonic crystal the problem is more difficult, requiring the use of many-body perturbation theory.[8-10] However, the results bear a strong similarity to those for a harmonic crystal, provided that the coupling of one-phonon states corresponding to the same wave vector but different polarization branches is neglected. This so-called polarization mixing effect vanishes for the zone-center modes, and elsewhere is usually assumed to be small. The expression for $S_1(\vec{Q},\omega)$ is then the same as that obtained in the harmonic case, except that the δ functions are broadened and shifted by the phonon-phonon interactions. The effect is to replace the spectral function (25) by

$$A_{\vec{q}\lambda}(\omega)=\frac{8\omega^2_{\vec{q}\lambda}\Gamma_{\vec{q}\lambda}(\omega)}{[-\omega^2+\omega^2_{\vec{q}\lambda}+2\omega_{\vec{q}\lambda}\Delta_{\vec{q}\lambda}(\omega)]^2+[2\omega_{\vec{q}\lambda}\Gamma_{\vec{q}\lambda}(\omega)]^2}\,. \tag{27}$$

$\Delta_{\vec{q}\lambda}(\omega)$ and $\Gamma_{\vec{q}\lambda}(\omega)$ are related to the anharmonic terms in the Taylor series expansion of the total potential energy of the crystal in powers of the displacements of the ions from their equilibrium positions.[8] To second order in the cubic anharmonicity and first order in the quartic anharmonicity the shift is given by

$$\begin{aligned}\Delta_1(\omega)&=12kT\sum_{2}\frac{V(1,-1,2,-2)}{\omega_2}\\&-18kT\sum_{2,3}|V(1,2,3)|^2\left(\frac{(\omega_1+\omega_2)^2}{(\omega_1+\omega_2)^2-\omega^2}\right)\\&\times\left(\frac{2}{\omega_1\omega_2}\right)\end{aligned} \tag{28}$$

and the width by

$$\begin{aligned}\Gamma_1(\omega)&=18kT\sum_{2,3}|V(1,2,3)|^2\left(\frac{\omega_1+\omega_2}{\omega_1\omega_2}\right)\\&\times\pi[\delta(\omega_1+\omega_2-\omega)-\delta(\omega_1+\omega_2+\omega)]\,,\end{aligned} \tag{29}$$

where $\omega_1\equiv\omega_{\vec{q}_1\lambda_1}$, etc., and the matrix elements V are essentially the Fourier transforms of the cubic and quartic terms in the potential energy.

In the case where polarization mixing is included, the Green's functions are obtained as a matrix with finite nondiagonal terms. The one-phonon contributions to $S(\vec{Q},\omega)$ can be expressed as a sum over all the elements multiplied by appropriate products of structure factors. The detailed formalism has been given by Cowley.[8]

In the actual lattice-dynamical calculations reported here, the $\omega_{\vec{q}\lambda}$ and $\vec{e}^{\kappa}_{\vec{q}\lambda}$ were calculated from the interionic potential (1) for lattice spacings identical to those used in the simulations. This is the so-called quasiharmonic (QH) approximation, use of which allows the harmonic frequencies to change with volume. Similarly, the anharmonic force constants V in (28) and (29) were evaluated for each volume. Coulomb contributions to the harmonic and cubic terms were evaluated using an Ewald transformation. All other types of force, and the Coulomb contribution to the quartic shifts, were summed in real space over enough shells of neighbors to give a converged result.

The principal parts and δ functions appearing in expressions (28) and (29) were replaced by analytic functions corresponding to a finite width. In the usual applications[10] of this technique the expectation is that when a sufficiently fine mesh of wave vectors is used the results become independent of the width of the function used. In the present case the number of wave vectors is fixed at 108. We must then of necessity use quite a wide representation of the δ function. In fact the value used in all of the present calculations corresponds to a full width at half-maximum of 0.3×10^{13} sec^{-1}. The most obvious justification for such a smoothing function is that the intermediate phonons themselves have finite lifetimes, so that the necessary width can be estimated from a suitable average lifetime. The value we have used then corresponds quite well to what we should estimate for 300 K. For higher temperatures the phonons are less well defined and we can expect that the response functions may show even less structure than we have calculated. For lower temperatures, however, the use of a mesh of the size together with a reasonable estimate of the phonon widths leads to the response functions having a spiky appearance. This may possibly indicate that at the lowest temperatures (i.e., below 300 K) the size of sample we have used cannot be considered "large," but at the higher temperatures it should be satisfactory.

We have calculated the scattering function both with and without the inclusion of the polarization mixing terms. For the particular examples shown here the effect is not large, but we hope to consider other examples in later work. We whall refer to calculations based on Eqs. (27)–(29) as anharmonic perturbation theory (APT).

We wish to stress that our definition of the one-phonon approximation ensures that $S_1(\vec{Q},\omega)/[Qd(Q)]^2$ is independent of Q. Hence a study of the quantity $S(\vec{Q},\omega)/[Qd(Q)]^2$ provides a simple means of monitoring the effects of higher-order terms in the phonon expansion (18), i.e., effects due to interference and multiphonon processes.

IV. MOLECULAR-DYNAMICS RESULTS

A. Equilibrium properties

Results on selected equilibrium properties (pressure, internal energy and rms displacements) are shown in Table II. In the case of thermodynamics properties there is fair agreement with previous Monte Carlo calculations[7] based on the same potential model, and also with experimental measurements. The highest temperature studied is actually above the melting point (at atmospheric pressure) of real NaCl. In the simulation, however, the crystal is apparently still stable. In particular, the pressure is positive and the quasiharmonic normal-mode frequencies are all real. The rms amplitude of vibration of the ions is very large at high temperatures, but is everywhere in fair agreement with the results of quasiharmonic lattice-dynamical calculations. It is evident from these results that anharmonicity makes only a small contribution to the mean-square displacement of the ions. However, the molecular-dynamics results, which contain all anharmonic effects, are systematically larger than those obtained from the quasiharmonic calculations. In the harmonic approximation the mean-square displacement is given by

$$\langle u_\kappa^2\rangle = \frac{kT}{Nm_\kappa}\sum_{\vec{q}\lambda}\left(\frac{|\vec{e}^{\,\kappa}_{\vec{q}\lambda}|}{\omega_{\vec{q}\lambda}}\right)^2 . \tag{30}$$

The dominant contribution to this expression comes from the lowest frequency branch of the dispersion curve, i.e., the TA mode. In this branch the anharmonic frequency shift is negative, as we shall see below. Thus the effect of anharmonicity is to increase the mean-square displacement.

B. Temperature dependence of the phonons

In all four molecular-dynamics runs, $S(\vec{Q},\omega)$ was calculated for the following values of the wave vector $\vec{Q}$ (expressed in units of $2\pi/3a$):

(1, 0, 0), (1, 1, 0), (1, 1, 1), (2, 0, 0),

(2, 2, 0), (3, 0, 0), (3, 3, 0), (0, 2, 4),

(0, 1, 5), (3, 3, 3), (0, 6, 1), (0, 6, 2),

(0, 6, 3), (7, 0, 0), (9, 9, 9), (10, 10, 10).

Additionally, at 302 and 954 K, we studied the points (2, 2, 2), (4, 4, 4), and (8, 8, 8), while at 80 and 1153 K we studied also the points (4, 2, 2), and (6, 0, 0).

Figure 1 shows the calculated phonon frequencies along the ⟨100⟩ direction at 80 K. For each of the four modes, the molecular dynamics results agree well with those of quasiharmonic theory. On the other hand, agreement with the experimentally measured[3] phonons is not especially good, the main failure of the simulation being the overestimation of the frequencies of the LO phonons. This is a straightforward consequence of the neglect of ionic polarization in our model, as we shall show in Sec. IV F. It should be noted that our use of a finite system means that we are unable to observe the $q=0$ LO phonon.

At sufficiently low temperatures the phonons are all very well defined, the only feature of the spectrum being a sharp peak close to the quasiharmonic frequency. With increasing temperature the phonons shift and broaden, a behavior illustrated in Figs. 2 and 3 for selected longitudinal phonons propagating in the ⟨100⟩ direction. Figure 2 shows the neutron cross section for the zone-boundary LA phonon at the lowest and highest temperatures studied. The observed frequency shift is large, approximately 20%, but is nonetheless considerably smaller than that predicted by quasiharmonic theory. The other obvious effect of increasing the temperature is the growth in intensity at low frequencies. Figure 3 shows the neutron

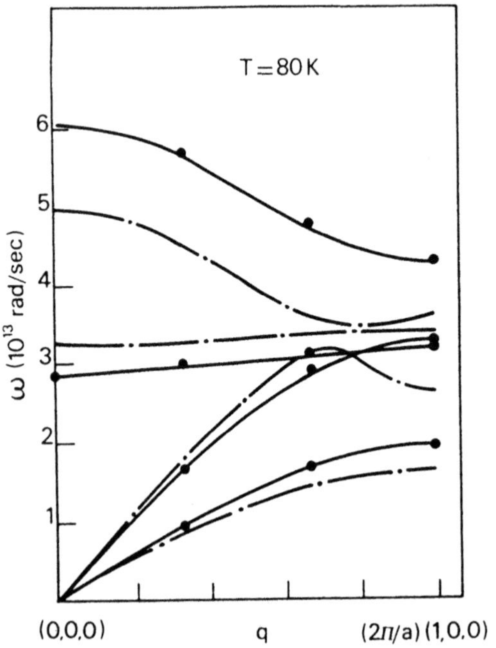

FIG. 1. Phonon frequencies for the ⟨100⟩ direction at 80 K. The dots are molecular-dynamics results and the solid lines give the quasiharmonic results for the same model. The dash-dot lines represent experimental neutron scattering data (Ref. 3).

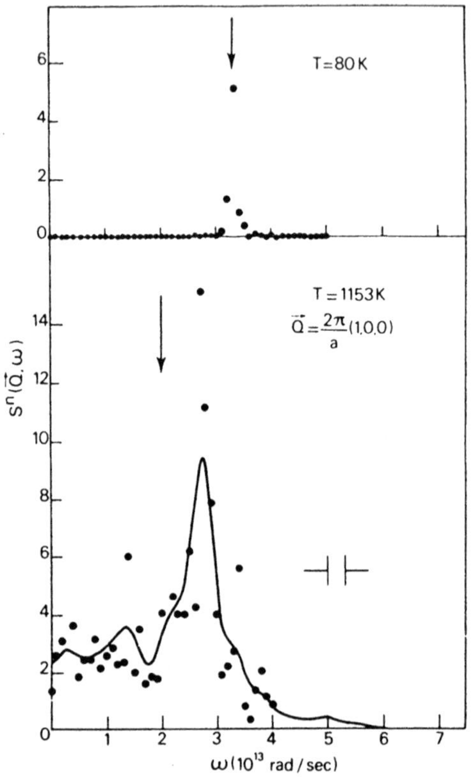

FIG. 2. Zone-boundary LA phonon (in arbitrary units) at two temperatures. The dots are molecular dynamics results and the arrows mark the location of the quasiharmonic frequencies. The curve represents the results of the smoothing procedure described in the text.

cross section at 302 and 1153 K for the smallest wave vector we can study in our periodic system of 216 ions, i.e., $\vec{Q}=(2\pi/3a)(1,0,0)$. The main peak is the LA phonon. This remains well defined as the temperature increases, and simultaneously a peak centered at $\omega=0$ develops, so that the spectrum qualitatively resembles a Rayleigh-Brillouin triplet. The weak response at high frequency is shown in the inset diagrams; the peak near 5×10^{13} rad sec^{-1} corresponds to the LO mode. The dashed curves in the insets show the charge-weighted spectra, plotted on a different relative scale; this comparison between $S^n(\vec{Q},\omega)$ and $S^z(\vec{Q},\omega)$ aids the identification of the optic-mode frequency at high temperatures. As is well known, the relative intensity of the acoustic and optic peaks for the same $\vec{q}$ can be very different in different Brillouin zones, and this behavior can also be exploited in identifying a mode frequency.

In Fig. 3(a) the dots on the figure are the direct output from the molecular-dynamics calculations. At 954 and 1153 K the direct output is somewhat noisy, particularly at large momentum transfer, because our method of computing $S(\vec{Q},\omega)$ necessarily includes the long-time statistical errors in $S(\vec{Q},t)$. The noise level can be reduced by convoluting the calculated spectrum with a Gaussian filter of given width, say δ. This in turn is equivalent to truncating correlations in $\rho_{\vec{Q}}(t)$ beyond a time $\tau\simeq2/\delta$. The solid lines in Figs. 2 and 3 show results obtained by this procedure; the width used for the filter is shown on each graph and is usually equivalent to 600 time steps. The correctness of the method can be checked by transforming the raw $S(\vec{Q},\omega)$ data to yield $S(\vec{Q},t)$, truncating the long-time tail at the appropriate value of τ, and the transforming back to obtain the smoothed $S(\vec{Q},\omega)$.

C. TO ($q=0$) phonon

The $q=0$ TO phonon is of particular interest because it can be studied by infrared spectroscopic methods as well as in neutron scattering experiments. For this reason it is the only phonon on which experimental data are available at relatively high temperatures (up to approximately 700 K). In our calculations we have studied this particular

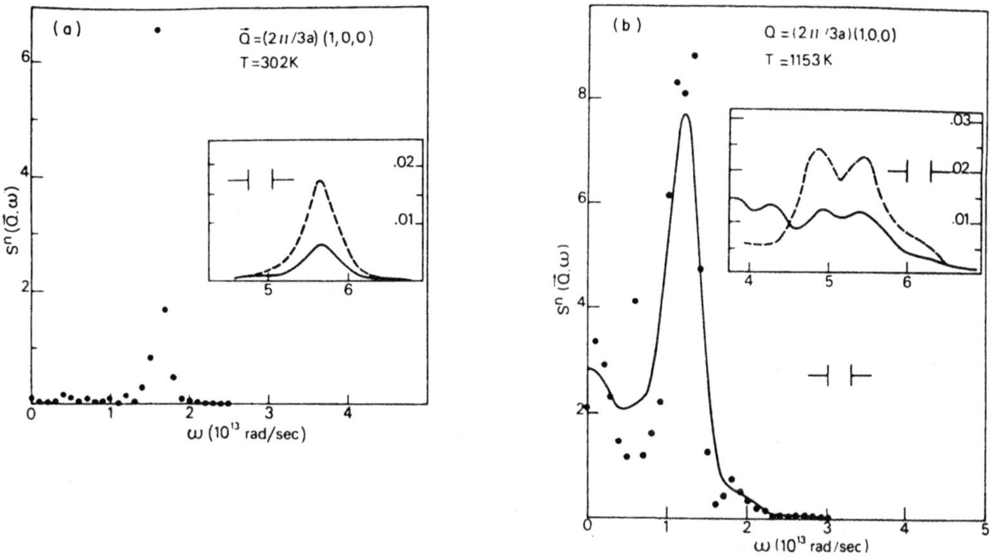

FIG. 3. (a) Neutron cross-section (in arbitrary units) for the smallest accessible wave vector. The curves in the inset show the high-frequency response in $S^n(\vec{Q}, \omega)$, plotted on the same relative scale (full line), and in $S^z(\vec{Q}, \omega)$, plotted on an arbitrary relative scale (dash line). (b) As (a) but at a higher temperature. The full curve in the main figure shows the effect of smoothing the molecular dynamics results.

phonon for a momentum transfer corresponding to $\vec{Q} = (2\pi/3a)(3, 3, 3)$ at temperatures of 302, 954, and 1153 K. At the lowest temperature the spectrum is sharp, as for all the phonons we have measured. At 954 K the peak is still clearly defined but sits now on a broad background. There is also some evidence for a weak secondary peak at approximately one-half the frequency of the main peak. At 1153 K the subsidiary peak has grown somewhat in intensity and a further satellite line can be seen at $\frac{3}{2}$ the main peak frequency. It is possible, of course, that the detailed structure of the spectrum is related to the small size of our system, but it could also represent a contribution from real multiphonon effects. The frequencies of the main peaks and their approximate full widths at half maximum height (with due allowance made for our resolution) are compared with experimental infrared results[16] in Fig. 4. The absolute values of the frequencies are too low, but the temperature dependence and the width are both in fairly good accord with the experimental data. The low value of the TO frequency reflects the inadequacy of the force constants in the Tosi-Fumi model, the question of polarization being of much less importance for this mode.

D. $\vec{Q}$ dependence and the one-phonon approximation

One of the original aims of this work was to investigate the $\vec{Q}$ dependence of the phonon spectra in an attempt to isolate the contribution from multiphonon processes. In Fig. 5 we show the charge-weighted spectrum $S^z(\vec{Q}, \omega)$ and its one-phonon approximation $S_1^z(\vec{Q}, \omega)$ for the LO phonon $\vec{Q} = (2\pi/3a) \times (7, 0, 0)$ at 302 K. In the one-phonon approximation the peak is asymmetric, whereas the full $S^z(\vec{Q}, \omega)$ is essentially symmetric. The influence of the interference term in the phonon expansion (18) is clearly visible, even at this low temperature and low value of $\vec{Q}$. The interference effect in lowest order contributes a term proportional to Q^3 which changes sign as the phonon wave vector $\vec{q}$ ($= \vec{Q} \pm \vec{g}$) crosses a Bragg vector $\vec{g}$. In the case illustrated, the result is to move intensity from the left-hand side to the right-hand side of the peak, incidentally making the peak more symmetric. At higher temperatures the effect is even more marked, as Fig. 6 shows. The one-phonon approximation for $\vec{Q} = (2\pi/3a)(7, 0, 0)$ is virtually identical with the full result for $\vec{Q} = (2\pi/3a)(1, 0, 0)$ when due allowance is made for the factor $[Qd(Q)]^2$. Thus Figs. 5 and 6 display directly the $\vec{Q}$ dependence of this particular phonon. We see that at 954 K the center of gravity of the spectrum is noticeably shifted by the interference effect. The increase in temperature has led to a substantial growth in the multiphonon background, but the peak remains easily identifiable.

In the left-hand part of Fig. 6 we show the $\vec{Q}$ dependence of the corresponding LA phonon at 954 K.

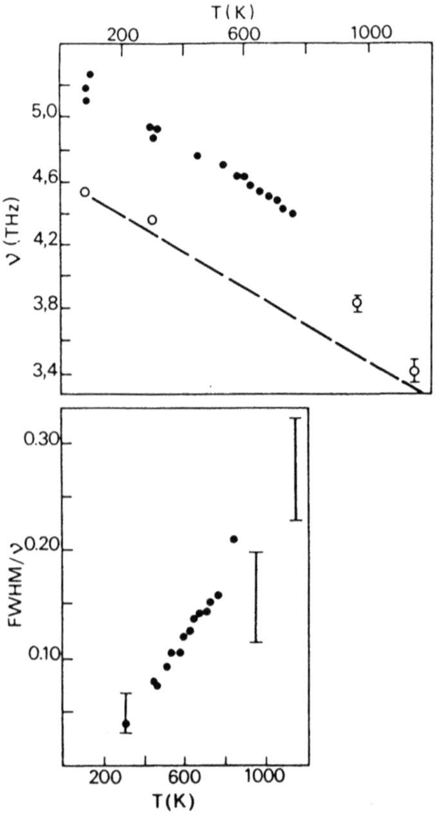

FIG. 4. $q=0$ TO mode. The upper half shows the peak frequency ν as function of temperature; circles are molecular-dynamics results, dots are experimental data (Ref. 15), and the dash line gives the predictions of quasiharmonic theory. The lower half shows the full width at half-maximum relative to ν; molecular-dynamics results are shown as error bars.

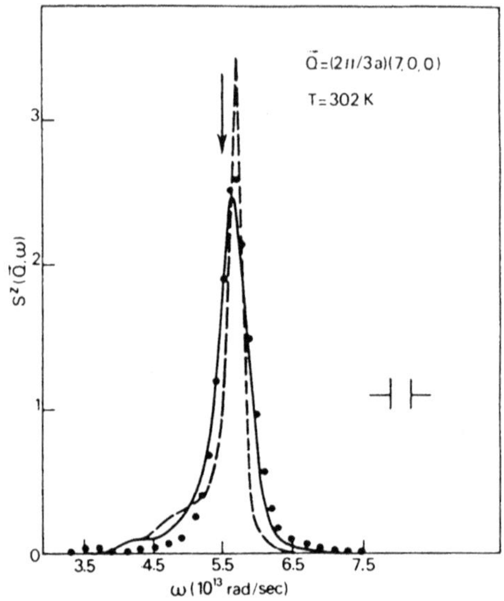

FIG. 5. Charge-weighted spectrum (in arbitrary units) for $\vec{Q}=(2\pi/3a)(7,0,0)$ at 302 K. The dots are the molecular-dynamics results for $S^z(\vec{Q},\omega)$ and the line shows the corresponding one-phonon approximation. The dashed line gives the results for $S_1^z(\vec{Q},\omega)$ from perturbation theory and the arrow locates the quasiharmonic frequency.

There is almost no evidence here of any interference effect, but the increase in the multiphonon background for $\vec{Q}=(2\pi/3a)(7,0,0)$ has caused a shoulder to appear on the side of the main peak. The one-phonon approximation has a peak at $\omega=0$, which is little different from that occurring in the full $S(\vec{Q},\omega)$ for the point $\vec{Q}=(2\pi/3a)(1,0,0)$ (shown as open circles). Thus the central peak is clearly not the result of multiphonon processes, lending support to our earlier suggestion that it is the remnant of a Rayleigh-type line.

Figures 7(a) and 7(b) show some of the phonons studied by Cowley and Buyers in their classic work[12] on the interference effect in KBr. To illustrate the importance of the effect we have plotted the neutron cross section $S^n(\vec{Q},\omega)$ and its one-phonon approximation for values of $\vec{Q}$ on either side of the Bragg vectors $\vec{g}=(2\pi/3a)(3,3,3)$ and $\vec{g}=(2\pi/3a)(9,9,9)$. For the smaller Bragg vector,

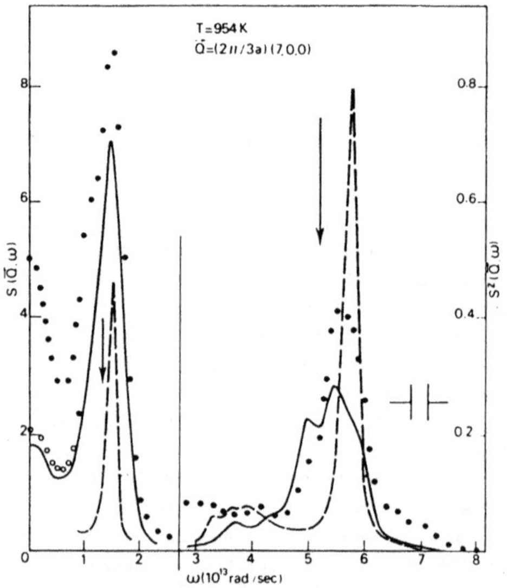

FIG. 6. Right-hand part same as in Fig. 5, but at a higher temperature. The left-hand part shows $S(\vec{Q},\omega)$ for the same wave vector $[\vec{Q}=(2\pi/3a)(7,0,0)]$. Results for $\vec{Q}=(2\pi/3a)(1,0,0)$, scaled by the factor $[Qd(Q)]^2$, are shown as open circles. For the sake of clarity, the perturbation results have been reduced in intensity by a factor of 4.

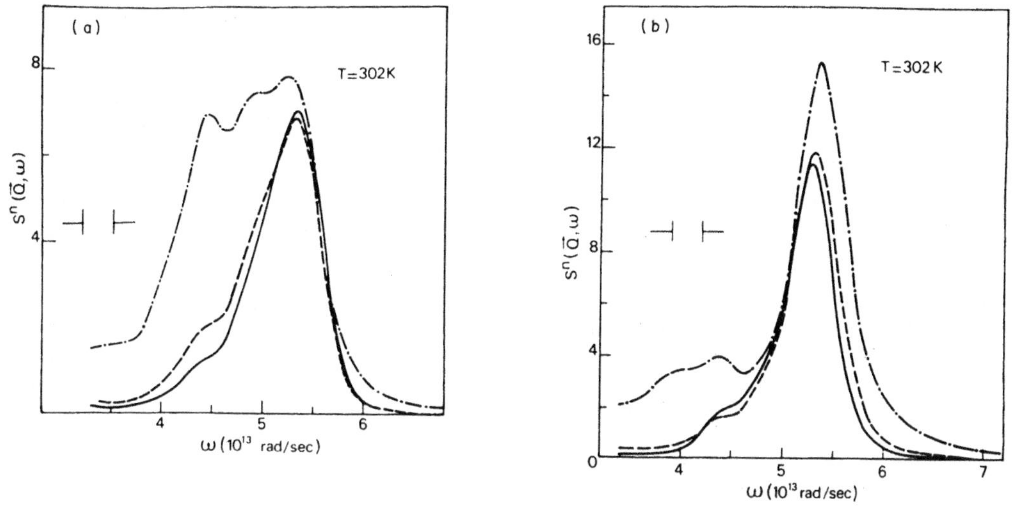

FIG. 7. (a) The interference effect in the neutron cross section at 302 K. The full curve is $S_1^n(\vec{Q},\omega)$ and the broken curves are $S^n(\vec{Q},\omega)$ for $\vec{Q}=(2\pi/3a)(2,2,2)$ (dashes) and $\vec{Q}=(2\pi/3a)(8,8,8)$ (dash-dot). The spectra have in all cases been divided by $[Qd(Q)]^2$. (b) As (a) but for wave vectors $\vec{Q}=(2\pi/3a)(4,4,4)$ (dashes) and $\vec{Q}=(2\pi/3a)(10,10,10)$ (dash-dot).

where the spectra are not complicated by multiphonon effects, the interference effect is clearly visible. For the larger one, both the multiphonon and interference effects have grown in magnitude and are not so readily separable. However, there is clearly a large enhancement of intensity on the low-frequency side of the spectrum for $\vec{Q}=(2\pi/3a)\times(8,8,8)$ and on the high-frequency side for $\vec{Q}=(2\pi/3a)(10,10,10)$. The relative effects seen here are not symmetrical. This arises from the necessity of both spectra (when divided by Q^2) having the same second moment; clearly the transfer of intensity from high to low frequency must be accompanied by a correspondingly greater enhancement of the spectrum relative to $S_1^n(\vec{Q},\omega)$ than in the reverse case.

E. Comparison with perturbation theory

Figures 5 and 6 show a comparison between the molecular-dynamics results for $S_1^s(\vec{Q},\omega)$ at the smallest accessible wave vector, i.e., $\vec{Q}=(2\pi/3a)\times(1,0,0)$, and the predictions of anharmonic perturbation theory. At 302 K (Fig. 5) there is reasonably good agreement. Both calculations give rise to a peak which is asymmetric (in the same sense) and shifted to a frequency higher than the quasiharmonic result. (Note that both spectra have been convoluted with the same smoothing function.) The simulation gives a broader peak and the shift from the quasiharmonic frequency is somewhat smaller. However, the perturbation theory used here is limited by the assumption that one phonon decays into two others, and therefore yields a lower limit on the width of the phonon or, equally, an upper limit on the lifetime.

At 954 K (Fig. 6) the molecular dynamics result for the same phonon is much broader and shifted noticeably less from the quasiharmonic frequency than in the perturbation calculations. In the case of the wave vector $\vec{Q}=(2\pi/3a)(7,0,0)$ there is somewhat better agreement with the full $S^s(\vec{Q},\omega)$ than with the one-phonon approximation. This, of course, is fortuitous, because the full result in-

TABLE III. Phonon frequencies, in units of 10^{13} rad $\sec^{-1}$ for $(\vec{Q})=(2\pi/3a)(3,0,0)$.

$T(K)$		TA	LA	TO	LO
80.3	QH	1.93	3.28	3.14	4.28
	APT	1.92	3.29	3.15	4.26
	MD [a]	1.90	3.30	3.15	4.30
301.7	QH	1.92	3.08	2.94	4.23
	APT	1.89	3.16	3.02	4.18
	MD	1.90	3.15	3.00	4.20
500.0	QH	1.92	2.91	2.78	4.19
	APT	1.85	3.05	2.91	4.14
700.0	QH	1.90	2.73	2.62	4.15
	APT	1.83	2.98	2.81	4.09
953.8	QH	1.89	2.54	2.45	4.11
	APT	1.82	2.9 [b]	2.70	4.04
	MD	1.75	2.85	2.75	4.05

[a] MD, molecular-dynamics results, with typical statistical uncertainty of ± 0.05.
[b] Very broad.

cludes a significant contribution from the interference effect. It is possible to extend the perturbation calculations to take account of the interference term, but we have not so far attempted this. The satellite peak in the molecular dynamics result is less separated from the main peak than in the perturbation calculation. Qualitatively this can be understood as resulting from the neglect of phonon renormalization in the basis set used in the perturbation theory. In the simulation, the phonon linewidth arises from interactions between phonons which already have a finite width and shift compared with the harmonic approximation. Hence any fine structure in the phonon spectral function will necessarily be shifted (relative to the main peak) and broadened when compared with that found in a simple perturbation calculation.

In Table III we make detailed comparison between the molecular-dynamics results and the theoretical predictions for the frequencies of the zone-boundary ⟨100⟩ phonons. Since zone-boundary phonons display no interference effect, the comparison is a meaningful one, even though the perturbation calculation of the neutron cross section is limited to the one-phonon approximation. For this particular value of $\vec{Q}$, the TA and LO phonons show only small negative shifts from the quasiharmonic results, whereas the LA and TO shifts are large and positive. (The shift in the LA phonon and its temperature dependence are also illustrated in Fig. 2.) Such a variety of behavior clearly constitutes a severe test of a theory of anharmonic effects. In fact, as the table shows, agreement between the molecular-dynamics and perturbation theory calculations is excellent at 302 K and remains good even at 954 K. Thereafter the perturbation theory rapidly breaks down and no useful comparison can be made at 1153 K. Nonetheless, at least insofar as the prediction of the phonon frequencies are concerned, the perturbation method used here remains adequate up to temperatures roughly 80% of melting. The range of validity of the theory is therefore substantially greater than in comparable calculations on rare-gas solids.[17]

F. Effect of ionic polarization

As we have already stressed, the Tosi-Fumi model takes no account of ionic polarization and for that reason is unable to account quantitatively for details of the dynamical properties of NaCl. The effect of polarization on the interionic forces can be incorporated explicitly in a molecular-dynamics simulation, though at the cost of a considerable increase in computing time. The technical problems involved in such a calculation have been discussed in detail in a recent paper,[18] both for a model consisting of point polarizable ions and for a simple version of the shell model. In the shell model the total ionic charge is assumed to be divided between a core and a shell, and polarization corresponds to a bodily displacement of the shell with respect to the core. The shell, which represents the outer electron cloud, is assumed to be bound to the core by a harmonic potential and the short-range repulsive interactions are assumed to act between the shells. In the molecular-dynamics "experiment" the equations of motion of the cores are solved in the usual way, whereas the shells are assumed to adjust themselves instantaneously so as to minimize the total potential energy of the system. The second step is considerably more complicated to carry out than the first and requires the use of iterative methods. Use of the shell model in molecular-dynamics calculations has also been discussed by Dixon and Sangster.[19]

In Fig. 8 we show a typical result obtained from a shell-model calculation in which polarization is superimposed on the Tosi-Fumi potential for NaCl. Only the Cl^- is treated as polarizable, the parameters of the model (the charge on the shell and the spring constant in the shell-core potential) being taken from the work of Sangster.[20] The phonon shown corresponds to the LO mode of smallest wave vector. There is a large negative shift with respect to the rigid-ion result, bringing the phonon frequency into good agreement with the experimental value (see Fig. 1). Such effects are, of course, already well known from lattice-dynam-

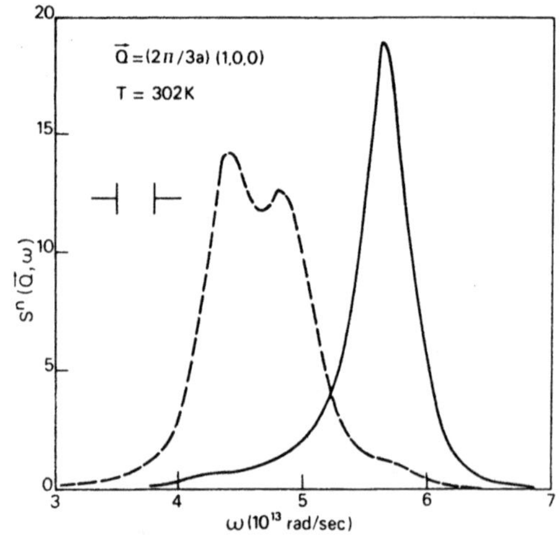

FIG. 8. Effect of polarization on the charge-weighted spectrum (in arbitrary units) for $\vec{Q} = (2\pi/3a)(1,0,0)$ at 302 K. The full curve shows results from the rigid ion "experiment," the dash curve those from a shell-model simulation.

ical calculations; we include the comparison here primarily to demonstrate that the molecular-dynamics simulation can be made substantially more realistic if this is particularly required. The broadening of the spectrum and the appearance of a double-peaked response is qualitatively similar to the behavior found by Cowley[10] in a perturbation calculation for the $q=0$ LO mode. Since polarization acts in such a way as to reduce the phonon frequencies (though the effect in other branches is less pronounced than for the LO mode), its inclusion necessarily leads to an increase in the amplitudes of vibration. This effect turns out to be very large, as shown by the results given in Table IV.

TABLE IV. Root-mean-square displacements of the ions. d $(=\frac{1}{2}a)$ is the nearest-neighbor separation.

T (K)	$100\langle u_+^2\rangle^{1/2}/d$ MD[a]	QH	$100\langle u_-^2\rangle^{1/2}/d$ MD	QH
(i) Rigid ions				
80.3	4.5	4.0	4.0	3.8
301.7	8.2	7.8	8.0	7.6
953.8	17.0	15.4	15.8	14.8
1153.0	20.6	19.0	19.5	18.5
(ii) Polarizable ions[b]				
306.3	12.6		11.0	
1153.5	24.3		23.6	

[a] MD, molecular-dynamics results.
[b] Shell-model results from Ref. 18.

V. CONCLUSIONS

The results reported here demonstrate that computer simulation can play a valuable role in analyzing the lattice vibrations of a simple ionic crystal such as NaCl. The fact that the one-phonon approximation to the dynamic structure factor can be evaluated in parallel with the calculation of the complete spectrum means that anharmonic effects can readily be isolated and their importance assessed. Our main quantitative result is that a simple perturbation treatment is adequate for the prediction of phonon frequencies at temperatures up to 80% of the melting temperature; on the other hand, the phonon linewidths are underestimated, particularly at high temperatures. In making this comparison it is sufficient to work with a much oversimplified potential model and most of our calculations have, in fact, been made for a system of rigid ions. The effects of polarization can be included, however, though only at considerable added expense in computing time, and fair agreement with experimental results can then be achieved. However, the qualitative features of our results are likely to be insensitive to details of the interionic potential, and hence should carry over to other ionic systems.

ACKNOWLEDGMENTS

We are grateful to C. Moser for providing the computing time which made the molecular-dynamics calculations possible. The work has been supported, in part, by a grant from the United Kingdom Science Research Council.

*Permanent address: G.N.S.M., Consiglio Nazionale delle Richerche, Instituto di Fisica, Università di Roma, Roma, Italy.
†Permanent address: Chemistry Division, National Research Council of Canada, Ottawa K1A 0R6, Canada.

[1] E. W. Kellermann, Philos. Trans. R. Soc. Lond. 238, 63 (1940).
[2] (a) F. G. Fumi and M. P. Tosi, J. Phys. Chem. Solids 25, 31 (1964); (b) M. P. Tosi and F. G. Fumi, *ibid.* 25, 45 (1964).
[3] G. Raunio, L. Almqvist, and R. Stedman, Phys. Rev. 178, 1496 (1969).
[4] B. G. Dick and A. W. Overhauser, Phys. Rev. 112, 90 (1958).
[5] J. R. Hardy, Philos. Mag. 7, 315 (1962).
[6] U. Schröder, Solid State Commun. 4, 347 (1966).
[7] D. J. Adams and I. R. McDonald, J. Phys. C 7, 2761 (1974); 8, 2198 (1975).
[8] (a) R. A. Cowley, Adv. Phys. 12, 421 (1963); (b) R. A. Cowley, Rep. Prog. Phys. 31, 123 (1968).
[9] E. R. Cowley, J. Phys. C 4, 988 (1971).
[10] E. R. Cowley, J. Phys. C 5, 1345 (1972).
[11] H. R. Glyde and M. L. Klein, Crit. Rev. Solid State Sci. 2, 181 (1971).
[12] (a) R. A. Cowley and W. J. L. Buyers, J. Phys. C 2, 2262 (1969); (b) R. A. Cowley, E. C. Svensson, and W. J. L. Buyers, Phys. Rev. Lett. 23, 525 (1969).
[13] H. Horner, Phys. Rev. Lett. 29, 556 (1972).
[14] D. Levesque, L. Verlet, and J. Kürkijarvi, Phys. Rev. A 7, 1690 (1973).
[15] V. Ambegoakar, J. Conway, and G. Baym, in *Lattice Dynamics*, edited by R. F. Wallis (Pergamon, New York, 1965), p. 261.
[16] J. E. Mooij, W. B. Van de Bunt, and J. E. Schrijvers, Phys. Lett. A 28, 573 (1969).
[17] J. P. Hansen and M. L. Klein, Phys. Rev. B 13, 878 (1976).
[18] G. Jacucci, I. R. McDonald, and A. Rahman, Phys. Rev. A 13, 1581 (1976).
[19] M. Dixon and M. J. L. Sangster, J. Phys. C 8, L8 (1975); 9, L5 (1976).
[20] M. J. L. Sangster, J. Phys. Chem. Solids 34, 355 (1973).

Molecular dynamics study of solid β-N_2

Michael L. Klein

Chemistry Division, National Research Council of Canada, Ottawa, Canada K1A OR6

Dominique Lévesque and Jean-Jacques Weis

Laboratoire de Physique Théorique et Hautes Energies, Université Paris-Sud, 91405 Orsay, France

(Received 22 October 1980; accepted 4 November 1980)

A computer simulation study of orientationally disordered solid β-N_2 is reported for the state condition $V = 26.1$ cm^3/mol, $T = 47$ K. We utilized an hexagonal system of 288 molecules interacting via a Raich–Gillis intermolecular potential. Particular attention is given to the dynamical structure factor $S(\mathbf{Q},\omega)$ and its dependence on the momentum transfer $\hbar\mathbf{Q}$. In this regard the present model agrees considerably better with experimental observations than previous work based upon a simple atom–atom potential with no electrostatic quadrupole–quadrupole interaction.

I. INTRODUCTION

At pressures below about 50 kbar liquid nitrogen freezes into an orientationally disordered structure (β-N_2) in which the molecular centers of mass form an hexagonal close packed lattice.

The properties of β-N_2 are well characterized and the literature up to 1976 has been reviewed by Scott.[1] In the same year the temperature and pressure dependence of the Raman spectra were reported[2] and Brillouin spectroscopy was used to measure the elastic constants near the triple point.[3] Coherent neutron scattering has been used to probe both the crystal dynamics[4] and its structure[5,6] as a function of temperature and pressure. Far infrared spectra have been reported[7] and the crystal structure has also been studied by x rays using a high pressure diamond-anvil cell.[8]

On the theoretical side a molecular dynamics (MD) calculation was carried out[9] using an atom–atom potential whose parameters were fitted to liquid state properties.[10] Many of the experimental findings could be understood on the basis of this calculation but quantitative agreement was often lacking. For example, comparison with the Brillouin scattering data[3] revealed that for the waves that propagate along the c axis, the theoretical value for the ratio of longitudinal to transverse sound speed (v_l/v_t) was 50% too large. The coherent inelastic neutron scattering data showed clear evidence not only for rotational diffusion but also for hindered librational motion. The latter features were less prominent in the MD data based upon the simple atom–atom model with no electrostatic quadrupole–quadrupole (EQQ) interactions. In view of the continuing interest in the properties of β-N_2[5-8] the present study addresses itself to these discrepancies between theory and experiment. We have therefore repeated our earlier MD calculations but this time we employ a more realistic intermolecular force model namely a Raich–Gillis potential.[11] This model and some details of the calculations are given next. Then in Sec. III we discuss the inelastic neutron scattering data. We will see that in many ways this more realistic intermolecular potential represents an improvement over our earlier work.

II. THE MODEL

We have employed the intermolecular potential number 5 from the work of Raich and Gillis.[11]

In this model the long range dispersion interaction between the molecular centers of mass has the form

$$v_{\rm disp}(R) = -\sum_{n=0}^{2} C_{2n+6}/(\delta + r^{2n+6}) ,$$

where $r = R/R_m$, $C_6 = 1.4647\epsilon$, $C_8 = 0.19293\epsilon$, $C_{10} = 0.013122\epsilon$, $\delta = 0.01$, $\epsilon/k_B = 85.48$ K, and $R_m = 3.9909$ Å.

The short-range valence interactions are written as the sum of interactions between three sites on each molecule. One of these, with hard core radius $R^c = 1.8615$ Å is the center of mass and the other two with $R^c = 1.6924$ Å are the atomic positions. The bond length was taken to be 1.11 Å. In detail

$$v_{\rm val}(R_{ij}) = \sum_{i>j}\sum_{n=0}^{5} A_n(r-1)^n \exp[-\alpha(R_{ij} - R_i^c - R_j^c)] ,$$

where $A_0 = 0.20188\epsilon$, $A_1 = -2.5069\epsilon$, $A_2 = -5.2835\epsilon$, $A_3 = 11.2490\epsilon$, $A_4 = 46.6470\epsilon$, $A_5 = 36.0040\epsilon$, $\alpha = 3.2$ Å^{-1}, and $r = R_{ij}/(R_i^c + R_j^c)$. The electrostatic interactions were represented by interacting fractional charges

$$v_{\rm eqq} = \sum_{i>j} q_i q_j / R_{ij} .$$

The charge q_i is $-Z$ for the N atoms and $+2Z$ for the center of mass. $Z = 0.481|e|$ reproduces the experimental quadrupole moment of -1.4 DÅ.

An investigation of the second virial coefficients of nitrogen using the family of Raich–Gillis potentials[12] showed that no model gave entirely satisfactory results. Model 1 performed best at high temperatures while model 5 was somewhat better at lower and intermediate temperatures. We have used the latter model in our work.

The MD calculation was carried out under the same state conditions as in our previous work ($V = 26.1$ cm^3/mol $T = 47$ K). We used an hexagonal system of 288 molecules and a time step of 5×10^{-15} s. The average energy per molecule was calculated to be -10.0ϵ and the

 0021-9606/81/042566-03$01.00

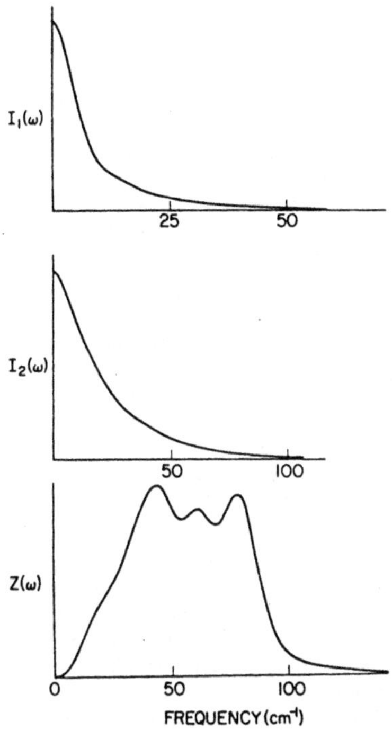

FIG. 1. Power spectra of the single molecule autocorrelation functions. From top to bottom the curves correspond to $P_1(\cos\theta)$, $P_2(\cos\theta)$, and the center of mass velocity v.

compressibility factor $pV/NkT = 9.5$. These values include interactions out to ~11.5 Å. The computed pressure is about 25% lower than the corresponding experimental value.[1] The mean square deviation of the molecular center of mass was found to be $\langle u_a^2\rangle = \langle u_c^2\rangle = 0.028$ Å^2 which is almost one half the value quoted in the recent neutron diffraction study at $p = 1.2$ kbar and $T = 40$ K (Ref. 5).

It is probable that these discrepancies are due, at least in part, to the neglect of quantum effects in our classical MD calculations. One structural quantity of interest[6] is $\langle\cos^n\theta\rangle$, where θ is the angle between the crystal c axis and the molecular symmetry axis. Within our statistical errors we find $\langle\cos^n\theta\rangle = 1/(n+1)$. As in our previous work, the molecules are undergoing rotational diffusion and are not precessing about the c axis.

III. CRYSTAL DYNAMICS

The MD calculations have been used to evaluate correlation functions whose power spectra can be related to experimental observations. Figure 1 shows the power spectra associated with the single molecule autocorrelation functions of $P_1(\cos\theta)$, $P_2(\cos\theta)$, and the center of mass velocity, v. The latter quantity, $Z(\omega)$ is the density of states of translational phonon in the solid. This is seen to have its main features around 40, 60, and 80 cm^{-1}, and these presumably correspond to the usual critical points that have become smoothed out partly because of our small system size and partly because of

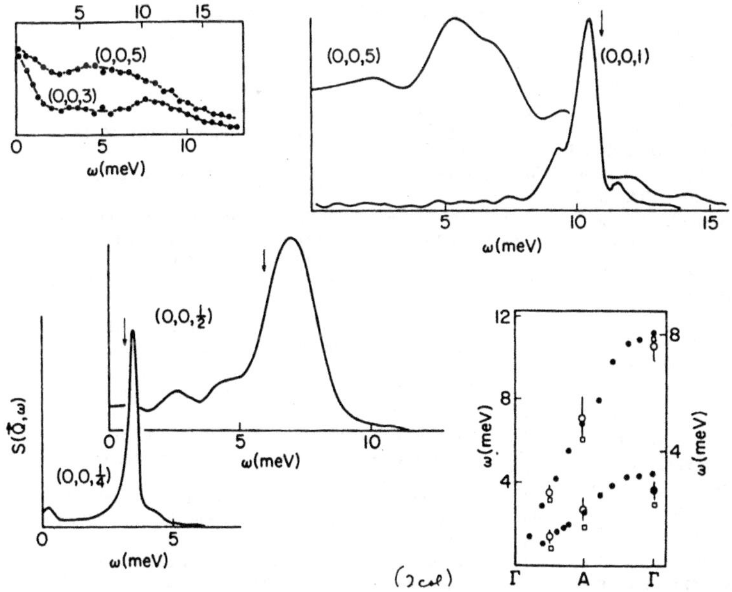

FIG. 2. The dynamical structure factor, $S(\mathbf{Q},\omega)$ for longitudinal phonons propagating along the c axis. The wave vector is indicated by $(0,0,\xi) = \mathbf{Q}c/2\pi$. The upper left shows the experimental data from Ref. 4. The lower right compares experimental and theoretical phonon dispersion curves for longitudinal and transverse modes (see text).

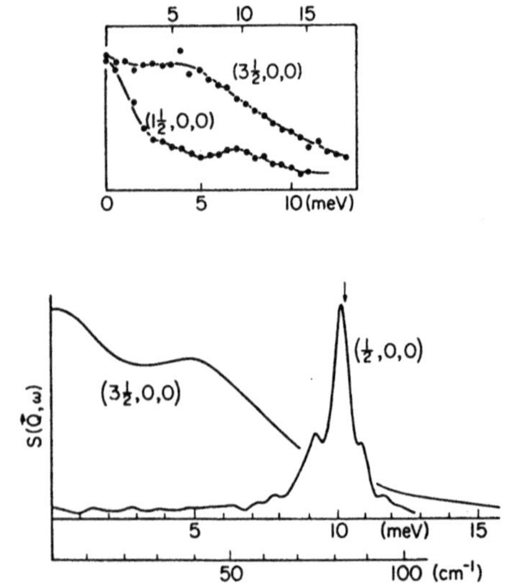

FIG. 3. The dynamical structure factor, $S(\mathbf{Q}, \omega)$ for the M point of the Brillouin zone. The wave vector is indicated by $(\xi, 0, 0) = \mathbf{Q}(a\sqrt{3}/4\pi)$. The upper part of the figure is taken from Ref. 4.

anharmonic effects. The single molecule contribution to the Raman spectrum $I_2(\omega)$ shows no structure but extends to around 100 cm^{-1}. We did not attempt a full calculation of the Raman scattering spectrum.[13]

Figure 2 shows the dynamical structure factor $S(\mathbf{Q}, \omega)$ for certain longitudinal phonons propagating along the c axis. The appropriate wave vector is indicated by $(0, 0, \xi) = cQ/2\pi$. For $\xi \leq 1$ these results are similar to our earlier work based upon an atom–atom potential.[9] The peak positions of the latter are indicated by arrows. In both cases the translational phonons are well defined with the broadest response at the A point. On increasing the momentum transfer $\hbar\mathbf{Q}$ the phonon peaks disappear [see the curve (0, 0, 5)] and an ill-defined peak around 6 meV appears superimposed upon a broader central peak due in part to rotational diffusion. The peak at 6 meV did not appear in our previous calculation (see Fig. 4 of Ref. 9). Inset in the upper left hand portion of Fig. 2 are experimental $S(\mathbf{Q}, \omega)$ measured at $V = 28.0$ cm^3/mol and $T = 36$ K (Ref. 4). The lower energy scale is the observed energy transfer $\hbar\omega$ while the upper scale is dilated to allow approximately for the difference in state conditions between our MD simulation and these measurements. The overall behavior of the experimental curves resembles quite closely the MD results. Inset in the lower right-hand portion of Fig. 2 are the dispersion curves for longitudinal and transverse phonons propagating along the c axis. The right-hand energy scale corresponds to that of the experimental measurements while the left-hand side corresponds to the MD calculations. The experimental data are shown by the full circles. The present MD data are shown as open circles with the error bars indicating the FWHM of the response functions. The squares are the old MD data using the atom–atom potential without the EQQ interaction.[9] A Γ-point value taken from the Raman measurements[2] is indicated by the large full circle. The main point to note is that the new MD data are in rather good agreement with the experimental data, particularly with regard to the transverse acoustic branch. Moreover, the ratio of v_l/v_t now agrees very well with that found in the Brillouin scattering measurements.[3]

Figure 3 shows another comparison between the experimental $S(\mathbf{Q}, \omega)$ as measured by coherent inelastic neutron scattering[4] and that calculated by MD using a Raich–Gillis potential. This time we show the longitudinal M point response. Once again the correspondence between the MD calculations and experiment is very good, once allowance has been made for the different energy scales. We have investigated $S(\mathbf{Q}, \omega)$ for values of $\mathbf{Q}$ other than shown in Figs. 2 and 3 and find results much the same as in our previous work.[9]

In summary, we have used molecular dynamics to calculate the dynamical structure factor $S(\mathbf{Q}, \omega)$ of solid β-N_2 for a Raich–Gillis potential. The main difference from previous work is that the restoring forces for the transverse acoustic response and the hindered librational motion are increased. The overall agreement between the experimental data and the MD calculations is now much better than found previously.

ACKNOWLEDGMENTS

We thank Maynard Clouter, Bill Daniels, Gerald Dolling, Harry Kiefte, Jørgen Kjems, and Brian Powell for helpful discussions and for supplying details of their various experiment before publication.

[1]T. A. Scott, Phys. Rep. C 27, 89 (1976).
[2]F. D. Medina and W. B. Daniels, J. Chem. Phys. **64**, 150 (1976).
[3]H. Kiefte and M. J. Clouter, J. Chem. Phys. 64, 1816 (1976).
[4]G. Dolling, *Dynamics of Molecular Crystals*, Conference on Neutron Scattering, Gatlinberg Tennessee, (1976), p. 263; also J. K. Kjems and G. Dolling (unpublished data).
[5]B. M. Powell, H. F. Nieman, and G. Dolling, Chem. Phys. Lett. **75**, 148 (1980).
[6]W. Press and A. Hüller, J. Chem. Phys. **68**, 4465 (1978).
[7]V. Buontempo, S. Cunsolo, P. Dore, and P. Maselli, Phys. Letters A 74, 113 (1979).
[8]D. Schiferl, D. T. Cromer, R. L. Mills, High Temps.-High Pressures **10**, 493 (1978).
[9]M. L. Klein and J.-J. Weis, J. Chem. Phys. **67**, 217 (1977).
[10]P. S. Y. Cheung and J. G. Powles, Mol. Phys. **30**, 921 (1975).
[11]J. C. Raich and N. S. Gillis, J. Chem. Phys. 66, 846 (1977).
[12]S. F. O'Shea (private communication); see also M. Whitmore and D. Goodings, Can. J. Phys. 58 (1980).
[13]J.-J. Weis and D. Levesque, Phys. Rev. A **13**, 450 (1976); D. Levesque and J.-J. Weis, Phys. Rev. A **12**, 2584 (1975).

RAPID COMMUNICATIONS

PHYSICAL REVIEW B VOLUME 32, NUMBER 6 15 SEPTEMBER 1985

Dispersion of surface phonons in xenon overlayers physisorbed on the Ag(111) surface

Gianni G. Cardini and Séamus F. O'Shea
Chemistry Department, University of Lethbridge, Lethbridge, Alberta T1K 3M4, Canada

Maurizio Marchese* and Michael L. Klein
Chemistry Division, National Research Council of Canada, Ottawa, Ontario K1A 0R6, Canada

(Received 19 June 1985)

Computer simulation and lattice dynamics have been used to investigate phonons propagating in monolayers, bilayers, and trilayers of xenon physisorbed on a Ag(111) surface. Agreement with the experimentally determined dispersion of the surface phonons has been achieved by the use of a realistic interatomic potential for the Xe adatoms and a new surface-adatom potential.

The inelastic scattering of helium atoms has proved to be a powerful method of studying surface phonons.[1] The extension of this technique to physisorbed overlayers[2] now offers an important complement to neutron scattering[3] as a means of probing the dynamics of such systems. Surface phonon dispersion curves have recently been measured for monolayer, bilayer, trilayer, and bulk xenon supported on a Ag(111) surface.[4] A dispersionless surface phonon branch for the monolayer was observed to evolve with increasing coverage to a Rayleigh wave behavior for the bulk Xe(111) surface.[4]

Previous experimental[5] and theoretical work[6] on such overlayers lead to a characterization of the relevant adatom-surface and adatom-adatom interactions. We have used a model based on this earlier work to calculate phonon dispersion curves which are compared with the He-beam data. Anticipating our results, we find that the existing surface-adatom potential[6] cannot account for the measured surface phonon dispersion curves. We are therefore led to propose a new softer adatom-surface potential which has enhanced dispersion interactions in the region of the potential minimum.[7] The revised model, coupled with a realistic Xe-Xe potential[8] which includes substrate mediation effects,[9,10] now gives a good account of the He-beam data.

We begin with a review of the salient facts. Experimental data indicate that a xenon monolayer is not in registry with a Ag(111) surface. At 20 K the monolayer Xe-Xe spacing (4.42 Å) is slightly larger than that of the bulk solid,[5] whereas the bilayer and trilayer have essentially the bulk value 4.33 Å.[4] At low temperatures the mean height of the monolayer above the Ag ion cores is 3.55 ± 0.10 Å.[11] The depth of the surface-adatom holding potential (172 meV) has been estimated[6] from the latent heat of adsorption of a monolayer (225 meV) by allowing for substrate mediated lateral interactions (65 meV) as well as the contribution from adsorption dipoles (6 meV), three-body forces (1.5 meV), and the zero-point energy (2.6 meV). At larger adatom-surface separations z the holding potential is known to have the form $-C_3/(z-d)^3$, where d is the height of the image plane above the Ag ion cores.[7] The original Debye-Waller factor measurements yield a value of 480 meV/Å^2 for the curvature at the minimum of the holding potential,[6] whereas a later determination[5] implies a curvature of 300 ± 60 meV/Å^2. We have used this information to parametrize surface-adatom potentials of the form $V(z) = A/z^m - C_3/(z-d)^3$, which, together with a realistic[8] substrate mediated Xe-Xe potential,[10] completes the characterization of the relevant interactions.

We have studied phonon vibrations in Xe overlayers physisorbed on a rigid Ag substrate using both conventional lattice dynamics[12] and molecular dynamics (MD) calculations.[13] In the latter case the Xe atoms were initially arranged (12×12) on triangular lattices with lattice constant $a = 4.33$ Å, periodic boundary conditions being used to simulate infinite layers. For the trilayer an hcp stacking sequence was adopted. The equations of motion were integrated using standard methods, the lateral potentials being truncated at 15 Å. An initial equilibration period was used to scale the temperature to the desired value (usually 20 K), and to relax the overlayer spacings normal to the surface. Phase-space trajectories were then collected for longer periods, typically 100 ps, in order to study the phonons.[14]

In the lattice dynamics calculations the required dynamical matrix was evaluated from the derivatives of the surface-adatom and adatom-adatom potential in standard fashion, with an option to optimize both the vertical and lateral lattice spacings if desired. If we neglect the dilation due to adsorption dipoles,[6] the zero-point energy, and three-body forces, the optimized monolayer spacing is 4.378 Å, which differs by only 1% from the experimental value.[5] We have ignored the contribution of these three effects to the phonon frequencies.

The modes propagating in the overlayers can be characterized by a two-dimensional wave vector $\bar{k} = (k_x, k_y)$, which is perpendicular to the surface normal $\bar{n}$; the plane defined by $\bar{k}$ and $\bar{n}$ is called the sagittal plane.[12] For a given $\bar{k}$ the modes are conveniently labeled by their dominant polarization characteristic: SH (shear horizontal) when the displacements are normal to the sagittal plane, $SP_\perp$ and $SP_\parallel$ when the displacements are in the sagittal plane, perpendicular and parallel to $\bar{k}$. With this classification in mind the appropriate correlation functions for probing the phonons via molecular dynamics have the form

$$F(\bar{k},t) = \langle f(\bar{k},t) f(-\bar{k},0) \rangle ,$$

with

$$f(\bar{k},t) = \sum_l W_l(t) \exp(i\bar{k}\cdot\bar{R}_l) ,$$

where $W(SP_\parallel) = (\bar{k}\cdot\bar{u}_l)$, $W(SP_\perp) = (\bar{n}\cdot\bar{u}_l)$, and $W(SH)$

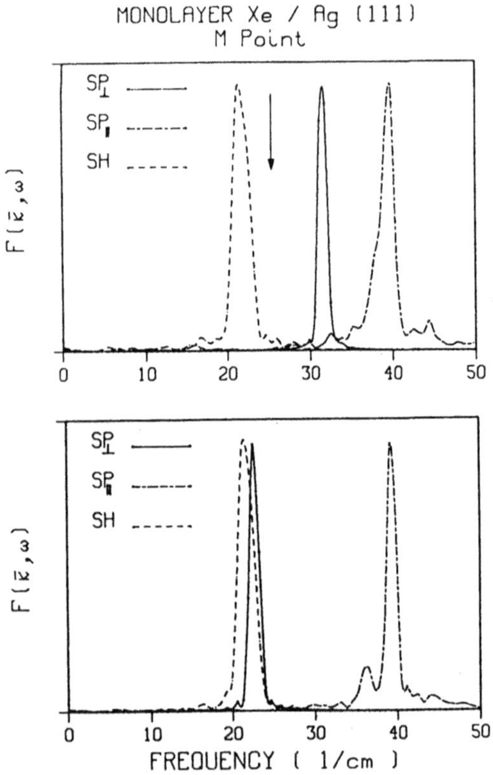

FIG. 1. Phonon response functions $F(\bar{k},\omega)$ in arbitrary units for the Brillouin zone M point calculated for a Xe monolayer on Ag(111) using molecular dynamics at 20 K. The measured $SP_\perp$ phonon frequency is 22.6 cm^{-1} (Ref. 4). The upper curves employ a surface-adatom potential fitted to the original Debye-Waller factor data (Ref. 6). The arrow indicates the $SP_\perp$ peak position based upon revised Debye-Waller data (Ref. 5). The lower curves are based on a surface-adatom potential fitted to the experimental $SP_\perp$ frequency.

$=(\bar{n}\cdot\bar{k})\cdot\bar{u}_l$. The summation index l runs over all the atoms in a given layer and $\bar{u}_l$ is the time-dependent displacement from the mean positions $\bar{R}_l$. The phonon frequencies are determined from the peaks in the response function $F(\bar{k},\omega)$ which has been evaluated using a direct method[14]

$$F(\bar{k},\omega)=\lim_{\tau\to\infty}|f(\bar{k},\omega)|^2/\tau ,$$

where

$$f(\bar{k},\omega)=\int_0^\tau \exp(i\omega t)f(\bar{k},\tau)d\tau .$$

In lattice dynamics the $\bar{k}$ vectors can be chosen at will but in MD our system size limits us to six distinct points on the branch $\Gamma\to M$, two for $M\to K$, and four for $K\to\Gamma$. Figure 1 shows $F(\bar{k},\omega)$ for the monolayer M-point phonons calculated using potentials of the type described above. A model fitted to the original Debye-Waller factor data, the adatom-surface separation,[11] and the holding potential (172 meV) has a $SP_\perp$ peak at 32 cm^{-1}, in poor agreement with

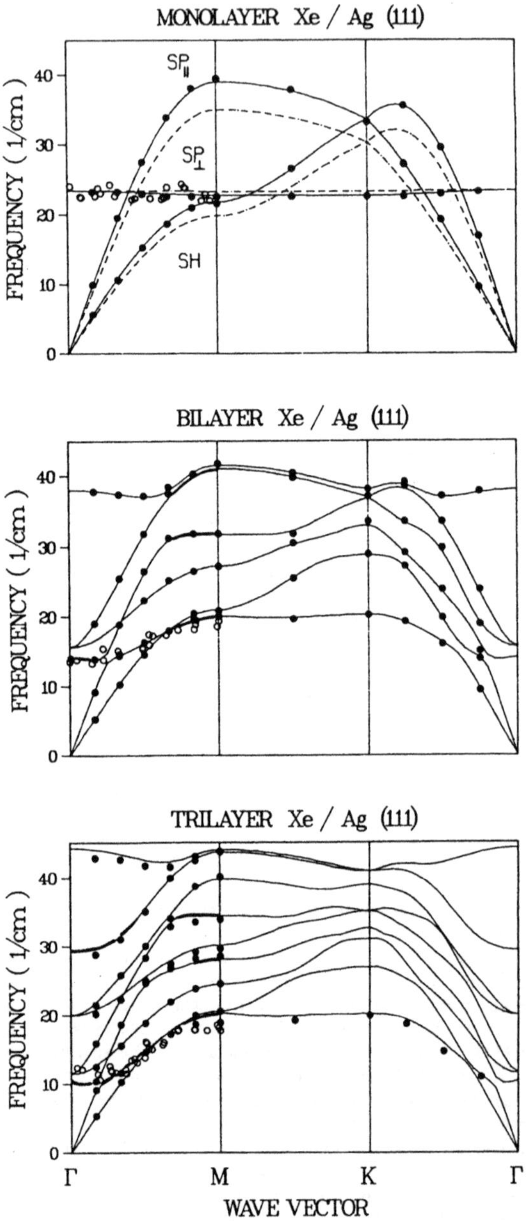

FIG. 2. Phonon dispersion curves for a monolayer, bilayer, and trilayer of Xe on Ag(111) based on a revised surface-adatom potential. The bold portions of the dispersion curves indicate that upper layer atoms have dominant motions with $SP_\perp$ polarization and are thus, in principle, probed by the atomic beam. The full circles indicate peaks in $F(\bar{k},\omega)$ derived from MD calculations at 20 K. The solid curves are lattice dynamics results with the same lateral spacing of 4.33 Å, whereas the dashed monolayer curves use a spacing of 4.378 Å. The bilayer curves were evaluated using the vertical spacings determined in the MD calculations, whereas for the monolayer and trilayer, statically optimized spacings were used. The softening of the trilayer M-point $SP_\perp$ phonon due to the vertical thermal expansion should be noted. The He atomic beam data of Ref. 4 are shown as open circles.

the experimental value[4] of 22.6 cm^{-1}. Use of the revised Debye-Waller factor data[5] yields a $SP_\perp$ peak at 25 ±3 cm^{-1}, which is consistent with the beam data. However, in view of the relatively large uncertainty in the revised Debye-Waller factor data[5] we prefer to constrain the surface-adatom potential to also fit the observed monlayer *M*-point frequency, rather than the Debye-Waller factor. The resulting potential ($m=8$, $d=1.80$, $A=12350$ eV Å,[8] $C_3=3.524$ eV Å^3) gave the dispersion curves for the monolayer, bilayer, and trilayer that are compared in Fig. 2 with the MD calculations and the beam data.[4] The monolayer lattice dynamics results show that lateral thermal expansion predominantly affects the SH and $SP_\parallel$ modes, whereas the $SP_\perp$ branch that is probed by the He-beam data is relatively insensitive to this effect.

The bilayer results shown in Fig. 2 indicate that provided the same interatomic spacings are used there is excellent agreement between the two independent methods of calculating the phonon dispersion curves. This fact plus the sharp response functions (Fig. 1) suggest that explicit anharmonic effects are small, at least at 20 K. It should be noted that in the trilayer case there is an important softening of the $SP_\perp$ branches due to thermal expansion normal to the surface. The agreement between the theory and the measurements (Fig. 2) was achieved via an enhanced surface-adatom attractive interaction, i.e., we employed $d=1.80$ Å, rather than the recommended value[7] $d=1.42$ Å. The need for additional attractive interactions accords well with theoretical expectation, since the term $-C_3/(z-d)^3$ is only the leading contribution of an asymptotic series.[15,16] Indeed, we have confirmed that a surface-adatom potential of the form $A\exp(-az)-C_3/(z-d)^3-C_5/(z-d)^5$, with d fixed at the theoretical value[7] 1.42 Å and both C_3 and C_5 taken from theory,[15] fits the beam data equally well.

In summary, it appears that inelastic scattering of the atoms from overlayers is a powerful means of characterizing adatom-surface interactions in systems where the heavy mass of the physisorbed atoms makes direct atom-surface scattering[1] very difficult.

K. Gibson, S. Sibener, M. Cole, and J. Phillips kindly allowed us access to unpublished data. We thank them as well as B. Williams, L. Bruch, J. Black, and D. Mills for useful discussions. This work was supported in part by the National Sciences and Engineering Research Council of Canada and NATO Grant No. 242.80.

*Permanent address: Dipartimento di Fisica, Università di Trento, Povo, Italy, 38050.

[1]G. Brusdeylins, R. B. Doak, and J. P. Toennies, Phys. Rev. Lett. **44**, 1417 (1980); **46**, 437 (1981); V. Bortolani, A. Franchini, F. Nizzoli, and G. Santoro, *ibid.* **52**, 429 (1984).

[2]B. F. Mason and B. R. Williams, Phys. Rev. Lett. **46**, 1138 (1981).

[3]H. Taub, L. Passell, J. K. Kjems, K. Carneiro, J. P. McTague, and J. G. Dash, Phys. Rev. Lett. **34**, 654 (1974); H. Taub, K. Carneiro, J. K. Kjems, L. Passell, and J. P. McTague, Phys. Rev. B **16**, 4551 (1977).

[4]K. D. Gibson and S. J. Sibener (unpublished).

[5]P. I. Cohen, J. Unguris, and M. B. Webb, Surf. Sci. **58**, 429 (1976); J. Unguris, L. W. Bruch, E. R. Moog, and M. B. Webb, *ibid.* **87**, 415 (1979); **109**, 522 (1981).

[6]L. W. Bruch, J. Unguris, and M. B. Webb, Surf. Sci. **87**, 437 (1979); L. W. Bruch and J. M. Phillips, *ibid.* **91**, 1 (1980); L. W. Bruch and M. S. Wei, *ibid.* **100**, 481 (1980); M. S. Wei and L. W. Bruch, J. Chem. Phys. **75**, 4130 (1981); L. W. Bruch, Surf. Sci. **125**, 194 (1983).

[7]E. Zaremba and W. Kohn, Phys. Rev. B **13**, 2270 (1976).

[8]J. A. Barker, M. L. Klein, and M. V. Bobetic, IBM J. Res. Dev. **20**, 222 (1976). Actually, we employed the Kr-Kr potential of R. A. Aziz, Mol. Phys. **38**, 177 (1979), scaled to the Xe well depth 282.35 K and minimum 4.3634 Å. This potential gives an excellent fit to the *bulk* phonon dispersion curves of solid Xe measured by N. A. Lurie, G. Shirane, and J. Skalyo, Phys. Rev. B **9**, 5300 (1974).

[9]A. D. McLachlan, Mol. Phys. **7**, 381 (1964).

[10]S. Rauber, J. R. Klein, M. W. Cole, and L. W. Bruch, Surf. Sci. **123**, 173 (1982).

[11]N. Stoner, M. A. Van Hove, S. Y. Tong, and M. B. Webb, Phys. Rev. Lett. **40**, 234 (1978).

[12]E. de Rouffignac, G. P. Alldredge, and F. W. de Wette, Phys. Rev. B **24**, 6050 (1981).

[13]F. E. Hansen, M. J. Mandell, and J. P. McTague, J. Phys. (Paris) Colloq. **38**, C4-76 (1977).

[14]J. P. Hansen and M. L. Klein, Phys. Rev. B **13**, 878 (1976); G. Jacucci and M. L. Klein, Solid State Commun. **32**, 437 (1979).

[15]X-P. Jiang, F. Toigo, and M. W. Cole, Chem. Phys. Lett. **101**, 159 (1983); Surf. Sci. **145**, 281 (1984); **148**, 21 (1984).

[16]P. Apell and C. Holmberg, Solid State Commun. **49**, 1059 (1984).

Thin Solid Films, 25 (1975) 65–70

MOLECULAR DYNAMICS CALCULATION OF THE ISOTOPE EFFECT FOR VACANCY DIFFUSION*

CHARLES H. BENNETT
IBM Thomas J. Watson Research Center, Yorktown Heights, N.Y. 10598 (U.S.A.)
(Received October 21, 1974)

The relative jump frequencies of tracer atoms with masses 1.00 and 1.05 times the host atom mass have been computed by molecular dynamics in a system of 255 Lennard-Jones atoms comparable with solid argon at 80 °K. The chief error of previous theoretical treatments of the isotope effect is found to be their neglect of unsuccessful saddle point crossings (*e.g.* U turns) which become more prevalent the lighter the diffusing atom. In the system studied, this effect lowers the ΔK factor from its harmonic value of 0.98 to an effective value of 0.89 ± 0.05.

INTRODUCTION AND THEORY

Because it can be measured experimentally with considerable precision and because it contains information on the diffusion mechanism, the isotope effect in tracer diffusion is an important target for theoretical calculation. Previous theoretical treatments[2–6] have equated the defect jump frequency to the crossing frequency through some hyperplane in configuration space, thereby tacitly assuming that all such crossings result in successful jumps. There is no reason why this must be so, and indeed recent molecular dynamics calculations[1] on the Lennard-Jones solid have revealed that in a significant number of cases the jumping atom reverses its direction one or more times while in the saddle point neighborhood. Allowing for the possibility of unsuccessful crossings, the spontaneous jump frequency Γ_j can be written as the product of two factors:

$$\Gamma_j = \Gamma' c \tag{1}$$

where Γ' is the crossing frequency through the designated hyperplane and c is the conversion coefficient, or fraction of successful crossings. Following Anderson[7] we count a crossing as successful if it is the *last* crossing on a success-

* *Proceedings Editors' note:* A very interesting cine film on vacancy diffusion in f.c.c. structures was shown by Dr. C. H. Bennett at the International Conference on Low Temperature Diffusion and Applications to Thin Films, Yorktown Heights, New York, U.S.A., August 12–14, 1974. This paper briefly outlines the computational methods that were involved in producing the cine film. For further details, the reader is requested to see ref. 1.

ful jump trajectory. (This convention avoids multiple counting of jumps such as the second one in Fig. 1.) It should perhaps be noted that, while Γ' and c are both dependent on the choice of dividing hyperplane, their product is not.

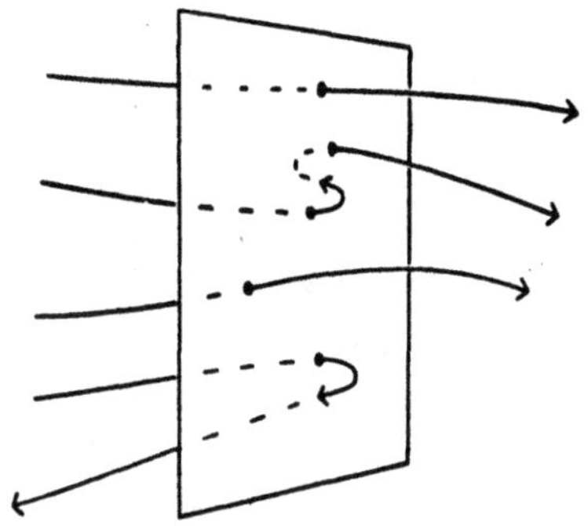

Fig. 1. Schematic representation of a hyperplane in configuration space and trajectories passing through it. Starting from a microstate on such a hyperplane, molecular dynamics can compute the trajectory forward and backward in time to determine its success. In a Lennard-Jones crystal comparable with solid argon at 80 °K, the fraction of successful crossings of the $Q_1' = 0$ saddle point hyperplane was found to increase with the mass of the jumping atom, thereby lowering the effective ΔK factor by about 0.1.

This paper describes molecular dynamics calculations aimed at measuring the mass dependence of c in a Lennard-Jones crystal near its melting point. Unsuccessful crossings are found to become slightly more prevalent the lighter the diffusing atom, thereby yielding an isotope effect about 10% lower than that calculated by rate theory, a result which is in agreement with most experimental data on f.c.c. materials (see, for example, ref. 8).

To determine the isotope effect we must find the change in the jump rate Γ_j caused by a differential change dm_1 in the mass of the jumping atom.

$$\frac{d\Gamma_j}{\Gamma_j} = \frac{d\Gamma'}{\Gamma'} + \frac{dc}{c} \tag{2}$$

The dividing hyperplane is conveniently described by the equation $\xi = 0$, where ξ is a linear function of the atomic coordinates:

$$\xi \equiv \text{const.} + \sum_{i=1}^{N} \boldsymbol{a}_i \cdot \boldsymbol{r}_i \tag{3}$$

Functions such as ξ are commonly called reaction coordinates because they measure the degree of completion of the jump. Any linear reaction coordinate can be characterized by a reduced mass μ:

$$\mu = 1/\Sigma(a_i^2\, m_i^{-1}) \tag{4}$$

and as m_1 is varied (while keeping $\boldsymbol{a}_i$ fixed) the crossing frequency Γ' varies simply as $\mu^{-1/2}$. The m_1 dependence of the jump frequency is then

$$\frac{\mathrm{d}\Gamma_j}{\Gamma_j} = -\frac{1}{2}\frac{\mathrm{d}\mu}{\mu} + \frac{\mathrm{d}c}{c}$$

$$= -\frac{1}{2}\frac{\mathrm{d}m_1}{m_1}\frac{a_1^2 m_1^{-1}}{\Sigma a_i^2 m_i^{-1}} + \frac{\mathrm{d}c}{c} \tag{5}$$

of which only the last term needs to be measured by molecular dynamics computer experiments.

Before discussing the measurement of the mass dependence of c, we shall point out the connection between eqn. (5) and previous theoretical treatments of the isotope effect. Vineyard[2], Mullen[3] and Le Claire[4] developed a theory based on the unstable normal mode Q_1' about the exact saddle point configuration of the many-body system. In the low temperature harmonic limit, where each normal mode (stable or unstable) moves independently of the others, this is indeed the correct reaction coordinate, and every crossing of the $Q_1' = 0$ hyperplane will lead to a successful jump. In this case the last term in eqn. (5) will vanish (dc must vanish where c attains its maximum of unity) and the coefficient of $-\frac{1}{2}(\mathrm{d}m_1/m_1)$ in the first term is simply the well-known ΔK factor, *i.e.* the fraction of the unstable mode's kinetic energy which resides in the jumping atom. The newer "dynamic" theory of the isotope effect[5,6] uses an arbitrary reaction coordinate involving only a few atoms, thereby obtaining a different value for ΔK; however, even at high temperatures there is no good reason to believe this reaction coordinate is "correct" in the sense of defining a hyperplane through which all crossings succeed. In the calculations described below the harmonic reaction coordinate $\xi = Q_1'$ was used, in order more clearly to separate the harmonic and anharmonic contributions to the isotope effect. The coefficients $\boldsymbol{a}_i$ defining this reaction coordinate were found by first locating the exact N-body saddle point configuration for the 255-atom system on which the dynamical calculations were to be performed, and then finding the direction of steepest descent (in $3N$ space) from this saddle point. The resulting reaction coordinate $\xi = Q_1'$ had a ΔK value of 0.98.

COMPUTATIONAL TECHNIQUE AND RESULTS

The molecular dynamics system studied was a periodic crystal of 255 atoms (*i.e.* $4\times4\times4$ f.c.c. unit cells with one vacancy), interacting via the Lennard-Jones pair potential

$$u(r_{ij}) = 4\varepsilon\left\{\left(\frac{\sigma}{r_{ij}}\right)^{12} - \left(\frac{\sigma}{r_{ij}}\right)^{6}\right\} \tag{6}$$

under conditions of energy and density ($E/N = -6.28\varepsilon$, $d = 1.134\sigma$ where d is the nearest neighbor lattice constant, and temperature $kT/\varepsilon = 0.67$) comparable with those of argon at 80 °K.

The anharmonic contribution to the isotope effect (*i.e.* the second term in eqn. (5)) was found by comparing the fraction c of successful $\xi = 0$ crossings in two systems, one with $m_1 = 1.00$ and the other with $m_1 = 1.05$ in units of the host atom mass. Unfortunately, the expected difference in c (a few tenths of a per cent) is too small to be measured directly by separate runs on the $m_1 = 1.00$ and $m_1 = 1.05$ systems. Therefore a differential sampling method was used to explore the very small subset of $\xi = 0$ crossings which lead to successful jumps in one system but not in the other. Such a one-to-one comparison of events in the two systems is possible because for every microstate of the $m_1 = 1.00$ system, $(\mathbf{r}_1, \mathbf{r}_2, \ldots, \mathbf{r}_N; \dot{\mathbf{r}}_1, \dot{\mathbf{r}}_2, \ldots, \dot{\mathbf{r}}_N)$, there exists a corresponding equiprobable microstate of the $m_1 = 1.05$ system, namely $(\mathbf{r}_1, \mathbf{r}_2, \ldots, \mathbf{r}_N; \dot{\mathbf{r}}_1/\sqrt{1.05}, \dot{\mathbf{r}}_2, \ldots, \dot{\mathbf{r}}_N)$. The differential sampling procedure is outlined below.

(1) A molecular dynamics calculation is performed on the $m_1 = 1.00$ system at equilibrium under the constraint $\xi = 0$. The microstates X generated by this calculation all lie on the $6N-2$ dimensional hyperplane $\xi = \dot{\xi} = 0$ and will be called *surface states*.

(2) Starting from the surface state X, a *crossing state* X_u is prepared by giving the reaction coordinate a positive thermal crossing velocity $\dot{\xi} = u\sqrt{kT/\mu}$, where u is chosen from a velocity weighted Maxwell distribution

$$p(u) = u \exp(-\tfrac{1}{2}u^2) \tag{7}$$

The states X_u so generated are representative of the forward $\xi = 0$ crossings which would occur in an unconstrained $m_1 = 1.00$ system at thermal equilibrium*.

(3) The trajectory through the crossing state X_u is computed forward and backward in time to determine its success. A crossing is considered successful if and only if it is the last $\xi = 0$ crossing on a trajectory segment which starts with $\xi < -0.2\ \sigma$ and ends with $\xi > +0.2\ \sigma$. For comparison, the stable defect configurations before and after the jump have ξ values of $\pm 0.6\ \sigma$.

(4) Empirically it is found that each surface state X has a threshold crossing velocity $\theta_{1.00}(X)$ such that the crossing X_u succeeds if and only if $u > \theta_{1.00}(X)$. The mean conversion coefficient for all $m_1 = 1.00$ crossings through the surface state X,

$$c_{1.00}(X) = \int_{\theta_{1.00}}^{\infty} p(u)\, du \tag{8}$$

can therefore be determined quite accurately by zeroing in on the threshold velocity $\theta_{1.00}(X)$ by successive trajectory calculations.

(5) From any crossing state X_u of the $m_1 = 1.00$ system, the corresponding crossing state X'_u of the $m_1 = 1.05$ system can be obtained by dividing the jumping atom's velocity by $\sqrt{1.05}$. The X'_u states so generated are representative of spontaneous crossings in an unconstrained $m_1 = 1.05$ system; in each, the jumping

* It should be pointed out that the implicit assumption of an equilibrium ensemble distribution for the spontaneously arising saddle point configurations and velocities in the unconstrained system is not an approximation. (It does not, for example, require that the typical jumping atom spend a long time in the saddle point neighborhood.) The equilibrium distribution in the saddle point neighborhood (or indeed in any part of phase space) follows rigorously as long as the system as a whole (including the N atoms and the occasionally jumping vacancy) is in a macrostate of thermal equilibrium. The use of constrained dynamics runs to sample equilibrium ensembles is discussed further in ref. 1.

atom has a greater momentum but a smaller velocity than in the corresponding X_u state. Like the X_u states, the X'_u states associated with a given X exhibit a threshold $\theta_{1.05}(X)$ such that X'_u succeeds if and only if $u>\theta_{1.05}(X)$. Typically the 1.00 and 1.05 thresholds differ by a few parts in 10^3; either may be higher, depending on X. The anharmonic contribution to the isotope effect arises from u values between the two thresholds, *i.e.* from crossings which succeed for one isotope but fail for the other (*cf.* Fig. 2).

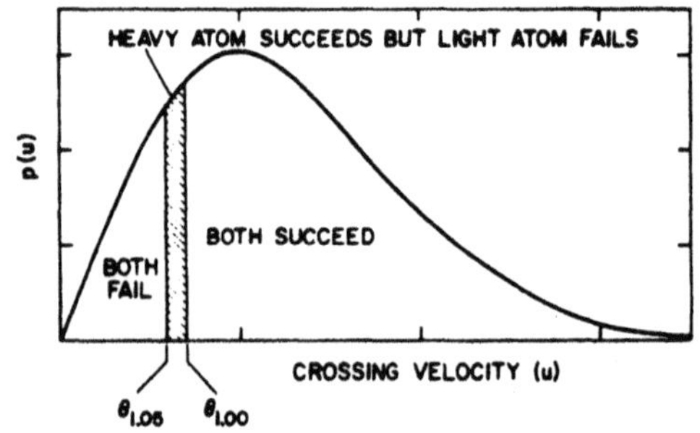

Fig. 2. Success of jump, as a function of normalized crossing velocity u, through a typical point X on the $6N-2$ dimensional saddle point hyperplane. The curve $p(u)$ is the Maxwellian spectrum of spontaneous crossing velocities. The shaded area is equal to $c_{1.05}(X)-c_{1.00}(X)$ and contributes to the anharmonic isotope effect.

(6) The difference in conversion coefficients, $c_{1.05}-c_{1.00}$, is obtained as an average over surface states X of the integral of $p(u)$ between the two thresholds:

$$c_{1.05}-c_{1.00} = \langle c_{1.05}(X)-c_{1.00}(X)\rangle_X$$
$$= \langle \int_{\theta_{1.05}(X)}^{\theta_{1.00}(X)} p(u)\,du \rangle_X \tag{9}$$

This may be found to the desired accuracy by measuring $\theta_{1.00}(X)$ and $\theta_{1.05}(X)$ for a sufficient number of representative surface states X, generated as described in paragraph (1) above.

The program outlined above has so far been carried out on 26 surface states of the 80 °K Lennard-Jones system, and the results are shown in Table I.

TABLE I
MASS DEPENDENCE OF THE CONVERSION COEFFICIENT

	$c_{1.00}(X)$	$c_{1.05}(X)-c_{1.00}(X)$
Minimum	0.006	−0.0028
Maximum	1.000	+0.0166
Mean	0.80	+0.0018
Standard error of mean	±0.06	±0.0009

Sample size = 26

The positive sign of $c_{1.05}-c_{1.00}$ means that saddle point crossings by the heavier isotope have a significantly higher chance of success, thereby reducing the effective ΔK factor

$$\frac{d(\ln \Gamma_j)}{d(\ln m_1^{-1/2})} = \Delta K_{\text{harm}} - \frac{40(c_{1.05} - c_{1.00})}{c_{1.00}} \tag{10}$$

from its harmonic value of 0.98 to an anharmonic value of 0.89±0.05, a finding which is in agreement with most experimental results on f.c.c. materials[8]. Anharmonicity of the jump mechanism is also indicated by the fact that 20% of all $Q_1' = 0$ crossings fail, whereas in a harmonic crystal all would succeed.

B.c.c. systems are an important target for future isotope effect calculations of the kind described here, because experiments (*e.g.* by Mundy[9] on sodium) suggest an isotope effect only half as large as reasonable harmonic estimates. A small and very mass-dependent conversion coefficient is quite plausible in b.c.c. systems due to the double barrier ring, which, it may be imagined, sometimes traps a diffusing atom (particularly a light one) in the saddle point neighborhood for some time, causing it to cross the $Q_1' = 0$ hyperplane several times without completing a successful jump.

ACKNOWLEDGMENTS

I wish to thank Paul Ho for helpful discussions.

REFERENCES

1 C. H. Bennett, in A. S. Nowick and J. J. Burton (eds.), *Diffusion in Solids: Recent Developments*, Academic Press, New York, to be published.
2 G. H. Vineyard, *J. Phys. Chem. Solids*, *3* (1957) 121.
3 J. G. Mullen, *Phys. Rev.*, *121* (1961) 1649.
4 A. D. Le Claire, *Phil. Mag.*, *14* (1966) 1271.
5 C. P. Flynn, *Phys. Rev.*, *171* (1968) 682.
6 M. D. Feit, *Phys. Rev. B*, *5* (1973) 2145.
7 J. B. Anderson, *J. Chem. Phys.*, *58* (1973) 4684.
8 N. L. Peterson, *Solid State Phys.*, *22* (1968) 409;
see also N. L. Peterson, in A. S. Nowick and J. J. Burton (eds.), *Diffusion in Solids: Recent Developments*, Academic Press, New York, to be published.
9 J. N. Mundy, *Phys. Rev. B*,*3* (1971) 2431.

Vacancy Double Jumps and Atomic Diffusion in Aluminum and Sodium

A. Da Fano and G. Jacucci
Gruppo Nazionale di Struttura della Materia del Consiglio Nazionale delle Ricerche, Istituto di Fisica, Università di Roma, Roma, Italy, and Centre Européen de Calcul Atomique et Moléculaire, Université de Paris XI, 91405 Orsay, France, and Section de Recherches de Métallurgie Physique, Centre d'Etudes Nucléaires de Saclay, 91190 Gif-sur-Yvette, France
(Received 27 June 1977)

The relevance of a high-temperature mechanism for vacancy migration, the double jump, has been established by molecular dynamics in Al and Na. Migration energies for single jumps are 0.42 ± 0.1 eV and 0.12 ± 0.02 eV, respectively, while for double jumps they are about five times larger. Second-near-neighbor jumps are found in Na near melting. Comparison with lattice statics and experiments indicates the adequacy of a monovacancy description. The inclusion of double jumps explains the anomalous diffusion of Na at high temperature.

One of the most remarkable experimental facts in solid-state diffusion is the linearity often exhibited by Arrhenius plots of the diffusion coefficient D, over wide temperature ranges.[1] This fact has led workers to regard the activation enthalpy H as constant with respect to temperature for a given defect type, since $-d(\ln D)/d(1/K_B T) = H$; and to ascribe eventual curvatures to the competition of two or more defect types. On the other hand, some authors have insisted that anharmonicity and lattice expansion can bring about variations for the activation parameters with temperature,[2] therefore causing the curvature observed in some cases. A controversy has thus originated about the interpretation of diffusion experiments in a number of metals.

The relative importance of various diffusion mechanisms and their temperature dependence can be investigated by computer simulation. While lattice statics has been widely employed, the relevance of dynamical simulation of atomic diffusion is perhaps insufficiently recognized. Dynamical correlations in the defect diffusion process have been revealed in this way.[3] Furthermore, when other than simple nearest-neighbor exchanges between atoms and defects contribute to diffusion, the description of the migration process is complicated by the presence of more than one activation energy and jump length.[4] In these conditions it is clear that lattice-statics evaluation of migration energy and entropy of the most obvious migration channel is not exhaustive.

The experimental Arrhenius plots of Al and Na show measurable curvatures. Their interpretation has received much attention in recent years.[2,5] In addition, Na shows a high-temperature anomaly, referred to as the premelting phenomenon,[2] also exhibited by other alkali metals (e.g., Li): The diffusion coefficient has a sharp rise above $T = 0.9T_m$ and the isotope effect abruptly decreases, in the same region, from the value apropriate to normal monovacancy diffusion.

The purpose of this Letter is to report a classical molecular-dynamics (MD) simulation of monovacancy diffusion in a 256-site fcc crystal model with an interaction pair potential appropriate to Al, and in a 250-site bcc model appropriate to Na, with periodic boundary conditions. Calculations were carried out at three temperatures within a 100-K interval in Al and four temperatures within a 45-K interval in Na. The system was followed for a time of the order of 10^{-9} sec to collect approximately 500 jump events for each run for Na and about a half of this for Al. We present here the results of this study. A comprehensive report will be published in due course, including a study on potassium.

The interaction pair potentials we have used are due to Dagens, Rasolt, and Taylor.[6] They are *ab initio* potentials, i.e., not fitted to empirical crystal properties. Previous studies of solid- and liquid-phase properties of Al,[7] K,[8] and Li[9] have shown that they give very good account of the experimental bulk behavior. Vacancy energies and volumes have been also computed with these potentials,[7,10] including anharmonic effects of lattice vibrations neglected in all existing calculations have given for the first time the correct positive sign[1] of the variation of defect formation and migration energies with increasing temperature. The formation energy for a vacancy in Na is found to be 0.25 and 0.28 eV at $T = 0$ and $T = T_m$ with formation volume ΔV_f of 0.6 atomic volume[10]; the migration energies in Al are found to be 0.42

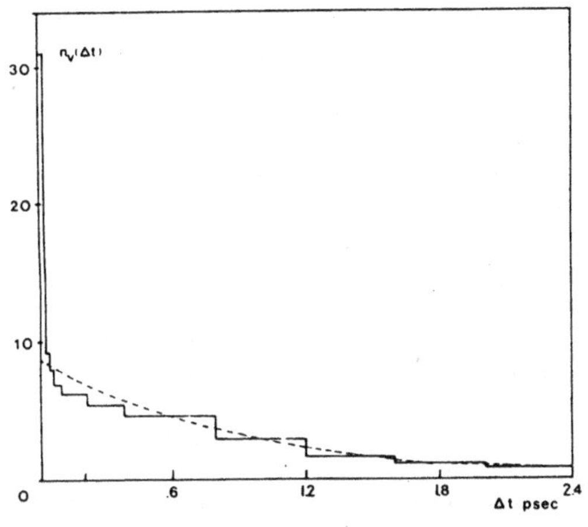

FIG. 1. Distribution of vacancy jumps in Al at melting vs their time delay Δt. The broken line represents the exponential distribution appropriate to a random process with the same overall jump rate. The histogram is much higher than the broken curve at very short times because of the double-jump events, and then becomes lower at intermediate times because of the suppression of backward jumps.

and 0.70 eV, with migration volume ΔV_m of 0.6 atomic volume.[7]

The analysis of the jump events of our MD simulation indicates the following:

(1) The distribution $n_V(\Delta t)$ of the time delays of successive jumps of the vacancy departs at short times (a few vibration periods of the lattice) from the exponential distribution appropriate to a random process (see Fig. 1).

(2) The huge peak occuring in n_V at short times grows rapidly with temperature: It is connected with simultaneous (within a vibration period of the lattice) forward-correlated jump events. The temperature dependence of their occurrence frequency indicates a migration energy much higher than for single-jump (SJ) events by about a factor of 5. Thus we must regard the double jump (DJ) as a second diffusion mechanism involving two atoms that jump at the same time collinearly, with the vacancy jumping to a non–nearest-neighbor site. Vacancy jumps to second near neighbors (SNNJ) have also been found in Na (5% at the melting point). They exhibit a migration energy also much higher than the SJ.

The defect jump rate Γ for the two mechanisms is obtained by counting the respective jump events observed in the various runs. For SJ we can write $\Gamma' = z\tilde{\nu}\exp(S_m'/K_B)\exp(-E_m'/K_BT)$, where z is the coordination number of the lattice. The slope of $\ln\Gamma'(T)$ versus $1/K_BT$ gives the value of E_m'. This quantity represents the migration enthalpy, that reduces to the migration energy at zero pressure. The migration enthropy S_m' is then deduced by inserting appropriate values of the attempt frequency $\tilde{\nu}$, which are 5×10^{12} sec^{-1} and 2×10^{12} sec^{-1} in Al and Na, respectively. Similarly, letting $\Gamma'' = \Gamma_0\exp(-E_m''/K_BT)$, a value of the migration energy for DJ is defined. We note that the value of Γ_0 is puzzlingly high if one tries to interpret it in the traditional way. Results are given in Table I, along with the overall migration energy E_m derived from the temperature variation of the asymptotic slope of the mean-square atomic displacement versus time.

TABLE I. Migration parameters of the vacancy. E_m', E_m'', and E_m (in eV) correspond to SJ, DJ, and atomic mean-square displacement, respectively.

Metal	E_m'	E_m''	E_m	S_m'
Na	0.12 ± 0.02	0.55 ± 0.1	0.18 ± 0.02	$\sim K_B$
Al	0.42 ± 0.1	~ 2	0.5 ± 0.1	$\sim K_B$

The values of the migration parameters need an important specification. The MD technique imposes constant energy (CEC) and constant volume (CVC) conditions on the atomic system in the atomic system in the periodic cell, in contrast to experiment, during the migration process—the volume changes, however, with temperature as that of the real solid at room pressure.

The CEC does not affect the results in a substantial way. The probability of occurrence of fluctuations of a given amplitude of the potential energy is somewhat reduced in our CEC microcanonical system with respect to the canonical distribution. Lower jump frequencies are then expected. An investigation of these distributions shows that the reduction is, however, negligible for SJ, but is nonnegligible for DJ and for SNNJ found in Na, since both processes show a migration energy of the order of 5 times E_m'. The occurrence frequency of these events is probably reduced by over 10%. Their migration energy should not be altered appreciably, however.

The Gibbs free energy of migration (and formation) of the defect is not altered by CVC, i.e., $H^P - TS^P = E^V - TS^V$ with P and V referring to constant-pressure or -volume conditions.[11,12] So that the value of Γ is not altered either, and (the constant) E_m' is directly comparable with the (high-temperature) value of H^P ($\sim E^P$) given by experiment. S_m' corresponds in turn to S^P. The effect of CVC must, however, be explicitly considered in static calculations.[7,10-12] Here E^P is obtained by adding to E^V the entropic contribution consisting essentially in the lattice entropy increase upon volume expansion[13]:

$$E^P \simeq H^P = E^V + T(S^P - S^V) \simeq E^V + T(\alpha/K_T)\Delta V^P.$$

For SJ in Al, at melting, static calculations give[7] $E^V = 0.3$ eV and $\Delta V^P = 0.6$ atomic volume Ω; the corresponding entropic term is 0.4 eV. The value of E_m' from Table I is then to be compared to the static result $E^P = 0.7$ eV and to the high-temperature experimental value $E^P > 0.62$ eV.[5] The difference between E^V and E^P for the migration of Na is instead very small. This is because ΔV^P must be small; in fact, the static evaluation of the formation volume[10] 0.6Ω is already larger than the measured activation volume 0.5Ω,[14] which is the sum of the two. In any case the entropic correction is only about 0.015 eV for each increase of 0.1 in $\Delta V^P/\Omega$.[10]

The dynamical result obtained for E_m' in Al is somewhat lower than the static evaluation E^P, which is close to the experimental value. The rather large variation of the static value E^P with temperature and a corresponding variation of the formation energy[7] suggest that there is no need to invoke divacancies to explain the curvature of the experimental Arrhenius plot in Al.

For Na we can add the migration energy E_m obtained in the present work, i.e., 0.18 eV at melting and 0.12 eV at low temperature (where only SJ contribute), to the static formation energies 0.28 and 0.25 eV to obtain 0.46 and 0.37 eV. These values are in excellent agreement with the temperature dependence of the measured self-diffusion coefficient.[14] By taking the slope of the best-fit formula of Mundy's self-diffusion data one obtains, in fact, 0.45 and 0.37 eV. Monovacancies, then explain, the curvature of the Arrhenius plot in Na.

At the melting point of Na, DJ contribute more than one fourth of the atomic migration processes. As a consequence near T_m the slope of Arrhenius plot, E_m, is substantially increased with respect to E_m'. Furthermore, the collective character exhibited by DJ causes a reduction of the isotope effect.[15] This is strong evidence that DJ are responsible for the premelting phenomenon observed in Na.

A preliminary study of K shows similar values for the migration parameters, and an accord with experimental data[16] just as good. The DJ in K are found to contribute a smaller amount to diffusion at melting than in Na; the same is true for SNNJ.

The DJ can be revealed by an accurate comparison of experimental results on mass diffusion with those of tracer diffusion. SNNJ can be instead revealed by incoherent neutron scattering. Their effect on the self-diffusion function $S_s(K, \omega)$ has been computed on the basis of the present data.[17] Experiments are indeed under way, and in preliminary runs it looks as if an interpretation of the data in term of nearest-neighbors jumps only is not possible.[18]

The overall agreement of the results of this work with static calculations and experiment indicate the following: (1) The pseudopotential pair interaction of Dagens, Rasolt, and Taylor is quite adequate for defect dynamics in Na and K. The situation is less conclusive for Al. (2) A description of atomic diffusion in these systems based on monovacancies only is able to give account of experimental results and explain the observed anomalies and curvatures.

The authors are indebted for many useful discussions to Y. Adda, C. H. Bennett, J. L. Bocquet, N. V. Doan, G. Martin, A. Seeger, R. D. Taylor, and A. Tenenbaum.

[1]A. Seeger and H. Mehrer, *Vacancies and Interstitials in Metals* (North-Holland, Amsterdam, 1969).
[2]M. Gilder and D. Lazarus, Phys. Rev. B 11, 4961 (1975).
[3]C. H. Bennett, in *Diffusion in Solids—Recent Developments*, edited by A. S. Nowick and J. J. Burton (Academic, New York, 1975).
[4]A. Da Fano and G. Jacucci, in Proceedings of the International Conference on the Properties of Atomic Defects in Metals, Argonne, Illinois, 1976 (to be published).
[5]A. Seeger, D. Wolf, and H. Mehrer, Phys. Status Solidi (b) 48, 481 (1971).
[6]L. Dagens, M. Rasolt, and R. Taylor, Phys. Rev. B 11, 2726 (1975).
[7]G. Jacucci, R. Taylor, A. Tenenbaum, and N. Van Doan, to be published.
[8]J. P. Hansen and M. L. Klein, Solid State Commun. 20, 771 (1976).
[9]G. Jacucci, M. L. Klein, and R. Taylor, Solid State Commun. 19, 657 (1976).
[10]M. S. Duesbery, G. Jacucci, and R. Taylor, to be published.
[11]G. F. Nardelli and N. Terzi, J. Phys. Chem. Solids 25, 815 (1964).
[12]C. P. Flynn, *Point Defects and Diffusion* (Clarendon Press, Oxford, England, 1972), Chap. 7.

[13]Note that while the correction due to CEC depends on the size of the system and vanishes at large N, the difference $E^P - E^V$ remains constant instead.

[14]J. N. Mundy, Phys. Rev. B 3, 2431 (1971).

[15]C. H. Bennett, in *Comptes-Rendus du Dix-Neuvième Colloque de Métallurgie, Saclay, 1976* (Centre d'Etudes Nucléaires de Saclay and Institut National des Sciences et Techniques Nucléaires, France, 1977), Vol. 2.

[16]J. N. Mundy, T. E. Miller, and R. J. Porte, Phys. Rev. B 3, 2445 (1971).

[17]A. Da Fano, G. Jacucci, and A. Rahman, to be published.

[18]A. Seeger, private communication; B. Alefeld, private communication.

Adequacy of lattice dynamics for high-temperature point-defect properties

G. De Lorenzi and G. Jacucci*
Department of Physics and Materials Research Laboratory, University of Illinois at Urbana-Champaign, Urbana, Illinois 61801
(Received 23 September 1985)

Comparison with existing Monte Carlo and molecular-dynamics data of novel quasiharmonic lattice-dynamics (LD) calculations for formation and migration of lattice vacancies in rare-gas crystals at melting shows that the effect of residual anharmonicity on atomic diffusion is not large. Inclusion of vibrational contributions to the free energy and pressure ensures, for the first time, thermodynamic consistency between LD results at constant volume, pressure, and lattice spacing. Lattice-statics predictions are badly in error. Constant-volume defect entropies are negative and surprisingly large.

Only recently the performance of modern computers has allowed the application of exact energy minimization and matrix diagonalization routines to perform lattice dynamics (LD) analysis of 100-site model crystals. Precise free energies of formation and migration of point defects can now be evaluated in the quasiharmonic approximation, while keeping under control the dependence of the results on the size cf the crystal.[1]

We employ such exact codes to compute normal-mode frequencies of 32- and 108-site cyclically repeating Lennard-Jones (LJ) crystals containing a vacancy, in relaxed static equilibrium or in relaxed saddle-point configurations, and in perfect lattice states. LD values for classical and quantum mechanical energies, entropies, and pressures are calculated for temperatures and volumes dictated by the equation of state in the domain of stability of the crystalline phase. At each state point, vacancy formation and migration parameters are evaluated while imposing on the crystal three different thermodynamic conditions: constant lattice spacing (CL), constant pressure (CP), and constant volume (CΩ). Each property is alternatively kept constant while comparing LD values for crystals with and without vacancy (formation), and with vacancy in equilibrium or saddle-point configurations (migration). This provides for the first time a stringent test of the thermodynamic consistency of the calculation.

The resulting complete picture of the thermodynamic properties of a point defect, in the quasiharmonic model, makes possible an accurate check of approximations currently employed in their estimation. In particular, our results unambiguously establish the inadequacy of the lattice statics (LS) description for the case of rare-gas LJ crystals at high temperatures. It is found to be strictly necessary to correctly include the contributions of thermal vibrations to the crystal (free energy and) pressure, in addition to the elastic term. In fact the LS estimate for the vacancy-formation volume close to melting exceeds the LD value by more than 50%. Use of this overestimate results in a *negative* value for the CP free energy of formation of lattice vacancies.

Fully anharmonic Monte Carlo (MC) calculations of vacancy formation[2] and migration[3] parameters are uniquely available for this same model system. Comparison with our LD results shows that the quasiharmonic theory overestimates both the free energy of formation and migration of lattice vacancies by about 10% close to melting. The effect of this residual anharmonicity turns out to be an increase, by a factor of about 4 only, of both vacancy concentration and jump frequency. Since anharmonicity is especially large in rare-gas crystals, these results establish that LD provides an excellent first approximation to point-defect concentration and jump rates, and hence to the atomic-diffusion coefficient, in high-temperature crystals. We briefly describe here results pertinent to classical mechanics, finite temperatures, and $N=108$. The results for $N=32$ show that the N dependence is small. An extensive account of the work is in preparation.

The free energy F of a vibrating crystal is expressed in terms of the potential energy Φ_0 and the normal mode frequencies ω_α calculated at the relevant minimum $\mathbf{R}_0$ in $3N$-dimensional configuration space at fixed volume V. At high temperatures F reduces to

$$F=\Phi_0+k_BT\sum_{\alpha}^{3N-3}\ln(\hbar\omega_\alpha/k_BT)\ ,$$
$$\Phi_0=\sum_{i<j}v(r_{ij})\ , \tag{1}$$

disregarding translation of the center of mass, and for pair potentials. The pressure is

$$P=-(\partial F/\partial V)_T=P_\Phi+P_\omega(T)\ , \tag{2a}$$

with

$$P_\Phi=-\partial\Phi_0/\partial V$$

and

$$P_\omega(T)=-\left[\frac{\partial(F-\Phi_0)}{\partial V}\right]_T .$$

Thermal volume expansion is automatically included in this description. This is because P from Eq. (2) depends on the temperature at a given V. Fixing the value of the external pressure P_{ex} and letting T vary produces the LD equation of state for the crystal, i.e., the volume V for which P equals P_{ex} at a chosen T. The thermal expansion of the crystal volume being one of the effects of anharmonicity of lattice vibrations, LD is also called a quasiharmonic treatment. It cannot be overemphasized that a thermodynamically consistent implementation of LD requires that when thermal volume expansion is considered and Φ_0 recalculated at different values of V corresponding to values of T derived from some equation of state (not necessarily the LD one) the variations of ω_α should also be considered, together with the resulting variations of S and P.

We see from Eq. (2a) that P is the sum of two terms, one coming from variations of Φ with V, the other from variations of ω_α with V. The first term is easily treatable analytically, giving the virial expression

$$P_\Phi = -\frac{N}{3V}\sum_{i<j}\frac{\partial v(r_{ij})}{\partial r_{ij}}r_{ij} . \quad (2b)$$

P_ω is best evaluated numerically instead, using incremental ratios built from recalculations of ω_α at two closely spaced values of V.

It is important to realize that the virial formula in Eq. (2b) differs from the usual formula based on the thermal average of the virial sum precisely by the term $P_\omega(T)$:

$$P = -\left[\frac{\partial F}{\partial V}\right]_T = -\frac{N}{3V}\left\langle\sum_{i<j}\frac{\partial v(r_{ij})}{\partial r_{ij}}r_{ij}\right\rangle_T$$

$$= P_\Phi + P_\omega(T) . \quad (2c)$$

As a result, the substitution of the thermal average of the virial sum with the value of the sum calculated at the most probable configurations, i.e., the configurations $\mathbf{R}_0$, may lead to disastrous results, because it essentially consists in using P_Φ alone, disregarding the volume dependence of ω_α. This is only correct at $T=0$ and for classical mechanics.

Our LD predictions of the perfect crystal properties coincide with previous calculations, already shown to deviate somewhat from high-T MC data.[4] Results for vacancy formation are obtained by comparing properties relative to the regular lattice crystal C_0 containing N atoms, with those relative to a crystal C_v containing $N-1$ atoms and one vacant site. In the CΩ calculation the lattice of C_v is squeezed so that its total volume V be reduced by the factor $(N-1)/N$ to leave the atomic volume $\Omega = V/N$ unaltered. In the CP calculation the lattice parameter of C_v is varied until the reading of the pressure coincides with that of C_0. The values are listed in Table I. Long-range corrections are not included in the values of Φ and P_Φ.

The formation volume at constant pressure is always within about 10% of unity. As a consequence, CL values are much closer to CP values than to CΩ ones. CΩ values are indeed remarkable: $(\Delta S)_\Omega$ is large and negative (as predicted by Flynn[5]), $(\Delta U)_\Omega$ decreases with temperature, and in $\Delta F = \Delta U - T\Delta S$ the temperature variations of the energy and entropy terms cancel to a large extent. CL and CP values for ΔU are much less dependent on T, and ΔS is much smaller and positive; ΔF is again only moderately dependent on T. $(\Delta G)_P$ is nearly equal to $(\Delta F)_\Omega$; it is also precisely equal to $(\Delta F)_L + P\Omega$. These last results are expected to order $1/N$. They constitute an important thermodynamic check and show that the model has been treated in a thermodynamically consistent way.

The static model, LS, neglects entropy related contributions, from lack of knowledge of ω_α. We see from Table I that in LG Ar at high T this produces disastrous results. From $(\Delta G)_P = (\Delta F)_\Omega$ one derives

$$(\Delta H)_P = (\Delta U)_\Omega + T[(\Delta S)_P - (\Delta S)_\Omega] . \quad (3)$$

Certainly, Eq. (3) does not hold without the entropy term: $(\Delta U)_\Omega$ is only about one-third of $(\Delta H)_P$. To correct LS, the entropy term can be estimated using a value for the thermal expansion coefficient α_P independently available (often from experiment) and the approximate relation[5]

$$T(\Delta S_P - \Delta S_\Omega) \cong T(\Delta V)_P\alpha_P/\kappa_T . \quad (4)$$

Equation (4) holds to order $1/N$ for our LD results. As a comparison, MC values of $P_\omega \equiv T\Omega\alpha_P/\kappa_T$ are some 15% lower than LD ones in LJ Ar at $T=80$ K (this discrepancy would be even larger were it not for a partial cancellation of differences in α_P and κ_T).

LS also neglects P_ω in Eq. (2). P_ω is by no means negligible in the perfect lattice. The formation volume $(\Delta V)_P$ predicted using $(\Delta P_\Phi)_L$, or $(\Delta P_\Phi)_\Omega$, comes out to be much too large. In fact, in the static approximation it is found $(\Delta V)_P = 1.24\Omega$ and 1.50Ω at $T \approx 60$ and 80 K instead of 1.00 and 0.92! Even the sign of the relaxation volume is wrong. This huge discrepancy of the static approximation to $(\Delta V)_P$ from the LD value is probably the most noticeable effect of neglecting P_ω and represents an important discovery of this work. Equation (3) holds exactly at $T=0$ in the classical static model, with $P \equiv P_\Phi$, for any given lattice parameter,

$$(\Delta U)_{P_\Phi} + P_\Phi(\Delta V)_{P_\Phi} = (\Delta U)_\Omega . \quad (5)$$

TABLE I. Variations of LD entropies (S), energies (U), Helmholtz (F) and Gibbs (G) free energies, and volume (V) of an fcc LJ periodic crystal with $N=108$ sites, upon formation of a vacancy. All quantities are in reduced units based on σ and ϵ of the LJ potential. The cutoff distance is 4.5σ. T (in kelvin) refers to Ar.

$\Omega = V/N$	k_BT	P	$(\Delta S)_L$	$(\Delta S)_\Omega$	$(\Delta S)_P$	$(\Delta U)_L$	$(\Delta U)_\Omega$	$(\Delta U)_P$	$(\Delta F)_L$	$(\Delta F)_\Omega$	$(\Delta G)_P$	$(\Delta V)_P/\Omega$
0.9734	0.3339($\approx$40 K)	−0.2084	2.3249	−6.7791	2.4658	8.3810	5.3525	8.4312	7.6047	7.6160	7.4019	1.0155
0.9974	0.4432($\approx$60 K)	0.3125	2.1451	−7.1558	2.1846	8.2929	4.1880	8.3111	7.2349	7.7172	7.5468	1.0043
1.0323	0.6678($\approx$80 K)	0.9503	1.7988	−7.8234	1.0130	8.0894	2.7451	7.6440	6.8882	7.9696	7.8685	0.9184

TABLE II. Variations of LD quantities of a LJ crystal with a vacant site upon raising it from equilibrium to the saddle for vacancy migration. Definitions of ΔS and $\bar{\omega}$ [in units of $(\epsilon/m\sigma^2)^{1/2}$] are in the text. Thermodynamic states same as for Table I.

$\bar{\omega}$	$(\Delta S)_\Omega$	$(\Delta S)_P$	$(\Delta U)_\Omega$	$(\Delta U)_P$	$(\Delta F)_\Omega$	$(\Delta G)_P$	$(\Delta V)_P/\Omega$
14.8043	−1.2284	3.1575	4.8035	6.3192	5.2137	5.1666	0.4847
13.7263	−1.2742	3.4404	4.3064	6.4588	4.9348	4.9209	0.5098
12.3219	−1.3506	3.9543	3.6314	6.6048	4.5333	4.5078	0.5542

If one desires to use the same data, obtained from knowledge of Φ_0, Φ_v, and their volume derivatives, to exploit Eq. (3) at finite temperature T, one must alter both sides to include entropy contributions. In the left-hand side, $P=P_\Phi+P_\omega$ will replace P_Φ throughout, and in the right-hand side the term $T(\Delta S_P-\Delta S_\Omega)$ will be added. Performing only one such modification will of course unbalance Eq. (5). An approximate treatment consistently neglecting P_ω but properly including S produces $(\Delta F)_\Omega \cong 8.06$ and $(\Delta G)_P \cong -1.86$ at $T\cong 80$ K, in complete disregard of thermodynamics and of the stability of the lattice.[6]

The free energy of formation of lattice vacancies in LJ crystals was first measured with MC by Squire and Hoover.[2(a)] Later calculations[2(b)] have recently been extended to include the formation volume.[7] The LD results exceeds the MC ones only by 6% at 60 K and 10% at 80 K. Almost all the explicit anharmonic contribution to $(\Delta G)_P$ comes from lattice relaxation around the defect. The effect is to lower its value, with an increase by a factor ~ 4 of the number of lattice vacancies at 80 K in Ar. An interesting result is that the MC value of the formation volume,[7] $(\Delta\Omega)_P/\Omega=1.21\pm0.05$ for $T\sim 60$ K/80 K, is intermediate between LD and LS values, LD being too high and LS too low by over 80%.[8]

The vacancy jump frequency Γ is given by classical harmonic rate theory as[9(a)]

$$\Gamma_0=(\bar{\omega}/2\pi)\exp(\Delta S/k_B)\exp(-\Delta U/k_BT)\ ,$$

with

$$(\bar{\omega})^{3N-3}=\prod_{\alpha}^{3N-3}\omega_\alpha\ ,$$

$$\Delta S=k_B\sum_{\alpha}^{3N-4}\ln\left[\frac{\bar{\omega}}{\omega_{s\alpha}}\right]\ ,$$

and

$$\Delta U=\Phi_s-\Phi_v\ .$$

The $\omega_{s\alpha}$ are the normal-mode frequencies at the saddle point, excluding the imaginary frequency relative to the reaction coordinate. Values of ΔU, ΔS, and $\bar{\omega}$ can be calculated by LD in the quasiharmonic approximation, as for vacancy formation quantities. Results are listed in Table II. Constant Ω and constant P conditions refer to saddle point versus equilibrium configurations, the values of P and Ω being appropriate to the formation of the defect at constant pressure. $(\Delta G)_P$ and $(\Delta F)_\Omega$ show again thermodynamic consistency.

The MC calculation of Γ, for LJ/Ar at $T\approx 60$ and 80 K, has been carried out by Bennett.[3] He finds $\ln[\Gamma(m\sigma^2/\epsilon)^{1/2}]=-7.5\pm1.0$ and -5.0 ± 0.5. The LD estimates from Table II are -9.2 and -6.1, showing an appreciable discrepancy between the quasiharmonic approximation and fully anharmonic rate theory predictions. Because in Bennett's work direct jump-frequency measurements were also made using molecular dynamics and gave the value of -5.3 ± 0.2 for the logarithm of the jump frequency at $T\approx 80$ K, in agreement with his anharmonic rate theory prediction, we conclude that (i) rate theory is applicable, and (ii) LD evaluations at high temperature may be low by a factor anywhere from 2 to 15. Previous evaluations of Γ_0 using lattice statics values of $(\Delta U)_\Omega$ and an effective attack frequency ν^* in the place of the LD factor $(\bar{\omega}/2\pi)\exp[(\Delta S)_\Omega/k_B]$ accidentally yielded better agreement with anharmonic jump rates.[3] The apparent accord is destroyed by the LD value for $(\Delta S)_\Omega/k_B$ that is larger than unity in modulus and negative in sign.[9(b)]

This is the first accurate comparison between quasiharmonic and anharmonic rate theory predictions. LD predictions are quite good. Representation of an imperfect crystal by a small quasiharmonic lattice is a useful approximate tool to use to get crystal properties concerning atomic diffusion. Slight LD overestimates of the barrier to atomic jumps may in fact be a general feature.[10] Work is in progress to extend LD to the first anharmonic term in a temperature expansion of the jump rate Γ,[11] and towards more accurate MC migration data.

We have benefited from discussions with C. P. Flynn. This research was supported by the National Science Foundation under Grant No. DMR-83-16981.

*Permanent address: Centro Studi del Consiglio Nazionale delle Ricerche e Dipartimento di Fisica, Universita di Trento, 38050 Povo, Italy.

[1]G. Jacucci, in *Diffusion in Crystalline Solids*, edited by G. E. Murch and A. S. Nowick (Academic, New York, 1984), Chap. 8.

[2](a) D. R. Squire and W. G. Hoover, J. Chem. Phys. **50**, 701 (1969); (b) G. Jacucci and M. Ronchetti, Solid State Commun.

33, 35 (1980).
[3]C. H. Bennett, in *Diffusion in Solids, Recent Developments,* edited by A. S. Nowick and J. J. Burton (Academic, New York, 1975), Chap. 2.
[4]A. C. Holt, W. G. Hoover, S. G. Gray, and D. R. Shortle, Physica **49**, 61 (1970).
[5]C. P. Flynn, *Point Defects and Diffusion* (Oxford University Press, London, 1972), Chap. 7.
[6]Neglecting P_ω in Eq. (5) seems to be a fair approximation in alkali metals. LS results [G. Jacucci and Roger Taylor, J. Phys. F **9**, 1489 (1979)] for energy and pressure, augmented by Eq. (4), met with precise agreement with careful experimental data [J. N. Mundy, Phys. Rev. B **3**, 2431 (1971); J. N. Mundy, T. E. Miller, and R. J. Porte, Phys. Rev. B **3**, 2445 (1971)] in those systems.
[7]G. De Lorenzi, G. Jacucci, and M. Ronchetti, unpublished.
[8]The experimental value of the formation volume for Shottky defects in ionic crystals is also intermediate between higher LS and lower LD results, as shown in recent CL calculations: H. Harding, Physica A (to be published).
[9](a) G. H. Vineyard, J. Phys. Chem. Solids **3**, 121 (1957); (b) G. H. Vineyard and J. A. Krumhansl, Phys. Rev. B **15** (1985).
[10]G. Jacucci, Roger Taylor, A. Tenenbaum, and N. Doan, J. Phys. F **11**, 793 (1981); G. De Lorenzi, G. Jacucci, and V. Pontikis, Surf. Sci. **116**, 391 (1982).
[11]M. Toller, G. Jacucci, G. De Lorenzi, and C. P. Flynn, Phys. Rev. B **32**, 2082 (1985); unpublished.

Jump rate of the fcc vacancy in the short-memory–augmented-rate-theory approximation. I. Difference Monte Carlo sampling for the Vineyard rate

G. De Lorenzi,* G. Jacucci,* and C. P. Flynn
Department of Physics and Materials Research Laboratory, University of Illinois at Urbana-Champaign, Urbana, Illinois 61801
(Received 20 April 1987)

The framework of short-memory–augmented-rate theory is employed to partition the vacancy jump rate into two factors related, respectively, to the Vineyard model and corrections for multiple barrier crossings inherent in the dynamics. The first of these two factors is treated in the present paper using methods of Monte Carlo difference sampling adapted to the vacancy jump problem. Both analytical and numerical methods are employed to derive the jump rate accurate to a few percent, starting from the interatomic potential and the atomic masses alone. Correction terms, and their important impact on the isotope effect, are discussed in the companion paper.

I. INTRODUCTION

This paper and the one by Marchese *et al.*[1] that immediately follows describe an accurately predictive treatment of the dynamical jump events by which atoms move through crystals in thermal equilibrium. In crystals, atomic migration typically occurs by defect mechanisms that define the atomic displacements caused by any given jump. Consequently, the defect density and jump rate together specify the rate at which intermixing takes place.[2] Monte Carlo methods have been devised to obtain the free energy of defect formation, and hence the defect density, with all necessary precision, given the many-body potential by which atoms of solids interact.[3,4] Comparable success in calculating the jump rate has been more elusive, and certain central phenomena have remained very poorly understood. Among these the problem of the isotope effect in diffusion stands out as the most important, because the isotope effect offers, in principle, the single accessible monitor of the dynamics during the atomic jump process itself.[5] In these two consecutive papers we describe a quantitatively successful treatment of the atomic jump rate, its isotopic dependence, and similar phenomena, for atomic migration by the vacancy mechanism in fcc crystals. The methods appear equally applicable to other crystal types and alternative defect mechanisms. It therefore appears likely that our methods can resolve many of the difficulties that have resisted description to the present time.

A central problem has been the lack of any precise criterion by which the likely occurrence or otherwise of an atomic jump from a given starting configuration can be assessed accurately. Any such assessment requires knowledge of the future evolution, and hence integration of equations of motion, for a many-particle system that is hopelessly nonlinear. In principle, large-scale computation could be employed with model crystallites to explore future trajectories and so, by sampling over initial states, to obtain an expected jump rate.[6] As the temperature is lowered, however, and jumps become progressively rarer, this approach quickly becomes inefficient. To overcome this difficulty, the sampling and the subsequent dynamic evolution may be used to evaluate the "success" of individual barrier crossings only for states on top of the barrier; the probability of "visiting" the top of the barrier follows separately from a numerical investigation of the distribution function. This is the principle of Bennett's[7,8] pioneering studies of atomic jumps by molecular dynamics, used also in more recent investigations of surface diffusion by Voter and Doll.[9] Others, particularly those concerned with large biological molecules in aqueous environments, have turned to stochastic methods, with simplified interparticle couplings, based on Kramers' method.[10] The present research improves on Bennett's procedure by using analytical methods[11] based on the theory of general dynamical systems to identify a precise criterion which selects trajectories that lead only to successful atomic jumps.

In this new theoretical framework, called short-memory–augmented-rate theory (SM-ART), certain invariant manifolds in the phase space of the crystal play critical roles.[12] Hopping events are, of course, inhibited by potential barriers that separate equilibrium positions in configuration space. The center manifold (CM) in phase space is the locus of all trajectories that linger indefinitely on this barrier; for a crystal of $N/3$ atoms this manifold is $2(N-1)$-dimensional. Two $(2N-1)$-dimensional manifolds known as the center stable (CS^+) and the center unstable (CS^-) manifolds intersect in CM. CS^- is the locus of all trajectories that depart infinitesimally from CM and hence dissociate from the barrier as time passes; CS^+ is the time-reversed analog, consisting of trajectories that converge asymptotically to CM. Figure 1 shows the relationships schematically, with q_1 representing the "direction of dissociation" (or reaction coordinate) and p_1 its canonically conjugate momentum. $\pm q_1$ thus lead from the barrier to the equilibrium sites. All directions perpendicular to q_1 lead to higher and eventually inaccessible energies (hatched regions). Trajectories of Hamiltonian flow are indicated schematically in Fig. 1 by lines with arrows.

At the center of the SM-ART treatment is the fact that CS^+ and CS^- divide phase space into four sectors, of which two contain all the flow across the barrier in one sense only, one sector for each sense. It follows that any surface Σ drawn from CM to remote regions of high

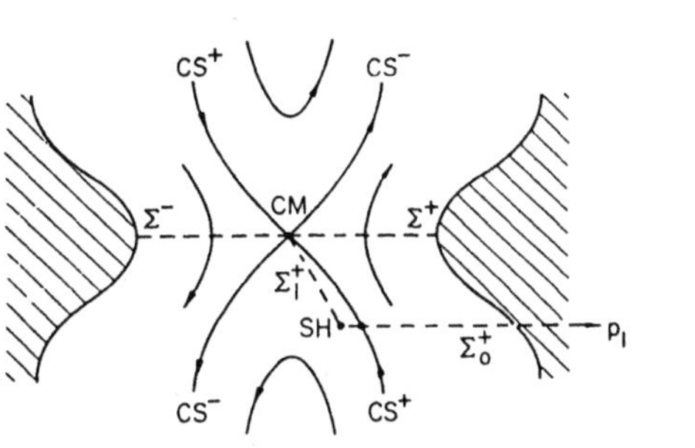

FIG. 1. Schematic representation of the flow through phase space for the evaluation of the net jump rate. The center stable (CS^+) and center unstable (CS^-) manifolds intersect in the center manifold CM. The total flux may be evaluated as an integral on any hypersurface extending from either CM^+, CM^-, or CM to regions of inaccessibly high energy (e.g., Σ^+ and Σ^-). Σ_0^+, which originates on the saddle hyperplane SH, cuts all positive flow and also some spurious flow from return jump trajectories. The flow through Σ_0^+ must be supplemented by a "correction" integral over the hypersurface Σ_I^+ that extends from CM to SH, thereby eliminating all and only the spurious flow.

energy inside one of these sectors cuts all the flux in one sense. The rate at which transitions occur is

$$R=\frac{1}{Z_\Gamma}\int_{\Sigma^+} e^{-\beta H}\mathbf{X}\cdot d\Sigma^+ , \tag{1}$$

or

$$R=\frac{Z_C}{\beta Z_\Gamma} , \tag{2}$$

in which H is the Hamiltonian, $\exp(-\beta H)$ is the distribution function, and Z_Γ is its normalizing partition function integral over the equilibrium site from which flow, velocity $\mathbf{X}$, originates. The second form, Eq. (2), follows from the Stokes theorem with Z_C the partition integral in CM. In writing Eq. (1) as the net flux we make the assumption that the motion is randomized during any passage around CS^+ and CS^-. This is the remaining "short-memory" approximation on which the SM-ART treatment focuses attention.

Two simple adaptations of this general result are of interest here. First, when the barrier is planar, CM is a "saddle hyperplane" that factors into separate space and momentum terms.[13] Then

$$R_0=\frac{Z_{SH}}{\beta Z_\Gamma} , \tag{3}$$

in which the momentum integrals can now be performed to yield the Vineyard result[14]

$$R_0=\left[\frac{kT}{h}\right]e^{-f/kT} , \tag{4}$$

with f the difference between $F_{SH}=-kT\ln Z_{SH}$ and $F_\Gamma=-kT\ln Z_\Gamma$, and h Planck's constant. Second, even when CM cannot be factored one can nevertheless still define a saddle hyperplane SH and write

$$R=\frac{Z_C}{Z_{SH}}\frac{Z_{SH}}{\beta Z_\Gamma}=cR_0 . \tag{5}$$

The full flux thus appears as the rate of saddle plane crossings R_0 multiplied by a correction factor c we term the "conversion coefficient," following Bennett's early concepts.[8]

Equation (5) provides the means by which we apply computational methods to the problem of vacancy jumps in fcc metals. It identifies the true flux as a product of two factors. One of these is the rate at which flow cuts an appropriate saddle hypersurface. The second factor is the ratio of true flux through Σ^+ to the larger flux through the surface connecting SH to remote regions of high energy. Accordingly, the second of these consecutive papers[1] considers the conversion coefficient c. The flux through SH is the main topic of the present paper. It will therefore be evident that, from one standpoint, the results described below bear no direct relationship to the SM-ART treatment but instead comprise an evaluation of the Vineyard model for the system under investigation. These results nevertheless form one basic ingredient of the complete flux when approached via Eq. (5), and require evaluation for a quantitative determination of the jump rate.

While the problem undertaken here has thus been clearly defined since 1957,[14] it has not been investigated in a fully comprehensive way by previous workers and, indeed, presents serious computational challenges. The analogous problem of the vacancy formation free energy was first treated at the required level by Squire and Hoover[3] using Monte Carlo methods and, more recently, by Jacucci and Ronchetti[4] using more accurate difference Monte Carlo procedures of the type pioneered by Bennett.[15] The vacancy jump calculation reported here is more difficult because the equilibrium and saddle-plane configurations to be compared have rather different geometries and relaxations. Vineyard's original work provided harmonic expressions for the partition function in f, Eq. (4), and these have since been investigated for explicit volume dependence, using the quasiharmonic approximation, in recent work by DeLorenzi and Jacucci[16] for Ar and by Harding[17] for diatomic salts. Parallel work by Voter and Doll[9] has concerned surface diffusion. In the present research we undertake a full calculation of f using difference Monte Carlo methods designed particularly for the jump problem. The principal goal was to obtain jump rates to a precision comparable with 1%, starting from the interatomic potential and masses alone. A preliminary description of the results has appeared elsewhere.[18]

As mentioned earlier, our evaluation of the jump rate R in Eq. (4) with the stated precision does not provide a comparably precise value of the true rate at which atoms change sites. In effect, the Vineyard approach miscounts the jumps.[19,7] A description of this problem is deferred to the second paper, where the correction terms that give the true jump rate are derived within the SM-ART approximation to an equal or higher accuracy. The overall effect of the work is to make true jump rates accessible at the 1% level. In addition, the available accuracy allows precise discussion of subtle phenomena like the isotope effect and the physical processes that determine it in real cases. A description of these matters is deferred to the second paper. In Sec. II of the present paper we take up the topic of Monte Carlo difference calculations applied to the vacancy jump problem for fcc crystals, and present the results of calculations in Sec. III. Section IV is a brief summary of the work.

II. THE FREE-ENERGY DIFFERENCE BETWEEN SH AND Γ

A. The difference method

In conformity with the discussion in Sec. I [cf. Eq. (3)] we wish to evaluate the ratio $Z_{SH}/\beta Z_\Gamma$ by a numerical procedure. In terms of the free energies $F_{SH} = -\beta^{-1}\ln Z_{SH}$ and $F_\Gamma = -\beta^{-1}\ln Z_\Gamma$, the quantity we seek is the free-energy difference $f = F_{SH} - F_\Gamma$. The free energies are extensive quantities of order $2N$, with $N/3$ the number of particles, while their difference f has a value of order unity, independent of N in the limit of large N. For this reason it is clear that a better statistical evaluation of f can be made by direct difference methods than from separate calculations of F_{SH} and F_Γ, since the statistical fluctuations of the extensive quantities would tend to mask their small difference. This problem is common to the evaluation of all the properties of point defects.[2]

Monte Carlo sampling procedures have been used to evaluate differences of free energy by many authors since the first suggestion by McDonald and Singer[20] (see, for example, the review by Frenkel[21]). The idea is to obtain the free-energy *difference* by sampling the internal-energy difference ΔU between the two systems. For this purpose the ratio of the partition functions may be written as an expectation value of $\exp(-\beta\Delta U)$, obtained from an ensemble average on one of the two systems. Thus

$$\frac{Q_2}{Q_1} = \frac{\int e^{-\beta(U_2-U_1)} e^{-\beta U_1} dq_1 \cdots dq_N}{\int e^{-\beta U_1} dq_1 \cdots dq_N} = \langle e^{-\beta\Delta U}\rangle_1 \, . \quad (6)$$

The accuracy in this "one-sided" evaluation of the average, when only a *finite* number of configurations is sampled using a Metropolis Monte Carlo algorithm, depends sensitively on the degree to which overlap exists between the density functions for the two systems. If the overlap is large, then a finite sample of configurations of system 1 identifies a representative sample of configurations of system 2 also, and the evaluation gains accuracy. If, conversely, the overlap is small, the configurations sampled in system 2 will not be satisfactorily representative of the distribution of ΔU.

A more accurate evaluation can be obtained for the same computational effort by dividing the sampling between the two ensembles. Normalized distribution functions

$$h_1(\Delta) = \int \frac{1}{Q_1}\delta(U_2 - U_1 - \Delta) e^{-\beta U_1} dq_1 \cdots dq_N \equiv \langle \delta(U_2 - U_1 - \Delta)\rangle_1 \quad (7)$$

and

$$h_2(\Delta) = \int \frac{1}{Q_2}\delta(U_2 - U_1 - \Delta) e^{-\beta U_2} dq_1 \cdots dq_N \equiv \langle \delta(U_2 - U_1 - \Delta)\rangle_2 \quad (8)$$

can then be constructed from separate samplings on the two systems. A direct evaluation of the ratio Q_2/Q_1 follows as

$$\frac{Q_2}{Q_1} = e^{\beta\Delta}\frac{h_1(\Delta)}{h_2(\Delta)} \, . \quad (9)$$

The statistical error of this procedure can be reduced considerably by the "acceptance ratio method" proposed by Bennett.[15] The method is based on the identity

$$\frac{Q_2}{Q_1} = \frac{\langle f(U_2 - U_1 + C)\rangle_1}{\langle f(U_1 - U_2 + C)\rangle_2} e^{C} \, , \quad (10)$$

with

$$C = \ln(Q_2 n_1 / Q_1 n_2) \, , \quad (11)$$

and where $f(x) = (1+e^x)^{-1}$ is the Fermi function, and with n_1, n_2, the number of independent configurations taken from ensembles 1 and 2, respectively. Bennett has shown that this choice of f and C minimizes the expected error σ of a finite-sample estimate of the free-energy difference. C is not known *a priori*, and in practice is evaluated by graphical methods. The overlap of the distributions $h_1(\Delta)$ and $h_2(\Delta)$ again determines the minimum uncertainty σ, which can be obtained as

$$\sigma^2 = \frac{1}{n_1}\frac{\langle f^2\rangle_1 - \langle f\rangle_1^2}{\langle f\rangle_1^2} + \frac{1}{n_2}\frac{\langle f^2\rangle_2 - \langle f\rangle_2^2}{\langle f\rangle_2^2} \, . \quad (12)$$

Before proceeding it is useful to examine more closely the partition functions whose ratio we wish to evaluate. A detailed description is given in Ref. 11 and is reviewed here in order to introduce the basic elements for the subsequent discussion. Our dynamical system consists of $N/3$ particles vibrating about the lattice sites of a face-centered-cubic crystal structure. One site is empty to simulate a lattice vacancy. The equilibrium configuration of the system is found by letting all particles relax towards the minimum of the total potential energy. In practice this is achieved by a damped molecular dynamics procedure. Similarly, the saddle-point configuration for migration is found by constraining one atom to lie halfway between two empty lattice sites and allowing all the other particles to relax towards the minimum of the potential energy. In these relaxation processes all the atoms of the system are free to move but only those closest to the vacancy, and for the saddle point those closest to the migrating atom, undergo significant displacements with respect to their perfect lattice positions.

The normal modes of the crystal with the vacancy are found by diagonalizing the dynamical matrix of the system in the equilibrium configuration. There are a total of N eigenvectors and eigenvalues, three of which correspond to the translational degrees of freedom and have zero eigenvalues (periodic boundary conditions eliminate rotations). In the discussion that follows these three translational degrees of freedom are always kept fixed. Similarly, the eigenvectors and eigenvalues for the saddle point can be evaluated. One of them has a negative eigenvalue which corresponds to the unstable mode or reaction coordinate q_1. The saddle plane in configuration space is defined as the hyperplane normal to q_1. If the jumping atom were completely decoupled from the other atoms of the lattice, q_1 would coincide with the jump direction of the jumping atom. In real cases the coupling between the jumping atom and the others produces nonzero components of q_1 along the coordinates of the other atoms.

We indicate by Γ the relevant part of phase space over which the distribution is normalized. The configurational part of Γ corresponds to the region around the equilibrium configuration. SH is a subspace of phase space, a "saddle hyperplane," normal both to the reaction coordinate q_1 and its conjugate momentum p_1. Its configurational part is the saddle plane defined above.

The two partition functions, Z_{SH} and Z_Γ, are thus defined on spaces of different dimensionality. For convenience of calculation we can make them congruent by introducing two extra degrees of freedom whose analytic contribution to the partition functions can be factored out exactly. To this end we introduce, in the ensemble corresponding to SH, two conjugate coordinates, q_1' and p_1', which are left uncoupled from the rest and whose contribution to the potential is taken to be harmonic with frequency ω_0, and with mass M equal to the mass of the other particles. The frequency ω_0 is chosen to match the square root of the curvature of the potential-energy surface at the equilibrium point in the direction parallel to the reaction coordinate, in order to maximize the overlap. The Hamiltonian function of this extended system SH' is

$$H'=\sum_{i=2}^{N-3}\frac{1}{2}\frac{p_i^2}{M}+\frac{1}{2}\frac{(p_1')^2}{M}+V(\{q_i\},q_1=0)+\tfrac{1}{2}\omega_0^2(q_1')^2\ . \tag{13}$$

The ratio $Z_{SH}/\beta Z_\Gamma$ can now be rewritten in terms of the new partition function $Z_{SH'}$ in the form

$$\frac{Z_{SH}}{\beta Z_\Gamma}=\frac{Z_{SH'}}{Z_\Gamma}\frac{Z_{SH}}{\beta Z_{SH'}}=\frac{\omega_0}{2\pi}\frac{Z_{SH'}}{Z_\Gamma}\ . \tag{14}$$

As the momentum part of the partition functions $Z_{SH'}$ and Z_Γ can now be factored out, we can write

$$\frac{Z_{SH}}{\beta Z_\Gamma}=\frac{\omega_0}{2\pi}\frac{Q_{SH'}}{Q_\Gamma}\ , \tag{15}$$

where $Q_{SH'}$ and Q_Γ are the corresponding configurational partition functions. It remains to implement a Monte Carlo procedure to evaluate the ratio $Q_{SH'}/Q_\Gamma$, which will give the jump rate as

$$R_0=\frac{\omega_0}{2\pi}\frac{Q_{SH'}}{Q_\Gamma}=\frac{\omega_0}{2\pi}e^{-\beta f'}\ . \tag{16}$$

The difference method by which we evaluate the migration free energy involves the comparison of two potential functions defined in the same configuration space. The ground and extended saddle surface configurations have equal dimensions, as described above, but their energies are different functions of position and they exhibit two well-separated energy minima. We need to fix a one-to-one correspondence between points in the two systems in order to proceed with difference sampling. We may do so by means of a transformation that leaves volume elements in configuration space unchanged and, at the same time, connects configurations of high statistical weight in the respective ensembles. Two transformations are discussed as candidates for this role in what follows.

If $\mathbf{P}_1$ is a configuration sampled in system 1, corresponding to a point in the equilibrium region, and its energy is $U_1=V(\mathbf{P}_1)$, where V is the sum of all pair interactions between the particles, we may identify a corresponding configuration in system 2, namely $\mathbf{P}_2$, such that

$$\mathbf{P}_2=T\mathbf{P}_1 \text{ and } \mathbf{P}_1=T^{-1}\mathbf{P}_2\ . \tag{17}$$

Here T is a transformation that maps points between the two systems. In general $\mathbf{P}_2$ will not lie on the saddle hyperplane; in this case we can use the projection operator Π to project $\mathbf{P}_2$ onto the saddle hyperplane, and separate off the out-of-plane component $q_1(\mathbf{P}_2)$. The energy U_2 may then be assigned according to

$$U_2=V(\Pi\mathbf{P}_2)+\tfrac{1}{2}\omega_0^2q_1^2\ , \tag{18}$$

where V is again the pair interaction between the particles in the configuration projected on the saddle hyperplane.

Various choices could be made for the transformation T. An appropriate choice will improve the overlap between the two distributions $h_1(\Delta)$ and $h_2(\Delta)$ of Eqs. (7) and (8). Consequently the transformation must be such that during the sampling of configurations in one system the corresponding set of configurations of the second system provides a satisfactory representation of a substantial fraction of the important configurations of the second ensemble.

One straightforward choice for the transformation is a simple translation

$$\mathbf{P}_2=\mathbf{P}_1+\mathbf{r}_1\ , \tag{19}$$

in which

$$\mathbf{r}_1\equiv(\Delta\mathbf{x}_1,0,0,\ldots) \tag{20}$$

is a displacement of the jumping atom alone through the distance from its equilibrium position in the lattice to its saddle-point position halfway between the two vacancies, all other atoms remaining fixed. This transformation is analogous to the method used by Jacucci and Ronchetti[4] to determine the free energy of vacancy formation, where one particle was "inserted" in the lattice and all the others remained fixed. In both cases, this procedure ensures that the interaction energies of all the remaining atoms, not affected by the transformation, exactly cancel in the potential-energy difference ΔU. This reduces ΔU to a nonextensive property and the desired free-energy difference f to a value comparable to the energy per atom.

A second useful choice for the transformation is a translation by the vector configurational displacement from the *relaxed* equilibrium configuration to the *relaxed* saddle point, namely

$$\mathbf{r}_N \equiv (\Delta\mathbf{x}_1, \Delta\mathbf{x}_2, \ldots, \Delta\mathbf{x}_{N/3}) , \qquad (21)$$

where

$$\Delta\mathbf{x}_i = \mathbf{x}_i^{\mathrm{SP}} - \mathbf{x}_i^{\mathrm{EQ}} . \qquad (22)$$

For this transformation the remote atoms are little affected and the advantage of major cancellations in the energy difference is again retained. In addition, the most probable configuration of one ensemble is directly mapped onto the most probable configuration of the other ensemble so that the relaxed migration energy subtracts out exactly and plays no role at low temperatures. This choice therefore guarantees a good overlap of the two ensembles, even at *low temperature.* The resulting procedures permit an accurate monitor of anharmonic deviations as the temperature is increased, and give a check of Monte Carlo results in the region where the anharmonic expansion and quasiharmonic lattice dynamics are expected to be valid.

The results of the two methods are compared in Fig. 2. "Right" and "left" distributions of ΔU are shown for a low temperature and for the melting temperature of the Lennard-Jones model. Results obtained with the transformation $\mathbf{r}_1$ are in Figs. 2(a) and 2(c), and with transformation $\mathbf{r}_N$ in Figs. 2(b) and 2(d). At low temperature the difference is dramatic. Almost no overlap occurs in the distributions obtained with $\mathbf{r}_1$, while good overlap is obtained with $\mathbf{r}_N$. At high temperature the distributions obtained with the two different methods look somewhat different but both exhibit an acceptable overlap. Thus use of the second method merely improves the overlap and therefore the efficiency of the free-energy estimate at high temperature, whereas at low temperature the second method is absolutely necessary to obtain a useful overlap at all. The relaxation energy released by displacing the system from equilibrium to the saddle point, no matter how large, is subtracted out exactly. The method can therefore be employed rather generally to defect formation and migration. A possible application is to ionic systems where the energy changes with configuration are often very large with respect to kT, and the distribution overlap prior to the transformation is therefore negligibly small.

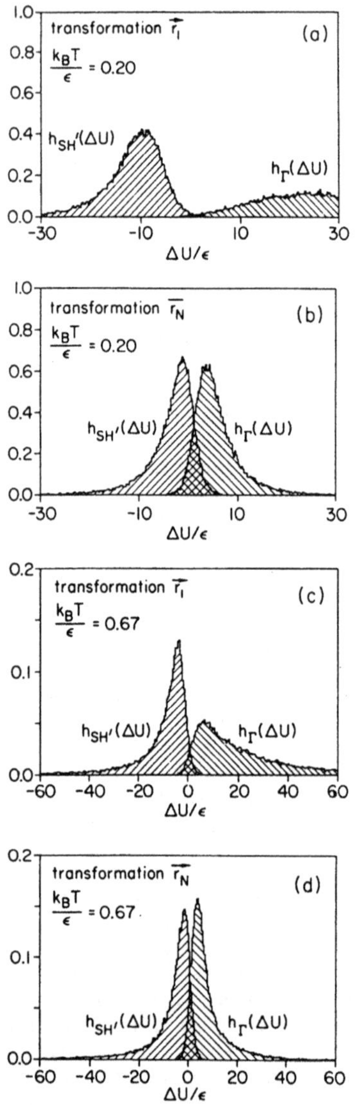

FIG. 2. Distributions of the energy difference $\Delta U = U_{\mathrm{SH'}} - U_\Gamma$ between the extended saddle plane SH' and the equilibrium state Γ, obtained from Monte Carlo samplings at two temperatures for the Lennard-Jones model at volume V_1 (see Table II) and for two different transformations $\mathbf{r}_1$ and $\mathbf{r}_N$ (see text).

B. Practical details

Difference Monte Carlo procedures outlined in Sec. II A were employed for a system of 31 particles in a 32-site fcc crystallite constrained by periodic boundary conditions. Transformation $\mathbf{r}_N$ was used at all temperatures.

To expedite sampling in configuration space, Monte

Carlo trial moves were constructed as small displacements of all normal coordinates, each scaled by its corresponding eigenvalue to mimic the probability distribution of the harmonic ensemble. After transformation to Cartesian coordinates the Metropolis algorithm was used to accept or reject the trial move according to the Boltzmann distribution of the system on which the sampling was performed. When sampling on the saddle plane (for the left distribution), 89 of the sampled coordinates [i.e., $(N-4)$, all except the three translational degrees of freedom and the reaction coordinate q_1] were transformed back to Cartesian coordinates to give a point on the saddle plane, and then used to evaluate the energy $V(\{\mathbf{x}_i\},q_1=0)$. To the coordinate q_1 a harmonic energy $\frac{1}{2}\omega_0^2 q_1^2$ was assigned and the sum of the two employed in the evaluation of the Boltzmann exponential.

Acceptance probabilities of $\sim 25\%$ were typically used at all temperatures. Correlation lengths for the energy difference, as measured by the number of correlated configurations along the Markov chain produced by the Metropolis algorithm with our chosen procedures, were monitored through the calculations. They provide estimates of the number of independent configurations from each ensemble for use in evaluating the expected error according to Eq. (12). Correlation lengths ranging from 100 to 150 were found for the equilibrium ensemble, while values ranging from 160 to 225 were found for the saddle-plane ensemble. Typically 2×10^5–10^6 Monte Carlo moves were used for each calculation.

The dynamics of the system in its ground state admit the possibility that a vacancy jump takes place. For this reason it is necessary to include an accurate check in the program to avoid jumps during the Monte Carlo sampling. To this end all the 12 saddles around the equilibrium configuration were monitored and their respective reaction coordinate checked at each Monte Carlo move. Any trial move in which any of the reaction coordinates passes the zero threshold was rejected. For sampling at the saddle point all the 23 neighboring saddles were checked; they would correspond to double jumps of the vacancy. The number of trial moves rejected for single jumps was of the order of 1 every 10^5 at the melting temperature and much less at lower temperatures. Only a few double jumps were found and rejected.

Boundary effects can be very large in small systems and special care is needed to avoid them. For the 32-site system employed in the present research these questions are of special concern. One delicate point concerns the potential. In our calculation the potential was set to zero at distances equal to or larger than the side of the simulation box. Periodic boundary conditions were used so that each particle interacted with more than one image of the same neighbor but did not interact with images of itself. Although such a large cutoff includes up to seven shells of neighbors (164 neighbors per particle in the bulk), and the Lennard-Jones potential there has a value of only 0.4% of the well depth, energy discontinuities were nevertheless observed during Monte Carlo runs, as particles passed through the cutoff limit. This problem was avoided by fixing the set of interacting neighbors from the outset and using the same set for the whole Monte Carlo calculation. Consequently, two sets of interacting neighbors were evaluated, respectively, in the relaxed equilibrium configuration and in the relaxed saddle-point configuration. The calculation ran faster when the same list of neighbors was retained. The average error in the energy for the Lennard-Jones model, introduced by always using the same list of neighbors instead of by a cutoff, is negative and increases in magnitude from zero at low temperatures to a maximum of 0.008ε at equilibrium and 0.015ε on the saddle plane at the melting temperature. Here ε is the well depth. In the free energy the error is again larger at the saddle point; it is positive and equal to 0.01ε at the melting temperature. This systematic error turns out to be about one-third of the statistical uncertainty found in our final Monte Carlo evaluations for 10^6 steps.

III. RESULTS

Values of the free-energy difference f were obtained for three different model potentials to simulate crystalline argon, silver, and copper. The same pair potentials used in earlier publications[11,13,18] were employed here: for Ar, the Lennard-Jones potential

$$V(r)=4\varepsilon\left[\left[\frac{\sigma}{r}\right]^{12}-\left[\frac{\sigma}{r}\right]^{6}\right] \tag{23}$$

was employed with $\varepsilon=119$ K and $\sigma=3.4$ Å; for Ag and Cu, the Morse potential

$$V(r)=D\left(e^{-2\alpha(r-r_0)}-2e^{-\alpha(r-r_0)}\right) \tag{24}$$

was taken, with $D=3823$ K, $\alpha=1.3939$ Å^{-1}, and $r=3.096$ Å for Ag, and with $D=3999$ K, $\alpha=1.3921$ Å^{-1}, and $r=2.838$ Å for Cu. These values of the parameters are designed to simulate the experimental values of the sublimation energy, the compressibility, and the lattice parameter at zero temperature. For purposes of useful comparison it will be convenient to report the results for the three models using the depth of the potential wells as units of energy, namely, ε for the Lennard-Jones potential and $\varepsilon\equiv D$ for the Morse potentials. In units of ε/k_B the experimental melting temperatures of the three substances are respectively 0.672 for Ar, 0.320 for Ag, and 0.339 for Cu. Similarly, the radii for the first zero of the potential functions can be employed as units of length, namely, σ for the Lennard-Jones potential and $\sigma\equiv r_0-\alpha^{-1}\ln 2$ for the Morse potentials.

Analytical predictions can be given exactly for the first three terms of the expansions of the free energy in powers of the temperature. The expansions are valid at low temperature and no estimates exist, to our knowledge, of the validity at higher temperatures in the range where actual diffusion experiments are carried out. Coefficients of the expansion

$$f_3=\Delta_1+\Delta_2\beta^{-1}+\Delta_3\beta^{-2} \tag{25}$$

have already been calculated in previous work[11,13,16] for the same models and thermodynamic states used in the present paper. Δ_1 is to the height of the barrier, ΔU_0, and Δ_2 is the classical harmonic entropy difference, $\Delta S = k_B \sum_{i=1}^{N-3} \ln(\omega_i^{EQ}/\omega_i^{SP})$. Here the ω_i^{EQ} are eigenfrequencies at equilibrium and ω_i^{SP} those for the system at the saddle point, including the fictitious frequency ω_0 introduced to extend the saddle hyperplane SH by one more dimension to SH'. Values of ω_0 for the systems studied in this work are reported in Table I. $f_H = \Delta_1 + \Delta_2\beta^{-1}$ is thus the free-energy difference produced by a harmonic approximation, i.e., the quadratic terms of expansions of the potential in powers of the atomic displacements from the equilibrium and saddle-point configurations. The coefficient Δ_3 is given by the third- and fourth-order terms of the anharmonic expansions. The complete expression in terms of coefficients of the potential expansions is derived in Ref. 11 and was used for actual calculations in Ref. 13. Values for Δ_1, Δ_2, and Δ_3 are reported in Table I for the cases relevant to the present calculation.

The main purpose of the present work is to evaluate the full anharmonic contribution to f and to compare it with the available analytical predictions of the power expansions. Calculations were performed for each model at densities close to the experimental values for the melting point at $P=0$, corresponding to lattice parameters of $1.6043\sigma_{Ar}$, $1.6165\sigma_{Ag}$, and $1.5795\sigma_{Cu}$, respectively. Figure 3 compares the results for the three model crystals. There the free energy f' [see Eq. (16)] is shown as a function of temperature. It is evident from the differences between the harmonic theory (solid lines) and the anharmonic results (points, Monte Carlo; dashed lines, analytical) that the Lennard-Jones model is very much more anharmonic than the Morse models of Cu and Ag. Despite these differences, the Monte Carlo and analytical evaluations of the jump rate agree in all three cases within the statistical uncertainties of the numerical methods. For the same amount of computer time, the uncertainties increase with temperature; larger error bars reflect shorter calculations for some of the reported data. It seems that precision within $\sim 1\%$ of the barrier height can be achieved by either method. The composite results give confidence that the jump rate within the present approximation can be predicted to within 1% or 2% at this level of effort. Further gains appear possible by means of more extensive computer runs.

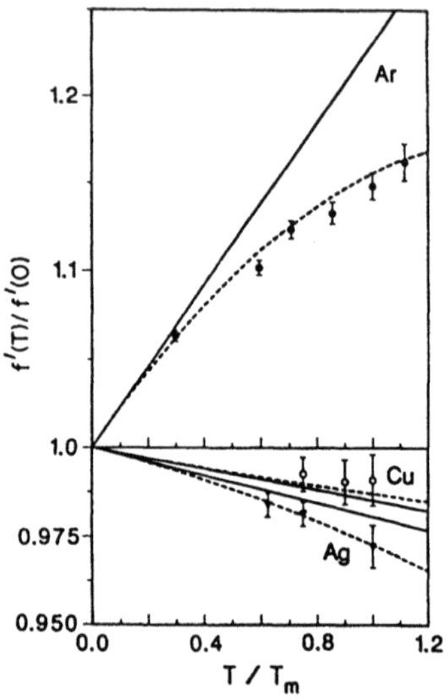

FIG. 3. Free-energy difference (in units of the barrier height) $f'(T)/f'(0)$ between the extended saddle plane SH' and the equilibrium state (see caption of Table I) as a function of the reduced temperature T/T_M, with T_M the melting temperature, for models of Ar, Cu, and Ag.

TABLE I. Coefficients of the anharmonic expansion $f'_3 = \Delta_1 + \Delta_2\beta^{-1} + \Delta_3\beta^{-2}$ of the free-energy difference between the extended saddle plane SH' (the saddle plane SH augmented by an extra degree of freedom with harmonic frequency ω_0: see text) and the equilibrium state Γ. Values are from Ref. 13. The jump rate through the saddle plane is $R_0 = (\omega_0/2\pi)\exp(-\beta f')$. The lattice parameters used in the calculations are $1.5861\sigma_{Ar}$ (V_1), $1.6043\sigma_{Ar}$ (V_2), $1.6165\sigma_{Ag}$, and $1.5795\sigma_{Cu}$.

	Ar V_1	Ar V_2	Ag	Cu
$\Delta_1\varepsilon^{-1}$	4.4004	3.7492	2.4897	3.9508
Δ_2	1.2668	1.2845	−0.1506	−0.0906
$\Delta_3\varepsilon$	−0.5540	−0.6209	−0.1922	0.0320
$\omega_0(m\sigma^2/\varepsilon)^{1/2}$	11.8844	13.2087	10.6128	11.0105

In order to investigate the volume dependence of anharmonicity we performed additional calculations for the Lennard-Jones model at a different volume corresponding to a lattice parameter of $1.5861\sigma_{Ar}$, equivalent to the experimental equilibrium density at temperature $T=0.75T_M$. The results are reported in Table II. We report there the differences between the Monte Carlo evaluations and the harmonic values, $f - f_H$, and also the analytical differences $f_3 - f_H$; these values are independent of any arbitrary choice for ω_0. At both volumes the agreement between the anharmonic expansion and the Monte Carlo results is excellent, and one cannot distinguish between the two sets of data within the statistical uncertainties.

It is interesting to compare the volume dependence of the anharmonic free energy determined here with that predicted by quasiharmonic theory. At the higher temperature $T=0.672$ and for a relative volume change of 3.4% we find a relative change in the free energy of 10%, to be compared to a relative free-energy change of 9% predicted by the quasiharmonic theory. As the Lennard-Jones potential exhibits the strongest anharmonicity among our models, we expect that the discrepancy between quasiharmonic and anharmonic predictions may be even smaller in other cases. We therefore conclude that the quasiharmonic theory is a reasonably satisfactory approximation. The deviation of anharmonic prediction from quasiharmonic results for argon is 6%, which corresponds to an error of 52% in the predicted vacancy jump rate at the melting tempera-

TABLE II. Comparison between anharmonic contributions to the migration free energy f of argon in units of the potential-well depth ε, from Monte Carlo and from the anharmonic expansion (anharm. exp.) for two different volumes, V_1 and V_2, corresponding to lattice parameters $d_1=1.5861\sigma_{Ar}$ and $d_2=1.6043\sigma_{Ar}$. The analytical predictions are taken from Ref. 13 (see Table I).

T (ε/k)	V_1 Monte Carlo $f-f_H$ (ε)	V_1 anharm. exp. f_3-f_H (ε)	T (ε/k)	V_2 Monte Carlo $f-f_H$ (ε)	V_2 anharm. exp. f_3-f_H (ε)
0.200	-0.020 ± 0.012	-0.025	0.200	-0.014 ± 0.019	-0.022
0.400	-0.133 ± 0.016	-0.099	0.300	-0.045 ± 0.028	-0.050
0.475	-0.148 ± 0.019	-0.140	0.400	-0.088 ± 0.040	-0.089
0.575	-0.242 ± 0.024	-0.205	0.521	-0.133 ± 0.021	-0.150
0.672	-0.307 ± 0.030	-0.280	0.600	-0.236 ± 0.070	-0.199
0.750	-0.355 ± 0.040	-0.349	0.800	-0.397 ± 0.041	-0.355

ture. Again, the errors for less anharmonic crystals are substantially smaller.

IV. SUMMARY

Methods of difference Monte Carlo sampling have been adapted to an evaluation of the difference of partition function integrals between the saddle-plane configuration and the equilibrium configuration of a model crystallite containing a vacancy. Our method provides accurate evaluations of the free-energy difference (a nonextensive property) between two many-particle systems by subtracting out exactly the difference of the relaxed energies, thereby reducing the free-energy difference to a value comparable to the energy per atom. The method can usefully be extended to defect and migration in other systems, particularly to those in which the energy change is very large. From applications of these procedures to models of Ar, Cu, and Ag we have obtained the rate R_0 at which diffusion jumps take place through a saddle plane that is assumed to be planar but otherwise is arbitrarily anharmonic. This result provides an initial estimate of the true jump rate for use with correction terms that are calculated in the paper that follows.[1] An important conclusion is that the temperature dependence of R_0 at constant volume is accurately described by the first three terms of a power-series expansion as a function of temperature. Thus $\ln R_0$ from Eq. (25) coincides with the Monte Carlo data for Ar to within their 1% statistical uncertainty. The first two terms, obtainable by the quasiharmonic treatment, provide a good first approximation with a maximum deviation of 6% at the melting temperature in $\ln R_0$ of Ar.

ACKNOWLEDGMENTS

We have benefited from discussions with M. Toller and from the advice and assistance of M. Marchese. This research was supported in part through the University of Illinois Materials Research Laboratory, with funds provided by the National Science Foundation, under Grant No. DMR-83-16981. One of us (G.J.) thanks the University of Illinois National Center for Supercomputing Applications for support during a visit.

*Permanent address: Dipartimento di Fisica, Università degli Studi di Trento, I-38050 Povo, Trento, Italy.

[1]M. Marchese, G. Jacucci, and C. P. Flynn, following paper, Phys. Rev. B 36, 9469 (1987).
[2]See, e.g, C. P. Flynn, *Point Defects and Diffusion* (Oxford University Press, London, 1972).
[3]D. R. Squire and W. G. Hoover, J. Chem. Phys. **50**, 701 (1969).
[4]G. Jacucci and M. Ronchetti, Solid State Commun. **33**, 35 (1980).
[5]N. L. Peterson, in *Diffusion in Solids*, edited by A. S. Nowick and J. J. Burton (Academic, New York, 1975).
[6]D. H. Tsai, R. Bullough, and R. C. Perrin, J. Phys. C 3, 2022 (1970).
[7]C. H. Bennett, in *Diffusion in Solids*, edited by A. S. Nowick and J. J. Burton (Academic, New York, 1975).
[8]C. H. Bennett, Thin Solid Films **25**, 65 (1975).
[9]A. F. Voter and J. D. Doll, J. Chem. Phys. **82**, 80 (1985); A. F. Voter, *ibid.* **82**, 1890 (1985); A. F. Voter, Phys. Rev. B **34**, 6819 (1986).
[10]H. A. Kramers, Physica **7**, 284 (1940).
[11]M. Toller, G. Jacucci, G. De Lorenzi, and C. P. Flynn, Phys. Rev. B **32**, 2082 (1985).
[12]R. Abraham and J. E. Marsden, *Foundations of Mechanics* (Benjamin, New York, 1967).
[13]G. Jacucci, G. De Lorenzi, M. Marchese, C. P. Flynn, and M. Toller, Phys. Rev. B **36**, 3086 (1987).
[14]G. H. Vineyard, J. Phys. Chem. Solids **3**, 121 (1957).
[15]C. H. Bennett, J. Comput. Phys. **22**, 245 (1976).
[16]G. De Lorenzi and G. Jacucci, Phys. Rev. B **33**, 1993 (1986).
[17]J. H. Harding, Physica **131B**, 13 (1985).
[18]M. Marchese, G. De Lorenzi, G. Jacucci, and C. P. Flynn, Phys. Rev. Lett. **57**, 3280 (1986).
[19]C. P. Flynn, Phys. Rev. Lett. **35**, 1721 (1975).
[20]I. R. McDonald and K. Singer, Discuss. Faraday Soc. **43**, 40 (1967); J. Chem. Phys. **47**, 4766 (1967); **50**, 2308 (1969).
[21]D. Frenkel, in *Molecular Dynamics Simulation of Statistical Mechanical Systems*, edited by G. Ciccotti and W. G. Hoover (North-Holland, Amsterdam, 1986).

Jump rate of the fcc vacancy in the short-memory–augmented-rate-theory approximation. II. Dynamical conversion coefficient and isotope-effect factor

M. Marchese, G. Jacucci,* and C. P. Flynn

Department of Physics and Materials Research Laboratory, University of Illinois at Urbana-Champaign, Urbana, Illinois 61801

(Received 20 April 1987)

The framework of short-memory–augmented-rate theory is employed to partition the vacancy jump rate into two factors related, respectively, to the Vineyard theory and to corrections for multiple barrier crossings inherent in the dynamics. This paper treats the second of the factors using molecular dynamics to locate critical trajectories that lie on invariant manifolds of the dynamical systems, starting from states of motion in the saddle plane. For models of Ar, Cu, and Ag the errors of the Vineyard treatment are negligible at low temperature and contribute a rate reduction of only $\sim 10\%$ at the melting temperature. These factors are, however, strongly mass dependent and give dominant contributions to the isotope effect. Our calculations reproduce the experimentally observed $\kappa \simeq 0.87$ near the melting temperature very well, and are similarly model insensitive; κ decreases nonlinearly from its harmonic value at $T=0$ as the temperature is increased. Detailed examination of trajectories shows that the isotope effect is largely determined by the deformation of the manifolds caused by core repulsive forces. In effect, the isotope dependence derives mainly from infrequent energetic collisions that take place in jump events.

I. INTRODUCTION

Classical statistical theory of diffusion jumps, as exemplified by Vineyard's[1] many-body formulation, has been unable to account for the isotope effects for diffusion observed in simple solids. The isotope-effect factor κ conveys the mass sensitivity of the jump rate R through the equation

$$\frac{R(M)}{R(M+dM)}=1+\kappa\frac{dM}{2M}\,, \qquad (1)$$

so that $K=1$ when $R \sim M^{-1/2}$, as is appropriate for an independent particle in thermal equilibrium. For fcc metals κ is typically 13% below unity,[2] and between 40% and 50% for bcc metals.[2,3] These are the consequences of cooperative motion by neighbors during the jump, which reduce the mass sensitivity. Regarded in this light the isotope effect offers a direct, and indeed the only, monitor of dynamics during the jump event itself. Calculations based on Vineyard's theory give reductions several times smaller than those observed.[2] It was suggested by Flynn[4] that the discrepancies must be temperature-dependent effects of anharmonicity that cause incorrect accounting for jump events. This was consistent with early molecular-dynamics simulations of a Lennard-Jones model for argon near its melting point by Bennett,[5,6] who computed for that model a mass sensitivity near the value actually observed in metals. A good deal more data[7–9] on various systems has accumulated since that time but no theoretical advances have occurred.

Because of its importance as a potential probe of jump dynamics, the isotope effect in diffusion warrants careful attention. Indeed, extensive experimental investigations were originally undertaken in the hope that the results could shed light on otherwise inaccessible phenomena that take place in the jump process. At the same time, the fundamental question of precise criteria by which jump events can be identified has come to the forefront in our own research,[10–12] in that of Doll and Voter[13–15] on surface diffusion and, in a different context, in the rate processes exhibited by macromolecular systems.[16] One of the main results of the present paper is that the mass dependence of the jump criterion plays a critical role in determining the isotope effect in solid-state diffusion. By means of a precise formulation of the jump problem we are able to obtain an accurately predictive treatment of the isotope effect and to understand the physical causes of its hitherto unexplained behavior. These topics form the subject matter of the present paper.

This paper, the second of two, is concerned with the accurate description of jump processes within the short-memory–augmented-rate theory (SM-ART).[11] As described in the companion paper by De Lorenzi *et al.*[17] that precedes this, certain invariant manifolds in phase space divide the Hamiltonian flow over a potential barrier in a convenient way. Figure 1 shows schematically the stable (CS^+) and unstable (CS^-) center manifolds intersecting at the center manifold (CM) itself. For a crystal of $N/3$ atoms CM is the $(2N-2)$-dimensional manifold of all trajectories that linger indefinitely on the barrier; CS^- is the $(2N-1)$-dimensional manifold obtained when CM is slightly disturbed so that this configuration decomposes, and CS^+ is its time-reversed analog. CS^+ and CS^- divide phase space into four sectors. Two of these pertain to flow over the barrier parallel to the reaction coordinate q_1. The important point is that each of the two contains all the flow in one sense, so that any surface Σ cutting the appropriate sector intersects all the flow. Consequently, the rate R at which jumps take place in one sense may be written as an integral of the

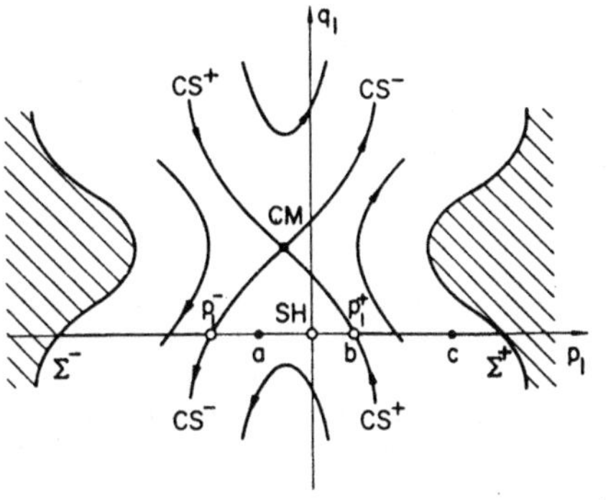

FIG. 1. Section of the tunnel in phase space connecting two equilibrium configurations A and B. The plane (q_1,p_1), with q_1 the reaction coordinate, is shown for a given choice of the remaining normal coordinate and momenta on SH. By definition (see text) SH is the origin while the intersection of this plane with the center manifold CM is the point specified by the crossing of the center stable (CS^+) and unstable (CS^-) manifolds; CS^+ and CS^- partition the flow of trajectories with different fates. For a given state on SH a change of the normal momentum corresponds to displacement along p_1. Only trajectories with initial normal momentum $p_1>\max(p_1^-,p_1^+)$ belong to the successful jump flow from A to B [$p_1<\min(p_1^-,p_1^+)$ to the one from B to A]. The full jump rate is obtained by integrating the flow through the surfaces $\Sigma^{\pm}$.

flow velocity $\mathbf{X}$ over the surface Σ:

$$R=\frac{1}{Z_\Gamma}\int_\Sigma e^{-\beta H}\mathbf{X}\cdot d\mathbf{\Sigma}\ , \quad (2)$$

with H the Hamiltonian, $\exp(-\beta H)$ the density distribution, and Z_Γ its normalization integral over the initial equilibrium configuration in phase space. When the surface Σ is chosen to extend from CM to regions of inaccessibly high energy, a more elegant formula, obtained by means of the Stokes theorem, displays the rate as a ratio between the partition function Z_Γ and that of the system constrained on the CM, namely Z_C:

$$R=\frac{Z_C}{\beta Z_\Gamma}\ . \quad (3)$$

This is the principal result of the SM-ART treatment of Toller *et al.*,[11] on which the work that follows is based.

We reiterate further, as in De Lorenzi *et al.*,[17] that it is often useful to work with a reference surface that has a simpler geometry than CM itself. This allows the total flow to be regarded as the flow with respect to the reference surface, together with a correction term that accommodates differences between the reference surface and the center manifold. A natural choice of reference is a "saddle hyperplane," SH, namely the product space of a saddle plane and the momentum coordinate it contains; the plane itself can usefully be defined by diagonalizing the potential function about the saddle point of the potential barrier. Then the rate follows from Eq. (3) as

$$R=R_0c\ , \quad (4)$$

with

$$R_0=\frac{Z_{SH}}{\beta Z_\Gamma} \quad (5)$$

and

$$c=\frac{Z_C}{Z_{SH}}\ . \quad (6)$$

Or, alternatively, from Eq. (1), we may write

$$c=\frac{\beta}{Z_{SH}}\int_\Sigma e^{-\beta H}\mathbf{X}\cdot d\mathbf{\Sigma}\ . \quad (7)$$

Here Z_{SH} is the partition function of the system constrained on SH. The separation of these terms in Eq. (4) is the basis on which the present calculations are undertaken.

Equation (5) for R_0 is essentially the Vineyard result[1] for the rate at which transitions take place through a planar saddle surface. Its precise evaluation for simple models of fcc solids is the task completed in the preceding paper. Our purpose in the present paper is to evaluate the dynamical "conversion coefficient" c to find the size of the error rate theory makes in counting the flux of transitions.

A point of special interest concerns the isotope effect which, from Eq. (1), is given by the expression

$$\kappa=-2\frac{\partial \ln R}{\partial \ln M}\ . \quad (8)$$

Using Eq. (4) for R, we can factorize κ into two terms:

$$\kappa=-2\frac{\partial \ln R_0}{\partial \ln M}-2\frac{\partial \ln c}{\partial \ln M}=\kappa_H-2\frac{\partial \ln c}{\partial \ln M}\ . \quad (9)$$

It has long been known[4] that the isotope-effect factor κ_H, for a barrier that is planar but otherwise arbitrarily anharmonic, is defined by the normal to the plane. For realistic models this gives predictions that are incorrect at high temperature, as first pointed out by Huntington *et al.*[18] In general the barrier is not planar,[4,10] however, and anharmonicity can bring mass-dependent terms into the conversion coefficient.[4,10,12] The consequent suggestion[4] that the experiments must probe the mass dependence of c is consistent also with the results of molecular dynamics simulation of Ar near its melting temperature.[5,6] Our interest in what follows therefore focuses on both the magnitude of c and its mass dependence.

The plan of the paper is as follows. In Sec. II we first address the topological structure of the invariant manifolds associated with the migration barrier, and then describe a numerical procedure for sampling the relevant manifolds to obtain the required correction factor. Section III presents the results of the calculations for models of Ar, Cu, and Ag. In the subsequent discussion of these data, and particularly their dependence on the interatomic potential, it is possible to identify without ambiguity the unexpected fact that energetic hard-core collisions play a major role in determining the isotope effect in diffusion. Section IV provides a brief summary of the results.

II. APPLICATIONS OF THE THEORY

The framework outlined in Sec. I can be applied to the conversion coefficient in different ways and with various levels of approximation. A principal purpose of this paper is to describe its exact, numerical evaluation within the SM-ART theory using the full interatomic potential. This is the main subject matter of Sec. II B. Section II A focuses on some analytical aspects that give useful insight into both the topology of the manifolds and the character of the trajectories around CM. In particular, one can calculate the partition functions of Eq. (3) for a truncated series expansion of the anharmonic potential energy and in this way obtain the jump properties as a power series in T. The first-order term in T has been determined and evaluated earlier[11,12] for particular model fcc systems. The results provide a theoretical framework on which the numerical methods may be based. A comparison among results from the various approaches in Sec. III gives new physical insight into the processes that determine the isotope effect.

A. Critical trajectories and the invariant manifolds

We are interested here in the trajectories available to a dynamical many-particle system near points at which the potential energy V possesses a simple saddle point P_0. The Hamiltonian of a crystal with $N/3$ atoms can be expressed in terms of mass-weighted normal coordinates q_i of this saddle point, together with their canonically conjugate momenta p_i, as

$$H=\sum_{k=1}^{N} p_k^2+V_0-\tfrac{1}{2}\eta^2 q_1^2+\tfrac{1}{2}\sum_{k=2}^{N}\omega_k^2 q_k^2+V_a(q_1,\ldots,q_N)\,, \tag{10}$$

where V_a is an infinitesimal of third order containing all anharmonic parts of V. These normal coordinates q_i are obtained by an orthogonal transformation of the N mass-weighted atomic Cartesian coordinates, $x_i M_i^{1/2}$, with associated velocities $\dot{x}_i$, $i=1,\ldots,N$, that diagonalize the dynamical matrix for the system at the saddle point $P_0\equiv(x_1^0,\ldots,x_N^0)$. Thus

$$\mathbf{D}_{rs}=(M_r M_s)^{-1/2}\left[\frac{\partial^2 V}{\partial \mathbf{x}_r \partial \mathbf{x}_s}\right]_{P_0}, \tag{11}$$

such that

$$\mathbf{D}_{rs}\cdot\mathbf{a}_r^i=\varepsilon_i \mathbf{a}_s^i\,, \tag{12}$$

where $\varepsilon_1=-\eta^2$, $\varepsilon_2=\omega_2^2,\ldots,$ and $\varepsilon_N=\omega_N^2$ are the eigenvalues and $\mathbf{a}_r^i$ the orthonormal eigenvectors which define the transformation

$$q_i=\sum_r M_r^{1/2}\mathbf{a}_r^i\cdot(\mathbf{x}_r-\mathbf{x}_r^0)\,. \tag{13}$$

We associate the position and velocities of the jumping atom with indices 1, 2, and 3 of the Cartesian coordinates. Further, q_1 and p_1 are the normal coordinate and momentum associated to the unstable mode at the saddle point; they define the "reaction coordinate" of the system.

For a dynamical system of this kind the center stable manifold CS^+ and the center unstable manifold CS^- are given by equations of the kind:[11]

$$p_1^{\pm}=\mp\eta q_1+F_{\pm}(q_1,q_2,\ldots,q_N;p_2,\ldots,p_N)\,. \tag{14}$$

CS^+ and CS^- intersect in the center manifold CM described by the equations

$$\begin{aligned} q_1&=f(q_2,\ldots,q_N;p_2,\ldots,p_N)\,,\\ p_1&=g(q_2,\ldots,q_N;p_2,\ldots,p_N)\,. \end{aligned} \tag{15}$$

The functions $F_{\pm}$, f, and g are defined for sufficiently small values of their arguments. Explicit analytical expressions for these functions, given in Ref. 11, and collected in the Appendix, are valid for a truncated power-series expansion of the anharmonic term V_a of the potential.

We wish to calculate the rate at which representative points pass through the surface Σ in Eq. (2). When this surface is defined by coordinates and conjugate momenta of a planar saddle surface, the behavior in the neighborhood of P_0 is simplified. This "saddle hypersurface," SH, defined by the equations

$$q_1=0,\quad p_1=0\,, \tag{16}$$

comprises a second $(2N-2)$-dimensional hypersurface analogous to the center manifold. The two coincide in general only for planar potential barriers. A given set of values for all the normal coordinates and momenta in SH are required to specify the Hamiltonian flow in the plane (q_1,p_1). In this plane SH is a point that locates the origin. The center manifold is also a point but, being determined by Eq. (15), is in general different from SH. The forms of CS^+ and CS^- remain determined by Eq. (14) and are represented in Fig. 1 as curves.

CS^+ and CS^- separate the flow paths through phase space of trajectories with different fates. This is the most important realization of the SM-ART treatment. All trajectories that cut the surface Σ that extends along the p_1 axis from SH to remote regions, may cut it again later any number of times as the system evolves. However, only those trajectories that cross Σ in this section with $p_1>\max(p_1^-,p_1^+)$ belong to the flow of successful jumps from the equilibrium in configuration A to that in B, while for $p_1<\min(p_1^-,p_1^+)$ they belong instead to the net flow from B to A.

By means of exact numerical evaluations of the dynamics we have verified that the monotonic behavior indicated in Fig. 1 does indeed occur in real systems. In Fig. 2 we show the time evolution from an initial state on SH, projected along the reaction coordinate, for the three values of normal momentum p_1 corresponding schematically to the points a, b, and c of Fig. 1. Those momenta larger than p_1^+, e.g., point c, give trajectories leading directly to transitions from A to B; upon reducing the normal momentum to p_1^+ (point b in Fig. 1) there occurs a critical trajectory that lies on CS^+. Smaller momenta, e.g., point a lead to return jump trajectories that reverse their perpendicular momentum. Further reductions would result in a trajectory that originated at a prior time from oscillations near the barrier (CS^- at p_1^-) and finally trajectories that correspond to jumps in the reverse direction. Note that the critical trajectories associated with CS^+ and CS^- approach and

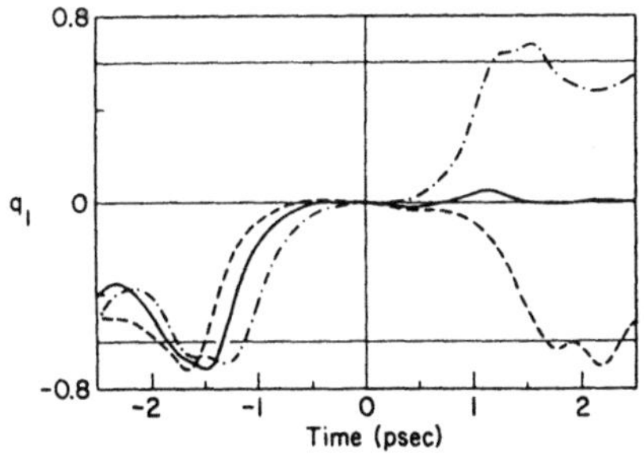

FIG. 2. Time evolution from an initial state on SH, projected along the reaction coordinate q_1, for a sequence of normal momenta, p_1, corresponding to points a, b, and c in Fig. 1. The monotonic behavior predicted by the theory is confirmed. The CM$^+$ trajectory (solid line, point b in Fig. 1) oscillates about the saddle plane in the forward time evolution. It separates the flow of return jumps having smaller normal momentum (dashed line, a in Fig. 1) from the flow of successful trajectories having larger normal momentum (dash-dotted line, c in Fig. 1).

linger indefinitely on top of the potential barrier, thus oscillating through the saddle plane. Similarly, all critical trajectories associated with the center manifold oscillate indefinitely about the saddle plane. For simple models of the anharmonic barrier, the nature and amplitude of such oscillations can be computed explicitly.[19] The fact that the critical trajectories do behave in this way plays a central role in their identification for the case of a general potential in Sec. II B.

The full jump rate is calculated by integrating the flow through the surface $\Sigma^\pm$ of equations

$$q_1=0, \quad p_1 \geq \max(p_1^-, p_1^+) \tag{17}$$

for Σ^+ and

$$q_1=0, \quad p_1 \leq \min(p_1^-, p_1^+) \tag{18}$$

for Σ^-. The dynamical conversion coefficient for the flow from A to B can then be evaluated by a thermal sampling of the hypersurface SH with analytical integration over all momenta larger than the critical one, $p_c(P) \equiv p_1^+$, be it positive or negative, for each sampling point P on SH; thus

$$c=\beta\left\langle \int_{p_c}^{+\infty} p_1 \exp(-\tfrac{1}{2}\beta p_1^2)\,dp_1 \right\rangle_{\rm SH} . \tag{19}$$

An analogous expression holds for the flow in the opposite direction. Equation (19) is a central result of the present treatment. The topological structure of the manifolds provides a rigorous definition for the critical momenta $p_c(P)$ as the momentum component perpendicular to SH required to place the trajectory on the center stable (or unstable) manifold for any given initial state on SH. The same structure then ensures that all larger momenta lead to complete jumps within the SM-ART formulation, much as assumed empirically in Bennett's calculation for a different definition of critical trajectory. As noted by Bennett,[6] the existence of a critical momentum allows sampling over the crossing momenta, still used in other approaches,[15] to be replaced by a more efficient integral from p_c to infinity.

We now turn attention to a determination of the isotope-effect factor. It is important to realize that, within the SM-ART treatment, the conversion coefficient c for a particular jumping atom and that, c', for a different isotopic mass constitute two entirely different problems. For both systems c is correctly represented by the ratio of partition functions in Eq. (6) or by the flux through a surface $\Sigma^\pm$ as defined in Eq. (7). However, both the partition functions and the definition of $\Sigma^\pm$ are complicated by the fact that the system with a different mass possesses a distinct saddle hypersurface SH' and center manifold CM' that are rotated and distorted with respect to those of the system with all masses equal. An accurate treatment of the isotope effect must therefore incorporate a correct description of the different jump conditions for the two systems.

For small changes in the mass of the jumping atom, $M_1 \to M_1(1+2\delta)$, the linear variation with δ of the surfaces of interest can be calculated by use of first-order perturbation theory. In order to obtain insight into the computational problems this complication brings, it is useful to consider the initial example of a system represented by its anharmonic expansion carried through to terms of third order. For this system the isotopically modified manifolds can be expanded by perturbation theory in terms of the coefficients of the unmodified system. The algebraic details are collected in the Appendix. It turns out that Eq. (19), when written for this system, can be evaluated analytically to first order in T, and this approach naturally leads to the result reported in Ref. 12. Of greater interest here is the fact that Eq. (19) can also be obtained by using a numerical sample of states on SH and assigning to each the critical momentum determined by the corresponding analytical expression of Eq. (14). We find immediately that the expected difference is too small to be measured directly by separate sampling runs on the two systems, since $c'-c \simeq 10^{-4}$ for $\delta=0.01$. A determination of $c'-c$ by means of separate numerical sampling on SH and SH' requires a number of sample points of the order of 10^6 to achieve an accuracy of 1% in κ for $\delta=0.01$. Such large samplings are not feasible, so the determination of κ for a general potential, where the critical momentum p_c must be determined by numerical methods, requires a different approach.

To overcome this problem we have incorporated into our calculation a difference sampling method analogous to that proposed by Bennett[6] to compute the mass dependence of c. We investigate the mass dependence of the critical momenta p_c in a chosen sample of SH states of the $M_1=M$ system by establishing a one-to-one correspondence between each state P of SH, $P=(x_1,\ldots,x_N,\dot{x}_1,\ldots,\dot{x}_N)$ and the equiprobable state of the $M_1=M'$ system, namely $\tilde{P}=(x_1,\ldots,x_N,\tilde{\dot{x}}_1,\ldots,\tilde{\dot{x}}_N)$. The transformation

$$\tilde{\dot{x}}_i=\begin{cases}\dot{x}_i(M/M')^{1/2}, & i=1,2,3\\ \dot{x}_i, & i=4,5,\ldots,N\end{cases} \tag{20}$$

ensures that the two states have the same statistical weight.

To employ this variance-reducing correspondence, one finds the critical normal momenta $p_c(\tilde{P})$ needed to place the trajectory of the new initial condition $\tilde{P}$ on CS^+ (or CS^-) for the $M_1=M'$ system. The difference in conversion coefficient is then obtained as an average over states on SH of the integral between the two critical momenta, just as in Bennett's work. Application of this difference sampling method dramatically reduces the number of states in SH needed to obtain an accurate determination for κ. In the case described above (see the Appendix) the difference method requires about 10^3 states to obtain the same accuracy achieved by a sample 3 orders of magnitude larger, in the evaluation by separate runs.

B. Sampling methods

For an arbitrary potential it is not, in general, possible to write down an analytical form for the CM. Numerical methods are therefore needed to locate CM, CS^-, and CS^+. For this purpose it is convenient to start from some specific locus in phase space. As explained in Sec. II, a natural choice that provides easy sampling of initial configurations is the saddle hyperplane SH.

The desired critical values of p_1 and q_1 are those for which the evolution of the representative point lies on the particular manifold of interest. These are determined by imposing the corresponding asymptotic behavior on the time evolution of the trajectory, namely, (a) the CM trajectory should oscillate indefinitely about the saddle plane; (b) the CS^+ trajectory should oscillate about the saddle plane only for the forward time evolution for the trajectory; and (c) the time-reversed behavior should occur for the trajectory CS^-.

Any determination of CM trajectories themselves would require finding critical values for both q_1 and p_1, and it would be necessary, in addition, to project the invariant measure of CM onto SH. We have therefore focused our effort instead on the determination of the critical momentum p_c required for an initial configuration on SH to obtain a trajectory on CS^+ or CS^- and so to evaluate c by means of Eq. (19).

An implementation of these principles for any given structure and model potential requires several distinct steps. The saddle-point configuration $P_0 \equiv (x_1^0, \ldots, x_N^0)$ must first be determined and the potential diagonalized at P_0 to yield eigenvalues and eigenvectors of Eq. (12) that define the saddle hypersurface SH. It is then necessary to sample representative points in phase space on the surface SH, and for each to iterate the evolution of trajectories for various normal momenta in order to locate the specific critical trajectory. The first step of the procedure has been carried out here using standard relaxation techniques such as those described in Ref. 10. The sampling on SH was performed by means of a molecular-dynamic (MD) run constrained to remain on SH. Details of the latter calculation are described in Sec. II B 1. The last step, namely the search for the critical trajectories, is the main topic of Sec. II B 2.

1. Sampling on a surface in phase space

Different methods have been widely used[5,14,15,17] to sample initial states on a surface in phase space for a given interatomic potential. In the present work we use an independent method to sample states on the saddle hyperplane. To maintain the dynamical system on SH we introduce an explicit dynamical constraint in the form of an external force $\bar{\phi}$, so that

$$q_1(t)=\sum_{i=1}^{N} a_i^1(x_i(t)-x_i^0)=0 \tag{21}$$

for every t. From Eq. (21) it follows directly that also $p_1(t)=0$ for every t. The force $\bar{\phi}$, applied in the direction of the constraint $q_1=(\{a_i^1\})$, has Cartesian components

$$\phi_i=-\left[\sum_{\alpha=1}^{N} a_\alpha^1 F_\alpha(t)\right] a_i^1 , \tag{22}$$

where $F_\alpha(t)=-(\partial V/\partial x_\alpha)_{x(t)}$ is the usual force acting on particle α in the absence of the constraint.

Because $\bar{\phi}$ acts along the direction of the constraint it does no work on the system. Likewise, all linear momenta are conserved when the initial configuration is chosen on SH, the natural choice for it being P_0. By starting a molecular-dynamics run from P_0 at a given temperature, subject to the Hamiltonian of the constrained system, we thus generate an ensemble of configurations on SH whose members are representative of configurations that would occur at thermal equilibrium for the unconstrained system of Hamiltonian H.

2. Determination of critical trajectories

Given an initial state on SH it is necessary to determine the critical value p_c of the perpendicular momentum. Our procedure uses a brief search of a generically broad interval of values of p_1 to locate the transition from the flow of returns and to the flow of successful jumps. With a specific initial value p^0 for the perpendicular momentum, chosen from within this interval (the natural choice being its arithmetic mean), the evolution of the system backward and forward in time is computed by MD for a period $\tau=n\tau_0$ with τ_0 the time step. In this run we monitor the quantity

$$q_1(t;p^0)=\sum_{i=1}^{N} a_i^1(x_i(t;p^0)-x_i^0) . \tag{23}$$

This projects the motion of the particles in the Cartesian space of the crystal onto the many-body reaction coordinate.

In addition, we calculate the time evolution of a perturbed trajectory, $q_1(t;p^1)$, for a slightly different initial value, p^1, of normal momentum. For small values of such perturbations the motion of the system falls in a linear-response regime where the magnitude of the mechanical response depends linearly on the strength of the applied perturbations.[20,21] Within this regime a general trajectory of normal momentum p can be obtained approximately as

$$q_1(t;p) \simeq q_1(t;p^0) + \frac{p-p^0}{p^1-p^0}\delta q_1(t;p^0,p^1) , \quad (24)$$

where

$$\delta q_1(t;p^0,p^1) = q_1(t;p^1) - q_1(t;p^0) . \quad (25)$$

In this way it is possible to study, within the linear-response approximation, the behavior of a family of trajectories, all originating from the same initial state on SH but corresponding to different values of the crossing momentum. An example of the resulting trajectories is given in Fig. 3.

The critical condition in which the system oscillates about the saddle plane can be expressed analytically by imposing an average of zero on the time integral of the reaction coordinate:

$$\int_0^\tau q_1(t;p)dt = 0 \quad (26)$$

for CS^+ and

$$\int_{-\tau}^0 q_1(t;p)dt = 0 \quad (27)$$

for CS^-. We can now estimate the critical normal momentum within the linear-response approximation by use of Eqs. (24) and (27) or (28) as

$$p_c^\pm = p^0 + p^1\Delta_\pm , \quad (28)$$

where

$$\Delta_+ = -\frac{\int_0^\tau q_1(t;p^0)dt}{\int_0^\tau \delta q_1(t;p^0,p^1)dt} \quad (29)$$

for CS^+

$$\Delta_- = -\frac{\int_{-\tau}^0 q_1(t;p^0)dt}{\int_{-\tau}^0 \delta q_1(t;p^0,p^1)dt} \quad (30)$$

for CS^-. By iterating the procedure, using successive estimates of p_c as initial momenta p^0, the critical momentum p_c can be obtained to any desired accuracy and for any interval τ of integration. Such an estimate of p_c depends to some extent on the particular choice of τ, but it must converge to a specific limit as τ is increased. In the case described in the Appendix an analytical value of p_c can be obtained exactly. For the iterative method just described this example suggests that the convergence can be considered complete as soon as τ approaches the characteristic time associated with the unstable mode frequency η (~5–6 Debye periods in the model studied, see Table I).

One further matter of computational significance warrants mention here. Rather than integrating $q_1(t;p^0)$ and $q_1(t;p^1)$ separately it has been possible, and computationally more economical, to solve an approximate differential equation for $\delta q_1(t)$ itself, valid within the same linear-response approximation. The advantage lies in the fact that the expression for the gradient required in the integration of the equation of motion for $\delta q_1(t)$ can be written at time t in terms of the second derivative of the potential function at the configuration of the unperturbed trajectory for the same time.[21] This result can be included at small cost in the same loop of the calculation of the gradient along the unperturbed trajectory. In practice the final search for p_c for any state on SH could be automated to require ~2.0 min of CPU (central-processing-unit) time on an FPS264 processor.

It is evident that the values of critical momentum can be obtained for each state on SH. One of these (on CS^+) is the critical momentum for jumps from A to B, and the other (on CS^-) for jumps from B to A. Both can be used, as appropriate, in the evaluation of c. This halves the required sample of initial states on SH.

To evaluate the isotope-effect factor by means of the differential method described in Sec. II A, precisely the same procedure can be applied to find the critical momentum $\tilde{p}_c$ for the corresponding state in the system with a jumping atom of different mass. The critical momentum $\tilde{p}_c$ corresponds in principle to the component along the unstable mode q_1, required to place the trajectory on the different CS^+ or CS^- of the new system, where it oscillates about the plane $q_1'=0$ rather than $q_1=0$. However, the computation effort is greatly reduced when the study is restricted to changes of critical momentum induced by a differential change in mass. In effect we confine attention to the mass derivative of c

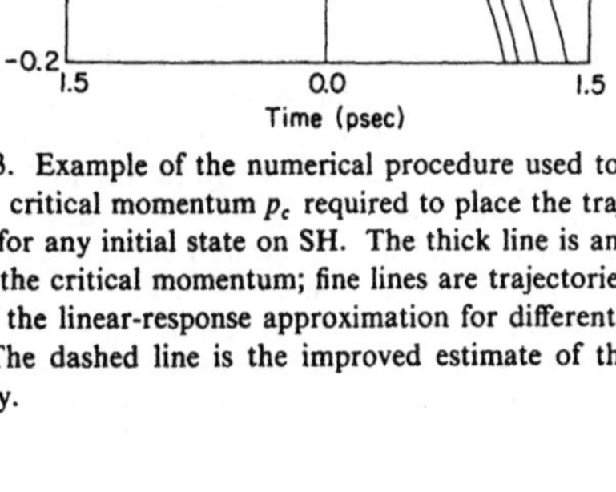

FIG. 3. Example of the numerical procedure used to determine the critical momentum p_c required to place the trajectory on CS^+ for any initial state on SH. The thick line is an initial guess at the critical momentum; fine lines are trajectories computed in the linear-response approximation for different values of p_1. The dashed line is the improved estimate of the CS^+ trajectory.

TABLE I. Numerical example of the dependence of the computed values $p_c(P)$ and $p_c(\tilde{P})$ at $T=0.1T_m$, $\delta=0.01$ [$T_m=0.67\varepsilon$; see De Lorenzi *et al.* (Ref. 17)] on the time of integration τ/τ_η ($\tau_\eta=2.2$ psec), for a state P on SH of the model potential described in the Appendix. Analytical values of p_c and $\tilde{p}_c$ reported in the last row for comparison correspond to an infinite integration time.

τ/τ_η	$p_c(P)$	$p_c(\tilde{P})$
0.40	0.050 644 563 4	0.050 038 766 3
0.80	0.058 150 133 3	0.057 494 734 1
1.20	0.058 152 309 0	0.057 508 694 4
∞	0.058 152 327 7	0.057 508 710 2

at $M_1 = M$. It is important to observe that a trajectory that lingers on the top of the barrier does so however one defines the plane of oscillation. In fact, the new reaction coordinate $q_1' = q_1 + \delta\sum_j \alpha_{1j} q_j$ (see the Appendix) oscillates with a phase that depends on $\mathbf{q}(t)$ so that any individual q_i appears with an essentially random phase. As it enters q_i' linearly it thus vanishes from $\langle q_i' \rangle$. The criteria $\langle q_i \rangle = 0$ and $\langle q_i' \rangle = 0$ therefore lead to identical results for $\tilde{p}_c$.

A second useful point is that, when the mass change is small, an excellent guess for $\tilde{p}_c$ is given by the value of p_c previously calculated for the corresponding state for $M_1 = M$, so that fewer iterations are needed to converge on the new critical value. The same program, modified only to treat the different dynamics associated with the new mass, gave the value of $\tilde{p}_c$ for the new state in less than 1 min of CPU time.

As examples Table II gives average values of c and κ at two different temperatures, obtained by the procedures just described, for the anharmonic expansion for Lennard-Jones (LJ) argon (see the Appendix) for comparison with the exact values for the same model evaluated analytically. There is excellent agreement between the two sets of data. These results provide a check on the procedures that are the basis for our subsequent exploration of conversion coefficients and isotope-effect factors for full interatomic forces. A summary of these latter calculations is provided in Sec. III.

III. NUMERICAL RESULTS

Our calculations follow earlier simulation of vacancy jumps by Bennett in which the conversion coefficient and nonharmonic isotope-effect factor were first computed for a Lennard-Jones fcc crystal near its melting point. Our work expands on these efforts in two directions. First, as explained above, we base our calculations on the SM-ART treatment in order to establish a rigorous theoretical framework for the results. Second, we have pursued these goals with much improved statistical precision. As will be apparent in what follows, the more accurate results allow new insight into the temperature, volume, and model potential dependences of the properties, and hence to a specific understanding of the anharmonic effects and their origins in atomic interactions.

As in the original SM-ART formulation by Toller *et al.*[11] our numerical results focus on particular models of Ar, Cu, and Ag. Details of the Lennard-Jones potential employed for Ar and the Morse potential used for Cu and Ag are provided in the paper by De Lorenzi *et al.*[17] that immediately precedes the present work, to which the reader is referred. For reasons of computational effort our calculations were confined to a 32-site fcc cluster with periodic boundary conditions, and with all sites but one occupied by atoms interacting through the chosen force laws. Details of the potential range and boundary effects are discussed by De Lorenzi *et al.*[17] It is our experience that these questions have little material effect on the conclusions drawn from the results reported in what follows.

In order to obtain a broad understanding of the processes under study we have completed model calculations at several levels of approximation. These include a full numerical investigation in which the real manifolds are sampled for the complete force law in order to obtain a complete and accurate determination of c and κ, together with their temperature and volume dependences. We refer to these calculations below as pertaining to the exact model. In addition, it has been valuable to examine the analogous results obtained by sampling methods when the exact model potential is replaced by a smoothed model that contains only the low-order terms of a truncated anharmonic expansion of the potential energy about the saddle-point configuration, as described in the Appendix. This system differs from the exact model mainly by the elimination of hard-core repulsive energy as two atoms approach each other closely. We refer to these results as deriving from sampling for a model with a truncated anharmonic expansion. Finally, from earlier work, we have available analytical results that express c and κ by a low-temperature expansion in powers of T, that also depends on an anharmonic expansion of the potential energy in the vicinity of the saddle-point configuration. These results are referred to as deriving from an analytical treatment of the anharmonic expansion.

Results for the conversion coefficient c and isotope-effect factor κ as functions of temperature for these three different calculations are given for fcc Ar in the upper panels of Fig. 4. Each point in Fig. 4 required a sample containing 400 independent configurations. The error bars, where visible, indicate that the final uncertainties are kept below 1% in both c and κ. Figure 4, together with Table II, shows that the analytical and numerical

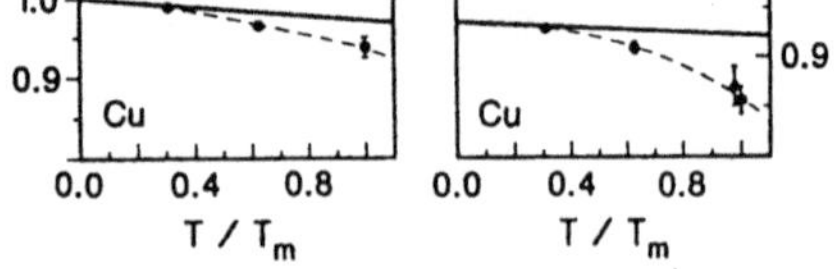

FIG. 4. The conversion coefficient c and the isotope-effect factor κ as functions of T/T_m for the exact models of fcc Ar, Ag, and Cu are shown by solid circles and connected by a dashed line as a guide to the eye. Analytical results for the anharmonic expansion (Ref. 11) (solid line) and sampling results for the same potential give identical results for argon (open circles). Points (△) are values for c and κ computed by Bennett (Ref. 6) for Ar. Points (▲) are experimental measurements of κ from Peterson (Ref. 2); they agree with the exact models but not with the anharmonic expansion (see text).

TABLE II. Values of the conversion coefficient and isotope-effect factor at two temperatures for the model potential of the Appendix, as computed by the numerical procedure (fourth and sixth columns) together with the corresponding analytical values (Refs. 11 and 12) (third and fifth columns). The second column gives the number of states on SH used in the sampling.

T/T_m	N	c Analytical	c Numerical	K Analytical	K Numerical
0.50	400	0.9868	0.9855±0.0016	0.9558	0.9545±0.0028
1.00	400	0.9790	0.9801±0.0020	0.9513	0.9482±0.0038

results for the anharmonic expansions are in essentially exact agreement. This verifies that each method provides an adequate account of the dynamical behavior for the smoothed potential. A startling discrepancy nevertheless exists between the exact model and the results for the truncated expansion. Whereas the latter exhibit only a weak linear temperature dependence in both c and κ, the results for the realistic potential exhibit a nonlinear dependence on T and a much larger deviation from unity. These effects are far outside the statistical uncertainties and must be regarded as firmly established by our calculation.

The lower panels of Fig. 4 compare the exact models with predictions for the truncated potentials of Cu and Ag. In both cases, and for both c and κ, there is a remarkable similarity between the calculated results and those for Ar. The calculated isotope effects obtained from the exact model for Cu and Ag agree well with the measured values, shown as solid triangles. More globally, all three exact calculations give an isotope effect near the melting temperature (T_m) of about 0.87, which is typical of the values more generally observed for fcc lattices. It seems apparent that the fcc lattice produces an isotope effect that is decreased by anharmonicity from the harmonic value[11] at low temperature to this fairly reproducible value near T_m, much as suggested earlier by Flynn.[4]

Analogous comments describe the calculated conversion coefficients. In each case the exact values decrease from unity at $T=0$ to about 0.9 at T_m, and with functional forms that have a remarkably similar dependence on T/T_m. The corrections to rate theory these results require are rather modest, and amount to about 10% at melting. This confirms earlier theoretical estimates[4,10,11,22,23] for dynamical corrections to rate theory in fcc systems. We note that some discrepancy exists between our LJ value of c near T_m and that given earlier by Bennett.[6] We believe that the differences are likely to be statistical in origin, rather than arising from differences of principle or from the different criteria employed for the critical trajectories. Numerical estimates of dynamical corrections to rate theory for surface diffusion reported by Voter and Doll[15,24] also amount to about 10% in the range of temperatures experimentally observed. Thus conventional rate theory as formulated by Vineyard gives an excellent approximation to the true jump rate for these systems with high barriers to atomic migration. However, this situation is not general and breakdown may occur for interstitial migration or even for vacancy migration in softer materials with lower potential barriers.

The precision with which our calculations distinguish the consequences of different interatomic potentials affords a unique opportunity to explore the physical processes that determine the isotope effect and conversion coefficient of realistically anharmonic crystals. For this purpose the isotope effect is the main focus of attention. In practice, the conversion coefficient itself cannot be measured, and amounts at most to 10% changes of a rate that varies exponentially with temperature by many orders of magnitude over the experimentally accessible range. In contrast, the isotope effect—mainly the mass dependence of the conversion factor—*is* experimentally accessible to 1% precision in favorable cases, given the diffusion mechanism, and remains the only available monitor of dynamical processes that take place during the diffusion jump. Our discussion of the detailed dynamics is therefore restricted to the isotope-effect factor κ.

The results in Fig. 4 suggest that closed-shell repulsion may be the critical ingredient in determining the observed isotope effects. One may reason that the anharmonic expansions all give similar but *incorrect* isotope effects, and that by construction they differ from the exact results, which all give similar and *correct* behavior, mainly by the omission of the hard-core forces. Therefore our purpose in what follows is to investigate the actual role of atomic size in general, and core repulsion in particular, in determining the isotope effect. To this end two distinct series of calculations have been made. The first involves the character of the trajectories and the second an alternative simplified simulation of atomic size.

We have followed trajectories of the exact and truncated systems from the same initial configurations, through the critical condition at which a jump does or does not occur, to identify the differences caused by the hard core of the exact potential. In general the trajectories of the two systems are remarkably alike. We find, however, that large departures occur in a small, temperature-dependent fraction of the trajectories. It turns out that these differences invariably originate at a point on the trajectory at which the diffusing atom undergoes a vigorous collision with one of its neighbors, such that the hard-core forces of the exact potential exert a significant effect and thus modify the subsequent evolution of the system. Moreover, we find that it is precisely this same subset of trajectories that have large changes of critical momentum, and that contribute the overwhelming portion of the calculated change of isotope effect. These results therefore establish beyond doubt that the observed isotope effects are determined in large part by infrequent, highly anharmonic processes that lie completely beyond the scope of available analytical theories. This is the reason the observed behavior has remained unexplained. From our present perspective we may observe that the repulsive core of the potential has a specific effect in deforming the center manifold, and to an extent that is strongly mass dependent. It is these changes that the isotope effect monitors.

Our conclusions are supported by the results of trajectory calculations summarized in Fig. 5. The figure shows, trajectory by trajectory, how the normalized fractional deviation Δc of the conversion coefficient from its truncated-potential value depends on the dynamics of the particles in the vicinity of the migrating atom. To obtain this concise information we have made use of the fact that motion along small principal radii of saddle-surface curvature (i.e., large curvatures) corresponds to relative moments localized at the migrating atom and its neighbors. These geometrical features of the saddle surface are described in an earlier report[10] and will not be reviewed here. For the present purpose we merely employ the components of the motion along these specific directions in configurational space to project out energetic collisions that involve the jumping atom. In Fig. 5 the fractional change of conversion coefficient is shown as a function of $Q=\sum_i \varepsilon_i/\rho_i^2$ for the trajectory. Here ε_i is the summed kinetic and potential energy owing to motion and displacements of the initial state along direction i (the harmonic value was used for the potential), and ρ_i is the principal radius i of saddle-surface curvature. Q is therefore a monitor of the behavior of energetic relative motion between the migrating atom and its neighbors. The behavior of time-reversed trajectories from the same initial state is also indicated in Fig. 5.

It is strikingly evident in Fig. 5 that the great majority of trajectories, and particularly those with Q small, have $\Delta c \simeq 0$. Particularly for Q large, however, there is a significant fraction of trajectories with Δc large and negative. This fraction decreases at lower temperatures, as shown at top left in Fig. 5 and as indicated by the histograms. The identifications of these changes with energetic local collisions is further confirmed by the fact that trajectories with initial velocities so directed as to *increase* the initial displacements (solid points) invariably have larger deviations than their time-reversed analog (open circles) that have the same value of Q but in which the velocity is directed to decrease the displacements. The velocity reversal gives a longer time interval for the energy to disperse before a collision. All these results therefore point clearly to the way in which the changes of c arise from local strong collisions during a potential jump event.

With these results in mind we have performed additional calculations to identify the explicit effects of atomic size. Rigid cores of radius σ' were superposed on the truncated expansion for the Lennard-Jones case in order to tune the potential from the smooth form to a more realistic form with a variable core size obtained by changing σ'. The results provide a convincing confirmation of the core effect discussed above. As seen in Fig. 6, both c and κ near T_m remain close to their truncated-potential values until σ' approaches 0.9σ, with

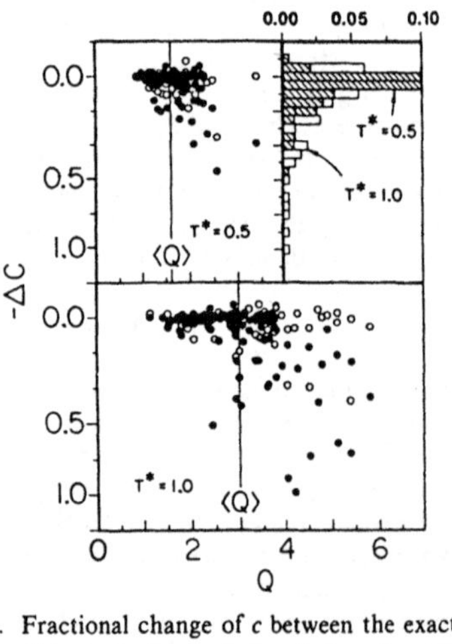

FIG. 5. Fractional change of c between the exact and truncated potential results for each sampled configuration on SH, shown as a function of $Q=\sum_i \varepsilon_i/\rho_i^2$ (see text) at $T=0.5T_m$ and $T=1.0T_m$ ($T_m=0.67\varepsilon$; see paper I). The faint lines indicate the mean value of Q for the sample. The inset histograms show the long, temperature-dependent tails of the distributions that result from energetic hard-core collisions for these two temperatures.

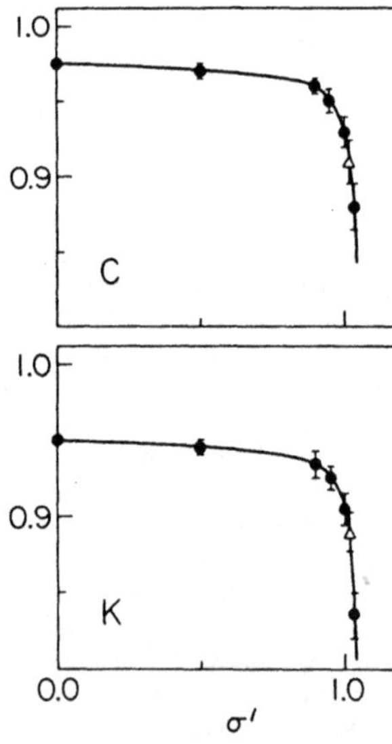

FIG. 6. Conversion coefficient and isotope-effect factor at fixed temperature $T=T_m$ and density $\rho=\rho_m$ when the anharmonic expansion is augmented by a hard-core interatomic repulsion of varying diameter σ' (in units of σ, σ being the usual Lennard-Jones diameter). Points (Δ) are results for the ordinary Lennard-Jones potential at $\sigma'_{LJ}=1.02$.

σ the Lennard-Jones radius, at which point marked reductions both of c and κ begin. For this temperature the appropriate hard-core radius of the model is 1.02σ, obtained by extrapolating Verlet's liquid-state data.[25] When σ' is increased to 1.02σ, the values $c = 0.90$ and $\kappa = 0.89$ are obtained from our hard-core calculations. Thus the superposition of an appropriate atomic size on the smooth potential brings the properties back into excellent agreement with the exact results of Fig. 4. The insensitivity of these phenomena to all except the core forces and size is once more clearly emphasized. It is worth noting in this connection that for $\sigma' = 1.06\sigma$ the saddle point itself is occluded by rigid-core overlap. The seriousness of this core radius constraint on saddle-surface dynamics has not been given sufficient attention in prior research.

If the jump characteristics depend so strongly on the core radius then there must exist a complementary sensitivity to the crystal density and hence pressure. We have carried through the relevant calculations to confirm that this is the case. Because the compressibility of solids is small, however, the resulting effects on c and κ turn out significant but not large in size. A simple estimate comes from the hard-core model and the results in Fig. 6. We may write

$$V\frac{\partial c}{\partial V} = -\sigma' \left[\frac{\partial c}{\partial \sigma'}\right]_{a_0} \left[\frac{V_0}{V}\right]^{1/3}, \quad (31)$$

$$V\frac{\partial \kappa}{\partial V} = -\sigma' \left[\frac{\partial \kappa}{\partial \sigma'}\right]_{a_0} \left[\frac{V_0}{V}\right]^{1/3}, \quad (32)$$

whence, from the derivatives in Fig. 6,

$$V\frac{\partial c}{\partial V} = 0.46 \pm 0.03 \quad (33)$$

and

$$V\frac{\partial \kappa}{\partial V} = 0.6 \pm 0.1 \quad (34)$$

near T_m. These results show that the fractional changes of c and κ have the same order of magnitude as the volume changes that cause them at the equilibrium density near melting. Accordingly, the pressures required to produce measurable isotope-effect changes are not readily achieved. Nevertheless, the calculated temperature and volume dependences of the isotope effect are predictions of the theory that remain for experiment to explore.

For Ar we have pursued these questions one step further by means of explicit calculations of c and κ for the exact model system. Table III compares the results for three lattice spacings with the estimates from the hard-core calculations. The exact and hard-core results for the changes of c and κ agree to within their statistical uncertainties. Thus Table III provides a final confirmation that these properties are largely determined by the dynamics of hard-core collisions.

IV. SUMMARY

We have shown in this paper that molecular dynamics may be used to incorporate exact dynamical behavior into an investigation of vacancy jump processes in solids. This permits a complete and accurate evaluation of the atomic jump rate within the SM-ART treatment. To the extent that long-term correlations appear negligible we obtain an exact prediction of the classical jump rate. In conjunction with the calculations of saddle-plane properties by De Lorenzi *et al.*[17] described in the preceding paper our results offer, for the first time, jump rate predictions for solids with 1% accuracy directly from the atomic masses and interatomic potential. For the fcc vacancy diffusion discussed here, the many-body rate theory of Vineyard is itself an excellent first approximation; it overestimates the jump rate by only 10% at the melting temperature and the error falls to zero as $T \to 0$.

We have used the available computational precision to study the dependence of conversion coefficient and isotope-effect factors for these systems on the details of their interatomic potentials. The isotope effects calculated for the realistic potentials agree very well with experimental results for fcc metals near the melting temperature, and the calculations give very similar values of κ from one system to the next, just as is observed. It has been possible to demonstrate unambiguously that the values of c and κ are mainly determined by the core repulsive forces, through infrequent hard-core collisions undergone by the migrating atom during the jump events. This accounts for the remarkable similarities among the conversion coefficients and isotope-effect factors of different model fcc solids when shown in reduced form as a function of T/T_m. The calculations predict a nonlinear decrease of κ from its harmonic value at $T = 0$, and also a significant volume dependence of κ, both deriving from these hard-core effects.

TABLE III. Comparison between predicted [Eqs. (33) and (34)] and computed values for c and K for the model of Ar at three different densities and at fixed temperature $T = T_m$.

a/σ	$\Delta V/V$	ρ/ρ_0	c Predicted	c Computed	K Predicted	K Computed
1.6403	0.00	1.00		0.905±0.005		0.891±0.006
1.5771	−0.05	1.05	0.890±0.010	0.885±0.010	0.865±0.016	0.871±0.015
1.5489	−0.10	1.10	0.872±0.010	0.874±0.015	0.830±0.016	0.848±0.018

ACKNOWLEDGMENTS

We have benefited from discussion with M. Toller and from the advice and assistance of G. De Lorenzi. This research was supported in part through the University of Illinois Materials Research Laboratory, with funds provided by the National Science Foundation, under Grant No. DMR-83-16981. One of us (G.J.) thanks the University of Illinois National Center for Supercomputing Applications for support during a visit.

APPENDIX

Our aim here is to introduce a useful model potential that we can use both for explicit analytical calculations and for testing the numerical procedures described in Sec. II B. In particular, we are interested in explicit expression for the center manifold (CM) and the center stable and unstable manifolds (CS^+ and CS^-) for this system, as well as their mass dependences. For this purpose we use a number of results found in earlier work[10–12] to which reference should be made for details.

The chosen potential is obtained from a general potential at a single saddle point P_0, such as that in Eq. (10), when the term contribution V_a is expressed in a power-series expansion in the displacements from P_0:

$$V_a=\tfrac{1}{6}\sum_{i,j,k=1}^{N} d_{ijk}q_iq_jq_k+\tfrac{1}{24}\sum_{i,j,k,r=1}^{N} e_{ijkr}q_iq_jq_kq_r+\cdots . \tag{A1}$$

The truncation leaves an anharmonic model potential which contains some of the anharmonic properties of interest in diffusion problems, including a curved saddle surface and some anharmonic characteristics of CM, CS^+, and CS^-, but which can still be treated analytically. The coefficients that appear in Eq. (A1) can be determined for a given interatomic potential, as in Refs. 10 and 11 for the Lennard-Jones potential, and the analytical results then used to identify the effects of higher anharmonic terms.

Here we employ the power series truncated after terms of third order in the q_i so that V_a contains only coefficients of the kind

$$d_{ijk}=\left[\frac{\partial^3 V}{\partial q_i\partial q_j\partial q_k}\right]_{P_0} . \tag{A2}$$

Then Eqs. (15) and (14) for CM, CS^+, and CS^- can be developed as power-series expansions in the q_i:[11]

$$f=\tfrac{1}{2}\sum_{i,k=2}^{N} u_{ij}q_iq_k+\tfrac{1}{2}\sum_{i,k=2}^{N} v_{ik}p_ip_k+\cdots , \tag{A3}$$

$$g=\sum_{i,k=2}^{N} w_{ik}q_ip_k+\cdots , \tag{A4}$$

$$F_\pm=\pm\frac{1}{6\eta}d_{111}q_1^2\pm\left[\sum_{k=2}^{N} d_{11k}\frac{2\eta q_k\pm p_k}{4\eta^2+\varepsilon_k}\right]\times q_1\pm\eta f+g+\cdots , \tag{A5}$$

where the coefficients u_{ik}, v_{ik}, and w_{ik}, fixed by the eigenvalues ε_i (note that $\varepsilon_1=-\eta^2$) and the third-order coefficients d_{ijk} of the potential V at P_0 are

$$u_{ik}=2h_{ik}d_{1ik} , \tag{A6}$$

$$v_{ik}=(\eta^2+\varepsilon_i+\varepsilon_k)h_{ik}d_{1ik} , \tag{A7}$$

$$w_{ik}=(\eta^2-\varepsilon_i+\varepsilon_k)h_{ik}d_{1ik} , \tag{A8}$$

$$h_{ik}=[\eta^2+\varepsilon_i+\varepsilon_k+2(\varepsilon_i\varepsilon_k)^{1/2}]^{-1}\times[\eta^2+\varepsilon_i+\varepsilon_k-2(\varepsilon_i\varepsilon_k)^{1/2}]^{-1} . \tag{A9}$$

The critical momentum that displaces the initial trajectory $(0,q_2,\ldots,q_N,0,p_2,\ldots,p_N)$ on SH onto CM^+ or CM^- is

$$p_c^\pm=\pm\eta f+g . \tag{A10}$$

This result can be used in Eq. (19) and the integration on SH carried out either analytically or numerically to yield the result for the conversion coefficient to first order in T given in Ref. 11.

When the mass M_1 of the jumping atom is changed, the system with changed mass has new orthonormal eigenvectors that are rotated with respect to the old ones. Moreover, the two systems possess two distinct CM, CS^+, and CS^-. For small changes in mass, $M_1\to M_1(1+2\delta)$, the linear variation with δ of the quantities that enter the equations for CS^+ and CS^- can be evaluated by first-order perturbation theory. The calculation proceeds by first expressing the new coordinates, momenta, eigenvalues, eigenvectors, and third-order coefficients (identified here by primes) in terms of the old ones:[12]

$$q_i'=q_i+\delta Q_i,\quad Q_i=\sum_{j=1}^{N}\alpha_{ij}q_j , \tag{A11}$$

$$p_i'=p_i+\delta P_i,\quad P_i=\sum_{j=1}^{N}\alpha_{ij}p_j , \tag{A12}$$

$$\varepsilon_i'=\varepsilon_i+\delta E_i,\quad E_i=-2\varepsilon_i\alpha_{ii} , \tag{A13}$$

$$d_{ijk}'=d_{ijk}+\delta D_{ijk} ,\qquad D_{ijk}=\sum_{r=1}^{N}(\alpha_{ri}d_{rjk}+\alpha_{rj}d_{irk}+\alpha_{rk}d_{ijr}) . \tag{A14}$$

Here

$$\delta=d\ln(M_1)^{1/2}=\frac{dM_1}{2M_1} , \tag{A15}$$

and

$$\alpha_{ij}=(\mathbf{a}_1^i\cdot\mathbf{a}_1^j)\frac{2\varepsilon_j}{\varepsilon_j-\varepsilon_i},\quad (i\neq j)\ \text{and}\ \alpha_{ii}=\mathbf{a}_1^i\cdot\mathbf{a}_1^i . \tag{A16}$$

From the above results, after a long but simple calculation, one can obtain all the coefficients that enter in the expression of the CS^+ and CS^-:

$$u_{ik}'=u_{ik}+\delta U_{ik} , \tag{A17}$$

$$v_{ik}'=v_{ik}+\delta V_{ik} , \tag{A18}$$

$$w_{ik}'=w_{ik}+\delta W_{ik} , \tag{A19}$$

where

$$U_{ik}=u_{ik}\left[A_{ik}-2\frac{t'_{ik}}{t_{ik}}\right]-\frac{t_{ik}D_{1ik}}{t_{ik}^2-4\varepsilon_i\varepsilon_k}\,, \qquad (A20)$$

$$V_{ik}=v_{ik}A_{ik}-\frac{2D_{1ik}}{t_{ik}^2-4\varepsilon_i\varepsilon_k}\,, \qquad (A21)$$

$$W_{ik}=U_{ik}-\varepsilon_i V_{ik}+2\varepsilon_i v_{ik}\alpha_{ii}\,, \qquad (A22)$$

and with

$$A_{ik}=\frac{4t_{ik}t'_{ik}-8\varepsilon_i\varepsilon_k(\alpha_{ii}^2+\alpha_{kk}^2)}{t_{ik}^2-4\varepsilon_i\varepsilon_k}\,, \qquad (A23)$$

$$t_{ik}=\eta^2+\varepsilon_i+\varepsilon_k\,, \qquad (A24)$$

$$t'_{ik}=\eta^2\alpha_{11}+\varepsilon_i\alpha_{ii}+\varepsilon_k\alpha_{kk}\,. \qquad (A25)$$

The critical perpendicular momentum p'_c along the new unstable mode q'_1 can thus be written by analogy with Eq. (A10), for any configuration $(q'_1,q'_2,\ldots,q'_N;p'_1,p'_2,\ldots,p'_N)$ of the new system, as

$$p'_c=\mp\eta' q'_1+F_\pm(q'_1,q'_2,\ldots,q'_N;p'_2,\ldots,p'_N)\,. \qquad (A26)$$

Equation (A26) can now be used to calculate the conversion coefficient c' for the system with the different mass by means of the equivalent of Eq. (19) for this new system.

In order to apply these results in the difference method required to compute the isotope effect we need to specialize the result of Eq. (A26) by explicitly introducing the correspondence between the state $P\equiv(0,q_2,\ldots,q_N;0,p_2,\ldots,p_N)$ of the system with all masses equal, and the equiprobable state $\tilde P\equiv(\tilde q'_1,\tilde q'_2,\ldots,\tilde q'_N;\tilde p'_1,\tilde p'_2,\ldots,p'_N)$ of the system with the mass of the jumping atom changed. We note here that the point $\tilde P$ does not lie on the new saddle surface SH′, but rather on the old SH. However, it is still possible by means of Eq. (A26) to find its critical momentum (note that one cannot now disregard terms involving the q'_1 coordinate). The correspondence given in the text in Cartesian coordinates can be expressed in the normal coordinate of the new system as

$$\tilde q'_i=q'_i\,,\qquad \tilde p'_i=p'_i-\delta\Delta'_i,\quad \Delta'_i=\sum_{j=1}^N(\mathbf{a}_1^i\cdot\mathbf{a}_1^j)p'_j\,, \qquad (A27)$$

where tildes are used to denote the component of $\tilde P$. By means of Eqs. (A11), (A12), and (A27) we obtain the new state $\tilde P$ in terms of the old coordinates and momenta, to the lowest order in δ:

$$\tilde q_i=q_i+\delta Q_i\,,\qquad \tilde p_i=p_i+\delta P_i-\delta\Delta_i,\quad \Delta_i=\sum_{j=1}^N(\mathbf{a}_1^i\cdot\mathbf{a}_1^j)p_j\,. \qquad (A28)$$

By substituting Eq. (A27) into Eq. (A26) the critical momentum along the new unstable mode $\tilde p'_c$ can now be evaluated for every $\tilde P$ in terms of the old coordinates and momenta. For our difference calculation we are interested, instead, in the component of the new critical momentum along the old unstable mode q_1 that is given to first order in the mass change by the equation

$$\tilde p_c=\tilde p'_c-\delta P_1\,. \qquad (A29)$$

On applying the above relation to Eq. (A26) and by repeated use of Eqs. (A11)–(A19) we finally obtain the expression for the corresponding critical momentum:

$$\tilde p_c=p_c-\delta A+\delta B\,, \qquad (A30)$$

with

$$A=p_c\alpha_{11}\pm\tfrac12\sum_{i,k=2}^N(v_{ik}p_i\Delta_k+v_{ik}\Delta_i p_k)+\sum_{i,k=2}^N w_{ik}q_i\Delta_k\,, \qquad (A31)$$

$$\begin{aligned}B=&\mp\eta Q_1-P_1+\left[\sum_{k=2}^N d_{11k}\frac{2\eta q_k\pm p_k}{4\eta^2+\varepsilon_k}\right]Q_1\\ &\pm\tfrac12\eta\sum_{i,k=2}^N(U_{ik}q_iq_k+u_{ik}q_iQ_k+u_{ik}Q_iq_k)\\ &\pm\tfrac12\eta\sum_{i,k=2}^N(V_{ik}p_ip_k+v_{ik}p_iP_k+v_{ik}P_ip_k)\\ &+\sum_{i,k=2}^N(W_{ik}q_ip_k+w_{ik}q_iP_k+w_{ik}Q_ip_k)\,. \end{aligned} \qquad (A32)$$

Term A comes from the correspondence of state points used in the difference method while term B comes from the changes of center stable or unstable manifold introduced by the mass change.

---

*Permanent address: Dipartimento di Fisica, Università degli Studi di Trento, I-38050 Povo, Trento, Italy.

[1]G. H. Vineyard, J. Phys. Chem. Solids 3, 121 (1957).
[2]N. L. Peterson, in *Diffusion in Solids,* edited by A. S. Nowick and J. J. Burton (Academic, New York, 1975), and references therein.
[3]J. N. Mundy, L. W. Barr, and F. A. Smith, Philos. Mag. **14**, 785 (1966); J. N. Mundy, Phys. Rev. B 3, 2431 (1971).
[4]C. P. Flynn, Phys. Rev. Lett. 35, 1721 (1975).
[5]C. H. Bennett, in *Diffusion in Solids,* edited by A. S. Nowick and J. J. Burton (Academic, New York, 1975).
[6]C. H. Bennett, Thin Solid Films 25, 65 (1975).
[7]J. N. Mundy, C. W. Tse, and W. D. McFall, Phys. Rev. B **13**, 2349 (1976); J. N. Mundy, H. A. Hoff, J. Pelleg, S. J. Rothman, L. J. Nowicki, and F. A. Schmidt, *ibid.* **24**, 658 (1981).
[8]M. J. Jackson and D. Lazarus, Phys. Rev. B **15**, 4644 (1977).
[9]C. Herzig, H. Eckseler, W. Bussmann, and D. Cardis, J. Nucl. Mater. 69/70, 61 (1978); see also the review of C. Herzig and U. Kohler, Mater. Sci. Forum **15/18**, 301 (1987).
[10]G. De Lorenzi, G. Jacucci, and C. P. Flynn, Phys. Rev. B **30**, 5430 (1982).
[11]M. Toller, G. Jacucci, G. De Lorenzi, and C. P. Flynn, Phys.

Rev. B **32**, 2082 (1985).
[12]G. Jacucci, G. De Lorenzi, M. Marchese, C. P. Flynn, and M. Toller, Phys. Rev. B **36**, 3086 (1987).
[13]J. D. Doll and A. F. Voter, Ann. Rev. Phys. Chem. (to be published).
[14]A. F. Voter and J. D. Doll, J. Chem. Phys. **80**, 5832 (1984).
[15]A. F. Voter and J. D. Doll, J. Chem. Phys. **82**, 80 (1985).
[16]H. Frauenfelder and P. G. Wolynes, Science **229**, 337 (1985).
[17]G. De Lorenzi, G. Jacucci, and C. P. Flynn, preceding paper, Phys. Rev. B **36**, 9461 (1987).
[18]H. B. Huntington, M. D. Feit, and D. Lortz, Cryst. Lattice Defects **1**, 193 (1970).
[19]C. P. Flynn, Mater. Sci. Forum, **15/18**, 281 (1987).
[20]G. Ciccotti and G. Jacucci, Phys. Rev. Lett. **35**, 789 (1975).
[21]G. Ciccotti, G. Jacucci, and I. R. McDonald, J. Stat. Phys. **21**, 1 (1979).
[22]C. P. Flynn and G. Jacucci, Phys. Rev. B **25**, 6225 (1982).
[23]G. Jacucci, in *Diffusion in Crystalline Solids*, edited by G. E. Murch and A. S. Nowick (Academic, New York, 1984).
[24]A. F. Voter, Phys. Rev. B **34**, 6819 (1986).
[25]L. Verlet, Phys. Rev. **165**, 209 (1968).

Surface Science 116 (1982) 391–413
North-Holland Publishing Company

DIFFUSION OF ADATOMS AND VACANCIES ON OTHERWISE PERFECT SURFACES: A MOLECULAR DYNAMICS STUDY

G. DE LORENZI and G. JACUCCI

Dipartimento di Fisica, Libera Università degli studi di Trento, e Istituto per la Ricerca Scientifica e Tecnologica in Trento, I-38050 Povo, Italy

and

V. PONTIKIS

Centre d'Etudes Nucléaires de Saclay, Section de Recherches de Métallurgie Physique, B.P. 2, F-91191 Gif-sur-Yvette Cedex, France

Received 14 July 1981; accepted for publication 22 January 1982

Using computer simulation by the technique of molecular dynamics, we have investigated the influence of the terrace structure on the type and the dynamical aspects of atomic mechanisms for surface diffusion in fcc structure crystals. On the (100) terraces, vacancies are much more mobile than adatoms, while the opposite is true for (111) terraces. On the latter, vacancies migrate through the creation in their vicinity of paired, adatom–vacancy, defects. On the (100) face, the adatom jump length incrases with increasing temperature and reaches a value equal to several times the nearest neighbour distance. Adatoms are also fully delocalized on the (111) face and spend much more time in flight over the surface than by vibrations in the equilibrium sites. Large dynamical correlations are present in the vacancy movement on the (100) face and have been identified as new mechanisms of the defect migration by multiple jumps. On the (110) terrace, despite its anisotropic structure, two-dimensional diffusion takes place by an original atomic exchange mechanism. This mechanism has been identified to be the same as the one proposed by Halicioglu to explain two-dimensional diffusion on (110) Pt terraces, and recently corroborated by the FIM experiments of Wrigley and Ehrlich.

1. Introduction

The aim of this paper is to report the results of a detailed computer simulation study on the atomic migration mechanisms, involving point defects on perfect fcc crystal surfaces, well below the bulk melting point ($T \lesssim 0.6 T_m$).

According to the terrace, ledge and kink (TLK) model of real surfaces, they generally consist of dense packed terraces, separated by steps of monatomic height. Along these, point defects may be generated or absorbed, their concentration being kept in a situation of dynamical equilibrium. The surface we choose to simulate is, however, an infinite perfect terrace; a point defect is initially created on it and cannot disappear. Thermally generated point defects must be, and actually are, formed in adatom–vacancy pairs. Furthermore, in real surfaces the presence of steps can enhance or hinder the diffusive motion [1].

These two aspects of surface diffusion are not considered by the present approach: what we are interested in are the various mechanisms of point defect migration on otherwise perfect surfaces and their relative importance for atomic diffusion.

Computer simulation by the method of molecular dynamics is the only technique which permits a close analysis of the nature of the diffusion processes and an assessment of the adequacy of jump diffusion models, rate theory and the random walk hypothesis.

To this end, moleuclar dynamics has been successfully employed to the case of bulk diffusion in rare gases [2], metallic crystals [3] and in superionic conductors [4].

Many aspects of the dynamics of the atoms are inaccessible to present experimental methos. Recently, the use of field ion microscopy (FIM) has allowed considerable progress to be made, by the observation of the migration of individual atoms on clean surfaces [5,6]. Unfortunately, many important details of the atomic migration dynamics such as residence and flight times, correlations in the direction of subsequent displacements relevance of non-nearest neighbour jumps, etc., are still out of reach.

In this work, we shall investigate the influence of the crystallographic structure of the surfaces for an fcc crystal consisting of particles interacting via the Lennard-Jones (12–6) potential. Simulation runs will be presented for the three low index (100), (110) and (111) surfaces. The dependence of the results on the temperature will also be investigated. For temperatures higher than roughly one half of the melting temperature, thermal generation of surface defects in large concentrations takes place and the description of the atomic migration processes becomes very complicated. As a consequence, our data will be limited to temperatures lower than this.

We stress that the present simulation employs the standard molecular dynamics technique, i.e. it correctly includes many-body and anharmonic effects of the system.

Full lattice relaxation at the surface and around the defects are of course allowed for. For this, the classical trajectory of the system is followed on the correct n-body potential surface, at the expense of following "uninteresting" surface and bulk atoms in addition to the mobile atoms in the surface.

Recently, an approximate method which greatly reduces the computation effort has been applied by Tully, Gilmer and Shugard [7] to simulate the motion of adatoms on the (100) face of a Lennard-Jones fcc crystal. The present study can be used as a test of the adequacy of this procedure and eventually of other approximate methods.

Preliminary results of this work have been the subject of a short communication [8].

2. The model

The model we used for the computations is an fcc microcrystallite bounded in the Z-direction by two parallel free surfaces, created by removing the usual periodic conditions, while keeping them in the other two directions. It contains about two hundred particles interacting via the pairwise additive spherical (12–6) Lennard-Jones potential

$$\phi(r) = 4\epsilon\left[(\sigma/r)^{12} - (\sigma/r)^{6}\right]$$

cutoff at radius r_c between 2σ and 2.5σ.

Reduced units have been used throughout (table 1), but when useful they have been converted into real units, using values of energy (ϵ), length (σ) and mass (m) appropriate to solid argon. The ratio of the actual and the melting point temperature of solid argon, $T_M = 0.7$ in (ϵ/k) units, is sometimes employed.

The temperature is chosen such as to ensure that appreciable surface migration takes place during the run, and yet not so high as to cause significant thermal generation of defects. Table 2 gives the orientation, temperature and all the other relevant information of the surfaces studied. The temperature dependent lattice spacings and values of the physical constants were taken from experimental data for rare gas crystals.

Newton's coupled equations of motion are solved using the central difference algorithm [9] with a time step $\Delta t = 0.01$. About five hundred steps are allowed for equilibration. Each run consists of one hundred thousand steps, requiring 7 h of CPU time on a IBM 370/168 processor.

Surface migration is studied by producing artificially two point defects, one on each free surface of the crystallite. This is achieved by removing one atom from the surface plane of the lower face and placing it on the upper face.

A few words are in order here about counting jump events. As usual in performing an analysis of atomic diffusion in terms of point defect migration [10], there is some arbitrariness in defining the jump event. The difficulty arises from the necessity of converting the continuous trajectory of the system in $3N$

Table 1
Units used in the present work; σ and ϵ are the Lennard-Jones potential parameters, m is the atomic mass, and k is the Boltzmann constant

Physical quantities	Unit
Distance (r)	σ
Energy (E)	ϵ
Temperature (T)	ϵ/k
Time (t)	$(m\sigma^2/\epsilon)^{1/2}$
Diffusion coefficient (D)	$(\epsilon\sigma^2/m)^{1/2}$

Table 2
Temperature regions and crystallite dimensions as a function of the surface orientation

Orientation	T_{min}	T_{max}	Crystallite dimensions	Number of particles	Cutoff	Crystallite thickness
(100)	0.24	0.34	4[100]×4[010]	256	2.0	8 planes
(110)	0.25	0.30	8[110]×3[001]	240	2.20	6 planes
(111)	0.16	0.45	4[110]×4[112]	192	2.20	6 planes

coordinate space, into the motion of the system being in one of the many equilibrium sites or crossing the boundary from one to another, and consists in defining the boundary surfaces. In the dynamical lattice, these are different from the surfaces bounding the Wigner–Seitz cells associated with the rigid lattice. A further complication is due to the existence of "unsuccessful" or "U-turn" trajectories, corresponding to uncompleted jumps. As usual in this type of analysis, a jump is not scored if the system goes back before having travelled most of the way towards the new equilibrium position. The residual arbitrariness in counting jump events gives rise to negligible differences. Also the "time" at which a jump should be scored can be determined only with a finite uncertainty. This is due to the finite duration of a jump event and the geometrical difficulties encountered in defining the crossing of boundary surfaces, as already mentioned. As a consequence, the order in which a jump event should be scored can be sometimes uncertain, and whether two or more jump events are "simultaneous" cannot be unambiguously determined. Given the limitations of the jump counting, it is nevertheless useful to devise a method for the investigation of the rate of occurrence and the delays between successive jumps. An automatic jump counting procedure appropriate to each surface and to each defect (adatom or vacancy) has been elaborated by using the Wigner–Seitz cells as an operational definition of the surface equilibrium sites.

In addition to the dynamic calculations, lattice statics calculations have been performed on identical model systems to evaluate defect migration energies. The method consists of a molecular dynamics program, modified to include a damping force [2].

3. Results

3.1. The (100) face

The analysis of the simulation runs has been carried out both by using the automatic jump counting procedure (AJC) previously mentioned (section 2), and by drawing the atomic trajectories in the real space. The latter can obviously not be obtained for the vacancy. Careful examination of the atomic positions as functions of time, leads to the conclusion that the vacancy is not well localized during most of the run, as a result of its high jump frequency.

In the AJC procedure, the adatom and the vacancy trajectories are considered to be composed of single jumps between nearest neighbour equilibrium sites. Table 3 summarizes the jump frequencies obtained in this way at various temperatures.

However, the adatom trajectories (fig. 1) clearly show that the AJC procedure fails, in part, due to the occurrence of multiple jump events characterized by displacements to non-nearest-neighbour sites. This failure consists in the counting of an N-length jump as N single-jump events, because the AJC is unable to distinguish between them.

Despite this disadvantage, the AJC procedure still remains very usefully especially in the case of the vacancy for which no trajectories can be drawn. Examination of figs. 2 and 3, obtained for the vacancy by AJC, shows an enhancement of the frequency of occurrence for successive "jumps", separated by short delay times, with respect to a random distribution.

This type of successive "jumps" exhibits also a strong directional correlation (fig. 4). In order to calculate $\langle \cos\theta \rangle$, i.e. the mean value of the cosine of the angle between successive jumps directions, all the jumps counted by AJC have been separated in two groups, for delay times shorter or longer than 2 ps. This value has been chosen in order to process separately all the events leading to the enhancement of the actual jump frequency compared to that of a random process (figs. 2 and 3). For the remaining jump events there is no longer any angular correlation (fig. 4). Therefore, we conclude that the short delay events detected by the AJC procedure correspond in fact to vacancy jumps between non-nearest-neighbour sites as for the adatom (fig. 1).

The multiple jump processes for the vacancy involve the simultaneous displacement of nearby atoms, all moving to nearest neighbour sites and leading to the defect hopping to non-nearest-neighbour sites. At the highest temperature, these processes account for a great part of the defect migration events (fig. 4).

Table 3
Summary of the temperatures at which the molecular dynamics simulation runs were performed for the (100) surface; the observed jump frequencies of the vacancy Γ_v and of the adatom Γ_{ad} are also reported

T	0.24	0.27	0.30	0.34
T/T_M	0.34	0.39	0.44	0.49
Γ_v	0.16±0.01	0.27±0.02	0.44±0.02	0.59±0.02
Γ_{ad}	0.04±0.01	0.71±0.01	0.11±0.01	0.12±0.01

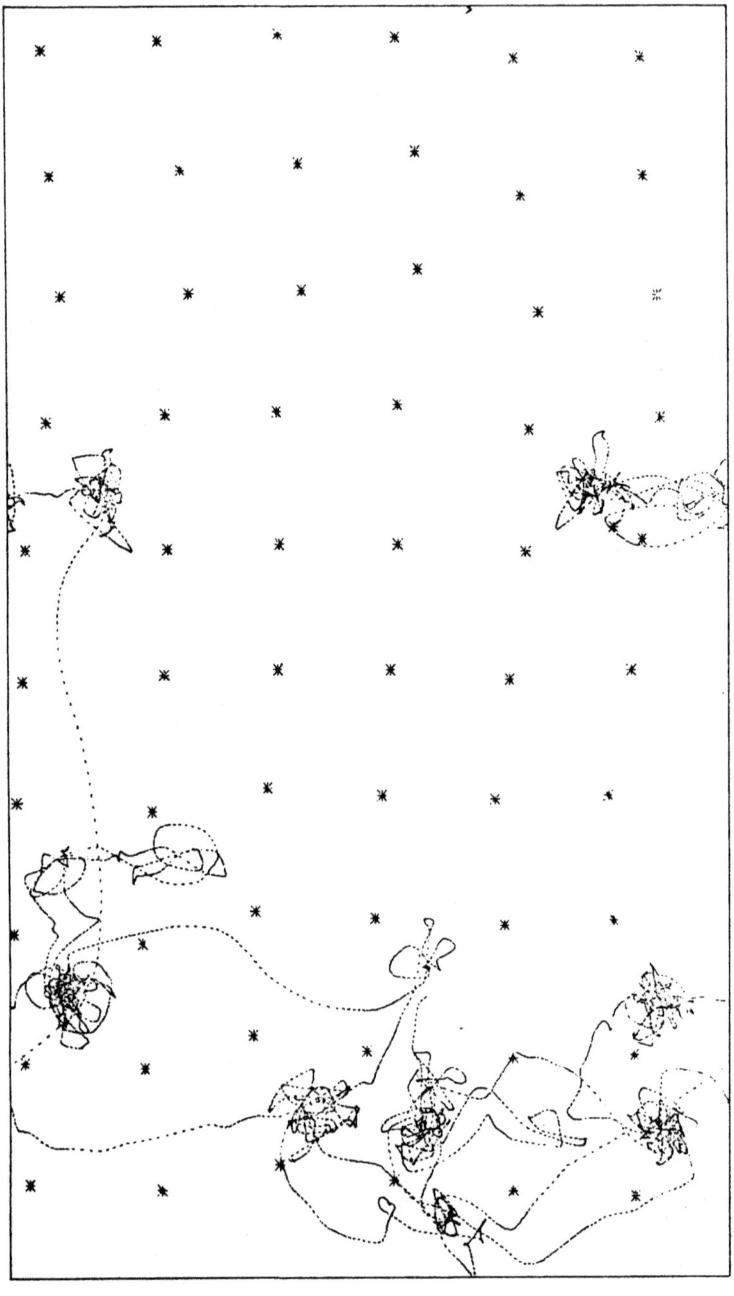

Fig. 1. Twenty-thousand steps trajectory of the adatom on the (100) surface showing non-nearest neighbour jump diffusion at $T=0.34$. The atoms on the surface are indicated by asterisks in their positions at the beginning of the run.

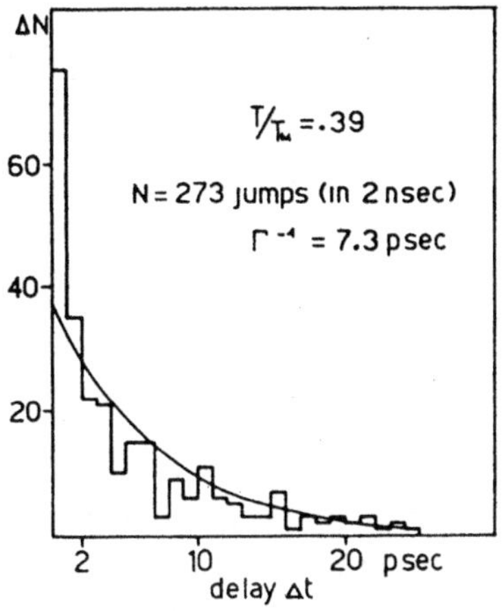

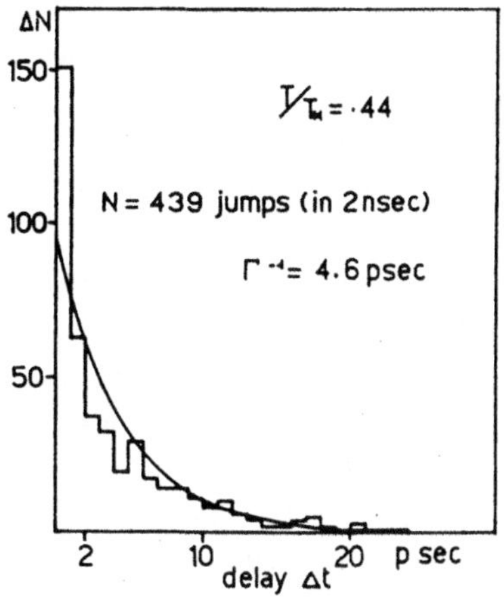

Fig. 2. Histogram of delay times between successive vacancy jumps on the (100) surface resulting from the AJC procedure. The solid line corresponds to the delay time distribution of a random walk process with a mean frequency equal to that obtained by the AJC procedure. $T/T_M = 0.39$.

Fig. 3. Same as fig. 2, but for $T/T_M = 0.44$.

Multiple jumps can be considered as new and different jump mechanisms, and the question arises as to whether the associated activation energies are different. The number of those events during the simulation runs is still rather small, yet, with some uncertainty, an answer can be attempted.

The operational definition used to count multiple jump events consisted of choosing a time interval of suitable length $\tau = 2$ ps (deduced from the AJC analysis) and looking for groups of two or more subsequent jump events, all

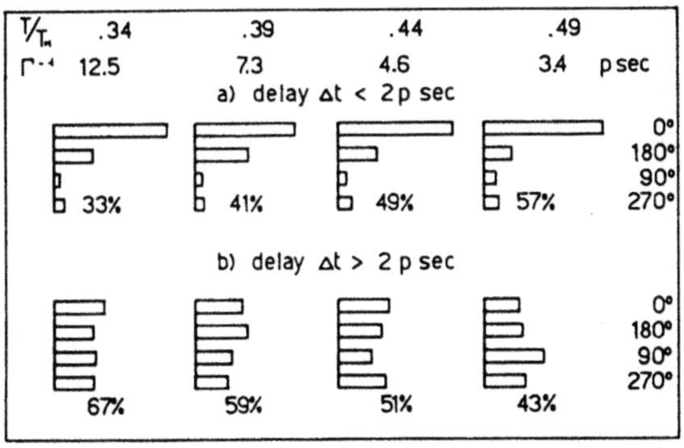

Fig. 4. Dynamical correlation exhibited by the histogram of the angles between the directions of successive vacancy jumps on the (100) surface, as obtained by the AJC procedure. Pairs of successive jumps are divided into two categories according to the value of the time delay between jumps.

occurring within a delay time τ. For the vacancy, the number of double (DJ) and triple (TJ) jumps, involving simultaneous motion of two and three atoms, respectively, increases strongly with temperature, and when plotted versus T^{-1} exhibit Arrhenius behaviour (fig. 5).

The activation energies obtained are $E^1 = 0.7$ for SJ, $E^2 = 1.3$ for DJ and $E^3 = 2.4$ for TJ.

A similar analysis is not feasible for the adatom, because the number of the corresponding jump events is much smaller than for the vacancy. Fig. 1 shows, however, that multiple jumps are present.

The temperature dependence of the adatom SJ number exhibits an Arrhenius plot (fig. 6), and the corresponding activation energy is equal to $E = 1.0$.

It should be noted that the operational definition employed to score multiple jump events will produce counts even in analysing the trajectory of a really single jump type random process. For a fixed choice of τ, the number of multiple jumps will increase with the overall process frequency Γ. If Γ is made to increase exponentially with the parameter $\alpha = E/kT$, then the frequencies Γ_2 of DJ and Γ_3 of TJ will also increase exponentially with the parameters $\alpha_2 = 2E/kT$ and $\alpha_3 = 3E/kT$, respectively, in the limit $\Gamma\tau \ll 1$. (Otherwise Γ_2 and Γ_3 will not be simple exponential functions of α_2 and α_3.)

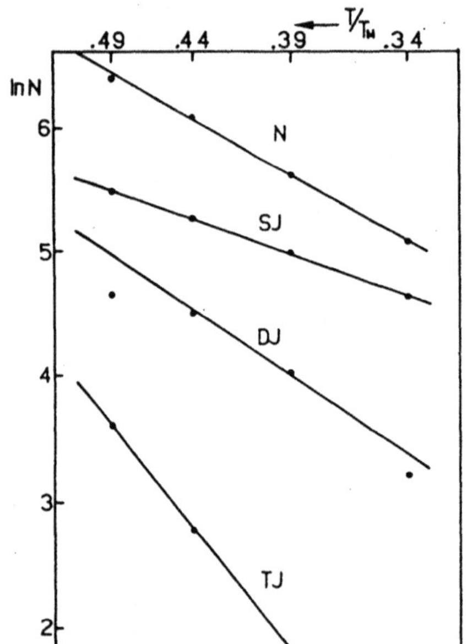

Fig. 5. Dependence of the number of jumps for the vacancy on the (100) surface versus the inverse temperature. Separate plots are presented for the total number of "jump" events furnished by the AJC procedure (N), and also for single (SJ), double (DJ) and triple (TJ) jumps (see text for definition).

We have determined analytically Γ_1, Γ_2 and Γ_3 for a random process having a mean occurrence frequency Γ equal to the overall jump frequency deduced from the AJC analysis for the vacancy. As a criterion to distinguish between SJ, DJ and TJ we chose the same as previously.

The results are compared in fig. 7 to the SJ, DJ and TJ frequencies obtained from the AJC procedure. We observe an enhancement of DJ in the simulation results with respect to those for a random process. No better agreement is found by changing the value of τ in the analysis of the random process or by introducing a finite duration for the jump event. Therefore, we conclude that DJ and TJ are not artifacts resulting from the particular analysis, but genuine and distinct migration mechanisms.

3.2. The (110) face

The (110) surfaces of an fcc crystal are made up of close-packed rows of atoms running in the [110] direction. They can also be considered as surfaces with a maximum step density. Contrary to what we expect, simulation runs have shown that adatoms move as easily perpendicular to the [110] channels as along them. A careful analysis of the defect trajectories, by a method similar to that previously used for the (100) face and by precise monitoring of the atomic

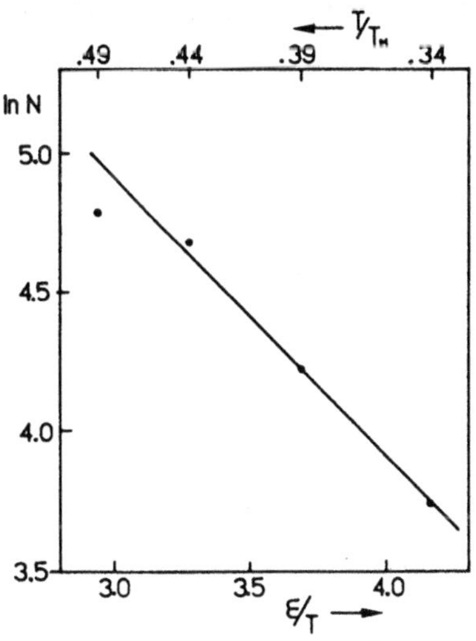

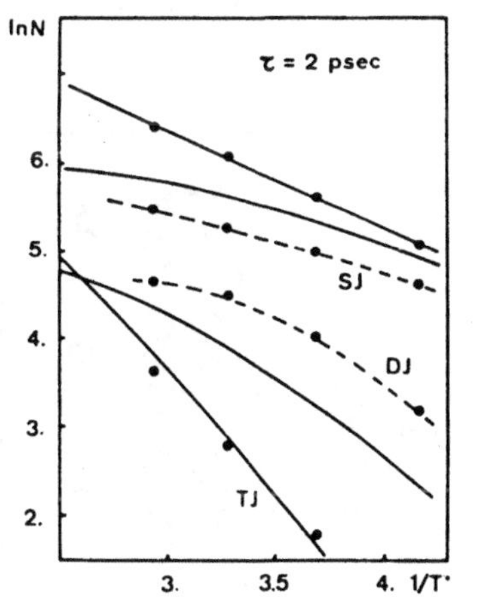

Fig. 6. Dependence of the number of single jumps for the adatom on the (100) surface versus inverse temperature.

Fig. 7. Comparison between the number of multiple jumps for the vacancy (filled circles, see also fig. 5) and the results obtained for a purely random process with the same mean frequency Γ as a function of temperature. The discrepancies observed show that the multiple jump events for the vacancy migration on the (100) surface are not due to an artifact of data analysis.

positions during each jump event, permits us to identify a variety of mechanisms for the adatom migration on this surface.

The adatom movement takes place mainly via a mechanism which produces either migration in the channel or the exchange between the adatom and one of the four adjacent row atoms. In this manner, diffusion takes place both along and across the [110] channels. Fig. 8 illustrates the saddle point and the four final positions for one adatom migrating via this replacement mechanism. The atomic trajectories obtained by molecular dynamics (figs. 9a and 9b) prove that this mechanism is identical to the one, first proposed by Bassett and Webber [11], and then refined by Halicioglu and Pound [12] to explain two-dimensional diffusion of adatoms on channeled (110) fcc crystal surfaces.

Simulation runs have also produced a few direct jumps of the adatom and the vacancy across the [110] rows, but their rate of occurrence is still negligible compared to that of the above exchange mechanism (table 4).

Both defects migrate along the [110] channels by direct jumps. The vacancy makes only single jumps, but for the adatom hopping to non-nearest-neighbour sites has been rarely observed in the highest temperature run (fig. 10).

3.3. The (111) face

3.3.1. The adatom

Comparison between the adatom trajectories on the (100) and the (111) planes (see figs. 1 and 11, respectively) shows that there is no hopping into the equilibrium sites on the latter surface.

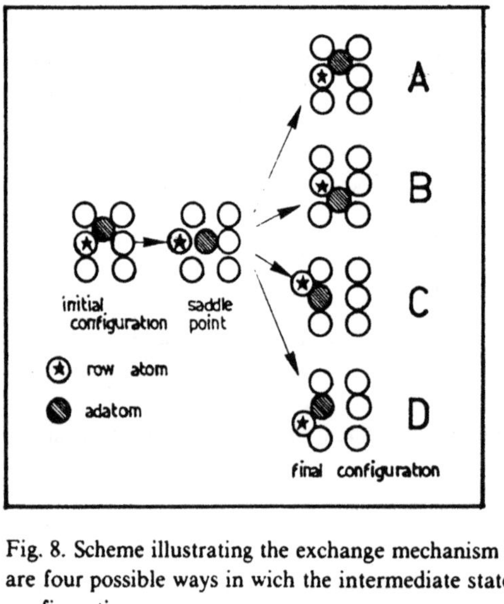

Fig. 8. Scheme illustrating the exchange mechanism for atomic motion on the (110) surface. There are four possible ways in wich the intermediate state (saddle pc int) can transform to a new stable configuration.

Table 4
Frequency of occurrence versus temperature, for adatom and vacancy single jumps along the [110] direction on the (110) plane and for adatom two-dimensional diffusion via the exchange mechanism

Mechanism	Γ	
	$T/T_M = 0.35$	$T/T_M = 0.41$
Exchange	0.014 ± 0.004	0.045 ± 0.005
Adatom direct parallel jumps	0.008 ± 0.003	0.029 ± 0.004
Vacancy parallel jumps	0	0.007 ± 0.002

The adatom movement is a nearly free translation over the surface without evaporation, and its mean velocity is of the order of $(2kT/m)^{1/2}$. During its motion, the defect is submitted to frequent collisions with the vibrating surface atoms. The mean free path, typical of the distance between two successive collisions, has been calculated using the expression:

$$\lambda = \tfrac{1}{2}(\lambda_x + \lambda_y),$$

where

$$\lambda_\alpha = (kT/m)^{1/2} f_\alpha^{-1} \quad (\alpha = x, y).$$

In this formula, k is the Boltzmann factor, T the temperature, m the atomic mass, and f_α the sign inversion frequency of the α component of the adatom velocity. In the calculation of f_α, those rare periods during which the adatom was observed to be localized at an adsorption site were not considered.

In the present results, f is seen to vary roughly as $T^{-0.4}$, so that the mean free path, λ, obeys the following empirical temperature law: $\lambda \propto T^{0.9}$ (fig. 12).

In the investigated temperature range, the diffusion coefficient of the adatom can be expressed as for a free particle by: $D = \frac{1}{4}\bar{v}\lambda$ ($\bar{v}$ is the mean square velocity). By substituting in this formula the values of λ and $\bar{v}$, we obtain:

$$D = \tfrac{1}{4}\lambda(2kT/m)^{1/2},$$

so the diffusion coefficient increases with increasing temperature with a power law $D \propto T^n$, where $n \propto \frac{3}{2}$ (fig. 13).

3.3.2. The vacancy

Unlike the results obtained for its migration on (100), the vacancy has a very low mobility on the (111) plane. For the three lower temperatures investigated (0.22, 0.27 and 0.35) the vacancy remained immobile during the entire simulation run. At $T = 0.40$, we observed 13 hops, three of which were to

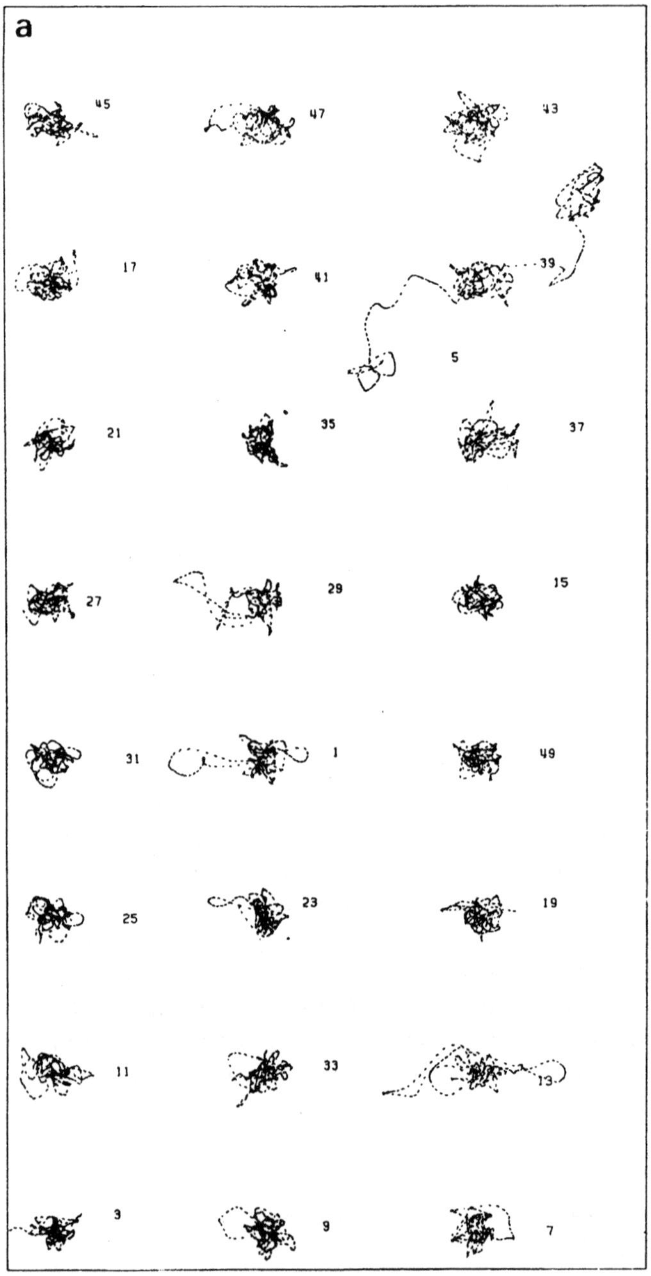

second neighbour sites. In all cases the vacancy motion was made possible by the thermal generation in its neighbourhood of an adatom–vacancy pair, which disappears after the jump was completed (fig. 14).

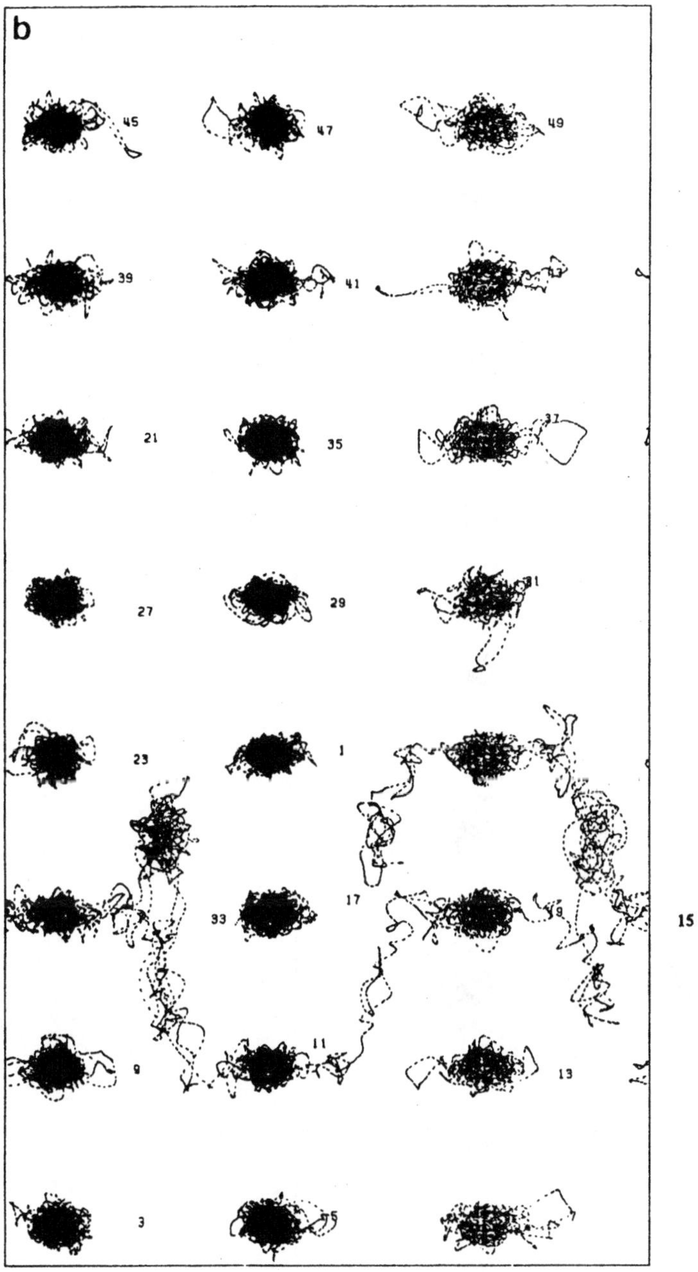

Fig. 9. Thousand steps trajectories for adatom diffusion on the (110) surface at $T=0.3$, illustrating the exchange mechanism. (a) The adatom starts from the indicated position and forces out the row atom ($T/T_M=0.41$). (b) Ten-thousand steps trajectory for a cascade of exchange events at $T/T_M=0.41$.

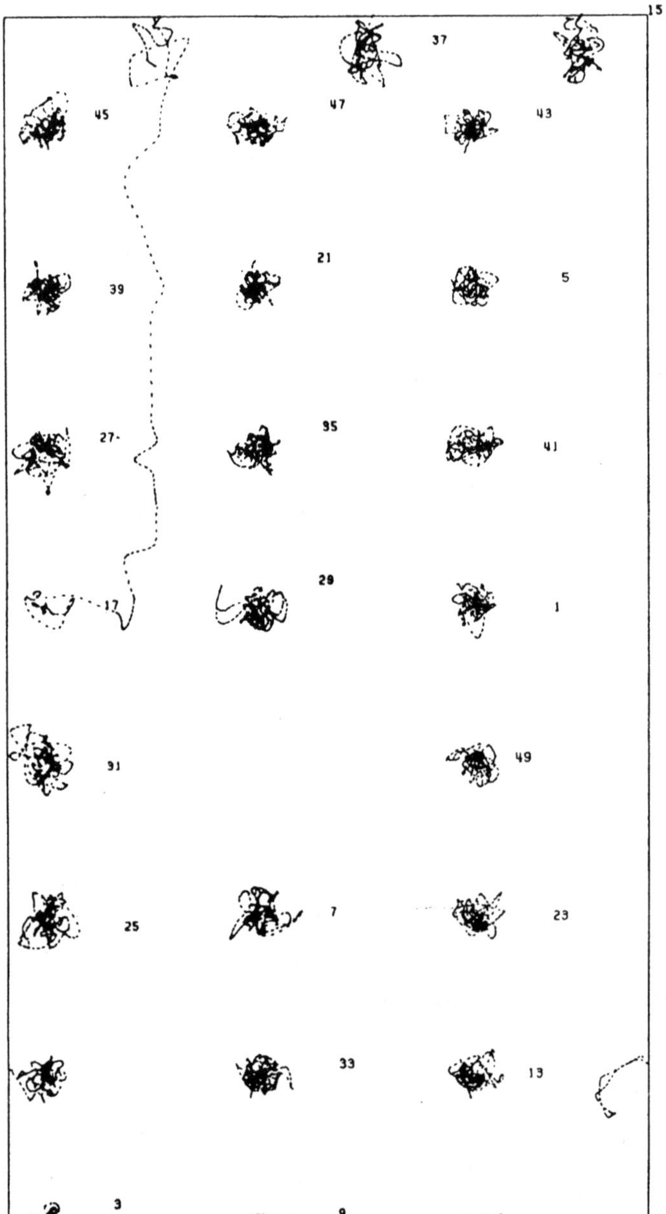

Fig. 10 Adatoms multiple jumps along the rows of the (110) surface. Trajectories obtained during thousands time steps at $T=0.3$.

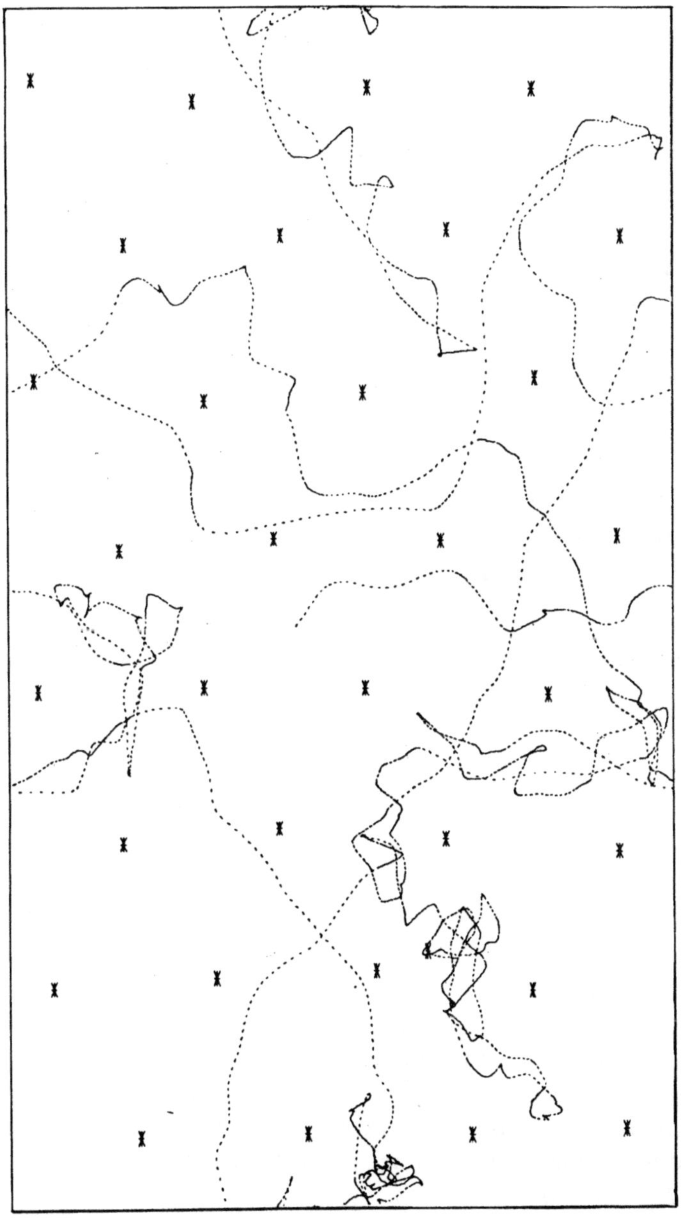

Fig. 11. Fully delocalized trajectory of the adatom on the (111) surface. Trajectories obtained at $T=0.3$ during ten thousand time steps. The asterisks represent the surface atoms at the start of the trajectory.

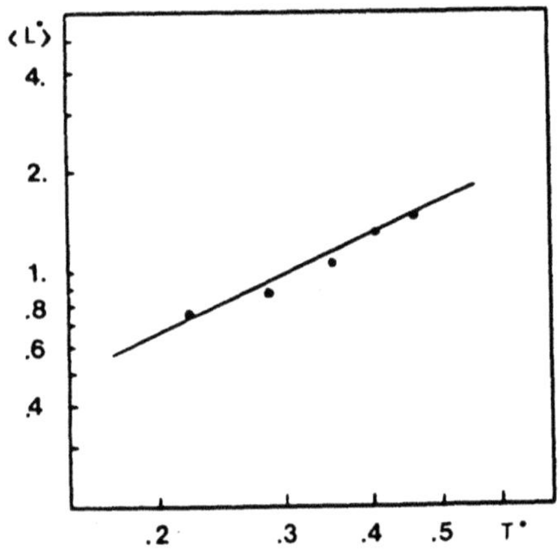

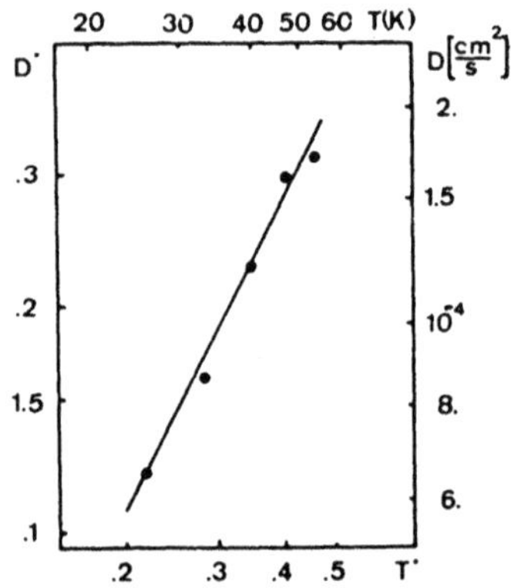

Fig. 12. Dependence of the adatom mean free path versus temperature on the (111) surface.

Fig. 13. Variation of the calculated adatom diffusion coefficient on the (111) surface versus temperature.

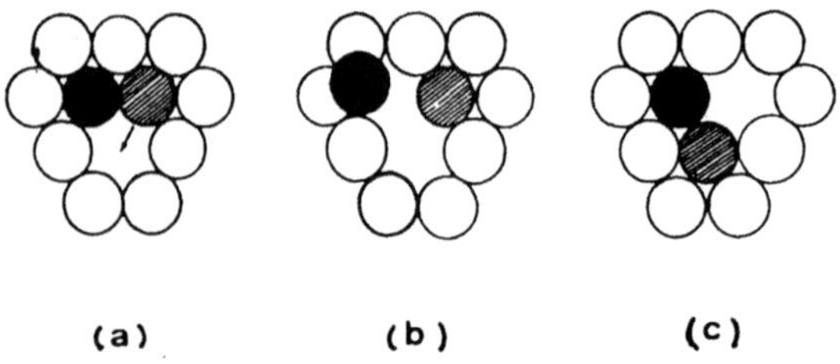

Fig. 14. Mechanism for vacancy migration on the (111) surface.

4. Lattice statics

4.1. The (100) surface

The migration energy of both adatoms and vacancies has been evaluated by lattice statics and found to be $E_m = 1.7$ for the adatom and $E_m = 2.1$ for the vacancy. We recall that the molecular dynamics simulation yielded $E_m = 1.0$ for the adatom single jumps. In the case of the vacancy we obtained $E_m = 1.1$ for the overall energy of migration (all nearest neighbour exchanges counted as completed jumps) and $E_m = 0.7$ for SJ (table 5).

These energies refer to the difference between the potential energies of the fully relaxed configurations with the defect at the saddle point (for single jumps!) and those with the defect at the equilibrium position at the bottom of the potential well.

Table 5
Migration energy for point defects on the (100) surface from static calculations and from the molecular dynamics simulation

Defect	E_m (static)	E_m (dynamic)
Adatom	1.7	1.0
Vacancy	2.1	0.7 (SJ)

4.2. The (110) surface

The static evaluation of the migration energy of adatoms along the channels and vacancies along the rows yields $E_m = 1.92$ and $E_m = 2.83$, respectively. The migration energy for the exchange mechanism (with the saddle point as defined by Halicioglu) is found to be $E_m = 2.30$. The molecular dynamics simulation gave $E_m = 2.0$ for direct parallel jumps of the adatom and $E_m = 1.8$ for the exchange mechanism (table 6).

4.3. The (111) surface

On the (111) surface, there are two types of sites, corresponding to fcc and hcp stacking, respectively. These yield two different values of the migration energy of the adatom, depending on the equilibrium site from which the jump

Table 6
Migration energy for point defects on the (110) surface; in this case the two principal jump mechanisms for the adatom are indicated

Defect	Mechanism	E_m (static)	E_m (dynamic)
Adatom	Exchange	2.30	1.8 ± 0.6
	Single parallel jumps	1.92	2.0 ± 0.7
Vacancy	Parallel jumps	2.83	–

Table 7
Migration energy for the adatom on the (111) surface from static calculations; hcp and fcc indicate the surface lattice site from which the jump was initiated

Defect	Equilibrium site	E_m (static)
Adatom	hcp	0.262
	fcc	0.394

is initiated. The migration energies for this defect obtained by static calculations are $E_m^1 = 0.262$ for jumps initiated from an hcp site, and $E_m^2 = 0.394$ for those initiated from an fcc site (table 7).

No static calculations have been performed for the vacancy in the (111) plane, since it is difficult to define a simple geometrical saddle point configuration for the observed migration mechanism.

5. Discussion

Before starting the discussion, it is appropriate to review briefly the main results on which it will be based or which it will try to explain:
(a) The mobility and the migration mechanisms of the adatom and the vacancy, are strongly dependent upon the structure of the surface on which they migrate.
(b) The jumps between non-nearest-neighbour equilibrium sites, seems to be a general feature for simulated surface diffusion.
(c) For both defects, the comparison between the migration energies obtained by molecular dynamics and those obtained by fully relaxed lattice statics shows the latter to be systematically higher.

5.1. Influence of the surface structure on defect mobility and the migration mechanisms

The structure of the three surfaces which are considered in this work is quite different. For fcc crystals, the (111) surface is the most densely packed, and this clearly influences the atomic mobility. The vacancy migrates unfrequently on this face, the opposite being true for the (100) surface, where it is the most mobile defect. The migration mechanisms are also different: the vacancy migrates on (111) mainly by generation in its vicinity of an adatom–vacancy pair and on the (100) by multiple jumps whose contribution could be very important at the higher temperatures. The above mechanism suggest that di-vacancies can be much more mobile than single vacancies on the (111) face.

This suggestion is also supported by a hard sphere model simulation realized by Pieranski et al. [13].

Now, the adatom is more mobile on the (111) than on the (100) face. In the investigated temperature range, the surface lattice structure of the former has little effect on its behaviour. It undergoes an essentially free two-dimensional motion on this plane, without evaporating, and, during its movement, it is scattered by the vibrating surface atoms. The mean free path of the defect exhibits a temperature dependence, indicating that the collision frequency decreases with increasing temperature ($\nu \propto T^{-0.4}$). This may be due to an adatom mean distance from the surface increasing more rapidly than the vibrational amplitude of surface atoms.

This "delocalized" behaviour of the adatom may disappear at lower temperatures, where the adatom may migrate by hopping between equilibrium sites.

On the (100) plane, the potential wells associated with the equilibrium sites are deeper than those on the (111) surface, and consequently the adatom spends a long time in such a site before jumping.

As for the vacancy, jumps occur between non-nearest neighbour sites and may enhance diffusion on this face. It should be noted that this situation is similar to that observed in vacancy bulk diffusion near the melting point in Lennard-Jones crystals [2] and for simulations using potential appropriate for metals [3].

The (110) surface must be considered apart from the two others, because of its anisotropic structure which is more realistic with respect to real crystal surfaces.

As can be expected, the vacancy migrates only along the [110] rows. The adatom migration takes place by the original exchange mechanism, which has been identified to be the same as the one proposed by Halicioglu and Pound [12]. The experimental observation of this mechanism by FIM [14], in the case of W diffusion on Ir(110), proves that the Lennard-Jones potential used for the simulation in this work correctly describes diffusion on metallic surfaces in the limit of weak, Van der Waals-like, atomic interactions.

Also the adatom migrates along the [110] channels either by single or multiple (i.e. non-nearest neighbour) jumps.

This behaviour seems to be a general feature of simulated surface diffusion and occurs for the three surfaces studied, despite their very different structure. On real crystal surfaces, if such mechanisms are operative, they will probably enhance difusion, especially at high temperatures, as has been assumed by Bonzel [15].

Simulation studies of surface diffusion have been attempted by Tully, Gilmer and Shugard (hereafter referred to as TGS) [7], and more recently by Garofalini and Halicioglu [16].

TGS have studied adatom migration on the (100) face of an fcc Lennard-Jones crystal with an approximate, but time saving simulation technique. They observed adatom migration by multiple jumps as we did in the present study. Unfortunately, their method did not permit the simulation of the vacancy migration on surfaces to be made. TGS have performed static calculations for the migration energy (E_m) of the adatom on the (100) plane, which is in good agreement with our results. On the other hand poor agreement is found between the two dynamic evaluations of migration energy (table 8).

The origin of the disagreement is probably related to the different definition employed to determine this quantity by TGS and in this work. TGS obtained an E_m value from the slope of the $\ln D$ versus $1/T$ plot. This is a questionable procedure because of the existence of variable length jumps.

Garofalini and Halicioglu [16] have used molecular dynamics to simulate diffusion on the (110) plane, and their results confirm the existence of the

Table 8
Comparison of migration energies for the adatom on the (100) surface given by TGS [7] and the present work

Method	Adatom on (100)	
	This work	TGS
Lattice statics	1.71	1.65
Dynamics: $\langle r^2 \rangle / 4\tau = f(1/T)$	–	1.55
Dynamics: $E_m = \partial \ln N_{SJ} / \partial(1/T)$	1.0	–

exchange mechanism, depending on the strength of the atomic bond and the size and mass of the migrating adatom. Qualitatively, there is agreement with our previous [8] and present results.

However, due to the short simulation runs performed by these authors, they have not observed multiple jumps for the adatom.

5.2. Diffusion anisotropy on the (110) face

The predominance of the exchange mechanism for atomic motion on the (110) surface leads to nearly isotropic diffusion. Otherwise, diffusion will be very anisotropic on this surface. We have investigated the mean square displacement of a tracer on such a surface, in a single encounter with an adatom, with the approximation that only the exchange mechanism operates, and by excluding direct hops parallel to the [110] channels. The method used is the same as the one proposed by Wolf, Differt and Mehrer [17] for diffusion of a vacancy in the bulk.

The results of 25 hops of the adatom in the encounter (table 9) shows that the anisotropy of the displacements distribution compensates in part the different lattice spacings in the [110] and the [001] directions. A small, temperature independent anisotropy still exists for this mechanism. Of course, if the other possible diffusion mechanisms, for the (110) plane, are taken into account, a diffusion anisotropy will appear, which should be temperature dependent because of the different activation energies associated with each of them. The superposition of all the possible contributions to diffusion on (110) planes should lead to a diffusion component along the [110] channels greater than the perpendicular component: $D_{\parallel} > D_{\perp}$. This is in agreement with recent experimental results for silver [19] and copper [21] diffusion on clean copper surfaces [20].

5.3. Lattice statics calculations

The defect migration energies from the statics calculation are consistently larger than those obtained by counting jump events occurring in the dynamical

Table 9
Mean square displacements of a tracer in a single encounter with an adatom, if only the exchange mechanism is taken into account

	$\overline{X}^2$	$\overline{Y}^2$		$\overline{X}^2$	$\overline{Y}^2$
1	0.125	0.125	14	0.360	0.898
2	0.203	0.329	15	0.365	0.914
3	0.240	0.462	16	0.370	0.929
4	0.265	0.555	17	0.374	0.943
5	0.282	0.627	18	0.378	0.957
6	0.297	0.681	19	0.381	0.968
7	0.309	0.725	20	0.385	0.976
8	0.320	0.761	21	0.388	0.986
9	0.329	0.793	22	0.392	0.996
10	0.336	0.819	23	0.394	1.007
11	0.343	0.841	24	0.398	1.016
12	0.349	0.863	25	0.401	1.023
13	0.355	0.882			

$N = 100{,}000$ random trajectories of the defect are followed for 25 exchange events, the first being between the defect and the tracer situated in a row site. The mean square displacements of the tracer are computed after each exchange event. The distances between atoms along the X([001]) and the Y([1$\overline{1}$0]) direction are both taken equal to unity (so that the distance row–channel is 0.5). Note that the exact values for $\overline{X}^2$ and $\overline{Y}^2$ after the first jump are 0.125 and 0.125, after the second jump 0.203 and 0.328.

simulation (tables 6 and 7). This is in analogy with molecular dynamics results on vacancy diffusion in the bulk for aluminium [3].

In order to identify the origin of this discrepancy we have tested for a dependence of the energy values of defect migration on lattice spacings, potential cutoff radius and stress applied to the free surfaces, to vary the relaxed interplanar distance normal to the surface. We did not find any essential difference in the lattice statics values. We, therefore, conclude that those discrepancies are inherent to the limitations of lattice statics. The real saddle point surface in configuration space is certainly very complicated, very different from the simple one used for the lattice statics calculations.

5.4. *A qualitative comparison with the experimental results*

We now attempt to make a comparison between our results, obtained by simulation, and those obtained by FIM experiments. This comparison is particularly interesting in the case of the (110) face on which, according to the results of this work, two-dimensional diffusion occurs, via the exchange mechanism. As already mentioned, this mechanism has been experimentally observed by Wrigley and Ehrlich by means of FIM [14] for the migration of W adatoms on the (110) Ir. In spite of the rough character of the present model, some qualitative aspects of the atomic diffusion processes are thus shown to be correctly described; on the other hand this is just what we basically expect from our study. Unfortunately, the range of investigation of the FIM is confined to low temperatures, and the limitations inherent to the method do not permit detailed information to be obtained on the dynamics of atomic migration.

There is also no any evidence for multiple jump events in surface diffusion arising from FIM experiments.

Yet, a few years ago Bonzel [15], in order to explain the temperature dependence of surface self-diffusion (Arrhenius plot) obtained by classical macroscopic experimental methods, has suggested that free translation of adatom dimers or trimers can occur on surfaces. His model predicts that for copper adatoms non-nearest-neighbour jumps occur and can reach, near the melting point, jump lengths of the order of ten lattice parameters.

Our results are clearly in qualitative agreement with this suggestion (figs. 1, 10 and 11).

We expect that for medium and high temperature surface diffusion those processes may be quite significant. Further experimental results in this area are needed to confirm the validity of the molecular dynamics simulation.

6. Conclusion

(1) The Lennard-Jones model used in this study can give a realistic description of surface diffusion, especially on the (110) plane. The two-dimensional diffusion of adatoms on that plane is due to the exchange mechanism, which seems to be the same as that recently observed by FIM.

(2) The mobility of adatoms and vacancies and the corresponding migration mechanisms depend strongly on the surface structure.

(3) New processes for atomic diffusion by multiple length defect jumps are operating on the (100) and (111) surfaces.

(4) A systematic difference exists between the values of defect migration energies obtained by molecular statics and by molecular dynamics calculations. This can be due to both the anharmonic features of the potential energy surface and the crude approximation of the saddle point surface used in the lattice statics calculations.

Acknowledgements

Very useful discussions with Y. Adda, J. Cousty, G. Martin, C. Moser and B. Perraillon are acknowledged. G. De Lorenzi and G. Jacucci are grateful for the kind hospitality they received at the SRMP during their repeated visit in the course of this work. This work has been supported by the Commissariat à l'Energie Atomique and the Consiglio Nazionale della Ricerche.

References

[1] J. Cousty, R. Peix and B. Perraillon, Surface Sci. 107 (1981) 586.
[2] C.H. Bennet, in: Diffusion in Solids - Recent Developments, Eds. A.S. Nowick and J.J. Burton (Academic Press, New York, 1975) p. 73.
[3] A. Da Fano and G. Jacucci, Phys. Rev. Letters 39 (1977) 950.
[4] G. Jacucci and A. Rahman, J. Chem. Phys. 69 (1978) 4117.
[5] G.L. Kellog, T.T. Tsong and P. Cowan, Surface Sci. 70 (1978) 485.
[6] G. Ehrlich, J. Vacuum Sci. Technol. 17 (1980) 9.
[7] J.C. Tully, G.H. Gilmer and M. Shugard, J. Chem. Phys. 71 (1979) 1630.
[8] G. De Lorenzi, G. Jacucci and V. Pontikis, in: Proc. ICSS-4 and ECOSS-3, Cannes, 1980, Vol. I, Eds. D.A. Degras and M. Costa, p. 54.
[9] L. Verlet, Phys. Rev. 159 (1967) 98.
[10] C.P. Flynn, Point Defects and Diffusion (Clarendon, Oxford, 1972) p. 306.
[11] D.W. Bassett and P.R. Webber, Surface Sci. 70 (1978) 520.
[12] T. Halicioglu and G.M. Pound, Thin Solid Films 57 (1979) 241.
[13] P. Pieranski, J. Malecki, W. Kuczynski and K. Wojcienchowski, Phil. Mag. 37 (1978) 107.
[14] J.D. Wrigley and G. Ehrlich, Phys. Rev. Letters 44 (1980) 661.
[15] H. Bonzel, Surface Sci. 21 (1970) 45.
[16] S.H. Garofalini and T. Halicioglu, Surface Sci. 104 (1981) 199.
[17] D. Wolf, K. Differt and H. Mehrer, Computer Phys. Commun. 13 (1977) 183.
[18] P. Benoist and G. Martin, Thin Solid Films 25 (1975) 181.
[19] R. Butz and H. Wagner, Surface Sci. 87 (1979) 85.
[20] J. Cousty and B. Perraillon, in: Proc. ICSS-4 and ECOSS-3, Cannes, 1980, Vol. I, Eds. D.A. Degras and M. Costa, p. 69.

Disorder at the Bilayer Interface in the Pseudohexagonal Rotator Phase of Solid *n*-Alkanes

Jean-Paul Ryckaert
Pool de Physique, Université Libre de Bruxelles, B-1050 Bruxelles, Belgium

Michael L. Klein
Chemistry Division, National Research Council of Canada, Ottawa, Ontario K1A 0R6, Canada

and

Ian R. McDonald
Department of Physical Chemistry, University of Cambridge, Cambridge CB2 1EP, Great Britain
(Received 30 October 1986)

Molecular-dynamics calculations are used to characterize the structure and dynamics of the two solid bilayer phases of the *n*-alkane tricosane ($C_{23}H_{48}$). In the crystalline orthorhombic phase at 38 °C, chains undergo translational, rotational, and torsional motions, but are otherwise essentially all-*trans* and perfectly ordered with a herringbone packing. By contrast, at 42 °C, in the pseudohexagonal (rotator) phase, there is a dramatic increase in longitudinal chain motion, each chain now has four possible orientations, and a significant number of conformational defects develop, predominantly at the chain ends.

PACS numbers: 61.50.Ks, 64.60.Cn, 64.70.Kb, 68.35.−p

The nature of the interface within molecular bilayers is not only of intrinsic interest but may, indirectly (i.e., through refinement of intermolecular force models) be of relevance to the understanding of both real and model biomembranes,[1] micelles,[2] and Langmuir-Blodgett films.[3] The simplest example of a bilayer system that is stable *around room temperature* is provided by the odd-chain-length normal (*n*-) alkanes (C_mH_{2m+2}) with m ranging from 19 to 25.

The complexity of the phase diagrams exhibited by these long-chain molecules near their respective melting points has only become apparent in recent years. Stable solid phases have been identified that contain both conformational and orientational disorder.[4,5] Moreover, the number and character of these phases vary with the molecular chain length.[5] Elegant experiments based on diffraction[6] and spectroscopic techniques,[4] plus calorimetric and NMR studies,[7] have established that when the crystalline phase transforms to the pseudohexagonal rotator phase, longitudinal diffusive motion of the chains increases dramatically and a significant concentration of conformational defects is generated, predominantly in the vicinity of the bilayer interface.[4]

We have used constant-volume molecular-dynamics calculations to study the structure and dynamics of the C_{23} *n*-alkane compound (tricosane) in both the crystalline and pseudohexagonal rotator phases close to the phase-transition temperature of 40 °C. The aim is to characterize further the nature of the bilayer interface, and the differences between these solid phases, making use of a microscopic model based on flexible (semirigid) chains and atom-atom intermolecular potentials.[8] We find that the model yields a stable crystalline phase with all-*trans* chains. In the rotator phase, each molecule is undergoing rotational jumps between four possible sites. These jumps are *independent* of enhanced chain longitudinal translational motion. A variety of intramolecular defects occur, predominantly near the chain ends, but these cause relatively little roughening of the bilayer interface. The most common defects are the so-called end-gauche defects and kinks.[4] In one example, we observed kink formation at the center of a chain and its subsequent migration down the chain backbone towards the bilayer interface. These findings complement experimental studies,[4-6] and give a microscopic picture of a stable, but highly disordered, bilayer system. The four-site orientational distribution function that we obtain for the bilayer phase confirms the structure inferred from an analysis of the x-ray and calorimetric data.[5]

In the crystalline phase, *n*-alkane chains adopt their all-*trans* most elongated conformation and pack in a lamellar bilayer structure. Within each layer the molecular longitudinal axes are mutually parallel and the planar zigzag chains have two possible orientations; thus there are four molecules in the unit cell. The experimental lattice constants for crystalline tricosane are $a = 7.47$ Å, $b = 4.98$ Å, and $c = 62.40$ Å.[9] However, around 40 °C, this crystalline phase transforms to a pseudohexagonal rotator phase with $a = 8.00$ Å, $b = 4.85$ Å, and $c = 63.22$ Å; this is sometimes referred to as the R_I or face-centered orthorhombic (fco) phase.[5,6] Just below its melting point, 46 °C, tricosane undergoes a further transformation to a rhombohedral (R_{II}) structure,[5] but, because this is a trilayer structure, it will not concern us here.

Molecular-dynamics calculations have been carried out on periodically replicated bilayer systems arranged as 3×5 unit cells in the (a,b) plane. The sample con-

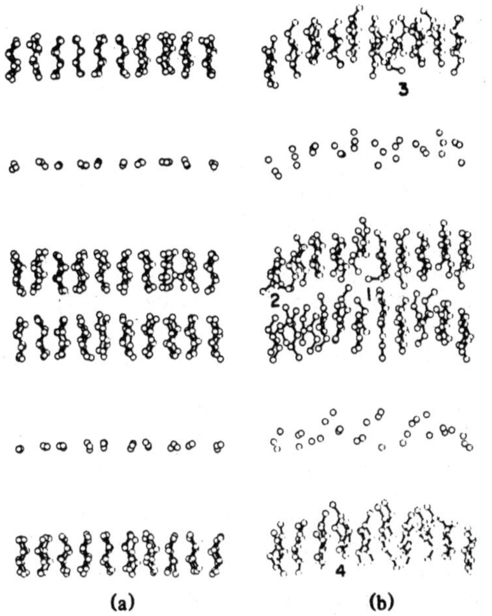

FIG. 1. Instantaneous configurations of the bilayer systems viewed down the a axis: (a) the crystalline orthorhombic phase at 38 °C, (b) the pseudohexagonal phase at 42 °C. For clarity, no H atoms and only the terminal plus the central C atoms are indicated. The disorder at the interface is dominated by longitudinal diffusion of the chains as is clearly indicated by the enhanced fluctuations in the position of the central C atom in case (b).

tained only one unit cell in the c direction so that the total system consisted of 60 chains (4260 atoms). Because of the boundary conditions there are two distinct interfaces. This was the largest system that could be conveniently studied with our available computing resources. The intermolecular interactions were of the atom-atom type and included C-C, C-H, and H-H contributions, truncated at around 8 Å. The chains were partially rigid in the sense that the H atoms follow the chain backbone, but are otherwise constrained to prevent the wagging, twisting, and scissoring motions of the methylene groups as well as the C-H stretching. The intramolecular potential consists of bending and torsional terms plus methylene-methylene group interactions starting from fourth neighbors. This potential model was used recently to study a monolayer of infinitely long, flexible chains.[8] In the present work we have added a torsional potential for the terminal methyl groups with a barrier height of 12 kJ/mol.[10] The equations of motion were solved with use of standard methods.[11]

The crystalline phase was brought to equilibrium at

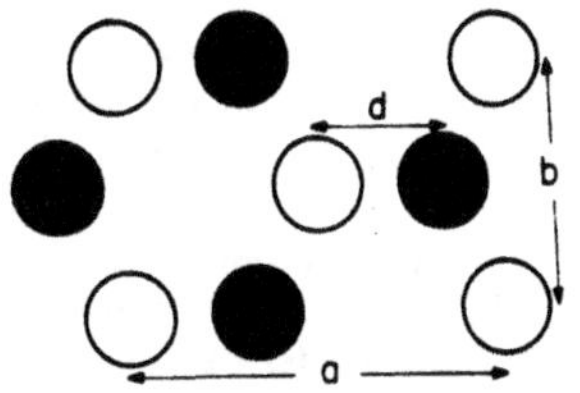

FIG. 2. A view down the c axis of an idealized bilayer system; d indicates the relative displacement of the two layers.

the desired temperature (3 °C) by a constant-temperature technique[11] and the phase-space trajectories were followed for 8 ps with a time step of 3.3 fs. A typical instantaneous configuration viewed down the crystal a axis is shown in Fig. 1(a). The crystal was very stable and the bilayer interface well defined, there being only modest longitudinal displacements of the chains, even though the temperature is very close to the melting point.

An analogous, but somewhat longer (19 ps), simulation was performed for the pseudohexagonal phase at 42 °C; the experimental a and b lattice constants were used, but c was held at 62.40 Å rather than the experimental value of 63.22 Å.[5] This modest compression was induced so that we could compare the crystalline and rotator phases at a constant bilayer thickness. We began with all-*trans* chains oriented parallel to the (b,c) plane and the chain centers of mass located as suggested by experiment, i.e., with d, the relative displacement along the a axis, equal to $\frac{1}{2}a$ (Fig. 2).[5] This configuration, which is consistent with the x-ray structure (*Fmmm*), immediately disordered; each chain rotated to adopt one of four possible orientations, located at $\pm 45°$ and $\pm 135°$ with respect to the crystal b axis. We therefore carried out the long calculation starting from a configuration in which all-*trans* chains were randomly distributed among these four possible sites. We note that this initial configuration is compatible with the *Fmmm* structure inferred from x-ray data[5] and is almost certainly also compatible with NMR data taken on deuterated samples of the n-alkane $C_{19}H_{40}$ in its pseudohexagonal rotator phase.[7] This structure is similar to the four-site orientational distribution found in molecular-dynamics[8] and Monte Carlo[12] studies on the rotator phases of n-alkane *monolayers*.

An instantaneous configuration of the bilayer rotator phase is shown in Fig. 1(b). Although the temperature is only 4 °C higher than in the crystalline phase, and the cross-sectional area per chain differs by only 6%, it is immediately evident there is now considerable disorder present. However, the system was stable as judged by the ease with which the temperature was controlled at its desired value and by the fact that the calculated internal stress-tensor components were all small (i.e., less than about 1 kbar). Other quantities, which will be discussed below, were also stable throughout the run.

It is worthwhile to comment on the relative lateral positions of the two monolayers which comprise the periodically replicated bilayer system. Figure 2 shows the centers of mass of an idealized bilayer system as viewed down the c axis. In the crystalline phase, the measured[9] relative displacement parameter is $d=b/3=2.5$ Å, whereas in the pseudohexagonal rotator phase it is $d=b/2=4.0$ Å. While the simulations agreed with the latter value, the calculated relative displacement for the crystalline phase was $d=3.3$ Å. We note also that in the crystalline phase the longitudinal chain axes are all parallel to the c axis, whereas in the rotator phase the chains have a mean canting angle of about 6°. Methyl-group rotation is active in both phases but is roughly twice as frequent in the rotator phase.

We now turn to a discussion of the rotational disorder. In the crystalline phase one rotational defect appeared and subsequently disappeared over a period of about 200 time steps but, apart from this isolated event, the system remained fully ordered. This behavior contrasts strongly with that found in the rotator phase where the four-site orientational distribution appeared to be quite stable (Fig. 3). The mean residence time between jumps was estimated to be about 13 ps which compares favorably with estimates based on neutron measurements.[13]

Even though the simulations were carried out at essentially the same temperature, the much larger dispersion in the positions of the chain centers of mass in Fig. 1(b) compared with that in Fig. 1(a) indicates the presence of enhanced longitudinal chain displacements in the rotator phase. The root mean square chain displacement along the c axis is calculated to be 1.2 Å, which compares rather well with the value 1.7 Å estimated from small-angle x-ray diffraction data.[6]

Figure 1(b) also indicates the presence of intramolecular defects in the rotator phase. The presence of a *gauche* defect, two of which are clearly evident, will cause a chain to shorten by about 1.1 Å. On average, we find about 6% of the chain ends to have *gauche* defects, a concentration which implies that the average chain length will shorten by $2\times0.06\times1.1$ Å $=0.13$ Å. The actual mean end-to-end chain length in the rotator phase was found to be shortened by about 0.30 Å, which is a considerably larger value than can be attributed solely to the presence of *gauche* defects. It is likely therefore that the chains are shortened predominantly via torsional motions.

Figure 4 compares the calculated percentage of *gauche* defects at various positions along the n-alkane chain backbone with that measured experimentally on a closely related system ($C_{21}H_{44}$).[4] Both the calculations and the experiments suggest an exponential growth of defects towards the chain ends, but the calculated number of defects seems to be too few, possibly because of inadequacies in the assumed form of the chain intramolecular potential.[8]

It seems natural to imagine that the bilayer interface is the most hospitable place for the birth of intramolecular defects; however, on one occasion we observed the spontaneous formation of a defect in the middle of a chain! This defect then proceeded to migrate along the chain. Such an observation suggests that the birth and death of intramolecular defects might profitably be modeled with use of theories of soliton dynamics.

In summary, we have used molecular-dynamics calculations to study the nature of the disorder present in the bilayer rotator phase of n-alkanes. We find that the dominant effect is due to longitudinal motion of the chains. Although intramolecular defects are present in

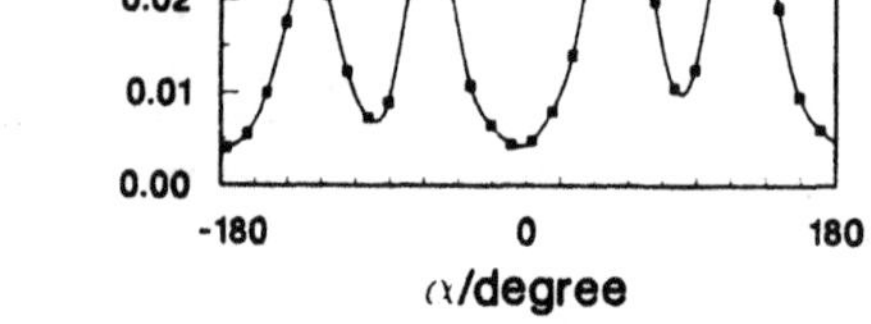

FIG. 3. Orientational probability distribution for chains in the rotator phase. Differences in peak intensities are a consequence of poor statistics (on average each chain has made only about 2 jumps in the total molecular-dynamics run).

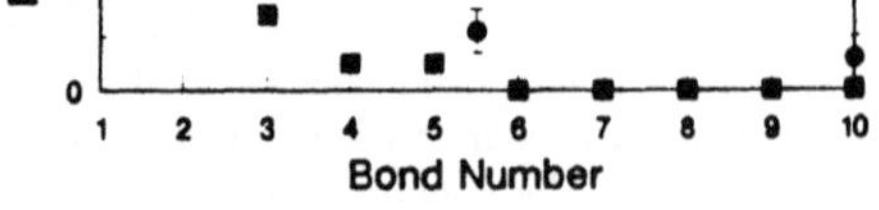

FIG. 4. Percentage of *gauche* defects along the chain backbone. The circles are experimental values taken from Ref. 4 the squares are the molecular-dynamics results with errors at least comparable to those of experiment.

abundance, and occur mainly at the bilayer interface, they have a relatively modest effect on the structure. These findings are in complete accord with inferences based on spectroscopic studies[4] and small-angle x-ray diffraction.[6] A prediction emerges from our calculations; it is that the pseudohexagonal rotator phase has a four-site orientational distribution function. It would be of interest to know whether or not this finding is supported by new experimental data.

Finally, our successful modeling of the bilayer interface required explicit inclusion of intermolecular interactions between hydrogen atoms. In previous attempts to model related membrane and micelle systems, such interactions have often been omitted.[14] It remains to be seen whether or not this simplification was indeed justifiable for these more complicated systems.

We thank the United Kingdom Science and Engineering Research Council for supporting a visit by one of us (J.P.R.) to the University of Cambridge, which enabled this work to be initiated.

[1]K. A. Dill and P. J. Flory, Proc. Nat. Acad. Sci. U.S.A. **78**, 676 (1981).

[2]H. Wennerström and B. Lindman, Phys. Rep. **52**, 1 (1979); M. Kotlarchyk, J. S. Huang, and S.-H. Chen, J. Phys. Chem. **89**, 4382 (1985).

[3]Th. Rasing, Y. R. Shen, M. W. Kim, and S. Grubb, Phys. Rev. Lett. **55**, 2903 (1985); V. Vogl and C. Wöll, J. Chem. Phys. **84**, 5200 (1986).

[4]M. Maroncelli, S. P. Qi, H. L. Strauss, and R. G. Snyder, J. Am. Chem. Soc. **104**, 6237 (1982); M. Maroncelli, H. L. Strauss, and R. G. Snyder, J. Chem. Phys. **82**, 2811 (1985).

[5]G. Ungar, J. Phys. Chem. **87**, 689 (1983).

[6]A. F. Craievich, I. Delonico, and J. Doucet, Phys. Rev. B **30**, 4782 (1984); G. Strobl, B. Ewen, E. W. Fischer, and W. Piesczek, J. Chem. Phys. **61**, 5257 (1974); B. Ewen, G. Strobl, and D. Richter, Faraday Discuss, Chem. Soc. **69**, 19 (1980).

[7]M. G. Taylor, E. C. Kelusky, I. C. P. Smith, H. L. Casal, and D. G. Cameron, J. Chem. Phys. **78**, 5108 (1983).

[8]J.-P. Ryckaert and M. L. Klein, J. Chem. Phys. **85**, 1613 (1986).

[9]A. E. Smith, J. Chem. Phys. **21**, 2229 (1953).

[10]R. A. Scott and H. A. Scheraga, J. Chem. Phys. **44**, 3054 (1966).

[11]J.-P. Ryckaert and G. Ciccotti, Mol. Phys. **58**, 1125 (1986).

[12]T. Yammamoto, J. Chem. Phys. **83**, 3790 (1985).

[13]D. Bloor, D. H. Bonsor, D. N. Batchelder, and C. G. Windsor, Mol. Phys. **34**, 939 (1977); J. Doucet and A. J. Dianoux, J. Chem. Phys. **81**, 5043 (1984).

[14]B. Jönsson, O. Edholm, and O. Teleman, J. Chem. Phys. **85**, 2259 (1986); E. Egberts and H. J. C. Berendsen, to be published.

Particle motions in superionic conductors*

A. Rahman

Argonne National Laboratory, Argonne, Illinois 60439
(Received 8 June 1976)

The motion of ions in $Ca^{2+}F_2^-$ has been studied by molecular dynamics. The results show that, at 2.525 gcm^{-3} and 1590 K, the F^-'s have a constant of self-diffusion of 2.6×10^{-5} cm^{2}sec^{-1} and the Ca^{2+}'s form a stable lattice. It is found that half of the F^-'s occupy the octahedral sites of the fcc Ca^{2+} lattice. The details of the self-diffusion process is analyzed showing the presence of a wave-vector dependent long relaxation time in the decay of $\langle \exp(i\mathbf{k}\cdot\mathbf{u}(t))\rangle$, which is the Fourier transform of the displacement $\mathbf{u}(t)$ of the F^- ions.

I. INTRODUCTION

Materials like Ag^+I^- and $Ca^{2+}F_2^-$, which are examples of "superionic conductors", are characterized by the fact that while they are still solid they show a remarkable degree of ionic conduction; this is because Ag^+ in Ag^+I^- and F^- in $Ca^{2+}F_2^-$ are mobile even though the other ions form a stable lattice in each case. The following is a report on some molecular dynamics calculations on $Ca^{2+}F_2^-$ that have been undertaken to elucidate the structural and dynamical correlations in the motion of particles which constitute this important class of materials. $Ca^{2+}F_2^-$ was chosen for this study, because to a good approximation, a simple rigid-ion model is an acceptable starting point for this material. In Ag^+I^- important effects due to covalent binding, ionic polarizability, etc., may be present and hence it was considered advisable to make a feasibility study[1] on the less well known superionic conductor, namely, $Ca^{2+}F_2^-$.

A large amount of molecular dynamics work on molten and crystalline ionic salts has been done over the last few years[2]; the method of molecular dynamics is well known and we shall not elaborate on it here. The only relevant piece of information that needs to be stated is the model for the short range repulsion between F^-–F^- on the one hand and F^-–Ca^{2+} on the other. In view of the large Coulomb repulsion between two Ca^{2+}'s we need not include their short range repulsion in the Hamiltonian. The table of functions given by Kim and Gordon[3] for the F^-–F^- and the Ca^{2+}–F^- repulsive potentials have been used in the present work. Using their repulsive potentials, the same authors have successfully calculated[4] low temperature binding properties of several ionic crystals.

Our purpose here is to use molecular dynamics calculations to test how well the same potentials give a reasonably good picture of an important high temperature solid state property of $Ca^{2+}F_2^-$, namely, the flow of F^-'s through a stable lattice of Ca^{2+}'s.

The calculation was made in a system of 108 cations and 216 anions initially forming a fluorite lattice; periodic boundary conditions were applied in the usual way. The density of the system was 2.525 g/cc corresponding to a molecular dynamics cell size of 17.7 Å or a lattice constant of 5.9 Å for the fcc cation lattice. After a considerable amount of "aging" a system in equilibrium at 1590 °K was studied in detail. The integration step was 1.843×10^{-15} sec and the results reported here were obtained out of a molecular dynamics run of 6000 Δt or about 10^{-11} sec. A run of about the same length was discarded to allow for equilibration, i.e., the "aging" mentioned above.

II. PAIR CORRELATIONS

Figures 1a, b, c show the three pair correlations $g_{ij}(r)$ between the two kinds of particles. The cation–cation pair correlation clearly shows that the Ca^{2+}'s form a thermally agitated fcc lattice with an rms deviation of about 0.3 Å from the mean lattice positions. The cation–anion pair correlation shows that each Ca^{2+} has a very well defined shell of eight F^-'s. The anion–anion pair correlation has much less clarity in its peak structure than the two mentioned above.

Figure 1(d) shows a new "pair" correlation which clarifies the dynamical structure of the system. The O sites along the cube edges formed by the cations can be defined unambiguously because, apart from developing a thermal cloud, the cations never abandon their permanent lattice sites. Looking out from the O sites for the density of anions one gets the O–F^- pair correlation shown in Fig. 1(d). In low temperature fluorite this should be very similar to $g(Ca^{2+}-F^-)$; however, at the high temperature we are dealing with the interstitial O sites are occupied on the average by one F^- [Fig. 1(d)] which implies, obviously, a vacancy in half the F^- positions of the regular fluorite lattice.

It is obvious that in the presence of the firm Ca^{2+} fcc lattice the analysis of the pair correlation should be made as function of vector $\mathbf{r}$ to further clarify the correlations. This will be part of a more detailed study which is now under way.

III. SELF-DIFFUSION

If $\mathbf{r}(t)$ denotes the displacement of a particle in time t, the time dependence of the mean square displacement $\langle \mathbf{r}^2(t)\rangle$ of the cations and anions is shown in Fig. 2(a). It is clear that the Ca^{2+} have zero constant of self-diffusion [in other words they form a stable structure and the limiting value of the mean square displacement of Ca^{2+}'s is consistent with the width of the first peak in the $g(Ca^{2+}-Ca^{2+})$ shown in Fig. 1(a).] However the F^-'s have $D = 2.6\times10^{-5}$ cm^2 sec^{-1} which is a typical liquid-like value of the constant of self-diffusion. It is a characteristic property of superionic conductors that the mobile species has liquid-like constants of self-diffusion.

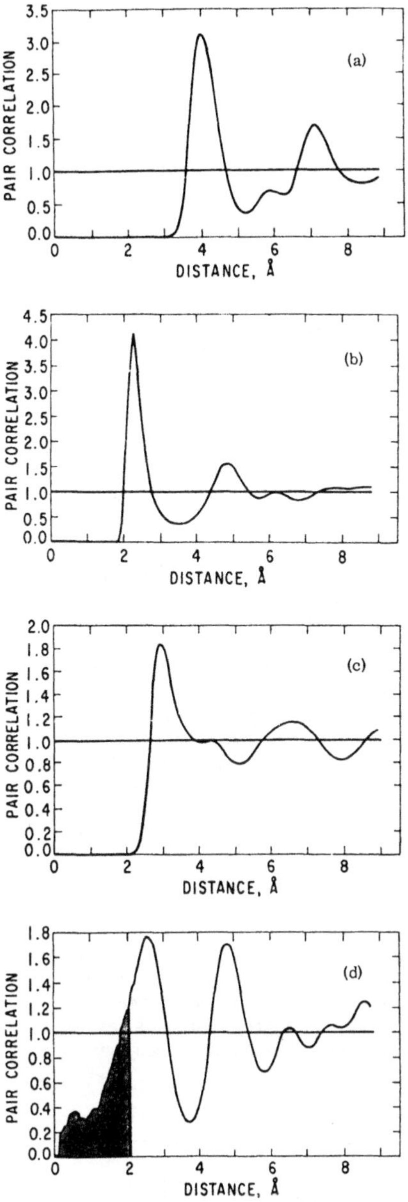

FIG. 1. (a) Pair correlation between Ca^{2+} ions. The population under the three peaks are 12, 6, 24, respectively. The Ca^{2+}'s preserve the face centered cubic lattice structure throughout the molecular dynamics run. (b) Pair correlation between Ca^{2+} and F^- ions. The population under the first peak is 8; the peaks appear at positions expected in the fluorite lattice. (c) Pair correlation between F^- ions. It indicates that the F^- ions have considerable disorder in their mutual arrangements. However, the shoulder and the second peak still correspond to positions in the fluorite structure. (d) Pair correlation between the octahedral site (of the Ca^{2+} fcc lattice) and the F^- ions; it shows a population of one under the shaded region. Thus half the anions have moved into regions which are unoccupied in a fluorite lattice.

To throw further light on the process of self-diffusion a function relevant to neutron inelastic scattering was calculated. It is denoted by $F_s(\mathbf{k}, t)$ and is defined as $\langle \exp[i\mathbf{k}\cdot\mathbf{r}(t)]\rangle$, where $\mathbf{k}$ is obviously a wave vector; if $\langle \mathbf{r}^2(t)\rangle$, see Fig. 2(a), happens to be sufficient to describe F_s completely, then $F_s(\mathbf{k}, t) = F_s^g(\mathbf{k}, t) \equiv \exp[-\frac{1}{2}\langle[\mathbf{k}\cdot\mathbf{r}(t)]^2\rangle] = \exp[-\frac{1}{2}\mathbf{k}^2\langle x^2(t)\rangle]$; this is true in the uninteresting limiting case of a system of noninteracting particles with a normal distribution of velocities (the ideal gas) and in the case of a harmonically vibrating crystal lattice, see Fig. 2(b). In monatomic liquids (e.g., argon) measurable differences between $F_s(\mathbf{k}, t)$ and $F_s^g(\mathbf{k}, t)$ have been found,[5a,b] and much larger effects are expected to occur for example in the diffusion of hydrogen in metals.[6]

For a magnitude of $\mathbf{k}$ ranging from 2 to 5 $Å^{-1}$, $F_s(\mathbf{k}, t)$ has been calculated for vector $\mathbf{k}$ in 3 special directions, namely, 100, 110, and 111 as well as in a few asymmetry directions. The results are displayed in Figs. 3(a)–3(d). Using $\langle x^2(t)\rangle$ from Fig. 2(a), we immediately see that $F_s^g(\mathbf{k}, t)$ is indeed quite different from $F_s(\mathbf{k}, t)$. From the remarks already made about interstitial occupancy by F^-'s away from their normal fluorite sites it is clear that the F^-'s must diffuse in such a way that they go from an O site to an F^- fluorite site. The dynamical characteristics of this process are seen through

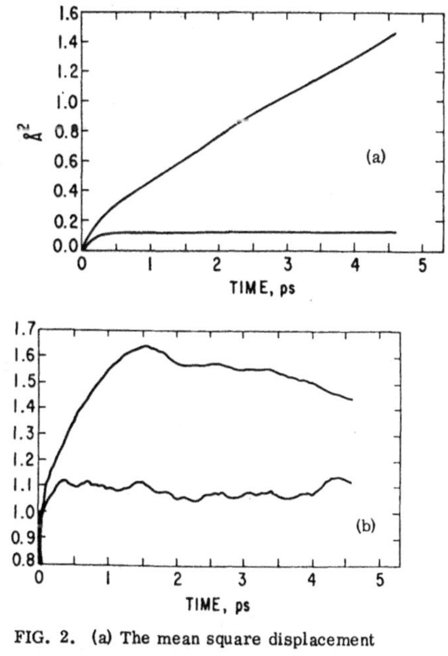

FIG. 2. (a) The mean square displacement

$$\langle \mathbf{r}^2(t)\rangle \equiv N^{-1}\sum_{j=1}^{N}\langle(\mathbf{r}_j(s+t)-\mathbf{r}_j(s))^2\rangle$$

for the two types of ions. The Debye–Waller thermal cloud of the Ca^{2+} has a width of 0.12 $Å^2$. The F^- have a constant of self-diffusion $D = 2.6\times10^{-5}$ $cm^2\,sec^{-1}$. The values shown are for $\langle \mathbf{r}^2(t)\rangle/6$. (b) Variation of $3\langle \mathbf{r}^4(t)\rangle/5\langle \mathbf{r}^2(t)\rangle^2$ with time; departure from unity shows that the simple approximation $F_s(\mathbf{k}, t) = F_s^g(\mathbf{k}, t)$ (see text for definition) is invalid for the motion of F^- but is not too bad for Ca^{2+}.

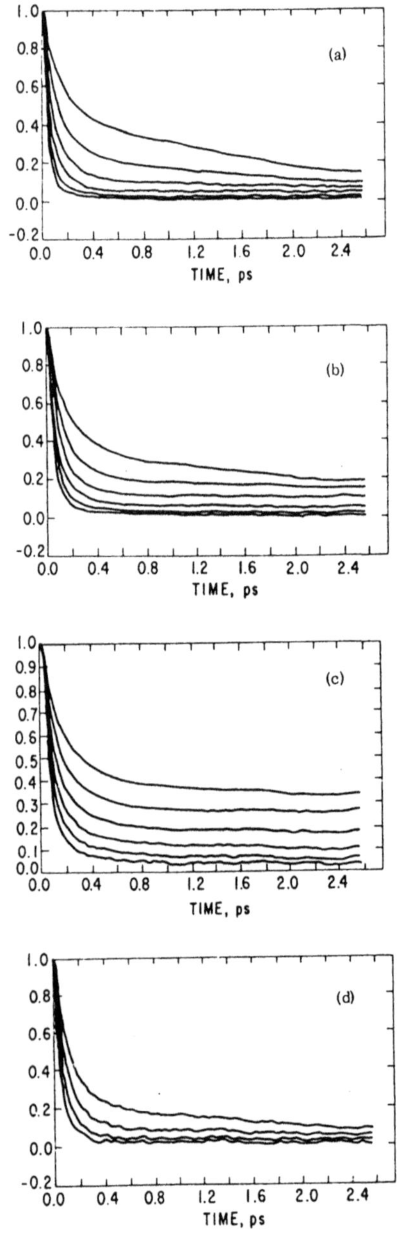

FIG. 3. (a) $F_s(\mathbf{k}, t)$ for F^- motion, with k in 111 direction; the magnitude of k is given by $\sqrt{3} \times 0.355$ Å$^{-1} \times n$, where $n = 8, 7, \ldots 3$ for the six curves shown; the fastest decaying curve corresponds to $n = 8$ and the slowest to $n = 3$ with the others situated monotonically in between. (b) Same as Fig. 3a but with k in 110 direction. Magnitude of k is $\sqrt{2} \times 0.355$ Å$^{-1} \times n$ where $n = 9, 8, \ldots, 4$. (c) Same as Figs. 3(a) and 3(b) but with k in 100 direction. Magnitude of k is 0.355 Å$^{-1} \times n$ where $n = 10, 9, \ldots, 5$. (d) Same as Figs. 3(a), 3(b) and 3(c), but the four k's are 0.355 Å^{-1} (l, m, n), where (l, m, n) are (8, 7, 6), (7, 6, 5), (6, 5, 4), and (5, 4, 3). The larger the magnitude of k the faster the decay.

$F_s(\mathbf{k}, t)$. From Figs. 3 one concludes that the decay of $F_s(\mathbf{k}, t)$ consists of an initial fast decay with a characteristic time of $\lesssim 10^{-12}$ sec and a subsequent slow decay which is hard to characterize by looking at $F_s(\mathbf{k}, t)$ itself.

Figure 4 shows $-\log F_s(\mathbf{k}, t)/\mathbf{k}^2$ and brings out the fact, more clearly than in Figs. 3, that there are two time characteristics, the slow one being of the order of 40 picosec. Thus neutron inelastic scattering from the diffusion of F^-'s will lead to a clearly defined inelastic region, usually referred to as the "phonon part", and a quasielastic region of essentially a Lorentzian shape with a width dependent on the direction and magnitude of k; however, the value of the width will be an order of magnitude less than $\mathbf{k}^2 D$ for values of $|\mathbf{k}|$ mentioned above.

IV. CONCLUSIONS

This preliminary report indicates that molecular dynamics calculations can be of considerable value in the understanding of the dynamical events underlying the motion of ions in superionic conductors. It also pinpoints the fact that the quasielastic region of neutron inelastic scattering data can be of use in increasing this understanding. More detailed calculations at a variety of densities and temperatures will be necessary to decide the extent to which a particular force model (in the above case, the rigid ion picture with Kim–Gordon repulsion) is a valid description. The calculation reported here gave a large negative pressure ($pV/NkT \sim -3$) for the $Ca^{2+}F_2^-$ system at 2.525 g cm^{-3} and 1590 °K. Together with the transport properties a thorough study of the equation of state using the Kim–Gordon potential will be a valuable point from which to suggest improvements in the model.

The electrical conductivity of the system described above is now being investigated using methods developed by Ciccotti and Jacucci.[7] A few sample runs at higher densities show that the diffusion constant is very sensitive to density; it is to be expected that this sensitivity will manifest itself in the electrical conductivity

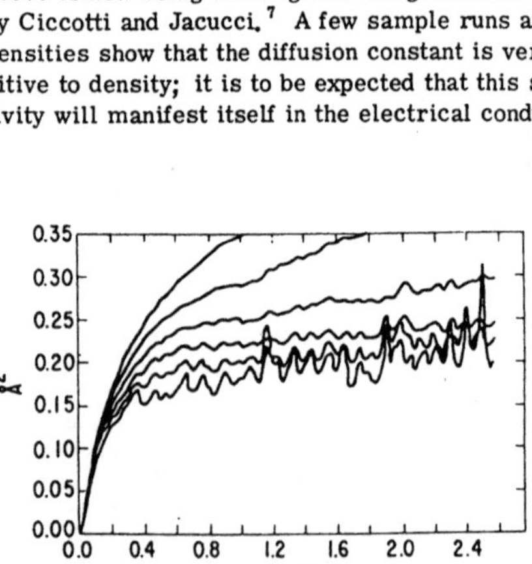

FIG. 4. $-\log F_s/k^2$ from Fig. 3(a). It shows the presence of a second long relaxation time arising from the large departure from unity of $\langle r^{2n}\rangle / C_n \langle r^2\rangle^n$ [which was shown in Fig. 2(b) for $n = 2$].

as well. This suggests that experimental investigation, at constant temperature, of the pressure (and hence density) dependence of the electrical conductivity in superionic conductors will be useful in understanding these materials.

From the point of view of the theory of the dynamics and structure of liquids, new problems are posed by the mobile species in superionic conductors; from the above report it is clear that the "liquid" of F^-'s is in equilibrium with an "external" field provided by the cations; this field excludes the anions from certain regions of configuration space and restricts them to regions which have a high degree of anisotropy (the octahedral and the F^- sites in fluorite in our case); in addition the anions are strongly interacting with each other.

Lastly, it will be desirable to work with larger systems of particles (e.g., 256 Ca^{++}'s and 512 F^-'s) to investigate possible size effects.[8] Previous molecular dynamics experience shows that even the "small" system we have considered above is capable of manifesting a wealth of realistic detail.

*Work performed under the auspices of the U. S. Energy Research and Development Administration.

[1]This suggestion was made by Professor G. B. Mahan, Indiana University.

[2]L. V. Woodcock, *Advances in Molten Salt Chemistry*, edited by J. Braunstein, G. Mamantov, and G. P. Smith (Plenum, New York, 1975), Vol. 3.
[3]Y. S. Kim and R. G. Gordon, J. Chem. Phys. **60**, 4332 (1974), see Tables VI and VII.
[4]Y. S. Kim and R. G. Gordon, Phys. Rev. B **9**, 3548 (1974).
[5](a) K. Sköld, J. M. Rowe, G. Ostrowski, and P. D. Randolph, Phys. Rev. A **6**, 1107 (1972). (b) B. R. A. Nijboer and A. Rahman, Physica (Utrecht) **32**, 415 (1966).
[6]J. M. Rowe, K. Sköld, H. E. Flotow, and J. J. Rush, J. Phys. Chem. Solids **32**, 41 (1971).
[7]G. Cicotti and G. Jacucci, Phys. Rev. Lett. **35**, 789 (1975).

[8]M. J. Mandell has investigated and explained the systematics of the dependence of pressure on system size in molecular dynamics calculations using systems consisting of Lennard-Jones particles; J. Stat. Phys. **15**, 299 (1976).

Diffusion of F^- ions in CaF_2[a]

G. Jacucci

Department of Physics, University of Trento, Trento, Italy

A. Rahman

Argonne National Laboratory, Argonne, Illinois, 60439

(Received 7 June 1978)

It is shown that in a recently reported model of the dynamics of CaF_2, the decay with time of the Fourier transform, $\langle \exp(i\mathbf{k}\cdot\mathbf{u}(t))\rangle$, of $\mathbf{u}(t)$, the F^- displacement, for various wave vectors $\mathbf{k}$ can only arise from diffusive motion occurring between the simple cubic anion sites of the fluorite structure. Most of the jumps (~80%) occur between near neighbor sites (i.e., they are in the 100 direction). The motion of F^- ions can be understood in terms of a spontaneous creation of anti-Frenkel pairs which diffuse and eventually get annihilated. At the temperature of this calculation (1590 K) the average number of such pairs at any instant is found to be 7.9 in an assembly of 216 anions. A detailed breakdown of the nature of various types of atomic jumps arising out of the dynamics of these pairs is also presented.

I. INTRODUCTION

In a recent paper[1] it was shown that the solid ionic conductor CaF_2 can be successfully simulated by the method of molecular dynamics. In the present paper we report on further work on this system; we have analyzed the diffusive motion of F^- ions and have tried to view this motion in the framework of a simple jump diffusion model. In addition we have attempted to analyze the dynamics of the system in terms of the motion of point defects.

Briefly, the salient features of the earlier work are as follows. A system of 108 Ca^{++} and 216 F^- ions in a cubic box of length 17.7 Å (mass density 2.525 gcm^{-3}), was studied at 1590 K. The potential functions given by Kim and Gordon[2] were used. Periodic boundary conditions of the usual type were applied. The system had the fluorite structure to begin with, however, in dynamic equilibrium, under the above stated conditions of temperature and density, the Ca^{++} ions continued to form (a thermally agitated) fcc lattice, while the F^- ions showed diffusive motion with a large, liquidlike, constant of self-diffusion (2.65×10^{-5} cm^2 sec^{-1}).

The nature of diffusive motion was further analyzed in Ref. 1 by calculating the function $F_s(\mathbf{k}, t) \equiv \langle \exp(i\mathbf{k}\cdot\mathbf{u}(t))\rangle$, where $\mathbf{u}(t)$ is a particle displacement [in Sec. II below we give more details regarding $F_s(\mathbf{k}, t)$]. The dependence of F_s on the wave vector $\mathbf{k}$ and on t was illustrated in Ref. 1. It was clear that its decay with time depends on a fast relaxation lasting only ~0.2 ps followed by a slow relaxation process; our aim here is to make the latter statement quantitatively precise in its dependence on k and on t.

The pair correlations shown in Ref. 1 indicated that the F^- ions, in spite of their fast diffusive motion, were situated predominantly at the F^- sites of the fluorite structure. This naturally suggests an analysis of the dynamics in terms of a jump diffusion model.

This suggestion is also strongly supported by the large deviation from Gaussian behavior exhibited in the picosecond region by the fourth moment of the distribution of the displacements of the F^- ions (see Ref. 1). These indications contradict the idea that the fluorine lattice is "melted," and that a liquidlike structural disorder of the F^- ions in the interstices of the fcc Ca^{++} lattice is responsible for the high value of their diffusion coefficient, a value which is appropriate for diffusion in the liquid phase.

In order to put on firm ground the relevance of a jump diffusion description to the present case, we have produced a visual display of the individual particle trajectories, by projecting the three-dimensional track onto *xy*, *yz*, and *zx* planes. This exercise readily confirms that the F^- ions generate well-defined Debye–Waller clouds, i.e., during a rather long time, the particle passes over and remains in one of well separated regions of space centered at the usual fluorine lattice sites. In many cases the sections of trajectories belonging to different clouds are connected by simple paths that the particle follows rather rapidly and decisively; in the connecting paths, the particles are often found to visit one of the empty octahedral sites of the crystal, describing a short wiggle or loop around it. Neighboring sites in the (110) and (111) directions are also found to be connected, in addition to the ones in the (100) direction.

Analysis of the F^- motions along the following two lines is thus indicated. On the one side the slow decay of $F_s(\mathbf{k}, t)$ can be interpreted with the Chudley–Elliott[3a] jump diffusion model, and best fit values for the jump frequency Γ and the relative proportions p_i of jumps in various directions can be derived. On the other side the trajectories of the F^- ions can be interpreted in a systematic way, with the help of a suitable code, directly in $\mathbf{r}$, t giving the "true" values of Γ and p_i and a great deal of useful quantitative microscopic insight.

In Sec. II we shall present the molecular dynamics results for $F_s(\mathbf{k}, t)$; then we shall briefly set down the equations of the jump diffusion model, and discuss how

---

[a]Work performed under the auspices of the U.S. Department of Energy.

 0021-9606/78/6909-4117$01.00

well this model describes the data; in later sections we shall see how the fitted parameters compare with the ones directly determined from the investigation of the trajectories in r, t space. We have not compared our results with recent more refined theories of jump motions is solids.[3b–d]

In Sec. III the direct r, t investigation provides additional detailed information about the jump motion of the particles, e.g., the duration of and the correlation between jumps. To our knowledge mathematical models for $F_s(\mathbf{k}, t)$ taking into account these elements of jump diffusion have not yet been formulated.

Once the jump diffusion description has been accepted, the question naturally arises as to what kind of highly correlated motion permits the fluorine ions to jump from site to site without closely approaching one another. In Sec. IV we show how the correlation between the jump motion of different F^- ions in our system can be precisely described in terms of formation and migration of point defects in the lattice.

The complexity and richness of structural and dynamical solid state properties is largely related to the presence of lattice defects. These are responsible for phenomena that are ruled out in the perfect solid by symmetry or by hindrance to atomic motion; thus defects in low concentration are capable of drastically affecting the properties of the solid.[4] The point defect picture that will develop in the present context is convincing enough that in the language of point defects it brings up new questions about the manner of using the computer simulation method for such studies. The question to ask at the beginning of any simulation study of solids is whether the model includes, by spontaneous generation or explicit introduction, the right kind of lattice defects, in the right concentration, with appropriate reaction rates, etc.

During any M.D. simulation with periodic boundary conditions the number density of particles is strictly conserved over the volume of the periodic box; thus point defects can be thermally generated only in pairs in the bulk of the system. The absence in the system of a free surface and of any other source or sink of point defects (e.g., dislocations) excludes the possibility of thermal generation of unpaired vacancies and interstitials.

Moreover, the generation process will take place at the same rate as it would under constant pressure conditions, and the same is true for the migration process. This is because the Gibb's free energies that determine concentration and rates at constant pressure are identical with the respective Helmholtz free energies that are relevant at constant volume. This fact has already been pointed out in connection with another M.D. study[5] (also see references therein).

The diffusion of F^- in CaF_2 has been shown experimentally to be supported by anti-Frenkel pairs,[6] i.e., pairs of vacancies and interstitials in the fluorine lattice. This conclusion has been drawn by Ure[6] already for temperatures much lower than the one of interest here; on theoretical grounds this "intrinsic" diffusion regime has to prevail at higher temperatures. The formula given by Ure[6] for the equilibrium concentration of anti-Frenkel pairs as function of the temperature predicts, through extrapolation over a rather wide temperature range, about six such pairs in our M.D. system of 216 F^- sites; this gives an indication of the feasibility of an M.D. simulation at the chosen temperature (1600 K). At a temperature of, say, 1000 K, this number would decrease by a factor $\sim e^{-6}$, and a simulation would be out of the question. This fact was implicit in the results presented in Ref. 1 albeit in different form, i.e., through the value of the diffusion coefficient.

Work is in progress in which we give attention to the electrical conduction properties of the system. The value of the relevant transport coefficient σ as well as the frequency dependent electrical conductivity $\sigma(\omega)$ are being computed using the appropriate Kubo formula. A small-amplitude-perturbation method[7] has provided independent check of these quantities, but the values so determined are affected by large statistical errors. Haven's ratio relates σ to the diffusion coefficient D of the F^- ions, and the value obtained is being analyzed in the light of the correlations induced in the jumps of F^- ions by the motion of the point defects.

II. FOURIER ANALYSIS OF F^- DIFFUSION

A. Molecular dynamics results

As in Ref. 1, the Fourier analysis of the diffusive motion of the F^- ions was made by evaluating the function $F_s(\mathbf{k}, t)$ defined by

$$F_s(\mathbf{k}, t) = \langle \exp(i\mathbf{k}\cdot \mathbf{u}(t))\rangle \ ,$$

where $\mathbf{u}(t) = \mathbf{r}_F(t+\tau) - \mathbf{r}_F(\tau)$ is the displacement of an F^- ion, $\mathbf{k}$ the wave vector of interest, and $\langle\cdots\rangle$ indicates averaging over all the F^-'s and a large number of initial conditions [i.e., the values of the time τ in the definition of $\mathbf{u}(t)$ given above].

In Figs. 1, 2, and 3 we show the dependence of $-\ln F_s(\mathbf{k}, t)$ on time and on the vector $\mathbf{k}$. It is seen quite unambiguously that the long time asymptotic form of $-\ln F_s(\mathbf{k}, t)$ is

$$-\ln F_s(\mathbf{k}, t) \xrightarrow{t\to\infty} g(\mathbf{k})t + c(\mathbf{k}) \ .$$

We note from Figs. 1–3 that the gradient $g(\mathbf{k})$ vanishes for values of $\mathbf{k}$ corresponding to the Bragg vectors of the simple cubic lattice of the fluorine sites in the fluorite structure. $c(\mathbf{k})$ is the analogue of the Debye–Waller exponent of crystal vibrations. Using the values of $c(\mathbf{k})$ to represent the quantity $|\mathbf{k}|^2a^2$, the root mean square extension a, of the thermal cloud in any direction is found to be ~ 0.5 Å; this corresponds to a distribution in space of the form $\exp(-r^2/4a^2)$ so that $2a \sim 1.0$ Å is to be compared with 1.28 Å which is $\frac{1}{2}$ the distance between an F^- site and the nearest octahedral site in the fluorite lattice. An overflow of the cloud into the region of the O sites is thus indicated.

We also see from Figs. 1–3 that the decay constant $g(\mathbf{k})$ shows the symmetry properties of the relevant reciprocal lattice: if d is the lattice constant of the simple cubic lattice ($d=2.95$ Å), $\mathbf{k}d/2\pi = (1/6, 0, 0)$ and $\mathbf{k}d/2\pi = (5/6, 0, 0)$ have the same $g(\mathbf{k})$ within the accuracy appar-

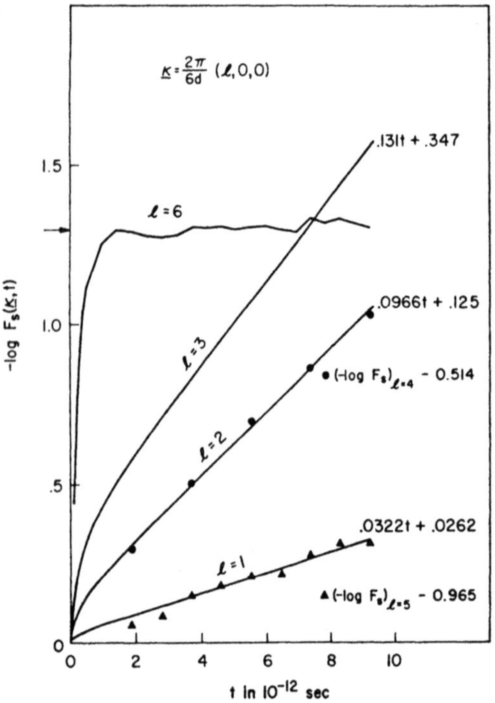

FIG. 1. Decay of $F_s(\mathbf{k}, t)$ for six equidistant **k** points in the 100 direction in reciprocal space. $l=6$ corresponds to the Bragg vector of the F⁻ simple cubic lattice, and for this point $F_s(\mathbf{k}, t)$ does not decay to zero. $l=3$ corresponds to the zone boundary. $d=2.95$ Å is the simple cubic lattice constant. The actual asymptotic forms are $l=1$: $0.0322t+0.0262$, shown as ——; $l=5$: $0.0322t+0.991$, shown as ▲; $l=2$: $0.0966t+0.125$, shown as ——; $l=4$: $0.0966t+0.639$, shown as ●; $l=3$: $0.131t+0.347$, shown as ——; $l=6$: $0t+1.30$, shown as ——. However, as shown in the figure, $-\log F_s$ for $l=4$ and $l=5$ have been shifted vertically by constant amounts (0.514, 0.965, respectively); this has been done to show the equality of the decay constants for the pair $l=2$ and $l=4$ and the pair $l=1$ and $l=5$.

ent from the data shown in Fig. 1. [The $c(\mathbf{k})$ for these two vectors, 0.0262 and 0.991, respectively, differ by 0.965, and the curve for $l=5$ in Fig. 1 has been shifted by this amount to show that $g(\mathbf{k})$ is the same for the two vectors.] The vectors (2/6, 0, 0) and (4/6, 0, 0) can be similarly bunched together on shifting the curve for $l=4$ in Fig. 1 by 0.514, which is the difference in $c(\mathbf{k})$ for $l=2$ and 4; however, the Bragg vector $\mathbf{k}d/2\pi=(1,0,0)$ has $g(\mathbf{k})=0$. Similar statements follow for the 110 direction (Fig. 2) and for the 111 direction (Fig. 3).

Using this information about $c(\mathbf{k})$ and $g(\mathbf{k})$ one can directly state that a simple cubic lattice with lattice constant 2.95 Å is the primary reference frame for the dynamics of the F⁻ ions; moreover, the sharpness of the localization (apparent from the value of a given above) and the value of the diffusion constant (2.65×10^{-5} cm^2 sec^{-1}) forces one to conclude that one is dealing with a jump diffusion mechanism with a jump rate $6D/l^2$ or ~ 0.15 psec^{-1}.

B. Jump diffusion model

A clear account of the Chudley–Elliott jump model, suitably generalized, is given by Rowe *et al.*[8] In the present case the input to the model consists in stating that

(i) Jumps occur between sites of a simple cubic lattice;

(ii) The jump rate from any site is τ^{-1};

(iii) Time in flight is negligible compared to τ;

(iv) Jumps occur to a prescribed set $\mathbf{L}+\{l, m, n\}$ of neighboring lattice sites from the site at $\mathbf{L}$, the probability of a jump to (l, m, n) being given by $p_{l,m,n}$;

(v) Successive jumps are uncorrelated in space and time.

Under (iv) above, we shall assume that only p_{100}, p_{110}, and p_{111} are nonzero (their sum being unity). There will be six neighbors available for a 100 jump, 12 for a 110 jump, but only four for the 111-type jump because the other four directions of the 111 type are blocked by the presence of the Ca^{++} ions; this makes it necessary to distinguish the totality of sites into two sets of inequivalent sites and to treat the problem in the manner of Rowe *et al.*[8]. However, if p_{111} is taken to be zero this complication does not occur; we shall henceforth assume this to be the case.

From Ref. 8 we easily get the result that

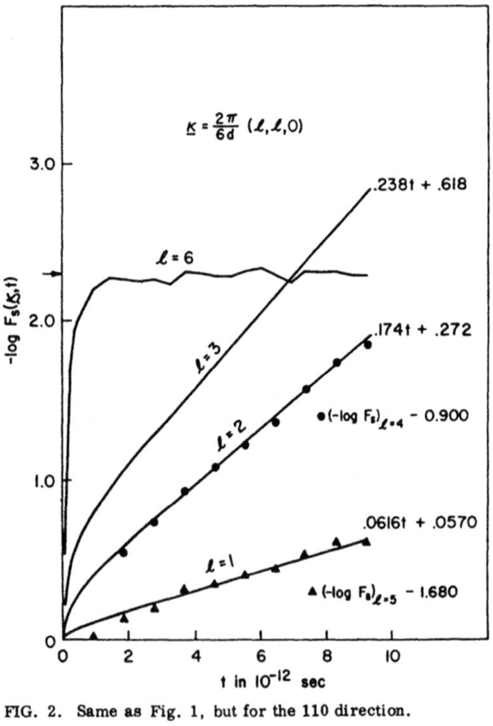

FIG. 2. Same as Fig. 1, but for the 110 direction.

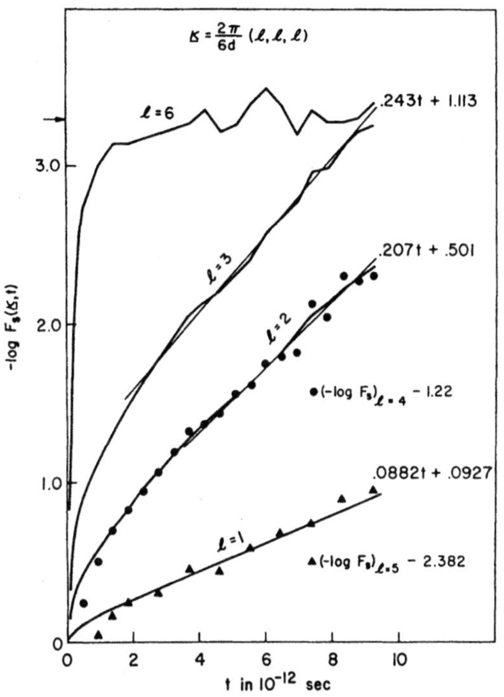

FIG. 3. Same as Fig. 1, but for the 111 direction.

$$F_s(\mathbf{k},t) = \exp\{-f(\mathbf{k})|t|/\tau\} ,$$

where

$$f(\mathbf{k}) = 1 - \sum_i \frac{p_i}{n_i} \sum_{l^{(i)}} \exp(-i\mathbf{k}\cdot\mathbf{l}^{(i)}) .$$

The summation over i contains two terms; p_1, p_2, are short for p_{100}, p_{110}, respectively. $\mathbf{l}^{(i)}$ is the vector displacement in going to a neighbor of the type i, and n_1, n_2, are 6, 12, respectively.

If d is the lattice constant of the simple cubic lattice of the F^- positions in the fluorite lattice (2.95 Å in our case), one gets for various vectors $\mathbf{k}$ the following $f(\mathbf{k})$:

$$\mathbf{k} = (k, 0, 0):\ f(\mathbf{k}) = [p_1 + 2p_2](1-\cos kd)/3 ,$$

$$\mathbf{k} = (k, k, 0):\ f(\mathbf{k}) = [(1-p_1/3) + (1-p_1-2p_2/3) \times \cos kd](1-\cos kd) ,$$

$$\mathbf{k} = (k, k, k):\ f(\mathbf{k}) = [1 + (1-p_1)\cos kd](1-\cos kd) .$$

In the limit $|\mathbf{k}| \to 0$ we get, in all three cases,

$$F_s(\mathbf{k},t) \to \exp(-\mathbf{k}^2 D|t|) ,$$

$$D = (p_1 + 2p_2)\, d^2/6\tau .$$

In Table I it is shown that the molecular dynamics calculations of the decay of $F_s(\mathbf{k},t)$ (Figs. 1–3) are in reasonable accord with the values $p_{100} = 0.79$, $p_{110} = 0.21$, and $\tau = 6.27$ psec. From these values we get $D = 2.8\times10^{-5}$ cm^2 sec^{-1}, the actual value being 2.65×10^{-5} cm^2 sec^{-1} (see Ref. 1).

The jump model [Assumption (v)] ignores all correlations between successive jumps. These, however, will be present if the motion of the atoms is a consequence of the motion of defects. For example, if a bias towards backward jumps makes them more frequent than they would be under Assumption (v), the jump model would ignore that the diffusive motion is inhibited to some extent by this bias; consequently in the jump model the residence time will appear to be longer than it actually is.

III. DIRECT ANALYSIS OF F^- JUMPS

Extensive molecular dynamics studies of single vacancy diffusion in metals have recently been made by DaFano and Jacucci.[5a] To follow the motion of the vacancy they made a sequential study in time of the entrance and exit of the particles in and out of subcells the totality of which makes up the volume of the periodic molecular dynamics cell. Dixon and Gillan[5b] have used suitably constructed spherical cells to make an analysis of particle trajectories in CaF_2; their analysis is very similar to the one presented here.

In a similar way we have used a program (to be referred to as QUOVADIS) to follow the motion of the 216 F^- ions in our system by dividing the molecular dynamics cell into 1728 ($=12^3$) cubic subcells (1.475 Å on the side) such that the 108 Ca^{++} mean positions and the corresponding 108 octahedral sites are centered in 216 of the 1728 subcells; obviously the 216 anion positions of the fluorite lattice will also be at the centers of another set of 216 of the 1728 subcells; the cells were labeled (l, m, n) with $l, m, n = 1, \ ,12$, such that all even l, m, n are the anion sites and all odd l, m, n are the cells with Ca^{++} and the O sites. We shall use the obvious notation *eee* or *ooo* for identification of the cells. Of the remaining 6×216 cells one-half are of the *ooe* type and the other half of the *eeo* type. It is clear that an *eeo* is between two adjacent *eee* cells along the 100 direction. Similarly, an *ooe* is along the 110 direction between two next nearest *eee* cells. The *ooo* cells are along the eight 111 directions emanating from an *eee* cell, four of which form a tetrahedral set and go to octahedral sites, the other four going to the cation sites.

TABLE I. Comparison of decay constants $g(\mathbf{k})$ observed in molecular dynamics ($g_{M.D.}$) and calculated from jump diffusion model, (g_{cal}) with $p_{100} = 0.79$, $p_{110} = 0.21$, $\tau = 6.27$ psec.

k/k_0[a] →	(100)		(110)		(111)	
l ↓	$g_{M.D.}$[b]	g_{cal}[b]	$g_{M.D.}$	g_{cal}	$g_{M.D.}$	g_{cal}
1	0.032	0.032	0.062	0.062	0.088	0.088
2	0.097	0.096	0.174	0.168	0.207	0.214
3	0.131	0.128	(0.238	0.213)[d]	0.243	0.252

[a] $k_0 = 2\pi/6d$; $d = 2.95$ Å.
[b] g is in psec^{-1}.
[c] Boxed-in entries used for calculating the model parameters given in the table caption.
[d] Worst discrepancy between $g_{M.D.}$ and g_{cal}.

Over the total molecular dynamics run the average population of the 216 F^- ions in the various types of cells was found to be *eee*: 131.3; *eeo*; 63.7; *ooe*; 16.3; *ooo*: (Ca^{++}) 0.02; *ooo*: (*O* site) 4.67. In fractional terms (i.e., dividing by the number of cells of each type) these numbers are 0.608, 0.098, 0.025, 0.0002, and 0.043, respectively. This constitutes a density map of the F^-'s in space using a grid of 1728 points in three dimensions.

Still another manner of observing the density of the F^- ions is to evaluate the pair correlation between the *O* sites and the F^-'s; in other words, the number of F^-'s found, on the average, at a distance r from an *O* site in a shell of thickness Δr, taking account of the volume of the shell and the number density of the F^-'s in the whole system, gives this "pair correlation." It is thus another manner of describing the density of F^-'s in space.

This pair correlation is shown in Fig. 4. It shows, in more detail than in the corresponding figure in Ref. 1, the degree to which the diffusing F^- ions penetrate into the octahedral regions of the mean fcc lattice of the cations. Figure 4 allows one to see in quantitative detail this degree of occupancy. The 1728 subcells already mentioned above are 1.475 Å on the side. A sphere of radius equal to half this value will contain (see Fig. 4) 0.02 F^-'s on the average, whereas a sphere of radius 1.3 Å ($=1.475\times\sqrt{3/2}$) has 0.15 F^-'s. The average population in a cube of side 1.475 Å around an *O* site is, as we have seen, 0.043 F^-'s. Thus, depending on the amount of space allotted to the *O* sites, out of the 216 anions a total of about two or five or even 16 F^-'s will be found on the average to be in the region of the *O* sites at any moment. However, approximately eight interstitials will be seen, in the analysis to follow, to be present on the average at any instant in the system and to play a role in the processes involved in tracer diffusion and in conductivity.

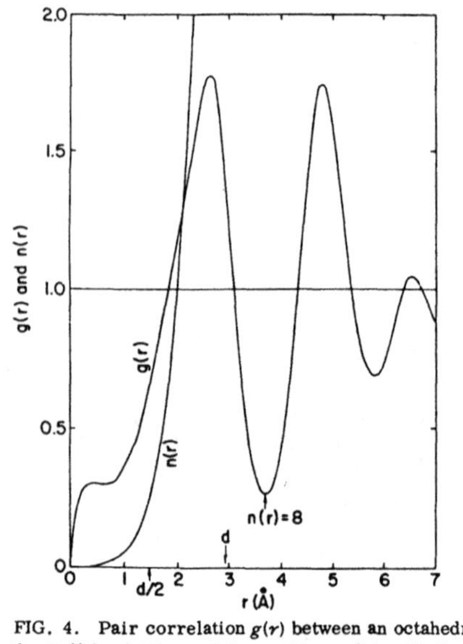

FIG. 4. Pair correlation $g(r)$ between an octahedral site of the Ca^{++} fcc lattice and the positions of the F^-; the coordination number $n(r)$ is also shown.

We note in particular that the pair correlation (Fig. 4) has a certain structure below 1 Å; in view of the central role played by the interstitial F^-'s present around the *O* sites in the dynamics of the F^-'s we believe that the extra piling up of the F^-'s around the *O* sites (as is seen in Fig. 4) is not a mere coincidence. We have derived above (Sec. II) a reasonable picture of the Debye–Waller cloud around each lattice site of F^-'s. Using $(2a\sqrt{\pi})^{-3}\times\exp(-r^2/4a^2)$ as the Debye–Waller cloud density centered at each F^- lattice site with $a=0.5$ Å we get the contribution of the cloud to $g(r)$ at the *O* site to be $8\rho_0^{-1}(2a\sqrt{\pi})^{-3}\exp(-2.56/2a)^2$, ρ_0 being the number density of F^-'s in the system. This value is 0.02. At 0.75 Å from the *O* site, using appropriate distances from eight F^- positions we get 0.056, at 1 Å we get 0.2, and at 1.7 Å we get 1.02. These values in comparison with those shown in Fig. 4 indicate that the piling up of the F^-'s at the *O* site is a genuine effect related with the presence of F^- ions at these sites during the lifetime of interstitialcy migration.

The dynamical information in QUOVADIS is obtained as follows. A jump is an event when a particle enters an *eee* cell having previously made a sortie out of a different *eee* cell (this being the *eee* cell to which it belongs). In other words, a large excursion of a particle out of its *eee* cell does not always end up as a jump according to this definition; the excursion has to lead to an entry into an *eee* cell different from the one out of which the excursion is taking place. When this happens QUOVADIS signals a jump and the new *eee* cell becomes the cell to which that particle belongs. 762 jumps were recorded in 17.02 psec giving an overall jump rate of 0.21 $psec^{-1}$, i.e., a residence time of 4.8 psec. Among the 762 jumps there were 638 100-jumps, 106 110-jumps, and 16 111-jumps (while two were of the type 210).

Thus QUOVADIS gives 0.84, 0.14, and 0.02 as the probabilities of various jumps and a residence time 4.8 psec. The directly observed spatial characteristics are thus in good accord with the parameters obtained from the simple jump diffusion model; the most useful comparison is between the jump rate (0.21 $psec^{-1}$) seen by QUOVADIS and that given by the jump model (0.16 $psec^{-1}$). We have already seen why in the jump model the residence time will appear to be longer than it actually is. The same effect has been observed in a study of the relevant correlations in the simulation of vacancy diffusion in metallic solids.[9] On the other hand, one cannot overlook the fact that the geometrical design used in QUOVADIS can also affect the manner in which the trajectories are sectioned. For example, if the region allotted to an *eee* cell is a sphere fitting into the cube rather than the whole cube then some of the sorties will fail to be recorded as jumps, giving a lower jump rate. From the inventory of "quick" return jumps to the original *eee* cell we conclude that about 3% of the jumps now recorded could be eliminated on allowing a more restrictive definitions of a jump, i.e., a more permissive one of an excursion.

It is pertinent to introduce the notions of vacancies and interstitials as they are defined by QUOVADIS. Each F^- *always* has an *eee*-type cell associated with it and the association is maintained till a jump is recorded when the association changes to the new *eee* cell. This allows one to count the *eee* cells momentarily not associated with any particle. This obviously defines a vacancy. This leads automatically to an equal number of doubly occupied *eee* cells which define an interstitialcy type defect. Using these criteria the average number of anti-Frenkel pairs in the system turns out to be 7.9.[10]

Given these definitions QUOVADIS provides a breakdown of the jumps of particles into four categories: (i) Generation of interstitials: g, (ii) Their recombination: r; (iii) Motion of vacancies: v; (iv) and the motion of interstitials: i. The 762 jumps are $119g+119r+282v+242i$.

Since the average number of (a–F) pairs is 7.9 and 119 g and r events were recorded in 17.02 psec we get $7.9\times 17.02/119=1.12$ psec as the (a–F) pair lifetime. This value is quite small and originates in the large number of generation and recombination events recorded (119 of each) in 17.02 psec. Similarly, the residence time of a vacancy and that of an interstitial is 0.48 psec and 0.55 psec, respectively; in other words these are the jump rates of successive ions along the path of a vacancy and an interstitialcy, respectively.

We have already given the overall count of the different jump directions for the 762 jumps. The breakdown of this count for the four categories of jumps are

$$119\ g=114_{100}+4_{110}+1_{111}\ ,$$

$$119\ r=91_{100}+23_{100}+3_{111}+2_{210}\ ,$$

$$282\ v=273_{100}+8_{110}+1_{111}\ ,$$

$$242\ i=160_{100}+71_{110}+11_{111}\ ,$$

$$762=638_{100}+106_{110}+16_{111}+2_{210}\ .$$

The pattern of behavior emerging out of these proportions will be discussed in the next section. Notice in particular that for the 242 i jumps the ratio is $\sim 15:6:1$ for the 100, 110, and 111 directions, respectively. Also, if the g and r jumps are excluded the *defect* jump rate is $(282+242)/(216\times 17.02\ \text{psec})=0.15\ \text{psec}^{-1}$, which is rather close to the value obtained by fitting the jump diffusion model to the decay of $F_s(\mathbf{k}, t)$.

IV. CORRELATION IN THE MOTION OF F^- IONS

The principal conclusion drawn in the previous sections is that F^- ions reside at the simple cubic lattice sites for a time long compared with the time spent in the displacement between two such sites. As a consequence, a jump model based on this geometric lattice has been found to be adequate to describe the self-diffusion process as seen through $F_s(\mathbf{k}, t)$. Jumps between first and second nearest neighbor sites have to be introduced. On the other hand, the existence of the O sites need not be taken explicitly into account. This is because the F^- ions reside at these sites for a rather short time during the displacement between *eee* sites.

We can now go one step further in developing a model description of mass and charge transport in the system. This will permit us to describe in a simple way the correlation in the motion of F^- ions and its effect on diffusion and electrical conduction.

If at a particular moment the F^- ions fill up completely the *eee* sites, then the displacement of an F^- ion from its lattice site into some interstice in the lattice will leave behind an empty *eee* site. This process certainly requires the expenditure of a large amount of free energy and we can expect it to be a rather rare event at temperatures not close to melting. However, when one such event has taken place somewhere in the bulk of the system, then the F^- ions can easily perform jumps by taking advantage either of the empty site or the interstitial ion. This is because the displacements of these "defects" from a lattice site to another one leaves the free energy of the crystal unchanged, and their activation free energy is certainly much smaller than that of the original jump.

As usual in these circumstances the simplest description of the correlated jumps of the F^- ions is in terms of point defect formation and motion. In CaF_2, F^- vacancies and interstitials are generated in pairs—anti-Frenkel pairs—by thermal fluctuations and then migrate individually till they encounter another defect of opposite "sign" and recombine. Along the path described by vacancies and interstitials F^- ions are displaced by one or more lattice spacings. In this respect the effect of interstitial and vacancy migration are similar in CaF_2; as is often the case, in addition to vacancy migration interstitial migration also changes the identities of the jumping ion at every jump. For this reason this type of defect is called indirect interstitial or interstitialcy: the interstitial particle forces one of its neighbors into another interstitial position, and replaces it at the normal lattice site. Let us consider this process in detail. The lattice of the F^- interstitial sites is a face-centered-cubic lattice. The interstitialcy can move from one interstitial site to another via two different lattice sites[11a,b] (see Fig. 5). While the interstitialcy performs a 110-type jump of length $\sqrt{2}\ d$ two F^- ions perform (111)-type jumps each of length $(\sqrt{3}/2)d$. Iteration of the interstitialcy jump motion results in displacements of the F^- ions in the simple cubic lattice along 100, 110, and 111 directions in the proportions of $3:3:1$. Other than nearest neighbor (n–n) jumps of the F^- ions are then a natural consequence of the interstitialcy migration process; this implies that a many body saddle point of the system is responsible for both n–n and non n–n jumps.

With the point defect picture in mind we have analyzed the motions of the F^- ions using QUOVADIS. Besides the information already mentioned we have also analyzed the role played by the ions reported in the O sites during the dynamical run. Since the time of the last sortie of a particle from its *eee* cell and the time of its entry into its new *eee* cell are recorded, it is possible to look up the population of the O sites during this time interval and check if the particle was reported as an inhabitant of an O site during this interval. Thus all jumps can be classified in this manner into two subgroups.

The numbers quoted at the end of Sec. III now have the following further breakdown:

$$119\ g = (54 + \underline{60})_{100} + (0 + \underline{4})_{110} + (0 + \underline{1})_{111}\ ,$$
$$119\ r = (31 + \underline{60}) \quad + (3 + \underline{20}) \quad + (1 + \underline{2}) \ + 2_{210}\ ,$$
$$282\ v = (161 + \underline{112}) + (3 + \underline{5}) \quad + (0 + \underline{1})\ ,$$
$$242\ i = (51 + \underline{109}) \quad + (7 + \underline{64}) \quad + (0 + \underline{11})\ .$$

The numbers underlined are the counts of jumps when the jumping particle was recorded as an O-site inhabitant during the interval mentioned above.

It is very clear that the O site plays an important role in i jumps changing the relative importance of 100, 110, 111 jumps from 160 : 71 : 11 to 109 : 64 : 11 (more reminiscent of the theoretical ratio 3 : 3 : 1 mentioned above). However, i jumps without O-site participation are not negligible and are mostly in the 100 direction. In addition we note that (see Sec. III) of the 4.67 F^-'s inhabiting the O sites at any moment only 2.50 participate in i jumps just described; the rest are not localized at the O site but represent "spillover" from the F^- lattice sites due to thermal agitation.

Using the label O for the old *eee* cell and N for the new *eee* cell a particle leaves and enters, respectively, in recording a jump we can identify the various jump types (g, r, v, i) as follows:

Description of event	Populations (old cell, new cell) (O, N) before	(O, N) after	Jump Type
Generation	(1, 1)	(0, 2)	g
Recombination	(2, 0)	(1, 1)	r
Vacancy mig. from N to O	(1, 0)	(0, 1)	v
Interstitialcy mig. from O to N	(2, 1)	(1, 2)	i

As exceptions to these four types of events there were also a total of 10 cases where three ions were assigned to the same *eee* site at the same time. This does not affect the discussions to follow.

From a g event to an r event the point defects (v's and i's) describe a migration path that is easily followed by reconstructing the sequence of events having one site in common. For example, $A(1)B(2)$ followed later by $B(1)C(2)$ means an i defect going from A to B to C. Similarly, $Y(0)Z(1)$ and later $X(0)Y(1)$ means that a v defect has gone from Z to Y to X (thus the ion jumps: X to Y then Y to Z, seemingly run "backwards" in time).

Chains of jumps connected in the way just described have been identified in the output of QUOVADIS. An example is given in Table II. The occurrence time of the jump events that constitute successive links of the chain clearly inverts its direction after a few jumps when g or r events occur. The very fact that the F^- motion can be ordered this way indicates the validity of using the point defect language in this context; in addition, as we have just seen, the space and time correlations between the jumping F^- ions are easy to visualize.

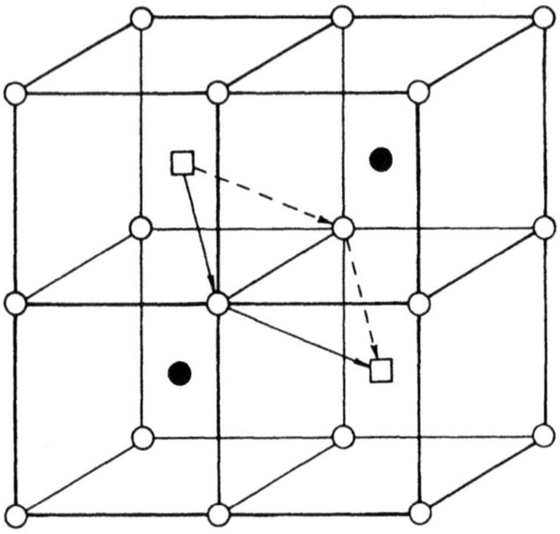

FIG. 5. ●: Ca^{++}; ○: F^-; □: Interstitial at an octahedral site of the Ca^{++} fcc lattice. Arrows denote two possible F^- jumps leading to an interstitial migration.

Forward correlated jump events are sometimes observed along a defect migration path, i.e., a collinear sliding of atoms in the 001 direction over five or six cells constitutes an i jump sequence; such sequences occur over an elapse of time of about 1 psec corresponding to a few vibration periods of the lattice. The O sites are not used in such a sequence. Similar events have also been observed in v jump sequences. No attempt was made to obtain precise statistics of the length and duration of such events, but from looking at many sequences of the type displayed in Table II the qualitative judgment is that such events are rather uncommon.

A small number of highly mobile and short lived defects are thus responsible for the high value of the diffusion coefficient of the F^- ions; the small value of the concentration makes it impossible to isolate the distortions of the density map of the F^- ions brought about by the lattice defects with the exception of the O sites which would be much less populated in the absence of defects. The high mobility of the interstitials reduces the residence time at the O sites, thus making it unnecessary to include them explicity in the jump diffusion model to account for the behavior of $F_s(k, t)$.

It should be pointed out that a large fraction of the jumps of the F^- ions are due to generation and recombination events. This fact will complicate the expression of transport coefficients in terms of defect dynamics. In particular, for ionic motion, including contributions from vacancy and interstitialcy migration alone will not suffice since generation and recombination events contribute significantly to the ionic transport.

The description of the diffusion process of the F^- ions according to the picture described above is not free from ambiguities. These ambiguities are caused by the fact that at high temperatures the ions frequently visit re-

TABLE II. A sample of jump events from QUOVADIS output ordered so as to show interstitialcy and vacancy motion.

Jump number	Jump time[a]	Cells involved[b] Old (O)	New (N)	(O, N)[c] Before	After	Particles in O and N Before jump	After jump[d]	Jump symbol[f]
532	5641	8 12 4	8 12 6	(1, 1)	(0, 2)	(93)(22)	(v)($\underline{93}$, 22)	g
540	5707	8 12 6	10 2 6	(2, 1)	(1, 2)	(93, 22)(23)	(93)($\underline{22}$, 23)	i[e]
541	5715	10 2 6	10 4 6	(2, 0)	(1, 1)	(22, 23)(v)	(22)($\underline{23}$)	r
521	5497	10 4 6	10 4 8	(1, 0)	(0, 1)	(45)(v)	(v)($\underline{45}$)	v
506	5253	10 4 8	12 4 8	(1, 0)	(0, 1)	(24)(v)	(v)($\underline{24}$)	v
504	5227	12 4 8	12 2 8	(1, 0)	(0, 1)	(164)(v)	(v)($\underline{164}$)	v
478	4891	12 2 8	2 2 8	(1, 0)	(0, 1)	(94)(v)	(v)($\underline{94}$)	v
473	4795	2 2 8	2 2 10	(1, 0)	(0, 1)	(142)(v)	(v)($\underline{142}$)	v
470	4773	2 2 10	2 2 12	(1, 0)	(0, 1)	(73)(v)	(v)($\underline{73}$)	v
464	4711	2 2 12	2 4 12	(1, 0)	(0, 1)	(201)(v)	(v)($\underline{201}$)	v
458	4605	2 4 12	12 4 12	(1, 0)	(0, 1)	(155)(v)	(v)($\underline{155}$)	v
454	4545	12 4 12	12 4 2	(1, 0)	(0, 1)	(17)(v)	(v)($\underline{17}$)	v
446	4453	12 4 2	2 4 2	(1, 0)	(0, 1)	(18)(v)	(v)($\underline{18}$)	v
435	4249	2 4 2	2 2 2	(1, 0)	(0, 1)	(1)(v)	(v($\underline{1}$)	v
418	3881	2 2 2	2 12 2	(1, 0)	(0, 1)	(71)(v)	(v)($\underline{71}$)	v
393	3671	2 12 2	4 12 2	(1, 0)	(0, 1)	(149)(v)	(v)($\underline{149}$)	v
391	3653	4 12 2	6 12 2	(1, 1)	(0, 2)	(145)(63)	(v)($\underline{145}$, 63)	g
399	3719	6 12 2	6 12 4	(2, 0)	(1, 1)	(145, 63)(v)	(145)($\underline{63}$)	r
392	3657	6 12 4	6 2 4	(1, 0)	(0, 1)	(15)(v)	(v)($\underline{15}$)	v
328	3055	6 2 4	6 12 4	(1, 0)	(0, 1)	(15)(v)	(v)($\underline{15}$)	v
314	2749	6 12 4	6 12 6	(1, 1)	(0, 2)	(59)(129)	(v)($\underline{59}$, 129)	g
331	3067	6 12 6	8 12 8	(2, 1)	(1, 2)	(59, 129)(212)	(59)($\underline{129}$, 212)	i[e]
339	3175	8 12 8	8 10 8	(2, 1)	(1, 2)	(129, 212)(134)	(129)($\underline{212}$, 134)	i
427	4049	8 10 8	8 10 10	(2, 1)	(1, 2)	(134, 212)(133)	(134)($\underline{212}$, 133)	i
428	4051	8 10 10	8 10 12	(2, 1)	(1, 2)	(133, 212)(205)	(212)($\underline{133}$, 205)	i
442	4309	8 10 12	8 10 10	(2, 1)	(1, 2)	(133, 205)(212)	(205)($\underline{133}$, 212)	i
459	4627	8 10 10	6 10 8	(2, 1)	(1, 2)	(212, 133)(135)	(133)($\underline{212}$, 135)	i[e]
462	4689	6 10 8	4 10 8	(2, 1)	(1, 2)	(135, 212)(198)	(212)($\underline{135}$, 198)	i
467	4765	4 10 8	4 10 10	(2, 1)	(1, 2)	(135, 198)(202)	(198)($\underline{135}$, 202)	i
471	4781	4 10 10	4 10 12	(2, 0)	(1, 1)	(135, 202)(v)	(135)($\underline{202}$)	r
469	4769	4 10 12	4 10 2	(1, 0)	(0, 1)	(176)(v)	(v)($\underline{176}$)	v

etc.

[a]In units of $\Delta t = 1.843\times10^{-15}$ sec.

[b]QUOVADIS defines events in terms of *eee*-type cell occupation by the F^- ions.

[c]See Sec. IV for explanation of O and N.

[d]The underlined number is the particle causing the event and (*v*) stands for an empty cell, i.e., a vacancy.

[e]110 jumps; others are all 100.

[f]See Sec. III.

gions in configuration space which are unfavorable in terms of potential energy resulting in a broad Debye–Waller cloud and in high jump rates. When the duration of a jump of an ion between sites and its residence time at one site are not different by several orders of magnitude and when, in addition, the Debye–Waller clouds spread onto neighboring cells, the identification of the arrival and departure time at the lattice sites with the time of crossing of the walls of the *eee* cells becomes an approximate procedure. The decision to assign an ion to a cell until it has moved into a different one certainly eliminates the possibility of overcounting jump events. However, some problems remain concerning the imprecision in assigning relative occurrence times of jump events that are close in time.

As an example consider the following jumps recorded as having occurred at $t_1 < t_2$:

$$\left.\begin{array}{l} g \text{ at } t_1: C(1)D(1) \rightarrow C(0)D(2) \\ r \text{ at } t_2: B(2)C(0) \rightarrow B(1)C(1) \end{array}\right] \text{ or } \left[\begin{array}{l} g \text{ at } t_1: D(1)C(1) \rightarrow D(0)C(2) \\ r \text{ at } t_2: B(0)C(2) \rightarrow B(1)C(1). \end{array}\right.$$

If $t_2 - t_1$ is short compared with the imprecision in the determination of jump "occurrence" times we can argue that, on reversing the order in which the two events occurred one would have obtained, instead,

$$\left.\begin{array}{l} B(2)C(1) \rightarrow B(1)C(2) \\ C(2)D(1) \rightarrow C(1)D(2) \end{array}\right] \text{ or } \left[\begin{array}{l} B(0)C(1) \rightarrow B(1)C(0) \\ D(1)C(0) \rightarrow D(0)C(1)\ , \end{array}\right.$$

which is an *i* going from *B* to *C* to *D* in one case and a *v* doing that in the other. Note that $B(1)C(1)D(2)$ in the former case and $B(1)C(1)D(0)$ in the latter continue to be the end result whatever the order of the events in each case. If during the interval t_1 to t_2 the cells *B*, *C*, *D* do not participate in other events and if $t_2 - t_1$ is less than 0.2 psec we have applied such inversions in analyzing the sequence of jump events and the counts and various other statistic includes the consequences of these inversions. In all 68 *v* jumps and 50 *i* jumps are consequences of such inversions.

ACKNOWLEDGMENT

We are grateful to K. Sköld for a critical reading of the manuscript.

[1]A. Rahman, J. Chem. Phys. **65**, 4845 (1976).

[2]Y. S. Kim and R. G. Gordon, J. Chem. Phys. **60**, 4332 (1974).

[3](a) C. T. Chudley and R. J. Elliott, Proc. Phys. Soc. London 77, 353 (1961); (b) W. Gissler and H. Rother, Physica (Utrecht) **50**, 380 (1970); (c) W. Gissler and N. Stump, Physica (Utrecht) **65**, 109 (1973); (d) R. Kutner and I. Sosnowska, J. Phys. Chem. Solids **38**, 741 (1977).

[4]P. F. Flynn, *Point Defects and Diffusion* (Clarendon, Oxford, 1972). In the introductory chapter is to be found a most elegant expression of these ideas.

[5](a) A. Dafano and G. Jacucci, Phys. Rev. Lett. **39**, 950 (1977); (b) M. Dixon and M. Gillan, J. Phys. C **11**, L165 (1978).

[6]R. W. Ure, Jr., J. Chem. Phys. **26**, 1363 (1957).

[7]G. Cicotti and G. Jacucci, Phys. Rev. Lett. **35**, 789 (1975); G. Cicotti, G. Jacucci, and I. R. McDonald, Phys. Rev. A **13**, 426 (1976); G. Jacucci, I. R. McDonald, and A. Rahman, Phys. Rev. A **13**, 1581 (1976).

[8]J. M. Rowe, K. Sköld, H. E. Flotow, and J. J. Rush, J. Phys. Chem. Solids **32**, 41 (1971).

[9]A. Dafano, G. Jacucci, and A. Rahman (unpublished).

[10]With considerable risk of going beyond the region of validity of experimental data we recall that a similar number is suggested by experiments of Ure.[6]

[11](a) A. B. Lidiard, Nuovo Cimento Suppl. **7** (X), 620 (1958); (b) K. Compaan and Y. Haven, Trans. Faraday Soc. **54**, 1498 (1958).

Ionic Motion in α-AgI

P. Vashishta and A. Rahman
Argonne National Laboratory, Argonne, Illinois 60439
(Received 14 March 1978)

A molecular-dynamics study of silver diffusion in superionic conductor α-AgI is performed. Interionic potentials are constructed using Pauling's ideas of ionic radii. The diffusion constant for silver and its temperature dependence are in good agreement with experiment. Good agreement is also obtained for the silver density map with the experiment of Cava, Reidinger, and Wuensch.

α-AgI has been extensively studied[1] because of its interest as a solid fast ion conductor (often called superionic conductor). In this Letter we present a molecular-dynamics (MD) study of this material. Our calculations show that in the computer model the diffusion constant of Ag^+ and their distribution in the interstices of the thermally agitated iodine lattice are faithfully reproduced. In addition, MD provides a detailed picture of the complete process of self-diffusion. The central issue in all such calculations is the construction of potential functions which, when used in the MD calculations, produce structural and dynamical correlations which compare favorably with those observed in the laboratory. The MD trajectories then porvide a wealth of microscopically detailed information which can be very hard to obtain by conventional experimental means. A study of CaF_2 along these lines has already been reported.[2]

For CaF_2 the ionic model given by Kim and Gordon[3] is sufficiently good.[2] In α-AgI this problem is not so straightforward. We have constructed effective pair potentials which provide a simple means of describing, with reasonable accuracy, the structural and dynamical properties of α-AgI. Schommers[4] had to use quite unphysical forces to keep the iodine lattice vibrating as a stable structure; in his model each iodine was attached to bcc lattice position by harmonic springs. In our model no such unphysical elements have been introduced. For AgI, we use

$$V_{ij}=\frac{A_{ij}(\sigma_i+\sigma_j)^n}{r^n}+\frac{Z_iZ_je^2}{r}-\frac{1}{2}(\alpha_iZ_i^2+\alpha_jZ_j^2)\frac{e^2}{r^4}-\frac{W_{ij}}{r^6}, \qquad (1)$$

where i, j describe the type of ions; A_{ij} the repulsive strength; σ_i, σ_j the particle radii; α_i, α_j the electronic polarizabilities. If σ's, α's, and W's are known, we need to determine five parameters (namely A_{ij}'s, n, and $|Z_i|=|Z_j|$). By assuming $A_{ij}=A$, the situation is considerably simplified. The repulsive term then implies that each ionic "contact" contributes energy A, i.e., the coefficient of r^{-n} is scaled according to the sum of particle radii. Low-temperature crystal structure, cohesive energy, and compressibility may be used to determine these three parameters.

However, AgI, and α-AgI in particular, pose special problems in this respect. It is certainly not purely ionic and the estimate of the cohesive energy from the Born-Haber cycle can be in considerable error. The γ-AgI compressibility, however, is known[5] ($\simeq 1\times10^{-11}$ cm^3/erg) and from phonon dispersion measurement[6] there is evidence of $|Z|\simeq 0.6$.

As to the σ's, Pauling's[7] concept of ionic radii is a means of expressing the bond length. Using the concept literally, we write $\sigma_{Ag}+\sigma_I=$ Ag-I distance and $2\sigma_I=$ I-I distance. Because of the large size differences between Ag^+ and I^- both Ag-I and I-I interactions together determine the crystal structure. We get $\sigma_I = 2.2$ Å and $\sigma_{Ag} = 0.63$ Å; σ_{Ag} is less than the ionic radius of Ag^+, presumably because of the presence of covalent interaction. It should be realized that we are imposing a pairwise additive effective-interaction scheme for the purpose of doing molecular-dynamics simulations with relative facility.

Thus, using the compressibility as a guide, we choose $n = 7$; with $|Z| = 0.6$ and with the condition that crystal-energy minimum occur at the lattice distance $= 5.06$, (with $\alpha_{Ag} = W_{AgAg} = W_{AgI} = 0$, $\alpha_I = 6.52$, $W_{II} = 6.93$), we determine $H_{AgI} = 17.893$, $H_{AgAg} = 0.062$, $H_{II} = 394.934$, where $H_{ij} = A(\sigma_i - \sigma_j)^n$. Here Å is the unit of length, $e^2/Å = 14.39$ eV is the unit of energy. The potential functions are plotted in Fig. 1(a).

In the final analysis, MD results for α-AgI themselves serve as justification for the validity of the chosen potential functions.

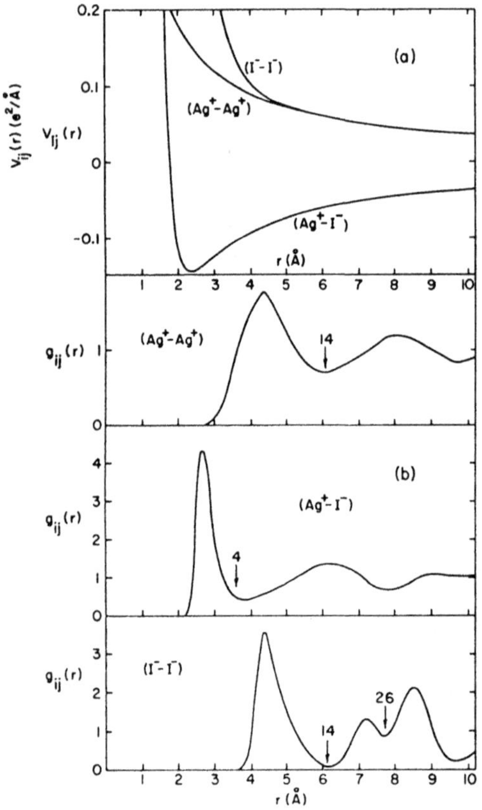

FIG. 1. (a) Interionic potentials in units of $e^2/Å = 14.39$ eV; (b) pair correlation functions.

The calculations were performed on a 256-particle system in a cubic cell of length 20.342 Å (mass density 5.928 g cm^{-3}) in which the iodines continue to form a thermally agitated bcc lattice with the lattice constant 5.0855 Å. Coulomb interactions were taken into account in the usual way.[8] The integration step was 2.5×10^{-14} sec and was quite satisfactory for energy conservation for several thousand steps. The calculation was initiated by placing the iodines in a bcc structure and the Ag ions in suitable tetrahedral sites.

After sufficiently long initial aging of the system (which was much longer than 10 psec), and because of large self-diffusion for Ag^+, the starting conditions have no effect on the results reported here. The calculations were performed at 430, 603, and 761 K, at the mass density mentioned above.

The correlation between the positions of the iodines clearly shows (as in all bcc structures at high temperatures) a coordination number of 14 (eight nn's + six nnn's) under the main peak of the radial distribution function. On the other hand, Ag^+-I^- correlation function gives a coordination of 4. The pair correlation functions are shown in Fig. 1(b). The main peaks in the I^--I^- and the Ag^+-Ag^+ pair correlation occur at about the same distance, but the former is much sharper. However, Ag^+-Ag^+ coordination number continues to be about 14, precisely the same as in I^--I^- structure. This structural information by itself cannot lead to any conclusion regarding the difference between the dynamical behavior of the two types of ions.

Since the iodines form a well-defined bcc structure, it is reasonable to use it as a reference frame to look at the presence of Ag ions in various positions. The most interesting region is a (100) face of the lattice and the presence of Ag ions in a plate of thickness 5.0855/16 Å was monitored. Using the density at the tetrahedral sites [($\frac{1}{4}$ 0 $\frac{1}{2}$) positions] as a reference value, Fig. 2 shows the density along various paths on the cube face. The subsidiary maximum at (0.4 0 0.4) when going along *I–S–C* is unambiguous and was found by Cava, Reidinger, and Wuensch.[9] It will be most instructive to use the density distributions of Ag and I ions to calculate the diffracted intensity patterns for comparison with the experiment The local maximum just mentioned[9] will be discussed below.

The constant of self-diffusion D was determined by calculating the mean-square displacement

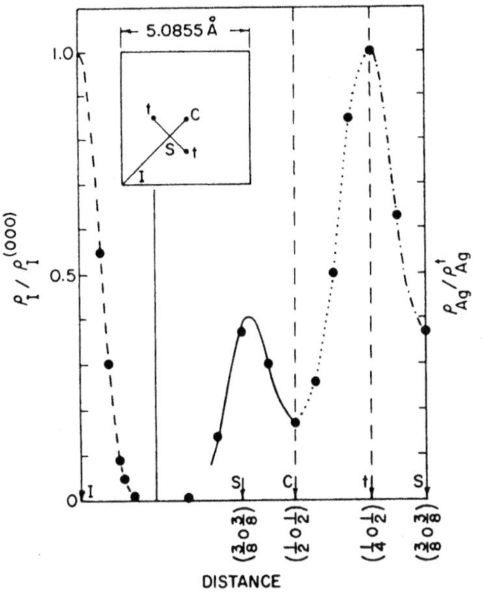

FIG. 2. Density distribution of I^- and Ag^+ on the (100) face at 157°C. In the inset I, C, and t mark the corner (iodine), center, and tetrahedral positions, respectively. Solid line is the only experimental curve (from Ref. 9) in the figure. Solid circles are MD calculations. Dotted, dashed, and dash-dotted lines joining solid circles are for clarity: dashed line, $\rho_{I^-}/\rho_{I^-}^{(000)}$ along line I–S–C; solid line $\rho_{Ag^+}/\rho_{Ag^+}^{t}$ along line I–S–C compared with experiment; dotted line, $\rho_{Ag^+}/\rho_{Ag^+}^{t}$ along line C–t; dash-dotted line, $\rho_{Ag^+}/\rho_{Ag}^{t}$ along line t–S. Point S is a local minima along line t–S–t.

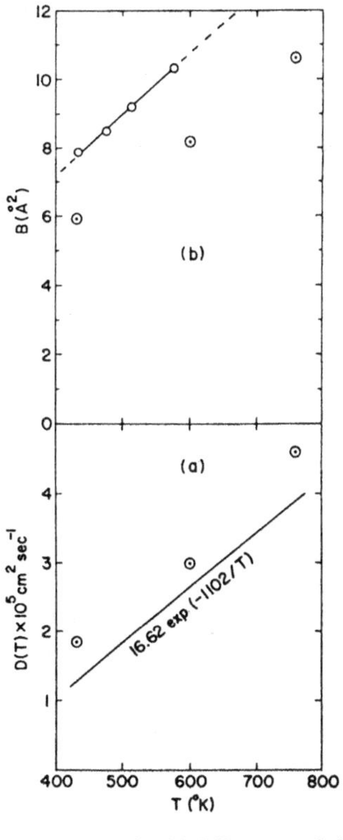

FIG. 3. (a) Constant of self-diffusion, $D(T)$, for Ag^+ in α-AgI. Circle with dot, MD calculation; continuous curve, experiment (Ref. 10). (b) Mean-square displacement, B, of iodine in α-AgI. Circles with dot, MD calculation; open circles connected by solid line, experiment (Ref. 9).

$\langle[\vec{r}(t+s)-\vec{r}(s)]^2\rangle$ for each ion type. As in the case of CaF_2,[2] the asymptotic time dependence of this quantity shows that Ag^+ has a large liquid-like D, whereas I^- has $D=0$. The asymptotic value of the function for I^- gives the extent of the Debye-Waller thermal cloud. In Figs. 3(a) and 3(b), calculated values of D and the mean-square displacement of iodine, B, are compared with the experiments.[9,10] For both quantities agreement is good, leading us to conclude that we are indeed dealing with a realistic model system for α-AgI.

Jacucci and Rahman[11] have analyzed the F^- diffusion in CaF_2 to show that the F^- motion is dominantly a jump in the [100] direction with a residence time of about 6 psec. It was already apparent from the behavior of the fourth moment of displacement, i.e., of $\langle[\vec{r}(t+s)-\vec{r}(s)]^4\rangle$, that the spreading of the probability distribution of the F^- ions [i.e., the Van Hove[12] function $G_s(\vec{r}, t)$] was quite unlike its behavior in simple liquids.

Using the method of Jacucci and Rahman, we find that the Ag ions diffuse by a jump process with jump frequency 0.34×10^{12} sec^{-1}; 82% of the jumps occur between nearest-neighbor tetrahedral sites (which can be characterized as [110] jumps), 9% in the [002] and 7% in the [112] directions from one t to the nearest t site in that direction. When successive [110] jumps occur, our analysis shows that there is no bias towards a rotation around the "t" sites which form a square on a (100) face of the bcc lattice. Each t site has four nearest-neighbor t sites (in [110] directions) and the analysis shows that when successive [110] jumps occur, there is a bias towards backward jumps in comparison with the other three possibilities which, so stated above, occur with equal probability. We find that among successive [110] jumps there are 40% in the backward direction and 20% each in the other three (in the unbiased case these would all be 25%).

Since the relative maximum in density at (0.4 0 0.4) in Fig. 2 is situated essentially along the line joining "*t*" sites, it is relevant to ask whether this site plays a role as a residence site. When the density is plotted from one "*t*" site to another, the location (0.4 0 0.4) is *a local minimum and not a maximum*; along this line the values of the density decrease from unity at ($\frac{1}{4}$ 0 $\frac{1}{2}$) to 0.63 at (0.31 0 0.44) and then to 0.37 at (0.375 0 0.375), point *S*. A concise manner of describing the situation is that the Ag ions reside at the "*t*" sites for about 3 psec, developing a thermal cloud of half-width 0.6 Å in the (100) plane and move between the "*t*" sites along rather narrow channels.

The work reported here makes it quite clear that using classic notions regarding the interparticle potentials in ionic materials it is possible to construct model systems which reproduce with fair precision the observed structural and dynamical behavior of α-AgI. We are therefore proceeding further on an analysis of the structure and properties of other forms of AgI as well as other materials like CuI. The analysis of density fluctuations in α-AgI is now in progress.

This work was performed under the auspices of the U. S. Department of Energy.

[1] *Superionic Conductors*, edited by G. D. Mahan and W. L. Roth (Plenum, New York, 1976); R. A. Huggins, in *Diffusion in Solids–Recent Developments*, edited by A. S. Nowick and J. J. Burton (Academic, New York, 1975).

[2] A. Rahman, J. Chem. Phys. 65, 4845 (1976).

[3] Y. S. Kim and R. C. Gordon, J. Chem. Phys. 60, 4332 (1974).

[4] W. Schommers, Phys. Rev. Lett. 38, 1536 (1977).

[5] G. Burley, J. Phys. Chem. Solids 25, 629 (1964).

[6] W. Buhrer *et al.*, Phys. Rev. B (to be published).

[7] L. Pauling, *The Nature of the Chemical Bond* (Cornell Univ. Press, Ithaca, New York, 1960).

[8] M. J. L. Sangster and M. Dixon, Adv. Phys. 25, 247 (1976).

[9] R. J. Cava, F. Reidinger, and B. J. Wuensch, Solid State Commun. 24, 411 (1977).

[10] A. Kvist and R. Tarneberg, Z. Naturforsch. 25A, 257 (1970).

[11] G. Jacucci and A. Rahman, to be published; M. Dixon and M. Gillan, to be published.

[12] L. Van Hove, Phys. Rev. 95, 245 (1954).

Structural Transitions in Superionic Conductors

M. Parrinello,[a] A. Rahman, and P. Vashishta
Argonne National Laboratory, Argonne, Illinois 60439
(Received 7 September 1982)

The $\alpha \rightleftharpoons \beta$ phase transition in AgI is studied with use of the new molecular-dynamics technique which allows for a dynamical variation of the shape and size of the cell. In the present model, upon heating of β-AgI, the iodine ions undergo a hcp → bcc transformation and silver ions become mobile, whereas the reverse transformation is observed on cooling of α-AgI. The calculated $\alpha \rightleftharpoons \beta$ transition temperature and structural and dynamical properties are in good agreement with experiments.

PACS numbers: 61.50.Ks, 64.70.Kb, 66.30.Hs

Superionic conductors are a class of systems in which phenomena observed in fluids and solids come together in an interesting manner.[1] Silver iodide is a superionic conductor which, as a function of density and temperature, shows a number of structural transitions.[2] Of particular interest to us is the $\beta \rightleftharpoons \alpha$ transition. The β phase has the wurtzite structure: Iodine ions form an hcp lattice and silver ions are tetrahedrally coordinated to each iodine and show no self-diffusion. At 420 K, the β-AgI undergoes a first-order phase transition into the superionic α phase in which iodine ions form a bcc lattice and silver ions are mobile, having a large constant of self-diffusion ($> 1\times10^{-5}$ cm^2/sec). Properties of α-AgI have been studied by a variety of experimental techniques by a number of workers.[1-3] A simple model for α-AgI was proposed[4] and the nature of ionic motions was studied by the usual molecular-dynamics (MD) technique in which both the shape and the size of the MD cell remain fixed.[4,5] Calculated results for the temperature dependence of the mean square displacement of iodine, the constant of self-diffusion of Ag, the density map of Ag in the unit cell, partial pair correlations, velocity autocorrelation functions, current-current correlation function, Haven's ratio, etc., were found to be in good agreement with experiments.[4-7]

However, a study of structural transitions is impossible within the framework of the usual MD technique in which both the shape and size of the cell are fixed. The iodine lattice, confined to a cubic cell and given a bcc structure, cannot undergo a structural transition; hence the $\alpha \rightleftharpoons \beta$ transition in AgI will be prohibited. The MD technique can be generalized to include changes with time in the volume of the cell, $\Omega(t)$.[8] However, the bcc lattice in a cubic cell cannot transform into another lattice structure even though the volume of the cubic cell can change with time. Using an appropriate Lagrangian one can set up a molecular-dynamics calculation in which both the volume *and* shape of the periodically repeating cell change with time.[9]

In this Letter we report the result of our studies of the $\alpha \rightleftharpoons \beta$ transition in AgI. Such studies have become possible because of the above-mentioned development in MD technique. Our aim is to construct a simple potential function for AgI which will make the iodines adopt a bcc structure at elevated temperatures with the silver ions jumping between the tetrahedral sites. On cooling of the system, the interaction potential, all on its own, should modify the bcc structure into a close-packed one, while at the same time the silver ions should settle down into a fourfold-coordinated nondiffusive configuration.

The problem of setting up a scheme to construct interaction potentials has been approached from a phenomenological point of view.[4] Since with the new MD technique we can study structural transitions, we are no longer confined to the α phase to determine the potential; hence we have used the true low-temperature structure of AgI to determine the constants in the potentials. This changes σ_{Ag} from 0.63 to 0.53 Å and leaves $\sigma_I = 2.2$ Å unchanged. It should be noted, however, that $\sigma_{Ag} = 0.53$ and the old exponents $n(\mathrm{II}) = n(\mathrm{AgI}) = n(\mathrm{AgAg}) = n = 7$ produced unsatisfactory results. The potential functions are therefore refined by using different values of the exponents n_{ij} in the repulsive terms. It is our hope that a single set of potentials of the form given below, where the constants are determined from a low-temperature structure, will be adequate to describe not only the phase transition $\alpha \rightleftharpoons \beta$, but also the high-pressure phases of AgI.[7] We take

$$V_{\mathrm{AgAg}}(r) = H_{\mathrm{AgAg}}/r^{n(\mathrm{AgAg})} + 0.36/r,$$

$$V_{\mathrm{AgI}}(r) = H_{\mathrm{AgI}}/r^{n(\mathrm{AgI})} - 0.36/r - 1.1736/r^4, \quad (1)$$

$$V_{\mathrm{II}}(r) = \frac{H_{\mathrm{II}}}{r^{n(\mathrm{II})}} + \frac{0.36}{r} - \frac{2.3472}{r^4} - \frac{6.9331}{r^6},$$

where $H_{ij} = A(\sigma_i + \sigma_j)^{n(ij)}$, with angstroms and $e^2/(1\ \text{Å}) = 14.39$ eV as units. We kept $n(\text{II}) = 7$ and changed $n(\text{AgI})$ toward a higher value [$n(x) < 7$ are uncommon and Pauling suggests $n(\text{Ag}) > n(\text{I})$]. Since the Ag-Ag repulsion term $r^{-n(\text{AgAg})}$ does not play an important role, we took $n(\text{Ag}) = 11$, $n(\text{I}) = 7$ and constructed $n(\text{II}) = 7$, $n(\text{AgI}) = 9$, and $n(\text{AgAg}) = 11$. In brief, the changes from old to modified potential result from taking σ_{Ag} and σ_{I} from the low-temperature structure and changing $n(\text{AgI})$ to 9 from 7. Using these $n(ij)$, we get $A = 0.010\,248$, $H_{\text{AgAg}} = 0.014\,804$, $H_{\text{AgI}} = 114.48$, and $H_{\text{II}} = 446.64$. The value of charge, $Z = 0.6$, and other constants were kept the same as in the earlier potential functions used to describe the α phase.[4,5] We shall refer to Eq. (1) as the "modified potential."

To study the structural transition of the iodine sublattice as well as the order-disorder transformation of silver ions, we take a neutral system of $N = 500$ particles (250 I^- and 250 Ag^+) obeying $3N + 9$ equations of motion.[9] Note that for a 500-particle system in a nonvarying cubic cell one can study only the properties of α-AgI and none of the other structures of AgI. Using the new MD method and the "modified potential," we made a calculation on α-AgI at 700 K. All the calculations reported here were carried out at a constant value of the pressure. After equilibration, a calculation extending over $\sim 5 \times 10^{-11}$ sec, i.e., 2500 MD steps, gave results in agreement with previous calculations.[4] The iodines form a thermally agitated, stable, bcc lattice; the Ag-I pair correlation gives a clear nearest-neighbor (nn) coordination of 4; the Ag ions diffuse with $D_{\text{Ag}} = 4 \times 10^{-5}$ cm^2 sec^{-1}.

Having thus confirmed that we have a satisfactory model and method of calculation for α-AgI we proceeded to study the structural phase transitions in this system.

On cooling of the above system from 700 to 350 K, in only about 2×10^{-11} sec the apparent "nn" I-I coordination started to drop from 14 (8 + 6) to 12 while the nn Ag-I coordination remained 4 throughout. In addition the MD cell changed to a noncubic shape. This already was indicative of a structural change for the iodines from bcc to close packed. After these secular changes had stopped, a 2500-time-step run was made to calculate various average properties. *Both* iodine and silver ions showed no diffusion, the mean square displacments being $B_{\text{I}} = 4.0$ Å^2 and $B_{\text{Ag}} = 6.3$ Å^2. The $g_{ij}(r)$ are shown in Fig. 1(a). The nn coordination shows that on cooling to 350 K

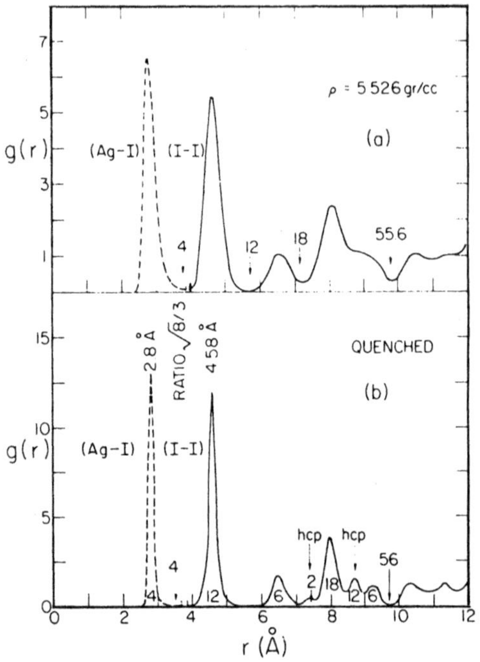

FIG. 1. (a) Partial pair correlation function, $g(r)$, after cooling from α phase at 700 K. The first peak for Ag-I shows a coordination of 4. Peaks for I-I are shown with coordination numbers. The first peak clearly shows a nn coordination of 12 and the arrow at 9.75 Å indicates a total coordination of 55.6 iodines. For a perfect fcc and hcp lattice this number is 54 and 56, respectively. (b) $g(r)$'s at 343.8 K on quenching to a very low temperature. The Ag-I peak at 2.8 Å is sharpened. The ratio between I-I and Ag-I nn distances is $(\frac{8}{3})^{1/2}$. The two peaks marked hcp containing two and twelve particles are clearly resolved. This shows that the pair correlations are for β-AgI with a wurtzite structure.

the α-AgI has been transformed to a zinc-blende or a wurtzite structure. From Fig. 1(a) by itself, one cannot resolve the situation unambiguously, but (i) the shoulder at 9 Å, and (ii) the value 55.6 of the number of I-I neighbors up to 9.8 Å indicate hcp character. Following the usual method in such cases, quenching the system to a very low temperature lifts the uncertainty. The result is shown in Fig. 1(b).

Ideally, for hcp structure, the $g_{\text{I-I}}$ should show 6 peaks and 56 neighbors up to 9.8 Å and 4 peaks and 56 neighbors for fcc. Figure 1(b) shows, firstly, an Ag-I and I-I coordination of 4 and 12, respectively, with a distance ratio of $\sqrt{\frac{8}{3}}$ between the peak positions. Secondly, 6 peaks in $g_{\text{I-I}}$, with 12, 6, 2, 18, 12, and 6 neighbors, are

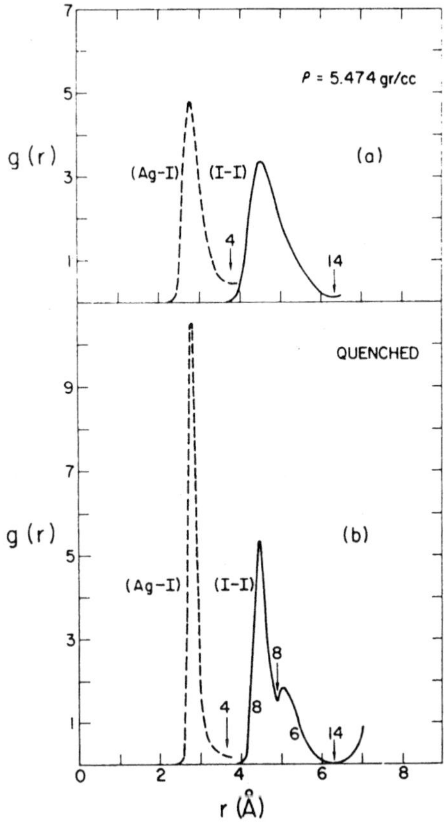

FIG. 2. (a) $g(r)$'s at 494.7 K obtained by heating from β-AgI at 343.8 K. The peaks for Ag-I and I-I show coordinations of 4 and 14, respectively. (b) $g(r)$'s at 494.7 K after quenching. The I-I correlation shows two peaks containing 8 and 6 iodines. This confirms that the figure described the pair correlation of α-AgI.

clearly seen. Thus α-AgI cooled from 700 to 350 K has transformed to β-AgI.

The $\beta \rightarrow \alpha$ transition was studied by heating of the β-AgI system from 343.8 to 495 K. After about 1000 time steps, the nn I-I coordination started to be more than 12. Pair correlation functions for a well thermalized system at 495 K are shown in Fig. 2(a). The first peak in $g(r)$ for I-I shows a nn coordination of 14—the characteristic value for a heated bcc lattice. Silver ions were found to be diffusing with $D_{Ag} = 2\times10^{-5}$ cm^2/sec and nn Ag-I coordination was tetrahedral. The result of quenching to very low temperature is shown in Fig. 2(b). The first two shells of the bcc iodine lattice, containing eight and six particles, respectively, are clearly identified. This

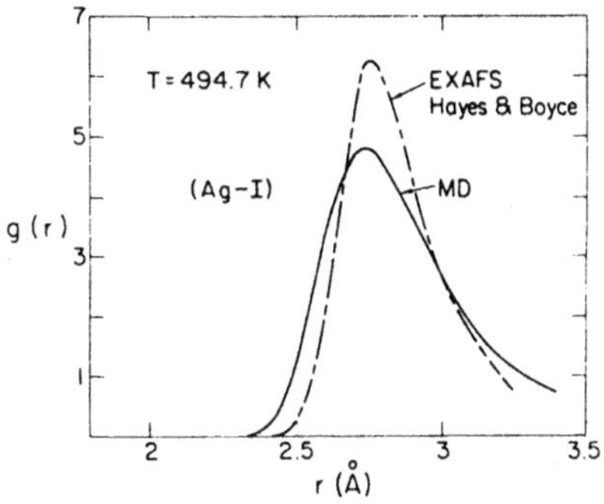

FIG. 3. The first peak of Ag-I pair correlation in α-AgI from MD calculations and EXAFS results (Ref. 10).

confirms that $\beta \rightarrow \alpha$ AgI transformation has been achieved.

By repetition of such calculations several times the $\alpha \rightleftharpoons \beta$ transition temperature was determined to lie between 472 and 495 K, in satisfactory agreement with the experimental value of 420 K.

The first peak of the Ag-I pair correlation function can be accurately measured by extended x-ray-absorption fine structure (EXAFS). In Fig. 3 we compare our MD results with the EXAFS results of Hayes and Boyce for α-AgI.[10] The agreement is indeed very good.

It is obvious that the "modified potential" given in Eq. (1), which has different exponents to describe Ag-Ag, Ag-I, and I-I repulsive terms, in conjunction with the MD technique which allows for a variation of the shape and size of the cell, successfully describes the $\alpha \rightleftharpoons \beta$ transition in AgI. Furthermore, the structural and dynamical properties of AgI in the superionic phase are in good agreement with experiments.

Very recently we have also studied structural transitions in AgI under hydrostatic pressure. In our computer model, AgI described by the potential function given in Eq. (1) successfully undergoes $\alpha \rightleftharpoons$(rock salt) and $\beta \rightleftharpoons$(rock salt) transitions. Results of this investigation will be reported elsewhere.

[(a)]Permanent address: University of Trieste, Trieste, Italy.

[1]*Superionic Conductors*, edited by G. D. Mahan and W. L. Roth (Plenum, New York, 1976); *Physics of*

Superionic Conductors, edited by M. B. Salamon (Springer-Verlag, New York, 1979); J. B. Boyce and B. A. Huberman, Phys. Rep. 51, 190 (1979); *Fast Ion Transport in Solids*, edited by P. Vashishta *et al.* (North-Holland, New York, 1979); *Proceedings of the Third International Meeting on Solid Electrolytes —Solid State Ionics and Galvanic Cells*, edited by T. Takahashi *et al.*, Solid State Ionics 3/4 (1981); *Fast Ionic Transport in Solids*, edited by J. B. Bates and G. C. Farrington, Solid State Ionics 5 (1981).

[2]K. Funke, Prog. Solid State Chem. 11, 345 (1976); B.-E. Mellander, A. Lunden, and M. Friesel, Solid State Ionics 5, 477 (1981).

[3]J. B. Boyce and T. Hayes, in *Physics of Superionic Conductors*, edited by M. B. Salamon (Springer-Verlag, New York, 1979).

[4]P. Vashishta and A. Rahman, Phys. Rev. Lett. 40, 1337 (1978).

[5]W. Schommers, Phys. Rev. Lett. 38, 1536 (1977); P. Vashishta and A. Rahman, in *Fast Ion Transport in Solids*, edited by P. Vashishta *et al.* (North-Holland, New York, 1979), p. 527; Y. Hiwatari and A Ueda, Solid State Ionics 3/4, 111, 115 (1981).

[6]R. J. Cava *et al.*, Solid State Commun. 24, 411 (1977); Y. Tsuchiya *et al.*, J. Phys. C 12, 5361 (1979); W. Andreoni and J. C. Phillips, Phys. Rev. B 23, 6456 (1981); I. Yokota, J. Phys. Soc. Jpn. 21, 420 (1966); H. Hyashi, to be published.

[7]A. Rahman and P. Vashishta, in *Physics of Superionic Conductors*, edited by S. W. de Leeuw and J. W. Perram (Plenum, New York, to be published).

[8]H. C. Anderson, J. Chem. Phys. 72, 2384 (1980).

[9]M. Parrinello and A. Rahman, Phys. Rev. Lett. 45, 1196 (1980).

[10]T. Hayes and J. B. Boyce, J. Phys. C Solid State Phys. 13, L731 (1980).

Structural and dynamic properties of lithium sulphate in its solid electrolyte form

Roger W. Impey and Michael L. Klein
Chemistry Division, National Research Council of Canada, Ottawa K1A 0R6, Canada
Ian R. McDonald
Department of Physical Chemistry, Cambridge University, Cambridge CB2 1EP, United Kingdom

(Received 19 December 1984; accepted 6 February 1985)

The properties of solid lithium sulphate have been studied by computer simulation. At sufficiently high temperatures, the simulated crystal behaves as a solid electrolyte with lithium ion (jump) diffusion and sulphate group rotation. The atomic radial distribution functions in the rotator phase are discussed in relation to the low temperature, fully ordered, monoclinic structure and the nature of the orientational disorder of the sulphate groups is characterized in terms of tetrahedral rotor functions. The crystal structure factor is found to be sensitive to the model adopted for the charge distribution of the anions; good agreement with experimental neutron diffraction data is obtained when a charge distribution consistent with *ab initio* quantum mechanical calculations is used. The phase transition whereby the low temperature monoclinic structure transforms to the disordered cubic phase has been investigated by the constant pressure molecular dynamics method. The nature of the lithium ion diffusive motion and its coupling to the anion reorientation, the relaxation of the orientational order and the lattice vibrations are all briefly discussed.

I. INTRODUCTION

At room temperature, solid lithium sulphate (Li_2SO_4) has a fully ordered monoclinic structure, but between 848 K and its melting point at 1133 K the solid exhibits extreme disorder of two types: orientational disorder of the molecular anions and a high mobility, diffusive motion of the lithium cations.[1-5] As in other solid rotator phases, the presence of orientational disorder restricts the amount of structural information that can be obtained from diffraction data. However, an analysis of neutron powder diffraction data collected at 908 K has revealed that the $SO_4^=$ ions lie on a face centered cubic (fcc) lattice with lattice parameter $a = 7.07$ Å, and that the Li^+ ions preferentially occupy $\pm(\frac{1}{4}\frac{1}{4}\frac{1}{4})$ positions in the unit cell.[2] The best fit to the intensities of the Bragg reflections was obtained for a model in which the distribution of anion orientations was assumed to be isotropic; in this model, scattering from the oxygen atoms was uniformly distributed on a shell of radius 1.49 Å from the sulphur atoms, whose root mean square displacement was $\langle u_-^2 \rangle^{1/2} = 0.81$ Å. The large value of the latter quantity, which corresponds to 16% of the mean S–S separation, is characteristic of a solid near its melting point. The analogous quantity for the Li^+ ions was even larger ($\langle u_+^2 \rangle^{1/2} = 1.1$ Å), presumably because of their high mobility. No evidence was found for significant occupation of the $\pm(\frac{1}{2}\frac{1}{2}\frac{1}{2})$ site in the unit cell, so that on average the crystal has an antifluorite-like structure. This model of the structure was consistent with x-ray data collected at 883 K, although the low x-ray scattering power of Li^+ means that the x-ray results, unlike those of neutron scattering, are insensitive to the positions of the Li^+ ions.

High temperature Li_2SO_4 is an example of a rotator phase solid that also behaves as a solid electrolyte. The observation that high cation mobility is maintained even for such large species as K^+ and Rb^+ has lead to speculation that the cation diffusion is coupled to the anion rotation.[2,5] The possible existence of a new diffusion mechanism in the antifluorite structure makes it worthwhile to compare the behavior of Li_2SO_4 with that of CaF_2: in the latter case, molecular dynamics (MD) studies have given important insights into the diffusion mechanism.[6-9] A study of Li_2SO_4 also follows naturally from our interest in the consequences of orientational disorder in ionic solids.[10-14] Apart from the structural and ionic mobility studies mentioned above, the results of Brillouin scattering have been used to estimate the elastic constants,[4] while Raman scattering experiments have provided information on the crystal dynamics of both pure and mixed crystals.[15]

The aims of the present investigation are as follows. First, in Sec. II, we describe the interionic potential models that have been used in the simulations.[16-18] On the basis of our earlier investigations we may expect that the model used for the charge distribution of the $SO_4^=$ ion will play an important role in determining the orientational order.[14] In computer simulation studies the latter quantity can be characterized in greater detail than is provided by experiments. Such a study will be described in Sec. V. However, before that, in Sec. III, we give some details of the MD calculations. In Sec. IV the MD results are analyzed to yield both the Bragg and diffuse scattering. We also discuss the real space atomic distribution functions and relate these to the low temperature (β-phase) monoclinic crystal structure and the mechanism of the monoclinic β- to cubic α-phase transition is investigated using constant pressure molecular dynamics.[19-22] The lattice vibrations are briefly discussed in Sec. VI and the question of the mechanism of Li^+ ion diffusion is taken up in Sec. VII. The article ends with a discussion and an indication of possible future avenues of research.

II. THE POTENTIAL MODEL

In all our calculations the lithium sulphate crystal consists of rigid Li^+ and $SO_4^=$ ions interacting via electrostatic

 0021-9606/85/104690-09$02.10

and atom–atom potentials. As in our earlier work,[13] the latter have been constructed from interionic potentials for the alkali ions[16] combined with potentials derived for molecular crystals.[17,18] The exp-6 form was used for the atom–atom contributions, i.e.,

$$V_{\alpha\beta}(R) = A_{\alpha\beta}\exp(-a_{\alpha\beta}R) - B_{\alpha\beta}/R^6, \quad (1)$$

where α,β = Li, S, O. Initially, the cross interactions were estimated from traditional combining rules but then the computed pressure is too high. We therefore softened the Li–O potential in an attempt to remedy this problem. The parameters used are listed in Table I.

While the potential energy of the solid is overwhelmingly dominated by the electrostatic interactions, it is the distribution of charge within the sulphate group which largely determines the orientational correlations in the crystal.[14] *Ab initio* calculations on the sulphate group[23,24] favor $Q_0 = -0.8e$ (model I). In addition, we have examined a model with $Q_0 = -0.5e$ (model II). Only the calculations based on model I employed the softer Li–O potential.

III. MOLECULAR DYNAMICS

We have used the models outlined in the previous section to carry out conventional (constant volume) and "new" (constant pressure) MD calculations on systems of either 96 or 324 ions. In the case of constant volume, the calculations were started by arranging $N = 32$ or 108 sulphate groups on a fcc lattice with parameter $a = 7.07$ Å.[2] The $2N$ lithium ions, all of atomic mass 7, were initially arranged on an interpenetrating simple cubic (sc) lattice with lattice parameter $a/2$. Periodic boundary conditions were used to simulate an infinite lattice and the electrostatic interactions were handled by the Ewald method. The equations of motion were integrated using standard techniques with a time step of either 2 or 3 fs[10–14]; the routine SHAKE[25] was used to maintain the rigidity of the $SO_4^=$ groups. A selection of results is given in Table II. We note that in general it proved quite difficult to achieve an equilibrium situation in which, simultaneously the $SO_4^=$ groups were rotating and the lithium ions were diffusing. The final structure invariably arose in a two stage process in which, first, the $SO_4^=$ ion disordered and then, eventually, the Li^+ ions began to diffuse. A pretransitional behavior involving the rotation of the $SO_4^=$ groups may well occur in real Li_2SO_4.[15] The MD program with $N = 108$ was rather slow in execution and this fact, combined with the length of time required for equilibration, is the reason for presenting results for the smaller system ($N = 32$). From Table II it appears that there is little difference between results for the $N = 32$ and 108 systems. Typically, the phase space trajectories were followed for 15 ps and stored on tape for subsequent analysis. While the softer Li–O potential used in the model with $Q_0 = -0.8e$ results in an improved pressure, there is clearly room for further refinement of the potential. In studying the α–β phase transition we used the constant pressure MD program previously developed to study ionic salts and polar fluids.[22] In this case the calculations were initiated in the ordered monoclinic β phase, the angular equations of motion being solved in quaternion form.[26]

TABLE I. Potential parameters for lithium sulphate $V_{\alpha\beta}(R) = A_{\alpha\beta}\exp(-a_{\alpha\beta}R) - B_{\alpha\beta}/R^6$.

$\alpha\beta$	$A_{\alpha\beta}$/kJ mol^{-1}	$a_{\alpha\beta}$/Å^{-1}	$B_{\alpha\beta}$/kJ Å^6 mol^{-1}
S–S[a]	3 210 000	3.80	9500
O–O[b]	331 850	4.07	1156
Li–Li[c]	1 812	2.33	4.4
Li–O[d]	24 522	3.20	71.3
Li–O[e]	22 116	3.34	71.3
Li–S[d]	76 266	3.07	204.5
S–O[d]	1 032 100	3.94	3314

[a] Reference 18.
[b] Reference 17.
[c] Reference 16.
[d] Traditional combining rules.
[e] Modified parameters used with $Q_0 = -0.8e$.

TABLE II. Constant volume molecular dynamics results for the cubic α phase of lithium sulphate at $a = 7.07$ Å.

N	Q_0 (e)	T (K)	U (kJ/mol)	P (kbar)	D (10^{-5} cm^2/s)
108[a]	−0.5	1153	−1954	71	3.1
32[a]	−0.5	1164	−1960	67	3.3
32[a]	−0.8	1180	−1943	74	4.6
32[a]	−0.8	861	−1964	74	0.9
32[b]	−0.8	991	−2055	10	1.5
32[b]	−0.8	1078	−2046	10	3.0
108[b]	−0.8	946	−2059	9	1.5

[a] Traditional combining rules used for Li–O potential parameters.
[b] Revised Li–O potential parameters (see Table I).

IV. THE STRUCTURE

A. α phase

The simplest way to study the structure of the simulated crystal is to examine the atom–atom radial distribution functions. Figures 1 and 2 compare the results obtained for models with $Q_0 = -0.8e$ and $-0.5e$. It is clear that the relatively modest change in the charge distribution of the sulphate ion has led to a significant rearrangement of the atoms. Other calculations[14] confirm that differences in the Li–O repulsions (recall Table I) are not primarily responsible for this effect.

The correlations between $SO_4^=$ groups is described by the S–S, S–O, and O–O distribution functions of Fig. 1. The position of the main peak in the S–S distribution function confirms that, on average, the S atoms form a face centered cubic lattice. The absence of sharp peaks in the S–O and O–O functions, whose general form closely resembles the corresponding curves for the cubic rotator phases of CH_4 and CCl_4,[27,28] confirms that the $SO_4^=$ groups are orientationally disordered. The arrows denote peak positions and coordination numbers corresponding to the low temperature β phase. In the case of the S–S distribution, the 12 equivalent nearest neighbors split into five subgroups in the β phase. The origin of these splittings will become somewhat clearer when we discuss the α–β phase transition. The main point to note here is that the broad peak in the α-phase S–S distribution encompasses the β-phase peaks. The S–O and O–O distributions

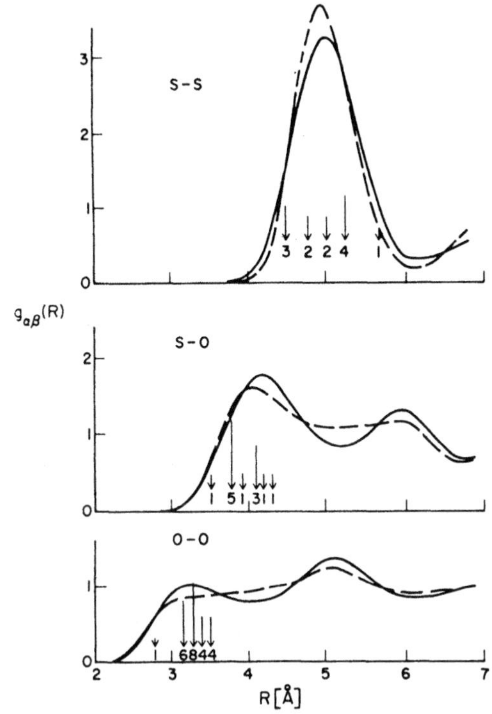

FIG. 1. Atom–atom radial distribution functions for the $SO_4^=$ ions in cubic (α-phase) lithium sulphate. The bold and dashed curves refer to models with $Q_0 = -0.8e$ ($T = 991$ K) and $-0.5e$ ($T = 1164$ K), respectively. The arrows indicate peak positions and coordination numbers in the low temperature, monoclinic β phase.

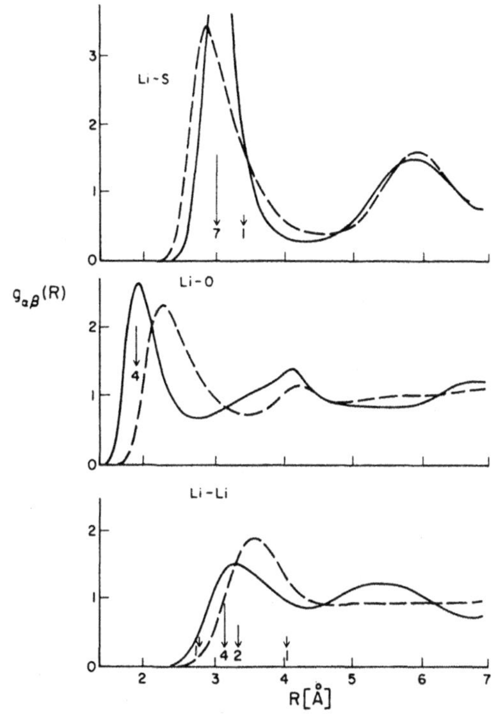

FIG. 2. Atom–atom radial distribution functions for cubic (α-phase) lithium sulphate. The bold and dashed curves refer to models with $Q_0 = -0.8e$ and $-0.5e$, respectively. The arrows indicate peak positions and coordination numbers in the low temperature, monoclinic β phase.

are complicated in the β phase, but the peaks again lie under the curves obtained for the α phase.

The first peak in the Li–Li distribution function in Fig. 2 occurs close to $R = d = a/2$, so that the lithium ions appear to form a simple cubic lattice. In the case of the model with $Q_0 = -0.8e$ the appearance of the second peak between $2^{1/2}d$ and $3^{1/2}d$, gives support to this interpretation. Moreover, in this case, unlike the model with $Q_0 = -0.5e$, the distribution nicely overlaps the β-phase peaks indicated by the arrows. The strong first peak in the Li–S function at $R \sim \sqrt{3}a/2$ confirms that the Li ions are preferentially located near the $(\frac{1}{4}\frac{1}{4}\frac{1}{4})$ sites of the crystal unit cell. Again, the results for model with $Q_0 = -0.8e$ accord better with the experimental β-phase structure. The well defined first peaks in the Li–O and Li–S functions suggest a strong correlation between the Li^+ ions and a specific site on the $SO_4^=$ tetrahedra.

B. Structure factor of the α phase

Experimental information on the structure of Li_2SO_4 has been obtained from the analysis of neutron powder diffraction measurements.[2] The experimental results discussed earlier are shown inset in Fig. 3. It is apparent that there is a large contribution from diffuse scattering and the absence of high order Bragg reflections indicates the presence of large center of mass displacements. The experimental structure factor is given by

$$S(\mathbf{k}) = \langle |\rho(\mathbf{k},t)|^2 \rangle, \tag{2}$$

$$\rho(\mathbf{k},t) = \sum_i b_i \exp[-\mathbf{k}\cdot\mathbf{r}_i(t)], \tag{3}$$

where $\hbar\mathbf{k}$ is the momentum transferred to the crystal in the scattering event and b_i is the scattering length of nucleus i whose position at time t is given by $\mathbf{r}_i(t)$. The scattering lengths of ^{7}Li, S, and O are, respectively, -2.33, 2.85, and 5.75 fermi. Bragg scattering is determined by the structure amplitude $\langle \rho(\mathbf{k},t) \rangle$ and hence, by difference, the diffuse scattering is related to the mean square fluctuation in $\rho(\mathbf{k},t)$.

We have used the MD data to evaluate $S(\mathbf{k})$ for the two models discussed earlier. The results for specific Bragg vectors labeled by the set of integers (lmn), where $\mathbf{k} = 2\pi(l,m,n)/a$, are shown in Fig. 3; it should be recalled that for Bragg vectors, $k = 4\pi \sin\theta/\lambda$, where λ is the neutron wavelength and 2θ is the scattering angle. The diffraction patterns predicted by the two models are seen to be radically different. The amount of diffuse scattering is very small for the model with $Q_0 = -0.5e$ even though the temperature is much higher than in the experiment and the relative intensities of the (111) and (200) Bragg peaks are incorrect. By contrast, the model with $Q_0 = -0.8e$ yields a diffraction pattern in good agreement with experiment. The results shown

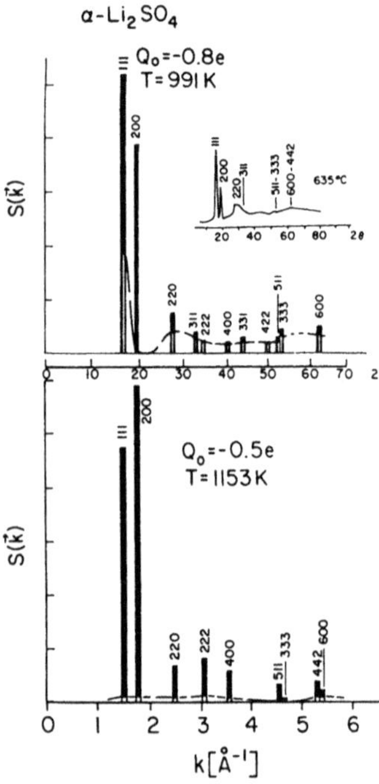

FIG. 3. The neutron diffraction structure factor $S(\mathbf{k})$ for α-Li_2SO_4 calculated for models with $Q_0 = -0.8e$ and $-0.5e$, respectively. At each wave vector $\mathbf{k} = 2\pi(l,m,n)/a$, labeled by the integers (lmn), the Bragg and diffuse contributions are indicated by the filled and open regions. The dashed line gives a rough sketch of the diffuse background. Inset is the experimental neutron diffraction pattern from Ref. 2.

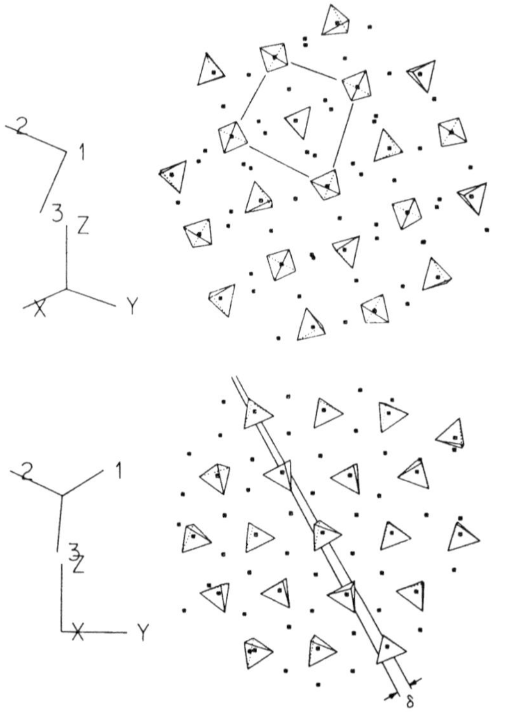

FIG. 4. An instantaneous configuration taken from the constant pressure MD calculation on the monoclinic β phase at $T = 300$ K; the crystal is viewed along the "cubic" ⟨001⟩ and ⟨111⟩ directions and the pseudo fcc cell of the β phase is indicated by the solid lines. At the phase transition, the displacement parameter δ goes to zero. The D_{2d} type order of the $SO_4^{=}$ tetrahedra should be noted.

in Fig. 3 were obtained with the revised Li–O potential of Table I but differ little from those previously published, which were calculated from potentials based upon traditional combining rules.[14] The new results presented here confirm that it is the charge distribution of the molecular anion which plays the dominant role in determining the structure.[13]

C. $\beta \rightarrow \alpha$ phase transition

Since the model with $Q_0 = -0.8e$ gave a good description of the structure of the disordered α phase we decided to attempt a simulation of the $\beta \rightarrow \alpha$ phase transition by means of the constant pressure MD technique.[19–22] The simulation was started from the β-phase room temperature structure, this is monoclinic, $P2_1/a$, with $a = 8.244$ Å, $b = 4.951$ Å, $c = 8.471$ Å, $\beta = 107.96°$.[2] The transition from the β to α phase is, at first sight, a complicated one. However, a simple transformation of the monoclinic unit cell vectors **a**, **b**, **c** can generate a pseudo face centered cubic structure with cell vectors $\mathbf{u}_1$, $\mathbf{u}_2$, $\mathbf{u}_3$.[21] At room temperature, the transformed cell has edges of length $u_1 = u_2 = 7.0$ Å and $u_3 = 6.7$ Å and angles $\theta_{12} = 92°$, $\theta_{13} = 89°$, $\theta_{23} = 89°$, while the α phase is characterized by $u_1 = u_2 = u_3$ and $\theta_{12} = \theta_{13} = \theta_{23} = 90°$. Thus the transition is not profound in character, involving as it does only modest changes in the cell lengths and angles. Figure 4 shows a view down the $\mathbf{u}_1$ axis in the room temperature structure in which the pseudo face centered cubic arrangement of the $SO_4^{=}$ tetrahedra is clearly visible.

The calculations were made for the model I with $N = 32$ and the modified Li–O potential parameters of Table I. In order to stabilize the MD cell with approximately the correct room temperature zero pressure lattice constants, it was necessary to impose an external pressure $P_{ex} \sim 7$ kbar. For an ionic crystal, this is a rather modest pressure, confirming that the potential model used is a reasonable one. Isobaric heating of the β-phase crystal caused the MD cell vectors to increase in length, following quite closely the experimentally measured expansion.[2] Below 900 K, we found that $u_1 \neq u_2 \simeq u_3$, with $\theta_{12} \sim 92°$, $\theta_{13} \simeq \theta_{23} \simeq 88°$ but at higher temperatures all u's and θ's become equal, indicating the formation of a cubic solid. Figure 5 shows phase space trajectories for Li and S atoms in both the α and β phases, plotted in the scaled reference frame of the constant pressure MD calculations.[21] Several points are apparent from this figure. In the β phase, the trajectories reveal the presence nonequivalent Li and S sites which coalesce in the cubic α phase to yield broad distributions. The pseudo cubic lattice of Li^+ ions is clearly visible at high temperatures, as is their diffusion between sites (we defer until later a discussion of the Li ion dynamics). The S atoms form a well defined face centered

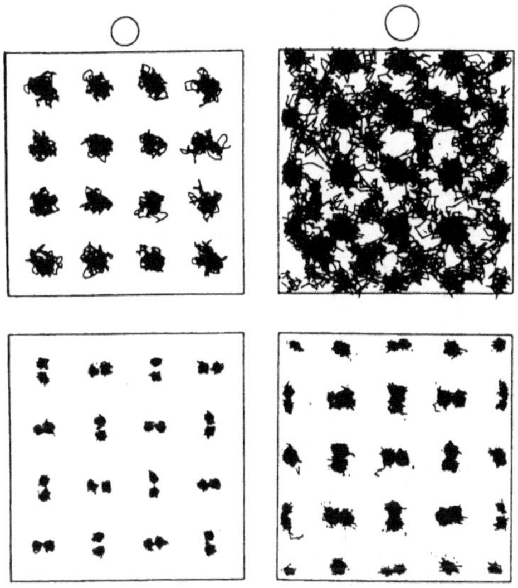

FIG. 5. Phase space trajectories for Li and S atoms in the scaled reference frames of the constant pressure MD calculations at $T = 814$ K (β phase) and $T = 1082$ K (α phase). For the β phase, the system is viewed down the pseudo ⟨001⟩ direction as discussed in the text. The circles represent the experimental root mean square amplitudes of the ions (Ref. 2).

cubic lattice, with very large center of mass motion. Overall the lattice appears to be simple cubic, because of the superposition of different layers of atoms; in reality, alternate sites are empty, as they should be in the antifluorite structure. The experimental amplitudes $\langle u_{\pm}^2 \rangle$ for the α phase are shown drawn to scale in the figure; there is quantitative agreement with the results of the simulation.

V. ORIENTATIONAL ORDER IN THE α PHASE

A. Statics

From the analysis of the experimental diffraction data it was concluded that the sulphate groups are spherically disordered, having no preferential orientation with respect to the crystal axes.[2] We can analyze the nature of the orientational order present in the simulations by exploiting the fact that the crystal is on average cubic. A natural method is one based on the behavior of the tetrahedral rotor functions M_μ ($\mu = 1,7$) associated with individual $SO_4^=$ groups[14]; these functions are analogous to the f orbitals of quantum chemistry. Under the influence of the cubic crystal field, the seven functions split into three groups corresponding, respectively, to $\mu = (1)$, (2,3,4), and (5,6,7):

$$M_1 = \frac{3}{4}\sqrt{3}\sum_{i=1}^{4} x_i y_i z_i, \qquad (4)$$

$$M_2 = \frac{3}{40}\sqrt{5}\sum_{i=1}^{4} (5x_i^3 - 3x_i r_i^2), \qquad (5)$$

$$M_5 = \frac{3}{8}\sqrt{3}\sum_{i=1}^{4} (y_i^2 - z_i^2)x_i. \qquad (6)$$

In these expressions, the summations are taken over the four arms of the $SO_4^=$ tetrahedra whose orientations are specified by the unit vectors $\mathbf{r}_i = (x_i, y_i, z_i)$. The $\mu = 1$ term monitors the presence of T_d orientations with respect to the cubic axes of the crystal; the group with $\mu = 2$, 3, or 4 monitors C_{3v} or C_{2v} configurations; and the group with $\mu = 5$, 6, or 7 monitors D_{2d} configurations. Figure 1 of Ref. 14 shows the typical orientations mentioned above and the same reference discusses their significance in more detail.

Since $\langle M_\mu \rangle = 0$ in an orientationally disordered crystal, the fluctuations $\langle M_\mu^2 \rangle$ serve as static order parameters. Because of the normalization condition $\Sigma_\mu \langle M_\mu^2 \rangle = 1$, a crystal with spherically disordered $SO_4^=$ groups would have $\langle M_\mu^2 \rangle = 0.143...$ for all μ. Departures from this mean value are a measure of the extent to which the sulphate ions adopt preferred orientations. Table III shows the computed order parameters for the two models. It can be seen that model I ($Q_0 = -0.8e$) shows a more isotropic distribution in better agreement with inferences based on experimental data.[2] In contrast, model II ($Q = -0.5e$) displays a definite preference for C_{3v} or C_{2v} order. The results of Sec. IV C (see Fig. 4) show that in the low temperature monoclinic solid (β phase), the local environment of a given $SO_4^=$ ion corresponds closely to D_{2d}-type configurations. The improved description of the α phase provided by the model with $Q_0 = -0.8e$ comes about precisely because of an increase in the amount of D_{2d} order.

B. Dynamics

The relaxation of the static order parameters is described by the time autocorrelation functions (acf's) $\langle M_\mu(t) M_\mu(0)\rangle$ plotted in Fig. 6. The main difference between the two models is that in model I, with $Q = -0.8e$, the D_{2d}-type order persists for much longer times, with a characteristic relaxation time in the picosecond range. This analysis is restricted to the dynamics of individual $SO_4^=$ ions. It is very likely however, that the reorientation of the $SO_4^=$ ions involves some form of cooperative effect. We have made no attempt to look for such behavior, which presumably would be monitored via the functions $\langle M_\mu^i(t) M_\mu^j(0)\rangle$, where i and j label different $SO_4^=$ groups. Our analysis of the collective dynamics has been restricted to a study of the van Hove function $S(\mathbf{k},\omega)$, to which we return in the next section.

In principle, reorientation of the $SO_4^=$ ions will influence the spectroscopic band shape associated with the intramolecular vibrations. The acf's which monitor the reorientation of a sulphate ion S_4 axis or S–O bond vector behave much like the C_{3v} or D_{2d} functions shown in Fig. 6, the relaxation time being again in the picosecond range. These results yield estimates for the Raman intramolecular vibrational bandwidths which vary from 15 to 30 cm^{-1} over the

TABLE III. Orientational order parameters $\langle M_\mu^2 \rangle$ for the α phase of lithium sulphate.

Q_0	$\langle M_1^2 \rangle$	$\langle M_2^2 \rangle$	$\langle M_5^2 \rangle$	T/K
$-0.5e$	0.06	0.20	0.11	1164
$-0.8e$	0.07	0.15	0.16	861
$-0.8e$	0.09	0.14	0.17	1085

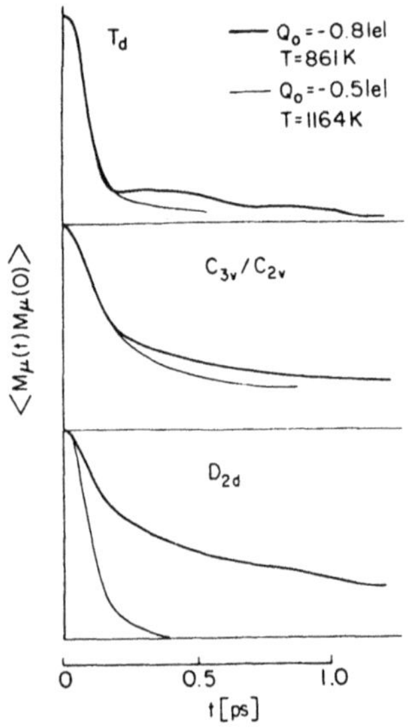

FIG. 6. Relaxation of the orientational order parameters for the models with $Q_0 = -0.8e$ (bold lines) and $-0.5e$ (thin lines), respectively. Note the slower decay of D_{2d} order for the model with $Q_0 = -0.8e$.

temperature range 860 to 1180 K. The corresponding experimental numbers are larger than this,[15] suggesting that rotational relaxation may not be the only mechanism contributing to the observed linewidths.

VI. LATTICE VIBRATIONS IN THE α PHASE

There are two possible ways to investigate the lattice dynamics of a crystal via computer simulation. The first consists in examining the velocity acf, defined as

$$Z_i(t) = \langle \mathbf{V}_i(t)\cdot\mathbf{V}_i(0)\rangle / \langle \mathbf{V}_i(0)\cdot\mathbf{V}_i(0)\rangle \qquad (7)$$

whose Fourier transform $Z_i(\omega)$, is the phonon density of states of species i. Figures 7 and 8 show the calculated $Z_i(t)$ for the Li^+ and $SO_4^=$ for the model with $Q_0 = -0.8e$. The associated power spectra, inset in the figures, reveal that the Li^+ ions participate in vibrations that range from 200 to 500

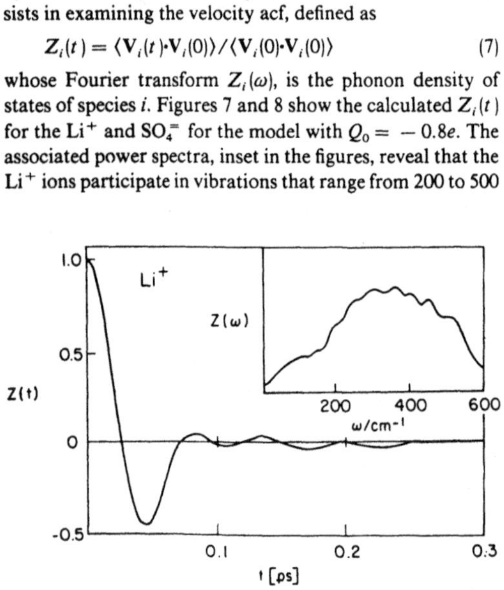

FIG. 7. Velocity autocorrelation function and associated power spectrum for Li^+ ions in α-Li_2SO_4 based on model I. The low frequency part of $Z(\omega)$ is uncertain due to errors in the Fourier transform.

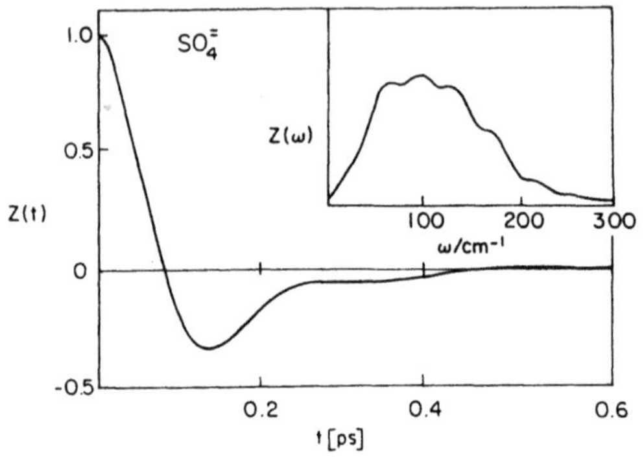

FIG. 8. Velocity autocorrelation function and associated power spectrum for $SO_4^=$ ions in α-Li_2SO_4 based on model I. The low frequency part of $Z(\omega)$ is uncertain due to errors in the Fourier transform.

cm^{-1} while the translational motion of the $SO_4^=$ groups is centered at aroung 100 cm^{-1}. The results on the Li^+ ion dynamics accord well with Raman data on the β phase, on the basis of which several Li modes have been assigned in the range 300 to 450 cm^{-1}.[15] The Li^+ results also confirm that the cations are diffusing, since the time integral of the velocity acf yields the diffusion coefficient. In the case of Li^+, this is significantly different from zero. The nature of the diffusive motion will be taken up in more detail in the next section.

The second way to study the crystal dynamics is the via the calculation of the van Hove function or dynamical structure factor $S(\mathbf{k},\omega)$ appropriate to the coherent inelastic scattering of neutrons. In particular, we have computed the so called intermediate scattering function $F(\mathbf{k},t)$ whose Fourier transform is $S(\mathbf{k},\omega)$:

$$F(\mathbf{k},t) = \langle \rho(\mathbf{k},t)\rho(-\mathbf{k},0)\rangle . \qquad (8)$$

Some typical examples for model II ($Q_0 = -0.5e$) are shown in Fig. 9 for wave vectors yielding phonons which propagate along the crystal $\langle 001\rangle$ direction; corresponding longitudinal and transverse modes are plotted side by side. Three types of correlation functions have been calculated. Those labeled N use the correct neutron weights and sum over all atoms when evaluating $\rho(\mathbf{k},t)$, those labeled CM set $b_{Li} = b_S = 1$ and $b_O = 0$; and those labeled OP employ $b_{Li} = +1, b_S = -1$, and $b_O = 0$. We have confirmed that model I (with $Q_0 = -0.8e$) gives essentially the same behavior, from which we conclude that Fig. 9 is representative of real Li_2SO_4. The arrows on the upper curves show the vibrational periods estimated from Brillouin scattering measurements.[4] In both cases the calculated periods are too short, indicating that the model crystal is too stiff. Thus, while the measured elastic constants in the α phase[4] are $C_{44} = 100$ kbar and $C_{11} = 240$ kbar, those derived from Fig. 9 are $C_{44} = 140$ kbar and $C_{11} = 440$ kbar. Similar calculations for model I yield $C_{44} = 115$ kbar and $C_{11} = 365$ kbar. While the latter results agree somewhat better with the experimental data, the high value of obtained for C_{11} is probably an intrinsic failing of the rigid ion model.

Figure 9 shows that the longitudinal acoustic (LA)

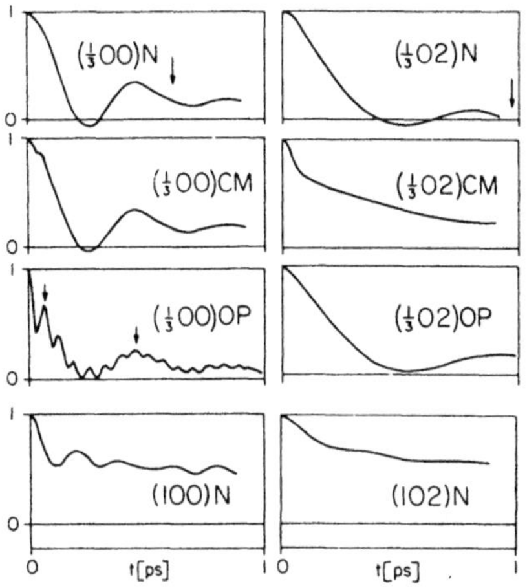

FIG. 9. Intermediate scattering functions $F(\mathbf{k},t)$ for various $\mathbf{k} = 2\pi(x,y,z)/a$. The curves are labeled by (xyz) and either N (neutron), CM (center of mass), or OP (optic). The left-hand column relates to longitudinal modes, the right-hand column to transverse modes. All results are for model II. The arrows on the upper most curves mark the vibrational periods predicted from Brillouin data (Ref. 4), while those on the lower OP curve indicate the presence of a high frequency LO and low frequency LA mode.

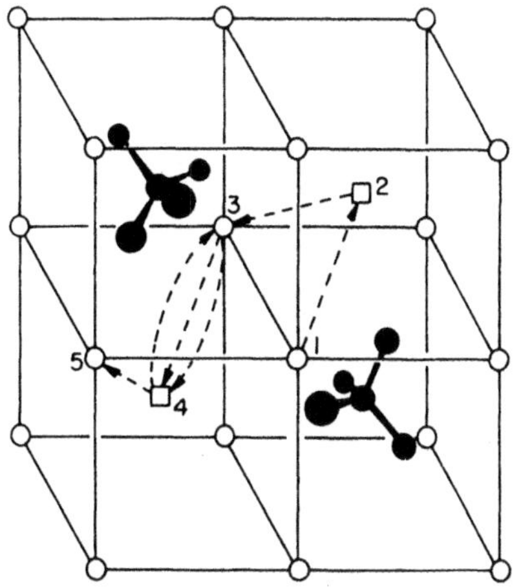

FIG. 10. Schematic diffusion pathway for Li^+ ions in an idealized cubic (α-phase) lithium sulphate crystal. The open circles are Li^+ ions, $SO_4^=$ tetrahedra are drawn explicitly, and the squares denote interstitial sites of the Li^+ lattice or, equivalently octahedral sites of the $SO_4^=$ lattice.

modes, as exemplified by the curves labeled ($\frac{1}{3}$00)N and (100)N, are well defined, whereas the transverse acoustic (TA) modes, ($\frac{1}{3}$02)N and (102)N, are heavily damped. There is clear evidence for the presence of central peaks, no doubt related to the extreme disorder in the crystal. The optic mode correlation function ($\frac{1}{3}$00)OP displays not only the high frequency LO mode but also the lower frequency LA mode.

The richness of the dynamical structure factor data suggest that inelastic neutron scattering could be usefully employed to unravel the crystal dynamics of Li_2SO_4. Unfortunately, it appears that no such data exist at present.

VII. LITHIUM ION DIFFUSION

A. Diffusion mechanism

Diffusion of the Li^+ ions is apparent in the behavior both of the phase space trajectories (Fig. 5) and the velocity acf (Fig. 7). In this section we discuss the Li^+ ion motion in more detail. We begin with some qualitative remarks. The idealized antifluorite structure of α-Li_2SO_4 can be viewed in a number of equivalent ways. In Sec. III we described the structure as a face centered cubic arrangement of $SO_4^=$ groups with lattice constant a and a simple cubic lattice of Li^+ ions with lattice constant $d = a/2$. Equivalently, we can regard α-Li_2SO_4 as being composed of two interpenetrating sc lattices but with only one half of $SO_4^=$ sublattice sites being occupied. These vacant sites play a key role in theories of diffusion in CaF_2[6–9] and are indicated by squares in Fig. 10. We note in the idealized structure shown in this figure, each $SO_4^=$ ion is surrounded by a sc cage of Li^+ ions. The Li^+ ions can also be regarded as occupying the tetrahedral interstitial sites of the fcc lattice of $SO_4^=$ ions, while the squares in Fig. 10 are octahedral sites of the fcc lattice.

As mentioned above, Fig. 10 is an idealized view of the structure. In Fig. 5 we showed a superposition of the phase space trajectories of Li^+ and $SO_4^=$ ions viewed down the ⟨001⟩ direction for the model with $Q_0 = -0.8e$. Since Fig. 5 contains a superposition of several crystal planes, the fcc structure of the $SO_4^=$ ions appears to be sc. When drawn on a graphics terminal, Fig. 5 provides a real time view of the Li^+ ion dynamics. Although the trajectories are complicated, it is clear that the Li^+ ions move via jump diffusion[7]; i.e., they are localized for considerable periods of time and then move rapidly to another localized state. The individual trajectories often appear to follow a curved pathway. An example is drawn schematically in Fig. 10, where the transition between sites 1 and 5 proceeds via the octahedral sites 2 and 4. As illustrated, the passage from the tetrahedral site 3 to the octahedral site 4 required two attempts. It is clear from Fig. 5 that the octahedral sites play a vital role in the diffusion process. In fact as observed in the study of $SrCl_2$ by Dixon and Gillan,[8] the diffusing ions actually skirt the octahedral site but nevertheless pass sufficiently close for the terminology to be useful. No quantitative analysis was performed, but from the study of individual diffusion events the residence time in an octahedral site such as 2 in Fig. 10 was estimated to be somewhat longer than the typical 0.2 ps jump time for $2 \rightarrow 3$. Roughly speaking, each Li^+ ion makes a jump between tetrahedral sites every 5 to 10 ps.

An attempt was made to establish whether or not the jump diffusion was coupled to reorientation of neighboring $SO_4^=$ groups as has frequently been argued.[5] For example, do the S–O arms of the tetrahedra "knock-on" the Li^+ ions into octahedral sites or vice versa? Unfortunately, in a given simulation there are relatively few diffusion events, and only one unequivocal example of a knock-on was observed. The difficulty inherent in attempting such an analysis can perhaps be appreciated by examination of Fig. 11, which shows a Li^+ ion in an octahedral site. Only the four coplanar neighboring $SO_4^=$ groups (in the [100] plane) are shown in this

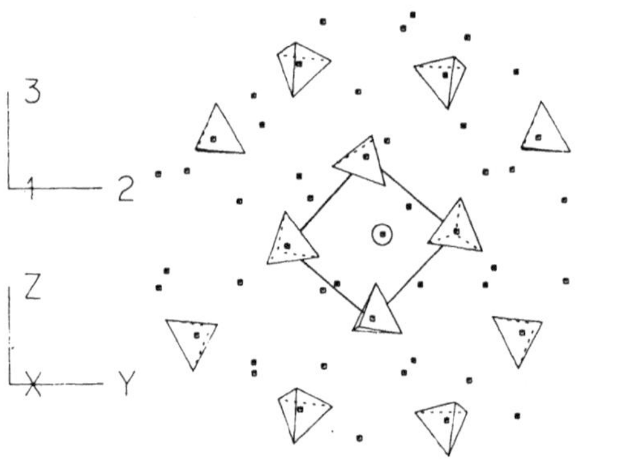

FIG. 11. An instantaneous configuration from the simulation of α-Li_2SO_4 showing a Li^+ ion (circled) in an octahedral site.

view; to characterize the configuration fully we also need to examine the [010] and [001] projections. A definitive answer as to whether or not the postulated coupled motion[5] plays an important role must be defered until we devise more automated search procedures, possibly akin to those used in elementary particle physics! It is also possible that a movie would be helpful. We are currently also exploring the diffusion of K^+ and Na^+ impurities[3] in Li_2SO_4 in an attempt to characterize the mechanism in more detail.

B. Diffusion coefficient

The lithium ion diffusion coefficient has been evaluated both from the velocity acf and the time derivative of the ionic displacements. The resulting values, derived from various MD runs, are collected in Fig. 12 where they are compared with the corresponding experimental results.[1] The calculated diffusion coefficients are seen to lie systematically below the experimental data, independent of the model. Other than softening the Li–Li repulsion, a possibility we have yet to explore, it is difficult to envisage what changes are required in the interionic potentials in order that theory and experiment be brought into better accord.

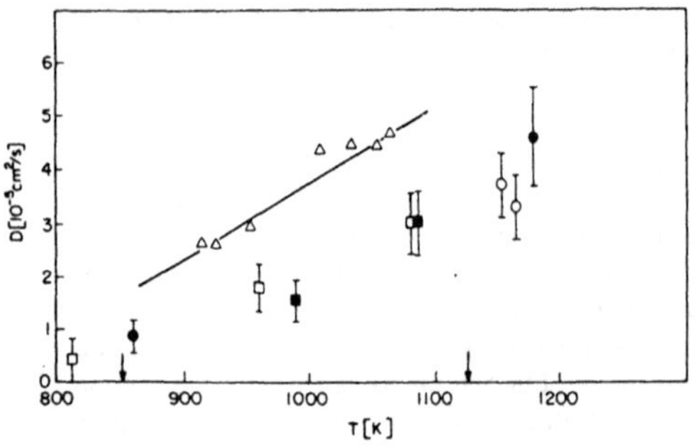

FIG. 12. Self-diffusion coefficients for Li^+ ions in α-Li_2SO_4. The experimental data for Ref. 1 are indicated by triangles. The full and open circles are the MD results for the models with $Q_0 = -0.8e$ and $-0.5e$, respectively, and Li–O potentials calculated from traditional combining rules. The squares are MD results for the model with $Q_0 = -0.8e$ and the modified Li–O parameters (see Table I); open squares are derived from constant pressure calculations and full squares from constant volume calculations.

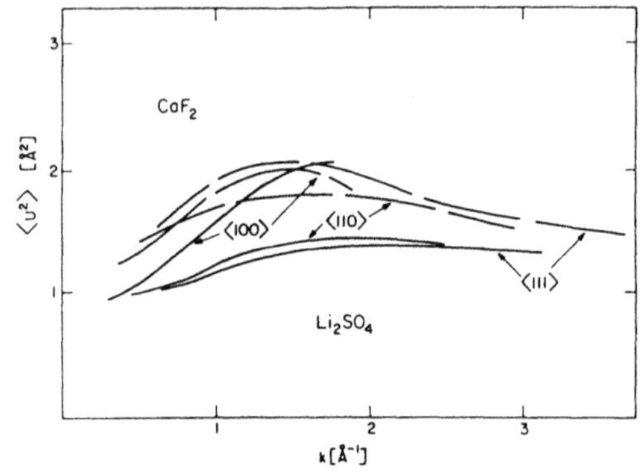

FIG. 13. Generalized mean squared vibrational amplitudes for Li^+ ions in α-Li_2SO_4, as derived from analysis of $F_s(\mathbf{k},t)$, and a comparison with analogous results for F^- ions in CaF_2 taken from Ref. 7 (see the text). The solid lines refer to Li_2SO_4 and the dashes to CaF_2.

C. Fourier analysis of the diffusive motion

A more quantitative approach to the Li^+ ion diffusion can be based on the work of Jacucci and Rahman[7] on CaF_2. These authors consider the behavior of the self-intermediate scattering function:

$$F_s(\mathbf{k},t) = \langle \exp[i\mathbf{k}\cdot\mathbf{u}_+(t)]\rangle. \tag{9}$$

where $\mathbf{u}_+(t) = \mathbf{r}_+(t) - \mathbf{r}_+(0)$ is the displacement (in our case) of a Li^+ ion, and $\mathbf{k}$ is the wave vector of interest. The long time asymptotic form of $-\ln F_s$ is

$$-\ln F_s(\mathbf{k},t) \to g(\mathbf{k})t + c(\mathbf{k}). \tag{10}$$

The Bragg vectors of the sc Li^+ lattice, with lattice constant $d = a/2 = 3.535$ Å, are given by $\mathbf{k} = 2\pi(l,m,n)/d$; for these values of $\mathbf{k}$, $g(\mathbf{k})$ vanishes. The quantity $c(\mathbf{k})$ can be related to the mean vibrational amplitude of the Li^+ ions from their lattice sites. In particular, for a harmonic oscillator, $c(\mathbf{k}) = k^2\langle u_+^2\rangle/6$. Thus a plot of $-\ln F_s$ vs t yields as its zero time intercept a mean squared amplitude, while the slope can be related via a model to the jump rate τ^{-1} and the diffusion coefficient D.[7]

The general behavior of the function $F_s(\mathbf{k},t)$ is analogous to that found by Jacucci and Rahman in their calculations on CaF_2. Accordingly, we do not show the individual $F_s(\mathbf{k},t)$ curves. In Li_2SO_4 the asymptotic long time behavior of Li^+ ion Bragg vectors takes longer to develop than in CaF_2. This dilation of time scale is perhaps related to the larger lattice constant of the Li^+ ions and hence the longer jump distances involved in the diffusion process.

In Fig. 13 we show a composite plot of $\langle u_+^2\rangle = 6c(\mathbf{k})/k^2$ for a selection of $\mathbf{k}$ vectors in the high symmetry directions ⟨100⟩, ⟨110⟩, and ⟨111⟩. In the long wavelength limit, $\langle u_+^2\rangle \sim 1$ Å^2 for all three directions, which is in good agreement with experimental diffraction data.[2] However, unlike the case of CaF_2, there is considerable anisotropy in the finite $\mathbf{k}$ results, the amplitude in the ⟨100⟩ direction being much the largest. This suggests that short wavelength phonons may be involved in the diffusion process. Analysis of the $F_s(\mathbf{k},t)$ data based upon the Chudley–Elliot model[7] yields values of D consistent with those shown in Fig. 12 and suggest that $\tau \sim 7$ ps, in agreement with the observations described earlier. Figure 13 also shows the analogous results for F^- ion diffusion in CaF_2.

VIII. CONCLUSION

We have shown that a simple rigid ion model is capable of exhibiting many of the observed properties of lithium sulphate. In particular, the high temperature, solid electrolyte phase can be obtained with a range of models. However, we have identified the charge distribution of the $SO_4^=$ ion as being an important quantity in determining the crystal structure.

A preliminary study of the monoclinic → cubic phase transition has been carried out by means of constant pressure molecular dynamics. The mechanism of the transition has been discussed and it has been found tht the $SO_4^=$ ions may disorder before the Li^+ ions begin to diffuse. The observed pretransitional behavior such as the anomalous Raman linewidths in the β phase,[15] may well have its origin in this phenomenon.

For the future, it remains to carry out more systematic studies on the diffusion mechanism, using better potentials and a larger system size. The role of polarization effects in the diffusion of the Li^+ ions remains to be investigated. A comparative study with the molten salt[29,30] and mixed crystals would also be of interest. On the experimental side, our results suggest that inelastic neutron scattering experiments could be very informative.

ACKNOWLEDGMENTS

We thank Professor A. Lundén, Lena Torrell, and Roger Frech for kindly providing us reprints and preprints, Austin Angell and Alan Chadwick for alerting us to this interesting problem, and Elizabeth Neusy for help with some of the data analysis.

[1] A. Kvist and A. Lundén, Z. Naturforsch. Teil A **20**, 235 (1965).
[2] L. Nilsson, J. O. Thomas, and B. C. Tofield, J. Phys. C **13**, 6441 (1980); B. E. Mellander and L. Nilsson, Z. Naturforsch. Teil A **38**, 1396 (1983).
[3] R. Aronsson, B. Jansson, H. E. G. Knape, A. Lundén, L. Nilsson, C. -A. Sjöblom, and L. M. Torrell, J. Phys. (Paris) **41**, C6-35 (1980).
[4] R. Aronsson, H. E. G. Knape, and L. M. Torrell, J. Chem. Phys. 77, 677 (1982).
[5] A. Kvist and A. Bengtzelius, *Fast Ion Transport in Solids*, edited by W. van Gool (North Holland, Amsterdam, 1973), p. 193.
[6] A. Rahman, J. Chem. Phys. **65**, 4845 (1976).
[7] G. Jacucci, and A. Rahman, J. Chem. Phys. **69**, 4117 (1978).
[8] M. Dixon and M. J. Gillan, J. Phys. C. **11**, L165 (1978); M. J. Gillan and M. Dixon, *ibid.* **13**, 1901 (1980); M. Dixon and M. J. Gillan, *ibid.* **13**, 1919 (1980).
[9] Y. Hiwatari and A. Ueda, J. Phys. Soc. Jpn. **49**, 2129 (1980).
[10] M. L. Klein and I. R. McDonald, Chem. Phys. Lett. **78**, 383 (1981).
[11] D. G. Bounds, M. L. Klein, and I. R. McDonald, Phys. Rev. Lett. **46**, 1682 (1981).
[12] M. L. Klein and I. R. McDonald, Proc. R. Soc. London Ser. A **382**, 471 (1982); M. L. Klein, I. R. McDonald, and Y. Ozaki, Phys. Rev. Lett. **98**, 1197 (1982).
[13] M. L. Klein and I. R. McDonald, J. Chem. Phys. **79**, 2333 (1983).
[14] M. L. Klein, I. R. McDonald, and Y. Ozaki, J. Chem. Phys. **79**, 5579 (1983); R. W. Impey, M. L. Klein, and I. R. McDonald, J. Phys. C **17**, 3941 (1984).
[15] D. Teeters and R. Frech, J. Chem. Phys. **76**, 799 (1982); Phys. Rev. B. **26**, 5897 (1982); E. Cazzanelli and R. Frech, J. Chem. Phys. **79**, 2615 (1983); R. Frech and E. Cazzanelli, Solid State Ionics **9/10**, 89 (1983); L. Börjesson and L. Torrell (to be published).
[16] M. P. Tosi and F. G. Fumi, J. Phys. Chem. Solids **25**, 45 (1964).
[17] M. L. Klein, D. Levesque, and J. J. Weis, Phys. Rev. B **21**, 5785 (1980).
[18] D. J. Tildesley and P. A. Madden, Mol. Phys. **42**, 1137 (1981).
[19] M. Parrinello and A. Rahman, Phys. Rev. Lett. **45**, 1196 (1980); J. Appl. Phys. **52**, 7182 (1981); J. Chem. Phys. **76**, 2662 (1982).
[20] M. Parrinello, A. Rahman, and P. Vashista, Phys. Rev. Lett. **50**, 1073 (1983); A. Rahman and P. Vashista in *The Physics of Superionic Conductors and Electrode Materials*, edited by J. W. Perram (Plenum, New York, 1983), p. 93.
[21] S. Nosé and M. L. Klein, Phys. Rev. Lett. **50**, 1207 (1983); J. Chem. Phys. **78**, 6928 (1983); Mol. Phys. **50**, 1055 (1983).
[22] R. W. Impey, S. Nosé, and M. L. Klein, Mol. Phys. **50**, 243 (1983); R. W. Impey and M. L. Klein, Chem. Phys. Lett. **104**, 579 (1984).
[23] H. Johansen, Theor. Chim. Acta **32**, 273 (1973).
[24] R. Nobes (private communication).
[25] H. J. C. Berendsen and W. F. van Gunsteren, in *The Physics of Superionic Conductors and Electrode Materials*, edited by J. W. Perram (Plenum, New York, 1983), p. 221.
[26] D. J. Evans and S. Murad, Mol. Phys. **34**, 327 (1977).
[27] D. G. Bounds, M. L. Klein, and G. N. Patey, J. Chem. Phys. **72**, 5348 (1980).
[28] I. R. McDonald, D. G. Bounds, and M. L. Klein, Mol. Phys. **45**, 521 (1982).
[29] L. Nilsson, Thesis, Department of Physics, Chalmers University of Technology, Göteborg, Sweden, 1981.
[30] H. Ohno and K. Furukawa, J. Chem. Soc Faraday Trans. **74**, 795 (1978).

THE INTERPRETATION OF COHERENT QUASIELASTIC NEUTRON SCATTERING IN SUPERIONIC FLUORITES

M. J. Gillan

Theoretical Physics Division, AERE Harwell, Didcot, Oxfordshire OX11 0RA, U.K.

ABSTRACT

New molecular dynamics simulations on superionic CaF_2 are reported. The results are used to calculate the dynamical structure factors, which can be measured by coherent quasielastic neutron scattering. We find a quasielastic peak whose intensity and width as a function of wavevector and temperature are in semi-quantitative agreement with experiment. The simulations are used to show that this peak is caused by the motion of the point defects responsible for diffusion, and by their associated lattice distortion.

1. INTRODUCTION

We report here the preliminary results of an extensive series of molecular dynamics simulations on superionic CaF_2 which, like other materials having the fluorite structure, becomes an anion superionic conductor at high temperatures. The transition to the superionic state is continuous, and in CaF_2 occurs at ~ 1430 K (Dworkin and Bredig 1968). Our work has two related aims: firstly to help interpret the results of coherent quasielastic neutron scattering due to Hutchings et al. (1984), and secondly to clarify the mechanism of diffusion in CaF_2. Since the ions in superionic conductors interact strongly with each other, it seems clear that diffusion must be a highly co-operative process, the motion of an ion from one site to another being accompanied by the correlated motion of other ions in its vicinity. Experimentally, one would expect to get a handle on such motions by coherent quasielastic scattering, which probes the dynamics of spatially varying density fluctuations. The double-differential cross-section for such scattering is given by (Marshall and Lovesey 1971):

$$(\partial^2\sigma/\partial\Omega\partial E)_{coh} = \frac{k_f}{\hbar k_i} \sum_{\alpha\beta} \sqrt{N_\alpha N_\beta}\,\bar{b}_\alpha \bar{b}_\beta S_{\alpha\beta}(\mathbf{q},\omega) \quad . \tag{1}$$

where $\hbar k_i$, $\hbar k_f$ are the initial, final neutron momenta. N_α, $\bar{b}_\alpha$ are the total number, coherent scattering length of ions of type α. The dynamical structure factors $S_{\alpha\beta}(\mathbf{q},\omega)$, which depend on the transfers of momentum $\hbar\mathbf{q}$ and energy $\hbar\omega$, are defined by:

$$S_{\alpha\beta}(\mathbf{q},\omega) = \frac{1}{2\pi} \int_{-\infty}^{\infty} dt\, e^{i\omega t} \langle \rho_\alpha(\mathbf{q},t)\rho_\beta(\mathbf{q},0)^* \rangle \quad . \tag{2}$$

Here, $\rho_\alpha(\mathbf{q})$ is the density of ions of type α at wavevector $\mathbf{q}$:

$$\rho_\alpha(\mathbf{q}) = \frac{1}{\sqrt{N_\alpha}} \sum_{i=1}^{N_\alpha} e^{i\mathbf{q}\cdot\mathbf{r}_{\alpha i}} \quad , \tag{3}$$

the sum going over all ions of type α.

Experimentally, measurements of the cross-section as a function of frequency ω at fixed wavevector $\mathbf{q}$ reveal a peak in $S_{\alpha\beta}(\mathbf{q},\omega)$ centred on zero frequency (Hutchings et al. 1984). The intensity of this 'quasielastic' peak increases smoothly with temperature, being first observable at ~100 K below the transition to the superionic state. This intensity is distributed in an extremely non-uniform way in **q** space, being particularly strong in the region just beyond the reciprocal lattice vector at $(2\pi/a_0)(2,0,0)$, where a_0 is the lattice parameter.

These results contain important information about the correlated motions associated with diffusion, and the simulation results we shall describe shed light on the nature of this information.

2. SIMULATIONS

Our molecular dynamics simulations are based on the rigid-ion model for CaF_2 developed by Dixon and Gillan (1980). The interionic potentials of this model give a good account of the low-temperature bulk and defect properties of CaF_2 and successfully reproduce the experimental diffusion coefficient in the high-temperature superionic state. The new calculations were performed for two sizes of simulated system, consisting of 96 and 324 ions, with the usual periodic boundary conditions (for a review of the molecular dynamics technique, see e.g. Sangster and Dixon 1976; simulation work on fluorites has been reviewed by Gillan 1985). We have made simulations at two different temperatures for the smaller system and at five temperatures for the larger one, but since our analysis is not yet complete, we report here only a selection of the results.

A molecular dynamics simulation produces a record of the ion trajectories over a certain time span, corresponding in the present work to typically 100 psec. The time-dependent densities $\rho_\alpha(\mathbf{q})$ and hence the dynamical structure factors $S_{\alpha\beta}(\mathbf{q},\omega)$ can be calculated directly from this record. At each temperature, we have made calculations for $\alpha, \beta = ++, +-$ and $--$ and for a set of 28 wavevectors **q** which cover some of the important regions investigated by Hutchings et al. (1984). It is important to note that, because of the periodic boundary conditions, only those **q** are accessible whose cartesian components are multiples of $2\pi/L$, where L is the periodic repeat length.

The dynamical structure factors of our simulated CaF_2 resemble those of the real material in showing a quasielastic peak in the superionic state but not at lower temperatures. We illustrate this in Fig. 1, where we show the anion-anion function $S_{--}(\mathbf{q},\omega)$ at temperatures well below and well above the transition temperature. The peaks at frequencies ~ 5 and 15 THz correspond to the longitudinal acoustic and optic phonons.

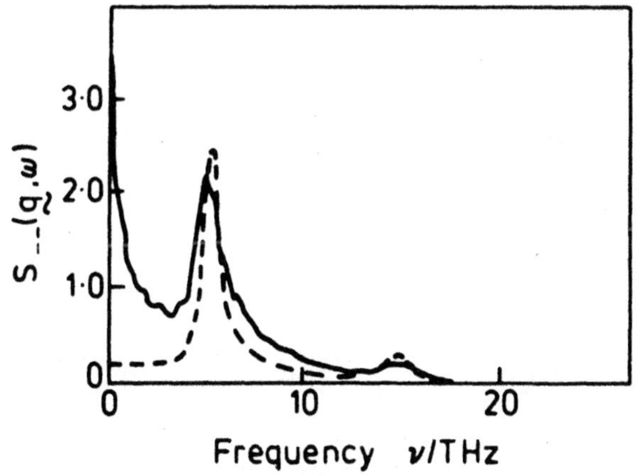

Fig.1. Anion-anion dynamical structure factor $S_{--}(\mathbf{q},\omega)$ in simulated CaF_2 (96 ions) at 1153 K (– – –) and 1657 K (——); wavevector **q** is $(2\pi/a_0)(1.5,0,0)$.

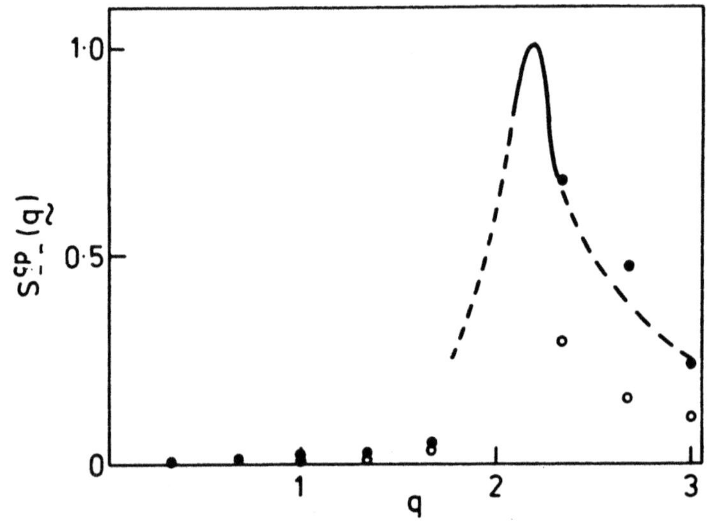

Fig. 2. Intensity of the quasielastic peak in the anion-anion dynamical structure factor from simulation of 324-ion system (○ 1390 K, ● 1609 K) and from experiments of Hutchings et al. (1984) at 1473 K (broken and solid lines).

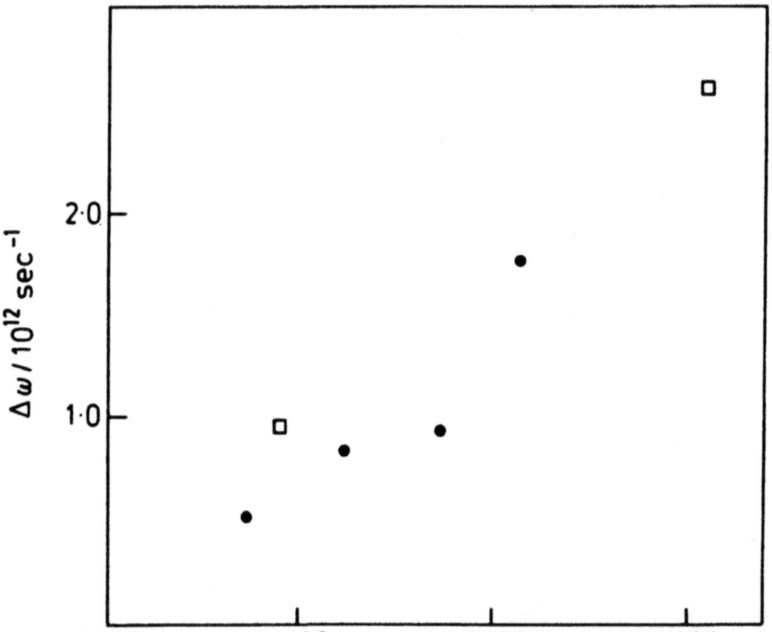

Fig.3. Simulation results from 324-ion system (open squares) for the half width of the quasielastic peak in $S_{--}(\mathbf{q}.\omega)$ at wavevector $\mathbf{q}=(2\pi/a_0)(7/3,0,0)$ compared with experimental results of Hutchings et al. (1984) (dots) at $\mathbf{q}=(2\pi/a_0)(2.3.0.0)$.

In order to quantify the intensity and width of the quasielastic peak, we have fitted the spectra to a sum of Lorentzian peaks at zero frequency and at the phonon frequencies, and a broad Gaussian background (Gillan and Dixon 1980a). The intensity of the fitted quasielastic peak can then be compared with the experimental intensity, for which Hutchings et al. obtained absolute results, assuming that the cation motions could be ignored (this assumption is supported by our results, for **q** values where the intensity is large). The comparison with experiment (Fig. 2) shows that the simulations reproduce, at least qualitatively, the experimental **q**–dependence and magnitude of the quasielastic intensity. Our analysis also gives the quasielastic half-width $\Delta\omega$, whose temperature dependence we compare with experiment in Fig. 3. The agreement is again qualitatively reasonable.

These comparisons provide strong evidence that the low-frequency co-operative motions in the simulated system are essentially the same as in real CaF_2.

3. DEFECT INTERPRETATION

We have used the detailed information produced by the simulations to show that these co-operative motions can be identified with the motions of point defects and their surrounding lattice distortion. Since some of the ideas are relatively unfamiliar, and space is short, we can do no more here than summarize the key points of our method.

We identify the vacancies and interstitials in the simulated system following the hopping analysis of Gillan and Dixon (1980b). This is based on two things: firstly on the fact, established both by simulation (Jacucci and Rahman 1978, Gillan and Dixon 1980b) and by neutron-scattering experiments (Dickens et al. 1983) that diffusion in superionic fluorites occurs by the hopping of anions directly between the anion regular sites; and secondly on the indication from simulation (Gillan and Dixon 1980b, Gillan 1985) that the anions do not reside on any sites except the regular sites. The implication is that both vacancies and interstitials should be regarded as occupying and hopping between only the anion regular sites. The procedure of Gillan and Dixon (1980b) allows us to analyze the simulation record so as to obtain for each time instant t the number n of interstitials and vacancies in the system, and a list of the sites $\{\mathbf{R}_s^i(t)\}$, $\{\mathbf{R}_s^v(t)\}$, $s=1,2 \ldots n$ occupied by the interstitials and vacancies respectively.

The idea that the quasielastic peak is due to the motion of these defects can be expressed by saying that the densities $\rho_a(\mathbf{q})$ can be decomposed into a slowly varying part $\delta_a(\mathbf{q})$ due to the defects and a rapidly varying part $\phi_a(\mathbf{q})$ due to the vibrations:

$$\rho_a(\mathbf{q}) = \delta_a(\mathbf{q}) + \phi_a(\mathbf{q}) \quad . \tag{4}$$

It turns out that these two parts are, to a good approximation, uncorrelated. This means that the spectra $S_{\alpha\beta}(\mathbf{q},\omega)$ separate into two parts, associated with defect motion and with vibrations:

$$S_{\alpha\beta}(\mathbf{q},\omega) = S^{\delta}_{\alpha\beta}(\mathbf{q},\omega) + S^{\phi}_{\alpha\beta}(\mathbf{q},\omega) \quad , \tag{5}$$

where $S^{\delta}_{\alpha\beta}$ and $S^{\phi}_{\alpha\beta}$ are defined as in equation (2), with δ_a and ϕ_a respectively instead of ρ_a. The meaning of this separation is that $S^{\delta}_{\alpha\beta}$ contains only the quasielastic peak and $S^{\phi}_{\alpha\beta}$ contains only the phonon peaks and the broad background.

It is an essential part of our description that the defects contribute to the density disturbance not only through their own presence, but through the strong lattice distortion which surrounds them. If we suppose that this distortion rigidly follows the motion of each defect, then the defect variables $\delta_a(\mathbf{q})$ will be expressible in terms of the defect positions $\{\mathbf{R}_s^i\}$, $\{\mathbf{R}_s^v\}$ alone, and it can be shown that they must take the form:

$$\delta_a(\mathbf{q}) = \frac{1}{\sqrt{N_a}}[f_a^i(\mathbf{q})\sum_s e^{i\mathbf{q}\cdot\mathbf{R}_s^i} + f_a^v(\mathbf{q})\sum_s e^{i\mathbf{q}\cdot\mathbf{R}_s^v}] . \quad (6)$$

The interstitial and vacancy quantities $f_a^i(\mathbf{q})$, $f_a^v(\mathbf{q})$ describe the distortion pattern around the defects, and in fact they are just the 'defect structure factors' well known from the theory of diffuse scattering from defects (Dederichs 1973, Bauer 1979).

The scheme we have summarized may perhaps seem rather abstract, but it can be put to a very concrete test. It can be shown that if the $\delta_a(\mathbf{q})$ constructed according to equation (6) are not the quantities whose fluctuations give the quasielastic peak, then no choice of the numbers f_a^i, f_a^v will effect the separation into quasielastic and vibrational spectra expressed in equation (5). In fact, our scheme does lead to an efficient separation for most $\mathbf{q}$, as we show in Fig. 4, where we give results for $\mathbf{q} = (2\pi/a_0)(7/3,0,0)$, which is close to the maximum in the quasielastic intensity (Fig. 2). The achievement of this separation establishes that the quasielastic peak can be attributed to the motion of the defects identified in our hopping analysis. We mention also that an analysis of the positions $\{\mathbf{R}_s^i\}$, $\{\mathbf{R}_s^v\}$ shows that the defects are almost uncorrelated. This implies that the strong $\mathbf{q}$–dependence of the quasielastic intensity comes mainly from that of the defect structure factors $f_a^i(\mathbf{q})$, $f_a^v(\mathbf{q})$, i.e. that it is a reflection of the lattice distortion.

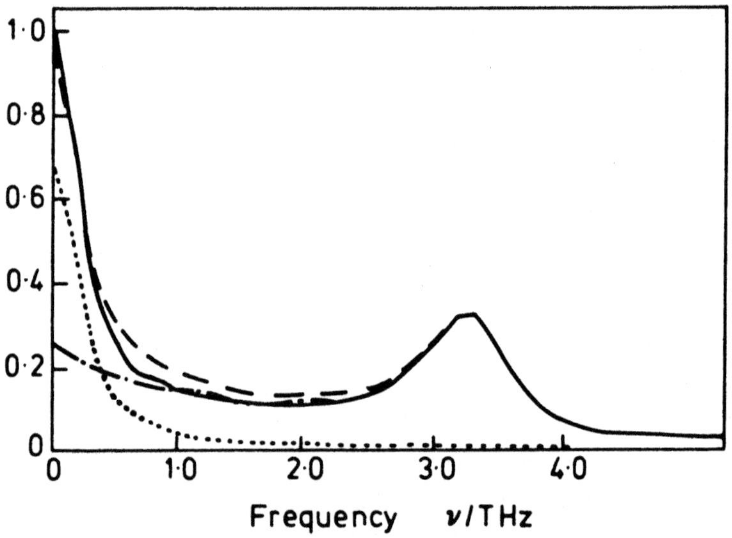

Fig.4. Separation of $S_{--}(\mathbf{q},\omega)$ (solid line) into defect fluctuation spectrum $S_{--}^{\delta}(\mathbf{q},\omega)$ (dotted line) and vibrational spectrum $S_{--}^{\phi}(\mathbf{q},\omega)$ (chain line); dashed line shows sum of S_{--}^{δ} and S_{--}^{ϕ}. Results are from simulation of 324–ion system at T=1390 K.

4. CONCLUSIONS

The dynamical structure factors of our simulated CaF_2 contain a quasielastic peak whose intensity and width are in reasonable agreement with experiment. The peak stems from the collective density fluctuations associated with the diffusion of the ions. These fluctuations may be identified with the motion of almost independent point defects and their surrounding lattice distortion, the defects being those defined by the hopping analysis of Gillan and Dixon (1980b).

REFERENCES

Bauer, G.S. (1979). Diffuse elastic neutron scattering from non-magnetic materials. In: Treatise on materials science and technology, Vol. 15. Edited by G. Kostorz (Academic, New York) 291–336

Dederichs, P.H. (1973). The theory of diffuse x-ray scattering and its application to the study of point defects and their clusters. J. Phys. F 3, 471–96.

Dickens, M.H., Hayes, W., Schnabel, P., Hutchings, M.T., Lechner, R.E. and Renker, B. (1983). Incoherent quasielastic neutron scattering investigation of chlorine ion hopping in the fast-ion phase of strontium chloride. J. Phys. C 16, L1–6.

Dixon, M. and Gillan, M.J. (1980). Computer simulation of fast ion transport in fluorites. J. de Physique (Paris) 41, Coll. C-6, 24–7.

Dworkin, A.S. and Bredig, M.A. (1968). Diffuse transition and melting in fluorite and anti-fluorite type of compounds. J. Phys. Chem. 72, 1277–81.

Gillan, M.J. and Dixon, M. (1980). Quasielastic scattering in fast-ion conducting $SrCl_2$: a molecular dynamics study. J. Phys. C 13, L835–9.

Gillan, M.J. and Dixon, M. (1980). Molecular dynamics simulation of fast-ion conduction in $SrCl_2$: I. Self-diffusion. J. Phys. C 13, 1901–17.

Gillan, M.J. (1985). The simulation of superionic materials. Physica, in press.

Hutchings, M.T., Clausen, K., Dickens, M.H., Hayes, W., Kjems, J.K., Schnabel, P.G. and Smith, C. (1984). Investigation of thermally induced anion disorder in fluorites using neutron scattering techniques. J. Phys. C 17, 3903–40.

Jacucci, G. and Rahman, A. (1978). Diffusion of F^- ions in CaF_2. J. Chem. Phys. 69, 4117–25

Marshall, W. and Lovesey, S.W. (1971). Theory of thermal neutron scattering (Oxford University Press, Oxford).

Sangster, M.J. and Dixon, M. (1976). Interionic potentials in alkali halides and their use in simulation of the molten salts. Adv. Phys. 25, 247–342.

Molecular-Dynamics Study of Ionic Motions and Neutron Inelastic Scattering in α-AgI

Guido L. Chiarotti
International School for Advanced Studies, Trieste, Italy, and Dipartimento di Fisica, Università di Trento, Povo, Italy

G. Jacucci
Centro Studi del Consiglio Nazionale delle Ricerche and Dipartimento di Fisica, Università di Trento, Povo, Italy

and

A. Rahman
Supercomputer Institute and School of Physics and Astronomy, University of Minnesota, Minneapolis, Minnesota 55455
(Received 7 August 1986)

Contrary to the current interpretation of inelastic–neutron-scattering data on α-AgI, molecular-dynamics calculations show that inelastic neutron scattering is dominated by coherent scattering from Ag^+ ions. The calculations agree with the available data. Ag^+ ions diffuse by jumps between tetrahedral sites, the consequences being in complete accord with the Chudley-Elliot model only if the full geometrical complexity of these sites is included. Phonon modes due to I^- motions are predicted for certain wave vectors.

PACS numbers: 66.30.Hs, 02.70.+d, 05.60.+w

Although α-AgI has been one of the most extensively studied superionic conductors, a direct comparison between calculations and experimental inelastic–neutron-scattering (INS) data has never been made. INS data are the only direct source of information about the motion of ions in its full anisotropic complexity. In α-AgI, Ag^+ ions are the mobile species which diffuse through interstitial spaces of the stable I^- lattice. We present here a comparison of molecular-dynamics computer-simulation results and INS experimental data, together with a theoretical interpretation of the motion of Ag^+ ions.

Using a large single crystal of α-AgI, Funke[1] and coworkers have reported measurements for two very different values of the scattering vector $\mathbf{k}$ (in the $\langle 522\rangle$ and $\langle 111\rangle$ directions) at 523 K. There are three coherent and one incoherent contributions to the intensity of the measured spectrum (the incoherent scattering from I^- ions is absent while that from Ag^+ ions is present[2]). This makes the interpretation of the data very difficult. Funke[1] interpreted the data as primarily due to the *incoherent* Ag^+ contribution; a comparison was then made between the data and the prediction of the Chudley-Elliot[3] (CE) jump model for diffusion and the conclusion was drawn that the CE model cannot explain the data.

A detailed molecular-dynamics (MD) study of α-AgI at 500 K has recently been completed and the full details are given elsewhere.[4] The conclusion drawn from molecular-dynamics calculations are simply stated as follows: (i) The computed INS spectrum agrees with the measured spectrum[1] for both values employed of the scattering vector $\mathbf{k}$ mentioned above. (ii) The most important contribution to INS comes from coherent Ag^+ scattering. (iii) The jump motion of Ag^+ ions is very well described by the CE model, provided that all the necessary geometric details[4] of the structure of tetrahedral (t) sites are taken into account.

The MD simulation was performed with 500 particles in a fixed cubic box with periodic boundary conditions, with use of a pairwise additive effective-interaction scheme based on a potential function,[5] constructed according to Pauling's ideas of ionic radii, which reproduces the $\alpha \rightleftharpoons \beta$ phase transition at the correct transition temperature. The lattice parameter of the I^- bcc sublattice was $d = 5.2512$ Å corresponding to a density $\rho = 5.384$ g/cm^3. The system was followed for 25000 steps ($\Delta t = 2.049 \times 10^{-14}$ s); an analysis of $\langle r^2(t)\rangle$, the mean square displacement of Ag^+ ions, gave a diffusion coefficient $D = 2.0 \times 10^{-5}$ cm^2/s.

With the hypothesis that the diffusion of Ag^+ ions takes place only via jumps between nearest-neighbor (nn) tetrahedral sites, the CE theory can be used to predict the time behavior of the self part of the intermediate scattering function $F_s(\mathbf{k},t)$.[6] Under the assumption that the vibrational (V) and diffusive (D) motions are decoupled this function can be factorized[3]:

$$F_s(\mathbf{k},t) = F_s^D(\mathbf{k},t) F_s^V(\mathbf{k},t). \tag{1}$$

The CE model provides an explicit expression for $F_s^D(\mathbf{k},t)$ on the assumption that the individual diffusion motion of the ions is a random walk consisting in subsequent instantaneous, directionally uncorrelated (Markoffian) jump events between nearest-neighbor geometrical sites on the chosen lattice. The only parameter in the model is the mean residence time τ.

The CE approach has to be generalized in order to take into account the fact that t sites do not form a Bravais lattice. To this effect we follow the approach of Rowe *et al.*[7] Because there are six *in*equivalent lattice sites the CE assumption leads to a system of six coupled differential equations[4] which are solved by a weighted sum of six decaying exponentials:

$$F_s^D(\mathbf{k},t) = \sum_{i=1}^{6} a_i(\mathbf{k}) e^{-M_i(\mathbf{k})t/\tau}, \quad (2)$$

where $a_i(\mathbf{k})$ and $M_i(\mathbf{k})$, given in terms of eigenvectors and eigenvalues of the matrix of coefficients appearing in the system of coupled equations, depend on the geometrical distribution of the sites, and $1/\tau$ is the frequency of jumps.

A real-space analysis of the molecular-dynamics trajectories showed in fact that, although various kinds of jumps could be observed, more than 80% occurred between nn t sites. Furthermore, the distribution of the angles between the directions of subsequent jumps was not far from random, albeit the probability of backward jumps was somewhat enhanced.

Vibrational motion contributes a simple Debye-Waller factor, as $t \to \infty$:

$$F_s^V(\mathbf{k},t \to \infty) = \exp(-k^2 a_k^2), \quad (3)$$

where a_k^2 depends on the direction of $\mathbf{k}$ to account for possible anisotropy of vibrations at t sites.

The most delicate point for the applicability of the CE model to this system is the fact that the residence time is (not much longer than the duration of a jump event, and) not much longer than a typical vibrational period. As a result, the asymptotic ($t \to \infty$) diffusive behavior of $F_s(\mathbf{k},t)$, as described by the CE model, sets in much later, in units of the residence time, with respect to what is observed in other systems, e.g., in CaF_2, where the time scales of vibrational and diffusional motion of the mobile species are more separated.[6]

Using the trajectories of particles generated by the molecular-dynamics computer simulation, we have computed $F_s(\mathbf{k},t)$ as[8]

$$F_s(\mathbf{k},t) = \frac{1}{N}\left\langle \sum_{j=1}^{N} e^{i\mathbf{k}\cdot[\mathbf{r}_j(t)-\mathbf{r}_j(0)]} \right\rangle. \quad (4)$$

The only parameters needed to make a comparison between the MD results and the CE model are the residence time τ and a_k^2. Their values are obtained by the fitting of CE-model predictions to MD results. The value for τ thus found is $\tau = 145\Delta t = 2.97$ ps. In Fig. 1(a) we show $F_s(\mathbf{k},t)$ for different values of k in the direction ⟨110⟩, and in Fig. 1(b) in the direction ⟨110⟩. The amplitude a_k is a measure of the extension of the Debye-Waller cloud. First neighbors of t sites are located along the ⟨110⟩ direction; nn jumps occur along those directions. Therefore, the fact that $a^2_{\langle 110\rangle}$ is larger than $a^2_{\langle 100\rangle}$ can be attributed to a Debye-Waller cloud which is (anisotropic and) more extended in the jump direction.

For both $\mathbf{k}$ vectors there is good agreement between the MD results and the CE model for $t \gg \tau$. With $\tau = 2.97$ ps and the assumption of nn jumps we get the diffusion coefficient to be 1.9×10^{-5} cm^2/s in accord with a direct calculation of the mean square displacement $\langle r^2(t)\rangle$ of Ag^+ ions. Good agreement is also found in other $\mathbf{k}$ directions, in particular the ⟨111⟩ and ⟨522⟩, which will be discussed in connection with INS experiments.

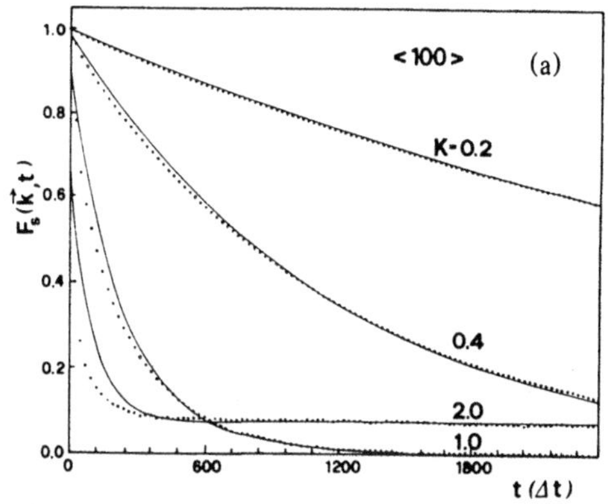

FIG. 1. Self part of the intermediate scattering function for Ag^+ ions in α-AgI, $F_s(\mathbf{k},t)$, as a function of time and for two different directions. Results from molecular-dynamics simulation are shown as circles and those from the Chudley-Elliot model are shown with lines. The parameters chosen for the CE model are $\tau = 145\Delta t$ ($\Delta t = 2.049\times10^{-14}$ s), $a_{\langle 100\rangle} = 0.05d$ and $a_{\langle 110\rangle} = 0.079d$ ($d = 5.25$ Å is the lattice parameter). Various values for the scattering vector are reported: $\mathbf{k}_{\langle 100\rangle} = (2\pi/d)(k,0,0)$ and $\mathbf{k}_{\langle 110\rangle} = (2\pi/d)(k,0,0)$ with $k = 0.2$, 0.4, 1.0, 2.0.

The inelastic scattering intensity $S_{\text{tot}}(\mathbf{k},\omega)$ measured by INS is made up of four different contributions:

$$S_{\text{tot}}(\mathbf{k},\omega) = \sigma_{\text{inc}}^{\text{Ag}} + \sigma_{\text{coh}}^{\text{AgAg}} + \sigma_{\text{coh}}^{\text{II}} + 2\sigma_{\text{coh}}^{\text{AgI}}, \quad (5)$$

where the scattering cross sections σ are proportional to the van Hove dynamical structure factors S through the scattering lengths b:

$$\sigma_{\text{inc}}^{a} = (b_{\text{inc}}^{a})^2 S_s^a(\mathbf{k},\omega)$$

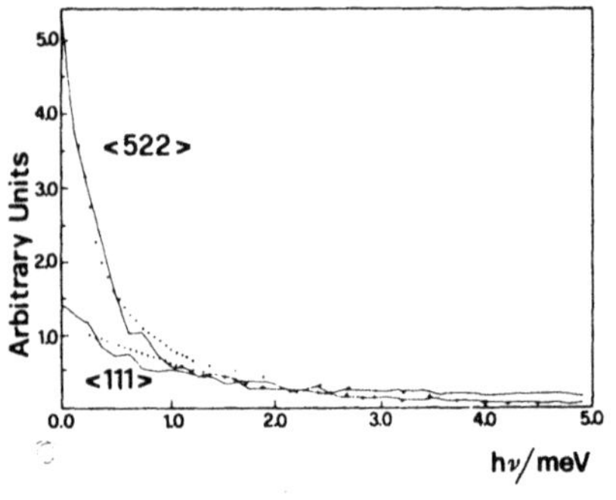

FIG. 2. $S_{tot}(\mathbf{k},\omega)$: solid line, MD; plusses, INS (Ref. 1) for the ⟨522⟩ and ⟨111⟩ directions of **k**. The experimental data have been scaled by the *same* constant for *both* reported values of **k**.

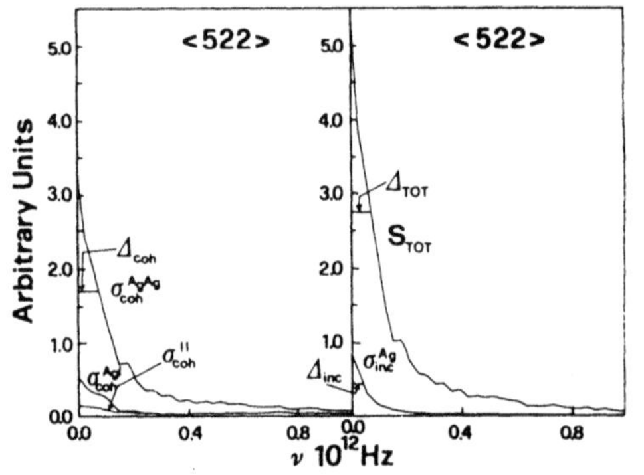

FIG. 3. All nonvanishing contributions to the INS spectrum computed by molecular dynamics, separately shown. The sum of all contributions is denoted by S_{tot}.

and

$$\sigma_{coh}^{\alpha\beta} = b_{coh}^{\alpha} b_{coh}^{\beta} S^{\alpha\beta}(\mathbf{k},\omega).$$

In order to compare our MD results with the experimental data of Funke,[1] we plot the inelastic scattering intensity $S_{tot}(\mathbf{k},\omega)$ in Fig. 2 for the two scattering vectors **k** for which data are available. The agreement is good for both scattering vectors. Because the size of the molecular-dynamics cell is finite (26.256 Å), not all the values of k can be used for the calculation. In the ⟨522⟩ direction the closest possible value was $k = 1.37$ Å^{-1}, to be compared with the experimental value of $k = 1.82$ Å^{-1}. In the ⟨111⟩ direction it was possible to use both $k = 1.66$ Å^{-1} and 2.072 Å^{-1} in the calculation, finding only a weak k dependence. The experimental value in this direction, namely, 1.82 Å^{-1}, lies in between.

The four contributions of Eq. (5) to the INS intensity $S_{tot}(\mathbf{k},\omega)$ cannot be separately measured, or isolated, without resorting to complicated (if available) isotopic substitutions. This fact very often hinders a reliable interpretation of INS experimental data. In contrast, MD calculations provide separate evaluations of the various dynamical structure factors, and hence are a unique basis for the interpretation of INS spectra. Figure 3 shows the four contributions, and their sum, for **k** in the ⟨522⟩ direction: $\mathbf{k} = (2\pi/d)(1.0, 0.4, 0.4)$. It is clear from the figure that the scattering is primarily due to the *coherent* contribution σ_{coh}^{AgAg}. This dominates the intensity and the width Δ of S_{tot}. We find in particular that $\Delta_{tot} \simeq \Delta_{coh} \simeq 2\Delta_{inc}$. Analogous conclusions are drawn from result of the MD calculations relative to the direction ⟨111⟩ of **k**.

An analysis of S_{tot} in the ⟨522⟩ direction was also made with $\mathbf{k} = (2\pi/d)(2.0, 0.8, 0.8)$ revealing a phonon peak, due to I-I coherent correlation, at a frequency $\nu = \omega/2\pi = 0.6 \times 10^{12}$ Hz. The fact that this phonon does not appear at $\mathbf{k} = (2\pi/d)(1.0, 0.4, 0.4)$ is probably because the frequency ν of the phonon for this **k** should be about 0.3×10^{12} Hz, which coincides with the frequency of jumps of silver ions which is 0.33×10^{12} Hz. This close equality should lead to a heavily damped phonon mode that disappears at the smaller value of **k** mentioned above. Thus we predict that an INS experiment with **k** in the ⟨522⟩ direction and magnitude in the range of 3 Å^{-1} should reveal the presence of a phonon mode.

In conclusion we have shown that coherent scattering from the mobile species Ag^+ dominates the INS spectrum in the superionic conductor α-AgI, at variance with current interpretations of experimental data. That this should happen far from the Bragg condition of the lattice is specific, and is probably general, to superionic conductors and may be related to the high degree of disorder of the sublattice of the mobile species.

Furthermore, we have shown that, even in the superionic conductor α-AgI where mobile ions diffuse extremely fast, the CE model, based on the interstitial lattice of t sites, adequately describes the diffusional motion of Ag^+ ions, at least for times long with respect to lattice vibrations. While this fact unambiguously identifies which is the lattice of interstitial positions used by Ag^+ ions in their diffusional motion, it also determines that diffusion occurs by jumps. Thus, many-body correlations in the motion of different ions in the solid produce, as the only consequence on the motion of individual Ag^+ ions, the determination of the value of the jump frequency that enters a simple random-walk model.

Those many-body correlations, however, are responsible for the shape of the dynamical structure factors $S^{\alpha\beta}$, related to coherent scattering cross sections, which dominate the INS spectrum in α-AgI, as well as in other cases.[9] A theory for σ_{coh}, analogous to the CE model for σ_{inc}, is at present still lacking.

Finally, the simulation has produced evidence against the "intermediate partially ordered phase" recently proposed[10] for superionic α-AgI. In particular we find isotropic mean square displacements of Ag^+ ions, in the course of time, in the x, y, and z directions.[10]

This work was started at the Materials Science and Technology Division of the Argonne National Laboratory. One of us (G.L.C.) is grateful for the hospitality during his stay at Argonne National Laboratory. We are grateful to Klaus Funke for helpful discussions and for making available to us unpublished experimental data, and to David Ceperley and Giulia De Lorenzi for a critical reading of the manuscript.

[1] K. Funke, Adv. Solid State Phys. **20**, 1 (1980).

[2] The scattering lengths are $b_{coh}^{Ag} = 4.55$, $b_{inc}^{Ag} = 0.49$, $b_{coh}^{I} = 3.50$, $b_{inc}^{I} = 0.0$ (in barns).

[3] C. T. Chudley and R. S. Elliot, Proc. Phys. Soc. London **77**, 353 (1961).

[4] Guido L. Chiarotti, Tesi di Laurea in Fisica, Università degli Studi di Trento A.A. 1984-85, unpublished.

[5] M. Parrinello, A. Rahman, and P. Vashishta, Phys. Rev. Lett. **50**, 1073 (1983).

[6] This type of analysis has already been shown to be valid for the superionic conductor CaF_2: G. Jacucci and A. Rahman, J. Chem. Phys. **69**, 4117 (1978).

[7] J. M. Rowe, W. Skold, H. E. Flotow, and J. J. Rush, J. Phys. Chem. Solids **32**, 41 (1971).

[8] M. Dixon, Philos. Mag. B **47**, 509 (1983).

[9] M. J. Gillan, "Collective Dynamics in Superionic CaF_2" (to be published).

[10] Gy. Szabo, J. Phys. C (to be published); Gy. Szabo and J. Kertesz, to be published.

Isothermal–isobaric computer simulations of melting and crystallization of a Lennard-Jones system

Shuichi Nosé and Fumiko Yonezawa
Department of Physics, Faculty of Science and Technology, Keio University, 14-1, 3 Chome, Hiyoshi, Kohokuku, Yokohama 223, Japan

(Received 26 August 1985; accepted 28 October 1985)

By means of constant-temperature, constant-pressure molecular dynamics techniques, we simulate the melting and crystallization processes of a model system composed of 864 Lennard-Jones (LJ) particles under periodic boundary conditions. On heating an fcc crystal of LJ particles, it is ascertained that melting takes place. On the other hand, a LJ liquid, when quenched slowly, crystallizes into a stacking of layers with stacking faults where each layer forms a close-packed structure with occasional point defects. The atomic configuration is not always nucleated into a completely ordered structure. A large hysteresis in the volume-temperature curve is observed. The volume contraction at the transition is characterized by two different growth rates, relatively slow at the first stage and relatively fast at the final stage. The critical cooling rate which separates the crystal-forming cooling rates and the glass-forming cooling rates is between 4×10^{10} and 4×10^{11} K/s for argon. On taking advantage of computer simulations, we analyze the microscopic atomic structure of our LJ system on the basis of the Voronoi and Delaunay tessellation.

I. INTRODUCTION

It is already a quarter of a century since Alder and Wainwright published the results of a molecular dynamics (MD) study[1] on the crystallization of a hard-core system which is now referred to as the Alder transition. Although quite a few investigations of crystallization have been reported since then,[2–10] the progress in this field has not been so very distinguished as would be expected from the remarkable improvement in computers themselves for the last few decades. One of the reasons is that the progress in computers is not profound enough to make it possible to increase the size of a simulated system by many orders of magnitude. But, it may be encouraging to recall how small and slow the computers used to be when the Alder transition was first discovered.

The MD studies of crystallization have been made using potentials (other than hard-core potential) such as soft-core potentials,[2,3] the Lennard-Jones potentials,[4–7,9] and metallic potentials.[8] Analyses have been made on the size of the critical nucleus,[5,6,8] the structural feature of nucleation,[2,3,8] the effects of the types of potentials,[7] and the effects of boundary conditions.[6] A point to make, however, is that most of the MD simulations of crystallization are performed for the (N,V,E) ensemble where the number N of particles, the volume V, and the energy E of the system are fixed. As is easily perceived, it is often desirable to perform the simulations of crystallization for the (N,P,T) ensemble where the pressure P and the temperature T are fixed as well as N, so that we can compare the results of the simulations with the experimental data. This is also the case for the simulations of melting.[9,11]

The Monte Carlo (MC) method has also been employed to carry out the isobaric–isothermal simulations of crystallization.[10] Although it is known that we can learn a great deal from the MC studies, the MD methods are suitable when the dynamical properties of systems are required.

With this situation in mind, it is the purpose of this paper to present the results of our constant-temperature, constant-pressure MD simulations on melting and crystallization. We also give some analysis of the microscopic structures using the Voronoi and Delaunay tessellation. Obviously, this kind of structure analysis is made possible only by computer simulations which give a full information of the structure at the molecular level.

In Sec. II, we describe our model and the simulation method. The results of the heating and melting processes are, respectively, given in Secs. III and IV, while the crystallization through different processes is discussed in Secs. V and VI. The structure analysis is made in Sec. VII, and a summary of our results is found in Sec. VIII.

II. MODEL AND METHOD

We study a model system composed of particles interacting with one another via (12-6) Lennard-Jones pair potential characterized by length σ and the minimum energy ϵ. The potential is truncated at 2.902 σ and the effect of this truncation is taken care of, as will be explained below. For our simulations, we use the combination of a constant-pressure MD method proposed by Andersen[12] and a constant-temperature MD method proposed by Nosé.[13] Some detailed description of our method is given in the Appendix. The equations of motion are solved by the fifth-order predictor–corrector algorithm, and periodic boundary conditions are assumed. We present the results for $N = 864$.

In what follows, the physical properties are expressed either in the reduced units or in the real units suitable for argon. For the former, we measure energy in ϵ, length in σ, temperature in ϵ/k with k being the Boltzmann constant, pressure in ϵ/σ^3, and time in $(m\sigma^2/\epsilon)^{1/2}$, where m is the mass of a particle. For the latter, we substitute $\epsilon = 125$ K, $\sigma = 3.446$ Å, and $m = 39.9$ g/mol. It is well known that the LJ potential serves as a good approximation for rare gas materials, and this is particularly the case for argon. A time step for integration is taken to be 0.002 34 $(m\sigma^2/\epsilon)^{1/2}$, which corresponds to 0.5×10^{-14} s for argon.

We are interested in systems under the atmospheric pressure for which the data of real experiments are available. Throughout our simulations, we choose the pressure P as 0.25 kbar $= 0.593\ (\epsilon/\sigma^3)$, which nearly compensates the effects of the potential truncation at 10 Å $= 2.902\sigma$ in such a way that the effective pressure substantially corresponds to the atmospheric pressure.

III. HEATING

We start with an fcc structure. The reason why we choose an fcc crystal as our initial structure is that argon is known to form an fcc crystal below 84 K. Note that a theoretical calculation does not predict an fcc structure as the configuration with the lowest energy for LJ particles. When the energies of closest-packed structures are compared at absolute zero, an hcp structure (a regular ABAB stacking of close-packed layers) takes the lowest energy; a completely random stacking of A, B, and C layers the second lowest; and an fcc structure (a regular ABCABC stacking) the highest of the three. But, the differences are less than 0.01%. The fact that an argon crystal in nature is of an fcc type indicates that, in an aggregation of argon atoms, there exists a part of interatomic interactions which is not described by a pairwise law of the LJ type, a part which is small but still efficient enough to favor an fcc structure.

Now, for our initial structure of an fcc type, we set temperature T at 40 K which is well below the experimentally observed melting temperature of 84 K. The pressure P is chosen as mentioned at the end of the last section. Then, in our constant-temperature, constant-pressure simulation, the system adjusts itself so that it can take the structure and the volume which it should take under given temperature and pressure. Since the energy barriers among the above-described three structures are fairly high, our system, once constructed in an fcc structure, does not transform to an hcp structure. From 40 K, we increase the temperature of this fcc crystal with each increment of 10 K up to 70 K and with each increment of 5 K up to 105 K. The volume-temperature relation is represented by filled circles in Fig. 1. Below 84 K, i.e., below the melting point of a real argon crystal, the volume of our system agrees well with that of an argon crystal measured in laboratories. The thermal expansion coefficient of our system is also in good agreement with the experimental data. These results indicate that our model simulates an argon system very well.

It is interesting to evaluate the ratio of the average deviation u of the particles to the interatomic distance a since Lindemann proposed that this ratio u/a could be used as a criteria for melting, asserting that $u/a = 0.1$ could be the critical value. For our system, the value u/a is about 0.1 at 70 K and about 0.2 at 90 K. This seems to be consistent with the fact that an argon crystal melts at 84 K. However, the volume of our system increases continuously and no catastrophic increase in volume is observed until T reaches 105 K.

The structure of our system in the temperature range between 80 K or so and 105 K could be checked by examining some appropriate physical properties. The first physical property which comes across our mind in this connection is the pair correlation function $g(r)$ of the atomic positions. This function for our system at 105 K is shown in Fig. 2(a) (bottom). From the figure, we can see that $g(r)$ shows a behavior similar to that for a liquid and there exists no peak structure characteristic of a crystal. This, however, does not mean that the system is in a disordered structure since, at the temperature as high as this, the effect of thermal vibrations is large enough to smear out any structure in $g(r)$. It would help to recall that usually $g(r)$ does not show any distinguishable change across the melting temperature. What is relevant in this case, therefore, is the pair correlation function $\bar{g}(r)$ defined by the time-averaged position vectors

$$\mathbf{r}_i = \frac{1}{\tau}\int_s^{s+\tau} \mathbf{r}_i\,(t)\,dt. \qquad (3.1)$$

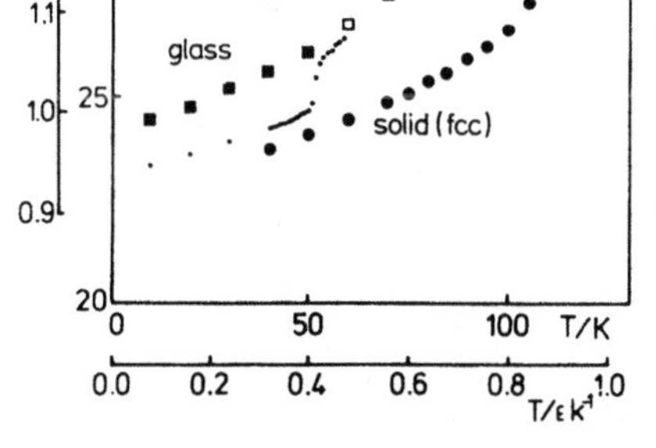

FIG. 1. Volume (V) vs temperature (T) for crystals (filled circles) and liquids (open squares). Small dots represent the cooling-crystallization process. The corresponding results are also shown for LJ glasses obtained by rapid quenching. A detailed discussion of the glass transition will be found elsewhere.

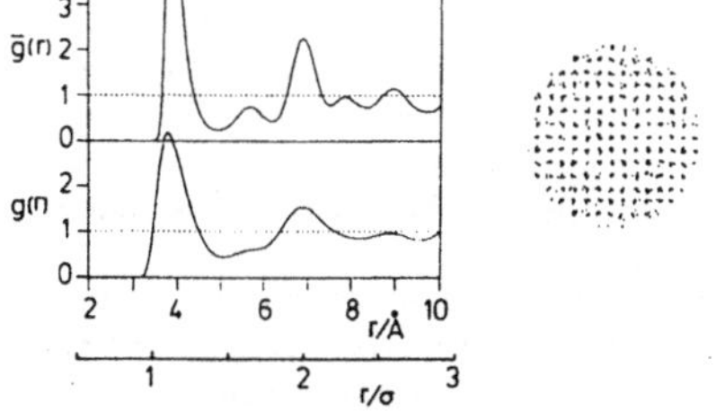

FIG. 2. (a) The pair correlation function $\bar{g}(r)$ defined by the instantaneous position vectors compared to the pair correlation function $g(r)$ defined by the average position vectors. Both are concerned with the LJ system at $T = 105$ K. (b) Projection of all particles onto the xy plane. Also for 105 K.

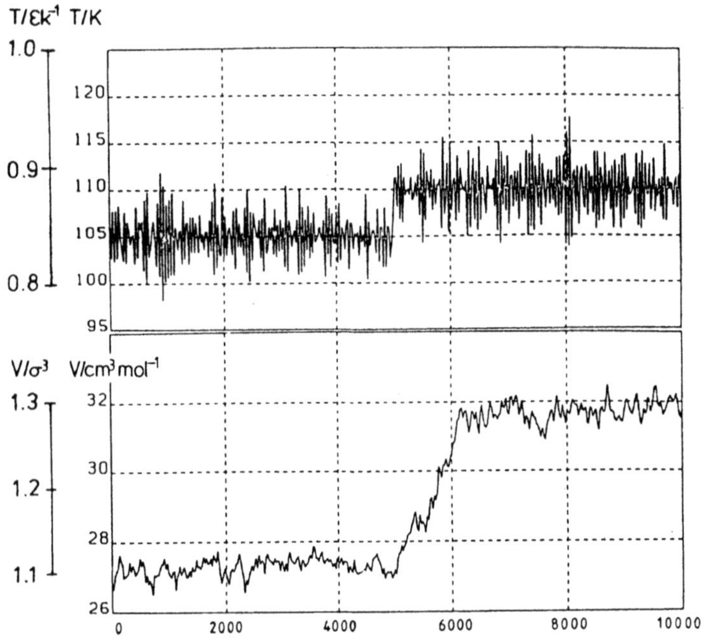

FIG. 3. Temperature and volume vs time steps after the switching of temperature from 105 to 110 K.

The idea of using $\mathbf{r}_i$ is that we can eliminate the effect of thermal vibrations by the averaging and, as a result, the essential structure is expected to show up. In Fig. 2(a), $\bar{g}(r)$ (the top figure) for the case of $\tau = 100$ time steps is compared to $g(r)$. We can clearly see that, at 105 K, $\bar{g}(r)$ has those peaks characteristic of an fcc structure. Thus, the system is ascertained to be in an fcc structure even at 105 K. This conclusion is also supported by a close analysis of the projections of the particles onto several appropriate planes. One example is shown in Fig. 2(b), and a clear-cut layer structure of the fcc type is evident.

IV. MELTING

A substantial expansion of the volume is observed when the temperature is switched from 105 to 110 K. The situation is clearly understood by referring to Fig. 3, in which the changes in temperature T and volume V are given as functions of time steps t. As is seen from the figure, the temperature is switched from 105 to 110 K at $t = 5000$. Following this temperature switching, the volume starts expanding, and the V-t relation indicates that the expansion ceases after about 1000 time steps.

Now, a question is: What is the structure of our system at 110 K? In order to find the answer to this question, we first study the trajectories of particles. They are presented in Fig. 4. The left three figures represent the particles which are in the first layer of the fcc crystal before the volume expansion. The right three figures represent those in the second layer. In either case, the top figure represents the trajectories between

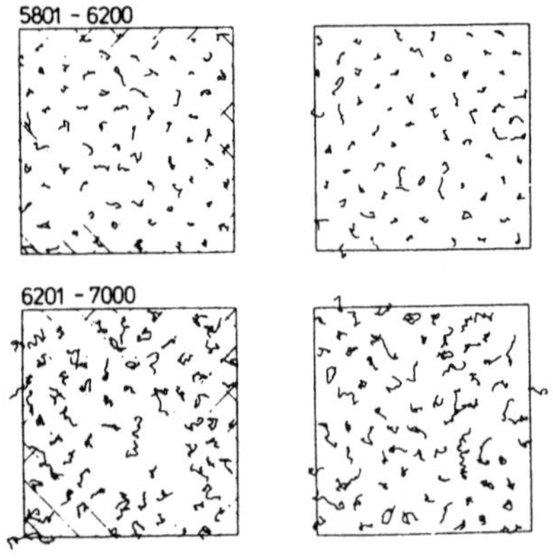

FIG. 4. Trajectories of particles in the first and second layers of the originally fcc structure. (a) Trajectories between 1 and 800 time steps after the increase of temperature from 105 to 110 K; (b) between 801 and 1200 time steps; and (c) between 1201 and 2000 time steps.

1 and 800 time steps after the temperature switching; the middle figure between 801 and 1200 steps; and the bottom figure between 1201 and 2000 steps. The figures are depicted in a normalized scale such that our MD cell with any given volume is reduced to a box of the same size. Needless to say, this normalized scale is adapted because we are interested in the relative positions of atoms alone, and the effect of the overall thermal expansion is not important here. In the left three figures, the square lattice typical of each layer in an fcc structure is drawn for the help of eye.

Let us start with the top figures for the first 800 steps following the temperature switching. It is suggested from Fig. 3 that, by the end of this first time interval, the increase in volume is more than halfway. In spite of this considerable expansion, most of the particles still stay in the vicinity of their respective original lattice points although there exist parts where the deviations of particles are appreciable. Accordingly, it is possible to find a one-to-one topological correspondence between the fcc lattice points and the particle positions. The middle figures show that the particles start leaving the original lattice points. Figure 3 denotes that the volume expansion is nearly complete by the end of this second time interval (801 and 1200 steps). Then, in the last time interval (1201 and 2000 steps) in which we no longer observe any substantial change in volume (see Fig. 3), the distribution of the particles is completely free from the fcc lattice points and has no topological order (see the bottom figures in Fig. 4).

The next question is then whether or not the disordered structure thus obtained is a fluid which is characterized by high fluidity. In order to study this problem, we evaluate the mean square displacement (MSD) of particles as a function of t, which is demonstrated in Fig. 5. The behavior of the MSD after 6000 steps is of fluid type in the sense that it increases linearly with time. From the slope of this linear relation, the diffusion constant D is calculated to be 3.7×10^{-5} cm^2/s at 110 K, which ascertains the fluid property of the disordered structure. Note that the diffusion constant of argon liquid at 84 K is 1.6×10^{-5} cm^2/s. Therefore, we can conclude that the melting does take place at 110 K. Here, it is interesting to note that the volume expansion at 100 K is about 14% which is comparable to the volume increase of about 16% at the melting point of argon. Our melting temperature 110 K is considerably higher than the experimental value of 84 K. This could be attributed to the finite size of our system and to the finite simulation time. The temperature range near our melting temperature of 110 K is considered to be the upper bound of the superheated region.

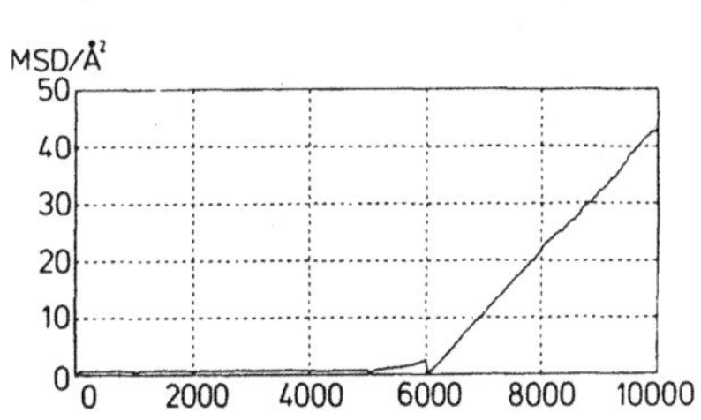

FIG. 5. Mean square displacement at $T = 110$ K.

V. COOLING AND NUCLEATION

In this section, we demonstrate what happens when a LJ liquid is cooled with relatively slow cooling rates. We start with a configuration of a liquid at 100 K. But, the physical properties of the system obtained therefrom are not influenced by the choice of the initial sample as long as we are concerned with temperatures which are high enough. Then, from the process of cooling the initial system, we derive the volume-temperature relation of our LJ liquid shown by open squares in Fig. 1. The V-T relation in this region is reproducible irrespective of the initial condition, the cooling rate, and the preceding process. This indicates that the system is in an equilibrium liquid state above 60 K or so. When the temperature is lower than 84 K or so, the system is expected to be in a supercooled region, but the mentioned reproducibility confirms that the system is still in equilibrium. On the other hand, the system seems to fall out from equilibrium below 60 K or so, since the state is strongly dependent on the process through which it is achieved. For instance, when the cooling rate is reasonably slow, nuclei grow and the system crystallizes, while a rapid cooling rate yields a glass. In this paper, we confine ourselves to the discussion of the slow-cooling simulations.[14] The rapid-cooling simulations and the problem of the glass transition will be discussed elsewhere.[15]

For the purpose of studying the slow-cooling processes, we first reduce the temperature of a supercooled liquid at 60 K with a step of 1 K down to 40 K. The system is kept at each temperature for 5000 time steps. The process is schematically illustrated by A in Fig. 6. The time development of the volume is shown in Fig. 7(a) where the temperature is fixed during each time interval of 5000 time steps and decreased by 1 K at the end of each interval. Five temperatures at the five time intervals in Fig. 7 are 54 K down to 50 K from left to right. They are indicated in Fig. 7(b). From the figure, we can see that the volume starts decreasing at 53 K and the contraction of the volume is nearly complete at 50 K. In Fig. 7(b), the potential energy is also shown as a function of time steps. Both for volume and energy, the decrease is relatively slow at the beginning while it shows a catastrophic tendency

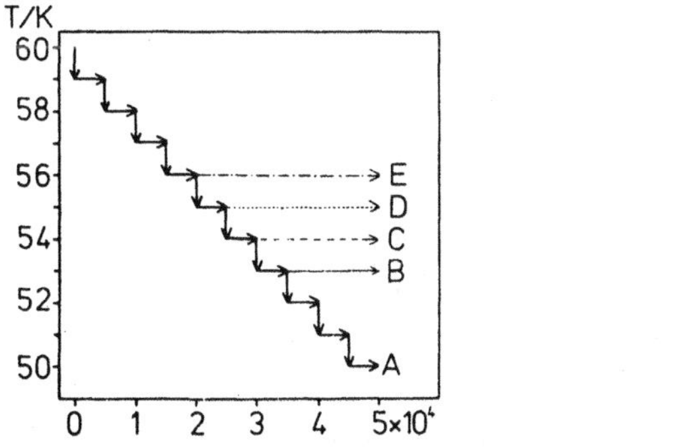

FIG. 6. Schematic explanation of cooling processes from A to E.

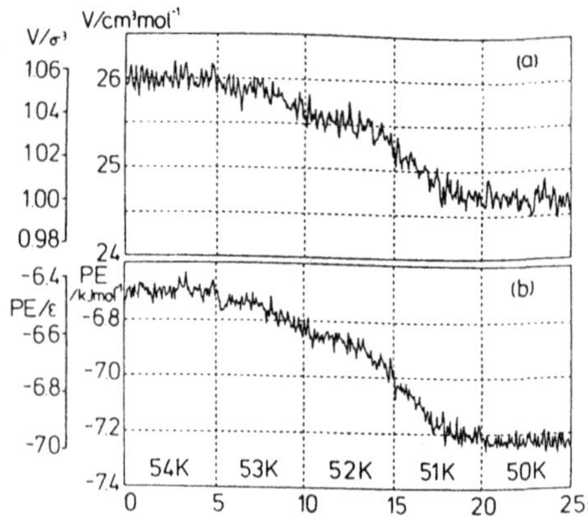

FIG. 7. (a) Volume vs time steps. Temperature (T) is fixed during each time interval of 5000 time steps. At the end of each time interval, T is reduced with a step of 1 K. (b) The corresponding potential energy vs time steps. The temperature at each time interval is indicated in the figure.

after the decrease proceeds to a certain extent. The corresponding MSD is given in Fig. 8. With each time interval of 5000 steps, the MSD for the first 1000 steps is measured relative to the configuration at $t = 0$, while the MSD between $t = 1000$ and $t = 5000$ is measured relative to the configuration at $t = 1000$. The purpose of this rescaling at $t = 1000$ is to eliminate the effects of the stepwise temperature reduction at $t = 0$ of each time interval. In our discussion, we use only the MSD after $t = 1000$. From the figure, it is evident that the lower the temperature, the lower the diffusion of particles. At $T = 50$ K, the slope of the MSD becomes extremely small indicating that the system is solidified. This is consistent with the fact that no substantial volume contraction is observed at 50 K as shown in Fig. 7.

Then, we investigate the structure of the *solid* thus obtained. Here again, it would be instructive to study the pair correlation function $g(r)$. In Fig. 9, $g(r)$ is given for the five different temperatures between 50 and 54 K. Even at 54 K at which the system is expected to be still in a supercooled liquid, the splitting of the second peak is already discernible. This fact may be regarded as indicating the existence of nuclei even at this temperature. Actually, this is not surprising since this temperature of 54 K is well below the melting temperature and accordingly the system is considered to be well supercooled. Accompanying the decrease in temperature, various peaks become marked, suggesting that the system is in some kind of a regular structure. In particular, the subpeak between the first and second main peaks is known as characteristic of a close-packed structure such as an fcc.

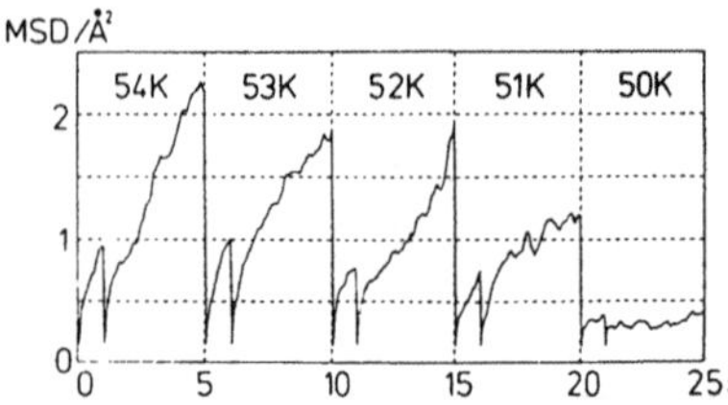

FIG. 8. Mean square displacement of atomic positions for the same process as described in Fig. 7.

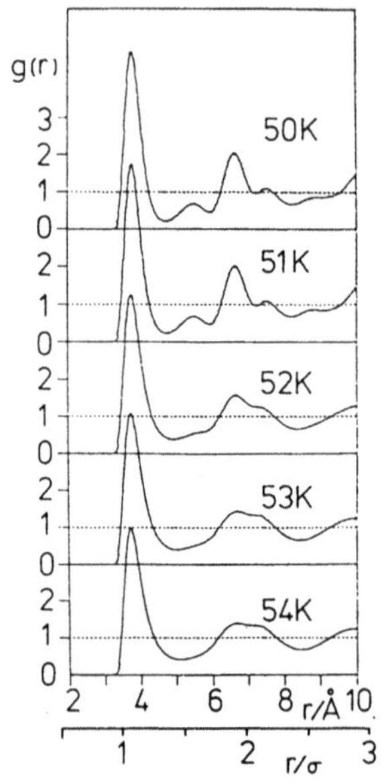

FIG. 9. Pair correlation function $g(r)$ at temperatures between 54 and 50 K.

To investigate the atomic structure more closely, we examine the behaviors of the projections of atoms onto several appropriate planes. One example is given in Fig. 10 where the final atomic configuration of our process, i.e., the configuration at 40 K is projected onto the yz plane where the particles within a sphere of radius 20 Å are picked up. A distinct ten-layer structure is observed. It must be noted that, because the diameter 40 Å of the sphere is larger than the edge length 32.7 Å of the MD cell, some of the layers appear twice in the figure. The atomic distribution in each of

FIG. 10. Projection of all particles onto the yz plane at 40 K.

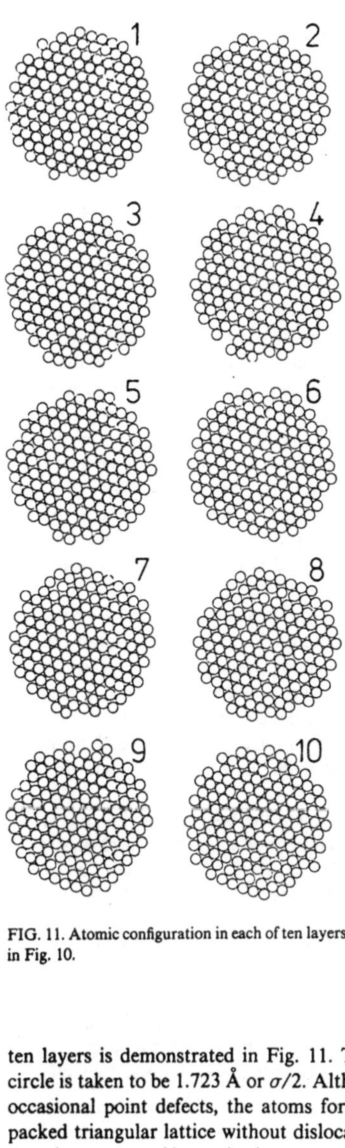

FIG. 11. Atomic configuration in each of ten layers in the system described in Fig. 10.

ten layers is demonstrated in Fig. 11. The radius of each circle is taken to be 1.723 Å or $\sigma/2$. Although there appear occasional point defects, the atoms form a regular close-packed triangular lattice without dislocations. It turns out that the number of layers in the structure attained through the process described in the above, or through other processes is not necessarily a multiple of 3. This indicates that the system does not form an fcc structure with the regular stacking of ABCABC type, but rather takes a close-packed layer structure with stacking faults. Since the differences in the minimum potential energies are negligibly small among an hcp, a random stacking, and an fcc (less than 0.01% as mentioned above), it is not surprising that the system does not show as much preference for a particular stacking. It is interesting to note that the atomic configuration is not always nucleated into a completely ordered structure, but some cooling processes yield the atomic distributions resembling polycrystals. A stereo figure of the nucleated structure is given in Fig. 12. We observe a point defect near the center

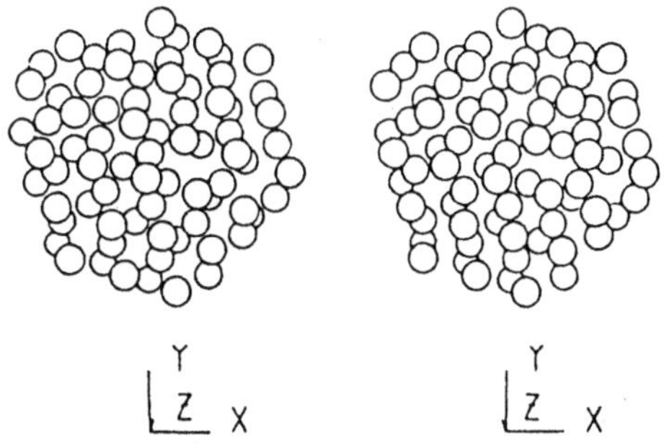

FIG. 12. A stereo figure of the atomic configuration.

of the figure. In this region, the particles form an ABC stacking.

The volume-temperature relation for the cooling process from 60 to 40 K with each decrement of 1 K is represented by small dots in Fig. 1 together with that for the cooling process from 40 to 10 K with each decrement of 10 K. The volumes in the nucleated phase are slightly larger than those of an original fcc crystal (expressed by the filled circles), which is accounted for by the existence of the point defects in the nucleated phase as shown in Fig. 11. In our simulations, no crystallization is observed at 60 K even after the annealing of 120 000 time steps (600 ps). All the simulations we have so far carried out guarantee that no tendency of nucleation appears at a temperature higher than 60 K, while the nucleation becomes observable at temperatures below 60 K. These facts indicate that the lower bound of the supercooled region of our system is in this temperature range. Obviously, this temperature range is lower than the freezing point experimentally observed. As can be seen from Fig. 1, the hysteresis found in the heating–cooling process is rather large.

The cooling rate of our process as described by Fig. 7 is 1 K/5000 time steps $= 4\times 10^{10}$ K/s. This cooling rate is actually higher, by two or three orders of magnitude, than the highest cooling rate attainable in laboratories. The results of our simulations of cooling and nucleation suggest that an argon liquid would crystallize even when quenched with the cooling rate as high as 4×10^{10} K/s. In other words, a quench rate higher than this is required to produce an argon glass through the so-called glass transition. This is notable in connection with the fact that an argon glass has never been obtained in laboratories. The poor glass-forming ability of argon is attributed to the isotropic pair potential. Then, it is interesting to clarify what is the critical cooling rate of a LJ system which separates the crystal-forming cooling rates and the glass-forming cooling rates. According to our simulations, the critical cooling rate is between 4×10^{10} and 4×10^{11} K/s.

VI. COOLING THROUGH OTHER PROCESSES

In a preceding section, we have learned that a LJ liquid starts contracting at 53 K when cooled with a cooling rate 4×10^{10} K/s. However, we show in the following that this temperature 53 K is not essential to the onset of the substan-

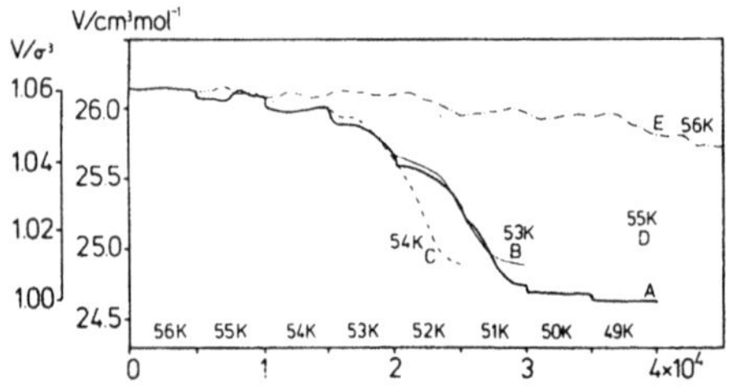

FIG. 13. Volume vs time steps for the annealing processes illustrated in Fig. 6. Each curve is a smooth interpolation of the average volume over 1000 steps at every 1000 step. The time steps are measured from the point of the temperature switching from 57 to 56 K.

tial volume contraction which eventually leads to homogeneous nucleation or crystallization. Actually, we can predict this without going into detail partly because the existence of nucleation is hinted even at higher temperatures as discussed in a preceding section, and mainly because the system of our concern here is in a nonequilibrium "transition" region where various behaviors of the system is strongly process dependent. In order to study the effects of processes, we present, in this section, the results of our cooling simulations through the processes other than that explained in a preceding section. In Fig. 6, the cooling process explained in a preceding section is denoted by A while the processes B to E are to be discussed in this section.

(i) With a view to confirming that the temperature 53 K is not a specific magic number, we first try and see what happens if we leave the temperature unchanged at the end of the first time interval in Fig. 7 and anneal the system at this temperature, 54 K. The process is schematically shown by C in Fig. 6. The volume-time step relation of this process is given in Fig. 13 together with other processes. The time steps are measured from the point of the temperature switching from 57 to 56 K, i.e., from the initial point of process E as shown in Fig. 6. Note from Fig. 13 that a stepwise volume contraction occurs whenever the temperature is reduced in a stepwise manner. This kind of the volume contraction has to be distinguished from the behavior characteristic of nucleation processes. The decrease in volume does take place even when the temperature is kept at 54 K without any further reduction. The time steps needed before the onset of the volume contraction is nearly the same for both cases A and C, i.e., it occurs at about the 18 000th time step (90 ps). This result indicates that the temperature reduction from 54 to 53 K at 15 000th step (75 ps) does not play an essential role in giving rise to the onset of the volume contraction at the 18 000th step (90 ps). In other words, we can regard this result as suggesting that the nucleation is prepared before the 15 000th step (75 ps), and possibly the critical nucleation may be already reached before this point.

Then, the next problem is to examine how early the nucleation is prepared in processes A and C. To solve this problem, we perform the annealing through process D as illustrated in Fig. 6. The volume-time step relation for this run is also given in Fig. 13. From the figure, we can observe that the onset of the volume contraction takes place at about the 26 000th step (130 ps) for process D. Obviously, much more time is required in process D than in processes A and C before the volume starts contracting. This ascertains that the nucleation is not prepared before the 10 000th step where all these three processes A, C, and D follow exactly the same path. Therefore, we are naturally led to the conclusion that, in processes A and C, the reduction of the temperature from 55 to 54 K at the 10 000th step yields a condition favorable to the realization of the critical nucleation. In Fig. 13, the volume-time step relation for process E is also given. In this case, the time required before the onset of the volume contraction is about 38 000 steps (190 ps) which is even longer. From these data, we can say that, the lower the temperature, the easier the nucleation.

It is also instructive to see the behavior of the system when annealed at 53 K. This annealing process is schematically explained by B in Fig. 6, while the *V-t* curve is given also in Fig. 13. The *V-t* curve for B is very similar to that for A. What is typical of these two curves A and B is the shoulder approximately between the 20 000th and 24 000th steps. This feature forms a remarkable contrast to curve C in which the shoulder of this kind is completely absent. This shoulder could be explained if we assume the formation of more than one nuclei at different points in the system which do not match one another, thus preventing the homogeneous nu-

TABLE I. The times necessary before the onset and completion of crystallization are presented for various cooling processes. (1) The cooling processes illustrated in Fig. 6. (2) The cooling processes discussed in step (iii) in Sec. VI.

	Cooling rate (K/s)		Temperature range	Crystallization onset	Crystallization completed
(1)	4×10^{10}	(A)	60 K → 40 K	53 K (90 ps)	50 K (150 ps)
		(B)	60 K → 53 K (anneal)	90 ps	140 ps
		(C)	60 K → 54 K (anneal)	90 ps	120 ps
		(D)	60 K → 55 K (anneal)	130 ps	180 ps
		(E)	60 K → 56 K (anneal)	190 ps	
(2)	2×10^{13}		100 K → 60 K	250 ps	
	4×10^{12}		100 K → 60 K → 55 K (anneal)	110 ps	180 ps
	4×10^{11}		100 K → 60 K → 55 K (anneal)	250 ps	290 ps

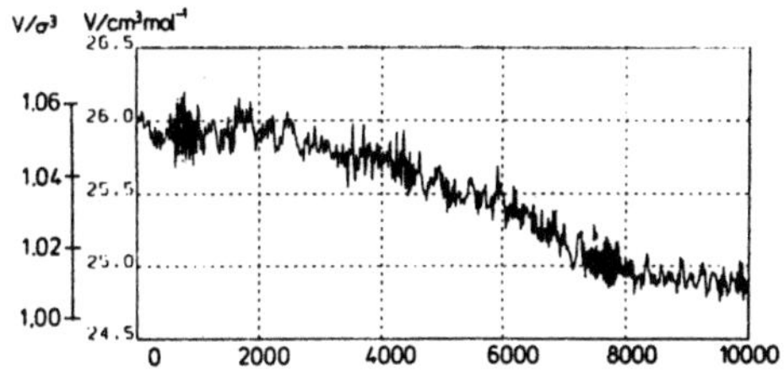

FIG. 14. Volume vs time steps at $T = 54$ K (process C).

cleation from growing. The formation and competition of more than one nuclei does not appear in process C where the system is annealed at 54 K. This annealing, in a sense, is considered to correspond to a cooling rate slower than processes A and B.

In the last column of Table I, the times at the onset and completion of the volume contraction are given where the time is measured from the switching of T from 57 to 56 K in the same way as in Fig. 13. As discussed above, we can learn, from the length of the time needed before the onset of the volume contraction, how well the system has been prepared for the nucleation. On the other hand, we can see, from the length of the time between the onset and the completion of the volume contraction, how easily the homogeneous nucleation grows. Naturally, this depends on the history of each sample. It is worth mentioning that even some of the glasses, obtained by rapid quenching, can crystallize when annealed long enough.

In all cases, we observe the tendency of a slow decrease at the beginning and a relatively faster decrease after a certain point. The abovementioned shoulder is, of course, excluded from the consideration here. As mentioned in a preceding section, this tendency is also observed in the cooling process described in Fig. 7. Actually, this tendency seems to be common to any cooling process, indicating that the growth rate is accelerated once the nucleation reaches a certain extent. Note that this is distinct from the critical nucleation at which the nucleation starts growing. We assert, in the above, that the critical nucleation is realized well before the macroscopic volume contraction of any kind is observed.

(ii) Now, in this section, let us study how the nucleation proceeds as the time develops while the temperature is fixed. For this purpose, we choose the annealing process represented by C. In Fig. 14, the volume is expressed as a function of the time steps which is measured from the step 30 000 in Fig. 6. The figure suggests that the decrease in volume stops at about 8000 steps.

In order to investigate the behavior of the atomic structure along this process, we first examine how the projection of particles (onto some appropriate plane) would change as the time develops from 0 to 10 000 steps. The results are presented in Fig. 15 where the particles are projected onto the xy plane. It would be interesting and instructive to study Fig. 15 with reference to Fig. 14. For instance, at the 4000th step where the decrease in volume is only about a quarter of the whole volume contraction, the layer-like structure is already discernible though partially. The spatial separation of

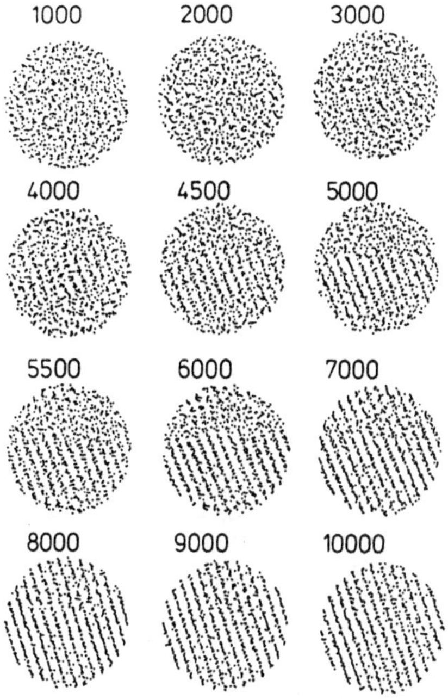

FIG. 15. Projections of particles onto the xy plane at various time steps at 54 K.

the layer-like region and the apparently disordered region is fairly clear at the 5000th step where the volume contraction is nearly halfway. The growth of the layer-like region is impressively demonstrated between the 5000th and 8000th steps. Thus, we can see clearly that, during these 10 000 steps, a transition takes place from a random distribution of atoms to a nearly complete layer structure. To study the atomic structure in each of the layers after 10 000 steps, we draw the figures like Fig. 11, from which it is ascertained that each layer forms a close-packed triangular lattice similar to that shown in Fig. 11. We also calculate the mean square displacement between 8000 and 10 000 steps, which proves that the diffusion is very small in the nucleated structure.

(iii) Since it has been shown above that the cooling rate is important below 60 K, it is interesting to see what significance would be borne by the cooling rate in the process above 60 K. In order to investigate this problem, we produce three samples by cooling the sample at 100 K all the way down to 60 K with three different cooling rates as follows:

(a) 2×10^{13} K/s;
(b) 4×10^{12} K/s;
(c) 4×10^{11} K/s.

Then, we switch the temperatures of the samples thus obtained from 60 to 55 K and let the runs continue without suffering any further change in temperature. The volume-temperature relations for these three samples are shown in Fig. 16 where the time is measured from the point of the

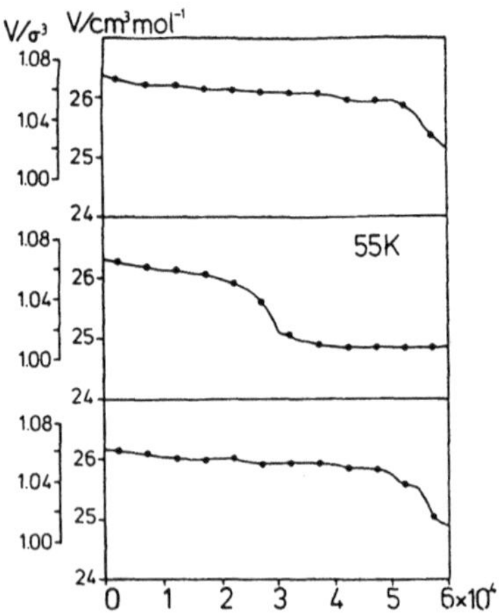

FIG. 16. Volume vs time steps of the annealing processes at 55 K for three different runs. Each dots represents the average volume over 5000 steps around the corresponding time step. The time steps are measured from the point of the temperature switching from 60 to 55 K.

temperature reduction from 60 to 55 K. The corresponding data are also found in Table I. Although the detailed behaviors of the V-t relations for the three samples are different from one another, all of them crystallize within the time steps of 20 000 to 60 000. Although it is difficult to draw any reliable conclusion from only three samples, we can at least say that we do not observe a systematic dependence of the volume-time step behavior on the cooling rate above 60 K. This indicates that the cooling rate in the process above 60 K is not a deciding factor as far as the nucleation is concerned. This conclusion is consistent with the aforementioned observation that the system is in equilibrium above 60 K.

VII. STRUCTURE ANALYSIS

One of the advantages of the MD simulations is that all structural information at the molecular level is available. This enables us to make a microscopic analysis of structure. In this section, we present some of our attempts on the structure analysis. Towards this end, we calculate several structure parameters of our MD systems, and discuss the behavior of the atomic distributions across the nucleation processes. The structure parameters we use in this section are as follows.

A. Distortion parameters

It is quite some time since the concept of the free volume was introduced to explain melting[16] as well as the glass transition.[17] In either case, the fluidity of a liquid is attributed to the existence of a relatively large free volume due to the loose packing of atoms in a liquid. The estimation of the size of the free volume is difficult[18,19] because the atomic distribution in a liquid is topologically disordered. One possible approach is to use the tessellation of the whole space by means of polyhedra which are characterized by the atomic distribution. When the free volume is large, the polyhedra must be distorted. The distortion in the atomic structure in turn can be estimated by the following quantities. (a) The dimensionless ratio of the surface area S to the volume V of the Voronoi polyhedron.[20]

$$w = [S/V^{2/3}]/[4\pi/(4\pi/3)^{2/3}] \tag{7.1}$$

and

(b) the fluctuation in the lengths of the edges of the Delaunay polyhedra[21]

$$\delta = \langle \delta_i \rangle = \langle \Sigma_\mu |l_\mu - \bar{l}\,|/6\bar{l}\,\rangle, \tag{7.2}$$

where $\langle \cdots \rangle$ denotes the average over all Delaunay polyhedra, l_μ the length of the μth edge of the ith Delaunay polyhedron, and δ is the average of the fluctuation δ_i in the ith Delaunay polyhedron. The summation is taken over all edges in the ith Delaunay polyhedron.

B. Bond-orientational parameters

It is well known in crystallography that the fivefold symmetry contradicts the translational symmetry of a crystal. On the other hand, small particles can take shapes having fivefold symmetry since the icosahedral structure, for instance, lowers the local energy owing to the large coordination numbers. Much the same, the local icosahedral configuration is expected in disordered systems such as liquids and amorphous solids. As a consequence, it seems reasonable to suppose that the degree of the fivefold symmetry could serve as a measure for the degree of amorphousness. Here, we use the word "amorphousness" as a synonym of "noncrystalline," thus indicating both liquids and disordered solids. Recently, Steinhardt *et al.*[22] have proposed bond-orientational parameters to discuss the relation between the bond-orientational order and the glass transition. Their parameters are defined as follows. With a bond whose midpoint is at r, the set of numbers

$$Q_{lm} \equiv Y_{lm}[\theta(r), \phi(r)] \tag{7.3}$$

are associated where the $[Y_{lm}(\theta, \phi)]$ are spherical harmonics, and $\theta(\mathrm{r})$ and $\phi(\mathrm{r})$ are the polar angles of the bond measured with respect to some reference coordinate system. The average

$$\bar{Q}_{lm} \equiv \langle Q_{lm}(r) \rangle \tag{7.4}$$

is defined where the average $\langle \cdots \rangle$ is taken over some suitable set of bonds in the sample. Then, the rotationally invariant parameters of second order and third order are introduced as

$$Q_l \equiv \left[\frac{4\pi}{2l+1} \sum_{m=-l}^{l} |\bar{Q}_{lm}|^2\right]^{1/2} \tag{7.5}$$

and

$$\hat{W}_l \equiv \frac{\displaystyle\sum_{\substack{m_1,m_2,m_3 \\ m_1+m_2+m_3=0}} \begin{pmatrix} l & l & l \\ m_1 & m_2 & m_3 \end{pmatrix} \bar{Q}_{lm_1}\bar{Q}_{lm_2}\bar{Q}_{lm_3}}{\left(\displaystyle\sum_m |\bar{Q}_{lm}|^2\right)^{3/2}}, \tag{7.6}$$

where the coefficients in the third order invariants (7.6) are the Wigner 3j symbols. Steinhardt *et al.*[22] assert that, when the average (7.4) is taken over all bonds in the sample, Q_6

and $\hat{W}_6$ can be used as the criteria for the existence of fivefold symmetry.

In this paper, we define the new parameters Q_6^V and $\hat{W}_6^V$ which are obtained by taking the average (7.4) over the bonds associated with each Voronoi polyhedron and calculate the mean of $Q_6(i)$ and $\hat{W}_6(i)$ over all Voronoi polyhedra in the sample. Moreover, when taking the average both for $(Q_6, \hat{W}_6)$ and for $(Q_6^V, \hat{W}_6^V)$, we employ the weighted average rather than the ordinary average. By the weighted average, we mean that the contribution from a given bond (connecting the atom under consideration with the nearest-neighbor atom defined by the Voronoi polyhedron) is weighted by the area of the corresponding face of the Voronoi polyhedron. The purpose of taking the weighted average is to eliminate the spurious effect from the bonds connecting distant nearest neighbors. As is readily expected, this effect is serious in Q_6^V and $\hat{W}_6^V$, and it is negligibly small in Q_6 and W_6.

C. Voronoi face parameter

The Voronoi polyhedra correspond to the Wigner–Seitz cells for a crystal, and both become identical in a crystal. Noting that no odd-edge faces except for triangles appear in the Wigner–Seitz cells in any crystalline structure, we predict that the portion of odd-edged faces can be used as a measure for the degree of amorphousness. With this in mind, we define n_i by the ratio of the number of i-edged faces to the number of all faces on the Voronoi polyhedra in the sample.

These structure parameters for processes C, A, and D are illustrated in Figs. 17(a), 17(b), and 17(c), respectively. We can understand the significance of the results better if we study them with reference to the V-t relation of the corresponding processes. In this connection, it is useful to notice that the double arrow at the bottom of each figure of Figs. 17(a) to 17(c) denotes the time interval during which the volume contraction is under way. Using these structure parameters, we can derive the following results.

(1) The distortion parameters δ and w show the behaviors analogous to that of the thermodynamic properties such as volume. To be more precise, they start decreasing at nearly the same time as the volume contraction begins, and their decrease is saturated at nearly the same time as the volume change stops.

(2) The third order invariant $\hat{W}_6$, determined from the average over all bonds in the sample, is denoted by filled circles in the third row from the bottom in Figs. 17(a) to 17(c). If the bond orientational order of fivefold type exists extensively in the whole MD cell, the parameter $|\hat{W}_6|$ is expected to take a value as large as 0.17, or, at least, of that order. As can be seen from Figs. 17(a) to 17(c), $\hat{W}_6$ does not show a systematic behavior. Even its sign is not definite, varying from time to time. This indicates that the extensive fivefold symmetry does not exist in our system.

(3) The third order invariant $\hat{W}_6^V$, determined from the average at each Voronoi polyhedron, is expected to reflect the local fivefold symmetry. This parameter is represented by open circles. In this connection, it is worth mentioning that $\hat{W}_6 = -0.17$ for an icosahedron while $\hat{W}_6^V$ is order of 0.01 for an fcc, bcc, hcp, and sc cluster. Now, as is seen from

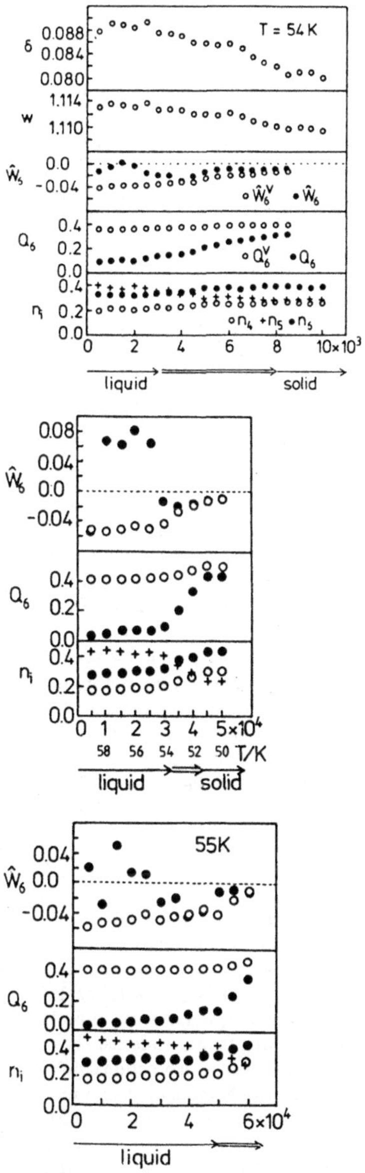

FIG. 17. Changes of structure parameters through the crystallization processes; (a) process C; (b) process A; (c) process D.

Figs. 17(a) to 17(c), $\hat{W}_6^V$ for a liquid is large, which indicates that the local symmetry varies from atom to atom, and yet the value -0.04 suggests local fivefold symmetry. This is a striking contrast to the case of a crystal for which $\hat{W}_6^V$ is nearly 0.01. An interesting point which we can study from Figs. 17(a) to 17(c) is that, in the transition region from a liquid to a crystal, $\hat{W}_6^V$ starts changing before the macroscopic volume contraction is observed. This fact means that the erosion of the local fivefold symmetry starts before the onset of the volume contraction. In other words, $\hat{W}_6^V$ serves as the precursor of the homogeneous nucleation or crystallization.

(4) The second order invariant Q_6, determined from the average over all bonds in the sample, is considered to be about 0.4 when the extensive bond-orientational order of *any kind* exists while it nearly vanishes otherwise. This parameter is denoted by filled circles in the second column from the bottom in Figs. 17(a) to 17(c). We can learn from the figures that Q_6 behaves just like $\hat{W}_6^V$. This is reasonable because the extensive bond-orientational order does not coexist with the substantial degree of local fivefold symmetry. Accordingly, Q_6 also plays a role of the precursor of the crystallization.

(5) The second order invariant Q_6^V, determined from the average at each Voronoi polyhedron, is expressed by open circles. We can see from figures that this parameter is always about 0.4. Actually, the value of this parameter for a cluster is in the range between 0.4 and 0.6 when the cluster has high symmetry. The fact that Q_6^V is about 0.4 even for a liquid indicates that local symmetry of some kind is sustained even in a liquid which has a disordered structure. This existence of local symmetry in a liquid is accounted for by geometrical restriction in the sense that the excluded volume effect does not allow a high degree of local distortion.

(6) The Voronoi face parameter is given in the bottom column of Figs. 17(a) to 17(c), where n_4, n_5, and n_6 are, respectively, denoted by open circles, crosses, and filled circles. The decrease in n_5 and the increase in n_4 and n_6 start a little earlier than the onset of the volume contraction. But the change is not so distinguished as in the case for $\hat{W}_6^V$ and Q_6.

VIII. SUMMARY

The observation of our simulations and some conclusions drawn therefrom are summarized as follows.

(1) When a LJ fcc crystal is heated, the melting takes place at 110 K. On the other hand, a LJ liquid, when cooled with an appropriate cooling rate, crystallizes below 60 K into a layer structure with stacking faults, while each layer forms a close-packed triangular lattice with occasional point defects, but free from dislocations. The atomic configuration is not always nucleated into a completely ordered structure.

(2) The hysteresis in the cooling–heating processes is considerably large as shown in Fig. 1. The upper bound of superheating is in the temperature range near 105 K, while the lower bound of supercooling is in the range near 60 K.

(3) The liquid state at temperatures higher than 60 K can be reproduced irrespective of the initial condition, the cooling rate, etc., thus confirming that the system is an equilibrium supercooled liquid.

(4) The difference in the cooling rates above 60 K does not seem to give an essential influence to the behavior of nucleation.

(5) On the other hand, the behaviors of nucleation sensitively depend on the cooling rate below 60 K. For instance, the smaller the cooling rate, the higher the crystallization temperature. This is interesting in connection with the fact that the smaller the cooling rate, the lower the glass transition temperature.

(6) Then, we can predict the following scenario. According as the increase of the cooling rate, the crystallization temperature $T_{\rm crys}$ becomes lower and the supercooled liquid stays persistent down to $T_{\rm crys}$. Then, at some critical cooling rate, a liquid ceases freezing into a crystal, but instead it freezes into a glass. When the cooling rate is increased still higher beyond the critical cooling rate, then the glass transition temperature becomes higher, though the dependence of the latter on the former is not very remarkable.[15] From our simulations, we can estimate the critical cooling rate q_c as follows:

$$4\times 10^{10}\ \mathrm{K/s} < q_c < 4\times 10^{11}\ \mathrm{K/s}.$$

(7) Concerning the transition from a liquid to a nucleated or crystalline phase, the transition cannot be halted but simply keeps proceeding once the volume starts contracting at a certain temperature below 60 K.

(8) The volume-time step curve as well as the energy-time step curve during the transition suggests that the process of the volume contraction is characterized by two stages defined by two different growth speeds, one relatively low and the other relatively high.

(9) The structure analysis of the nucleation process can be made in terms of the distortion parameter, the bond-orientational parameters, and the Voronoi-face parameter. Some useful information is obtained. In particular, Q_6 and $\hat{W}_6^V$ serve as the precursor of the crystallization.

APPENDIX

As our MD cell, we take a cube with an edge length $L = V^{1/3}$ where V is the volume of the cube. In order to cope with the constant-pressure method of Andersen,[12] it is convenient to introduce a reduced unit for position vectors so that they do not suffer the effect of the change in the volume. For this purpose, we employ a reduced unit for length in such a way that the position vector $\mathbf{r}_i$ of ith particle is scaled as $\mathbf{q}_i = \mathbf{r}_i/L$. Then, it is easily seen that each component of $\mathbf{q}_i$ varies only between 0 and 1 no matter how enormously L may change.

By analogy, we also introduce an additional variable s as a controlling factor of temperature.[13] Then, the equations of motion are written as

$$\ddot{\mathbf{q}}_i = -\frac{1}{m_i L}\frac{\partial\Phi}{\partial \mathbf{r}_i} - \left[\frac{\dot{s}}{s} + 2\frac{\dot{L}}{L}\right]\dot{\mathbf{q}}_i, \tag{A1}$$

$$\ddot{L} = \frac{s^2}{3L^2 W}\left[\frac{1}{3V}\left(\Sigma_i m_i L^2 \dot{\mathbf{q}}_i^2 - \Sigma_{i,j}\mathbf{r}_{ij}\frac{\partial\phi}{\partial \mathbf{r}_{ij}}\right) - P\right] + \frac{\dot{s}}{s}\dot{L} - \frac{2}{L}\dot{L}^2, \tag{A2}$$

$$\ddot{s} = \frac{s}{Q}\left[\Sigma_i m_i L^2 \dot{\mathbf{q}}_i^2 - gkT\right] + \frac{1}{s}\dot{s}^2, \tag{A3}$$

where k is the Boltzmann constant, $\phi(r)$ the interatomic potential, m_i the mass of ith particle, g the number of degrees of freedom of the system.

Parameters W and Q, respectively, serve as "masses" for the motions of L and s when these motions are regarded as those of particles. These parameters are determined in a more or less trial-and-error manner in the MD runs such that the convergence in the integration processes is optimum. It is interesting to note that these parameters thus determined usually fall in the ranges comparable to those estimated from the periods of the oscillations of L and s. In our simulations, W and Q are chosen from the ranges $8\sim16\times10^{-6}$ g mol^{-1} Å^{-4} and $8\sim64$ kJ/(ps)^2, respectively.

The pressure P and the temperature T are externally set to the desired values. The equations of motion, Eqs. (A1) to (A3), guarantee that, in the MD simulations based on them, the average of the internal pressure and that of the total kinetic energy respectively agree with P and T which are externally given as mentioned above. This can be clearly recognized when we express Eqs. (A1) to (A3) in the following forms:

$$\frac{m_i}{sL}\frac{d}{dt}(sL^2\dot{\mathbf{q}}_i) = -\frac{\partial\phi}{\partial\mathbf{r}_i}, \quad (A4)$$

$$\frac{W}{s}\frac{d}{dt}\left(\frac{\dot{V}}{s}\right) = \frac{1}{3V}\left[\Sigma_i m_i L^2\dot{q}_i^2 - \Sigma_{i,j}\mathbf{r}_{ij}\frac{\partial\Phi}{\partial\mathbf{r}_{ij}}\right] - P, \quad (A5)$$

$$Q\frac{d}{dt}\left(\frac{\dot{s}}{s}\right) = \Sigma_i m_i L^2\dot{q}_i^2 - gkT. \quad (A6)$$

The invariance in this method is the quantity

$$H = \Sigma_i \frac{m_i}{2} L^2\dot{q}_i^2 + \phi(L\mathbf{q}) + PL^3 + \frac{9}{2}\frac{WL^4}{s^2}\dot{L}^2 + \frac{Q\dot{s}^2}{2s^2} + gkT\ln s. \quad (A7)$$

It is a common knowledge in all simulation methods to use an invariance or invariances of the methods in order to check the precision of calculations, and we naturally use Eq. (A7) as such.

ACKNOWLEDGMENTS

The authors are grateful to Dr. M. Tanemura, Dr. N. Ogita, and Professor T. Ogawa for providing the program of the Voronoi tessellation. The simulations have been performed by using the HITAC S-810/M-280 system at Computer Center, University of Tokyo.

[1] B. J. Alder and T. J. Wainwright, J. Chem. Phys. **33**, 1439 (1960); Phys. Rev. **127**, 359 (1962).
[2] M. Tanemura, Y. Hiwatari, H. Matsuda, T. Ogawa, N. Ogita, and A. Ueda, Prog. Theor. Phys. **58**, 1079 (1977).
[3] J. N. Cape, J. L. Finney, and L. V. Woodcock, J. Chem. Phys. **75**, 2366 (1981).
[4] M. J. Mandell, J. P. McTague, and A. Rahman, J. Chem. Phys. **64**, 3099 (1976).
[5] C. S. Hsu and A. Rahman, J. Chem. Phys. **71**, 4974 (1979).
[6] J. D. Honeycutt and H. C. Andersen, Chem. Phys. Lett. **108**, 535 (1984).
[7] R. D. Mountain and A. C. Brown, J. Chem. Phys. **80**, 2730 (1984).
[8] C. S. Hsu and A. Rahman, J. Chem. Phys. **70**, 5234 (1979).
[9] R. D. Mountain and P. K. Basu, J. Chem. Phys. **78**, 7318 (1983).
[10] F. F. Abraham, J. Chem. Phys. **72**, 359 (1980).
[11] F. H. Stillinger and T. A. Weber, Phys. Rev. B **31**, 5262 (1985).
[12] H. C. Andersen, J. Chem. Phys. **72**, 2384 (1980).
[13] S. Nosé, Mol. Phys. **52**, 255 (1984); J. Chem. Phys. **81**, 511 (1984).
[14] S. Nosé and F. Yonezawa, Solid. State. Commun. **56**, 1005, 1009 (1985).
[15] S. Nosé and F. Yonezawa, Phys. Rev. A (submitted).
[16] See, for instance, J. Frenkel, *Kinetic Theory of Liquids* (Oxford University, New York, 1946).
[17] M. H. Cohen and D. Turnbull, J. Chem. Phys. **31**, 1164 (1959).
[18] T. Ishimura, N. Ogita, and A. Ueda, J. Phys. Soc. Jpn. **45**, 252 (1978).
[19] Y. Hiwatari, J. Phys. Soc. Jpn. **47**, 733 (1979); J. Phys. C **13**, 5899 (1980).
[20] M. Kimura and F. Yonezawa, in *Topological Disorder in Condensed Matter*, edited by F. Yonezawa and T. Ninomiya (Springer, Berlin, 1983).
[21] Y. Hiwatari, T. Saito, and A. Ueda, J. Chem. Phys. **81**, 6044 (1984).
[22] P. Steinhardt, D. R. Nelson, and M. Ronchetti, Phys. Rev. B **28**, 784 (1983).

PHYSICAL REVIEW B VOLUME 36, NUMBER 12 15 OCTOBER 1987-II

Melting and nonmelting behavior of the Au(111) surface

P. Carnevali
IBM European Center for Scientific and Engineering Computing, Rome, Italy

F. Ercolessi
International School for Advanced Studies, Trieste, Italy

E. Tosatti
International School for Advanced Studies, Trieste, Italy
and International Centre for Theoretical Physics, Trieste, Italy

(Received 13 April 1987)

A molecular-dynamics study of the melting behavior of Au(111) (both reconstructed and unreconstructed) has been carried out using a recently developed many-body force scheme. The reconstructed (denser) surface remains stable up to the bulk melting temperature T_M, showing no form of microscopic surface melting. By contrast, the two outermost layers of the unreconstructed surface "melt" $\sim$100 K below T_M. However, no more layers melt as T_M is approached. The nonmelting of Au(111) is contrasted with the gradual melting behavior of Lennard-Jones surfaces.

The idea that crystal melting could be a surface-initiated process is very old,[1] and some evidence has been provided long ago by macroscopic means.[2] Recently, interest in surface melting has been revived by qualitative ideas,[3] as well as by newly available microscopic surface tools,[4] and by the possibility to realistically simulate the warm crystal surface on the computer. The best simulation so far is that of Lennard-Jones (LJ) crystal surfaces, thoroughly characterized by Allen, De Wette, and Rahman[5] and by Broughton and co-workers.[6] They show clear evidence of surface-nucleated melting, down to temperatures as low as $\frac{3}{4}T_M$, in remarkable agreement with predictions based on simple qualitative models.[3,7] Experimentally, surface-initiated melting has been recently demonstrated on Pb(110)[4] as well as on Ar.[8] For Pb(110), a close correlation has been found[9] between anharmonic surface outwards relaxation and the onset of surface disorder, as was predicted.[7] The general situation is however still far from clear. In the case of Lennard-Jones crystals, the role of vacancy-related surface roughness could be important, and is as yet unclear.[10] Moreover, the thermodynamics of the warm surface is not well established in that one does not know if surface melting is or is not a well-defined phase transition, and of what type. Finally, metals are most commonly used for microscopic experiments, but a LJ crystal is not a good description for them.

We have undertaken a series of molecular-dynamics (MD) calculations to characterize the melting behavior of the Au(111) surface. We have used the many-body force scheme ("glue" potential) introduced by Ercolessi, Parrinello, and Tosatti.[11-13] This potential, though classical, reproduces many features due in reality to the bandlike d electrons. In particular, the Au surface reconstructions are modeled reasonably well by this potential. For the Au(111) surface, specifically, the many-body model predicts a distorted topmost (111) layer with higher density,[14] in qualitative agreement with experiment.[15]

We have used a slab geometry, with the two bottom layers assumed to be rigid in their bulklike positions to mimic the contact with a semi-infinite bulk. Periodic boundary conditions are used along x and y, and free motion is allowed along z (zero pressure). In order to minimize spurious slab effects, the lateral box size was adjusted to match the mean lattice parameter at $T\sim 1350$ K, as extracted from an independent bulk simulation.[13] Atom evaporation is a very improbable event in this system, and we have observed none during our simulations. On the other hand, we have directly checked that a surface vacancy or adatom has an extremely short lifetime ($\sim 10^{-13}$ and $\sim 10^{-12}$ s, respectively) before being annealed out. Therefore, the reconstructed surface is free of vacancies and adatoms at almost any time. Hence, we argue that a solid-vacuum interface—such as that realized in our simulation—should behave very similarly to the equilibrium solid-vapor interface.[16] For most of the calculations we have used slabs of 40 layers with 56 particles on each layer. Our (x,y) cell is defined by $L(a/2)(1,-1,0)\times M(a/2)(1,1,-2)$ with $L=7$ and $M=4$. In this approximately square cell, we can accommodate either an unreconstructed surface (56 top-layer atoms), or a denser unreconstructed surface (64 top-layer atoms). This corresponds to higher surface density $\delta\rho_s/\rho_s \sim 14\%$. This value is slightly higher than the value $\delta\rho_s/\rho_s \sim 9\%$, which is optimal for our potential,[14] but has the advantage of requiring a smaller size cell. The total number of particles with reconstruction is, therefore, $N=56\times 39+64=2248$, of which 112 belong to the rigid layers. This requires about 1 CPU second per MD step on an IBM 3090 with vector facility. Annealing of this system at low temperatures leads to an ordered surface structure, studied in detail elsewhere.[14] This reconstructed surface is, as it should be, much more stable and well packed than the corresponding unreconstructed surface.

To check for lateral size effects we have doubled, in some calculations, the lateral size in each direction (bringing in this way to 224 the number of particles of an unreconstructed layer), and decreased the number of layers to 12 (of which two are rigid) to limit the increase in the total number of particles. In these runs we have found no difference in behavior with the other runs, thus indicating that 56 particles per layer are sufficient for present purposes.

We have performed both microcanonical and canonical runs. Canonical runs have been realized by crudely rescaling the particles velocities at each time step to adjust the kinetic energy to conform to the desired temperature. Figure 1 summarizes the results of the canonical runs, for the reconstructed Au(111) surface, and presents the number n of molten layers at each temperature. There are several qualitative ways to define a molten layer: (a) the intralayer pair correlations have lost their crystalline shell structure; (b) diffusion is linear with time and large; (c) the average energy per atom is ~0.12 eV larger than in a typical bulk layer; (d) the in-plane orientational (hexatic) order parameter O_6 has dropped from close to 1 to close to 0. We define $O_6 = |\sum_{ij} W_{ij} e^{6i\theta_{ij}}| / \sum_{ij} W_{ij}$, where the sums run over first-neighbor pairs and θ_{ij} is the angle which the $i-j$ bond, projected on the xy plane, forms with the x axis. The weight function $W_{ij} = \exp[-(z_i - z_j)^2/2\delta^2]$, $\delta = 0.59$ Å, has the purpose of filtering out all "noncoplanar" neighbors. Figure 2 exemplifies the behavior of O_6 and of the (x,y)-averaged atomic density for a sample with $n = 18 \pm 2$ molten layers and $\langle T \rangle = 1350$ K. As a practical criterion, we call "solid" a layer with $O_6 > \frac{3}{4}$. We have checked that this definition generally fits well with the other criteria (a), (b), and (c) above. In particular, diffusion sets in rather sharply for O_6 smaller than this value.

Returning to Fig. 1, each arrow nn' represents a simulation, beginning with n and ending with n' molten layers. Initial configurations with any number n of molten layers are easily generated by a high-temperature run ($T > 1600$ K) where, starting from an initially solid slab, n grows

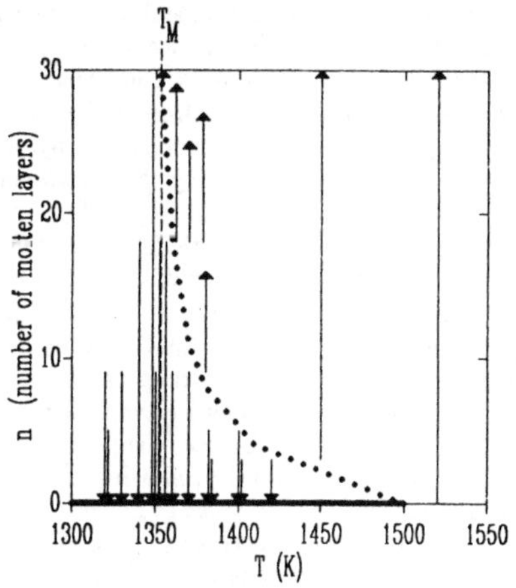

FIG. 1. Summary of the runs for the reconstructed Au(111) surface. The dotted line represents the curve of instability $\nu(T)$.

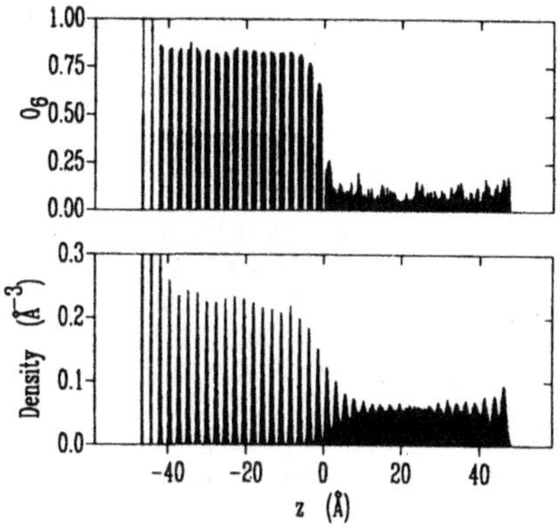

FIG. 2. Orientational order parameter O_6 and (x,y)-averaged density, for a microcanonical sample with $\langle n \rangle = 18$ molten layers and $\langle T \rangle = 1350$ K. The two leftmost layers are rigid. These data have been averaged over 2000 MD steps.

with time. Each run has typically required $\sim 10^4$ to $\sim 3 \times 10^5$ time steps, the larger times being required for runs close to the melting temperature. Since a time step $\Delta t = 7.14 \times 10^{-15}$ s has always been used, our equilibration times range from $\sim 10^{-10}$ s to $\sim 10^{-9}$ s. We generally find that the energy E of a sample is rather accurately related to temperature and to n by the simple relation $E = NC_V T + \Delta H n n_L$ where $C_V = 3.1 \times 10^{-4}$ eV/(K atom) is the solid-bulk specific heat and $\Delta H = 0.12$ eV/atom is the bulk heat of melting for our model potential. Here n_L is the number of particles in each layer. Up to a temperature $T_0 = 1357 \pm 5$ K, the only equilibrium configuration is crystalline ($n = 0$). Above T_0, we find two possibilities. If the initial liquid thickness is small enough, $n < \nu(T)$, the sample crystallizes, $n' = 0$. For $n > \nu(T)$, the sample melts completely, $n' \rightarrow \infty$ (really $n' \rightarrow 35$ due to our finite size and rigid layers). The "unstable line" $\nu(T)$ is oblique, and intersects zero at $T_1 \sim 1500$ K. Above T_1, any initial configuration, including $n = 0$, will melt.

We interpret the above as follows. The temperature T_0 is identified with the bulk melting (triple-point) temperature $T_M = T_0 = 1357 \pm 5$ K. This value is in fairly good agreement with the experimental value $T_M^{\rm expt} = 1336$ K, confirming the good accuracy of the glue potential, also at high temperatures. The crystalline reconstructed surface is stable below T_M, and remains metastable between T_M and T_1. Thus, microscopic surface melting does not occur on this surface. In principle, this does not imply that macroscopic surface melting, i.e., sudden formation of a thick liquid film extremely close to T_M, might not occur. We simply cannot address this question with our tools, due to size and time limitations. Within these limits, however, our surface is not only stable up to T_M, but can also sustain overheating by as much as ~100 K *above* T_M.

It is tempting to relate the lack of microscopic melting of the reconstructed Au(111) to its denser first-layer packing. To test this idea, and also to explore the more general possibility of a totally different behavior for a slightly different state of the surface, we have carried out a parallel study of the unreconstructed surface. Here, the topmost layer is taken to be simply identical to all other layers, i.e., no extra atoms have been added. This state of the surface might be experimentally accessible, in spite of its substantially higher surface energy σ (at $T=0$, $\sigma_{rec}=90.4$ meV/Å^2, $\sigma_{unrec}=96.6$ meV/Å^2).

Figure 3 describes our results for the unreconstructed Au(111) surface. Here, the first two layers melt simultaneously at $T^*\sim 1250$ K, with an energy increase $\Delta h \sim 0.03$ eV/atom, a value much lower than the bulk heat of melting. This may be due to the poor degree of packing of the unreconstructed surface layer and to the concurrent high quality of packing found on the double melted layer. This two-layer melting shows hysteresis, which could indicate a first-order character. Following this two-layer melting, one might have expected to observe the solid-liquid interface to propagate into the bulk, as T_M is approached further. However, this does not happen, and the double-melted layer state remains stable up to T_M. Moreover, in analogy with the reconstructed surface, the two-layer state can be overheated for about ~ 100 K above T_M. We conclude that indeed a situation of poorer surface packing can bring about some microscopic surface melting. Yet, this "nucleus" does not propagate into the bulk to give rise to a thick liquid layer as T_M is approached. In this sense, the lack of surface melting of Fig. 1 is confirmed.

We believe this nonmelting behavior to be an effect of the many-body forces. Specifically, the energetics of surface atoms, very poor in a system with two-body forces (such as LJ), becomes very much better once the many-body forces are included. For example, the (relative) excess energy of a surface atom with respect to a bulk atom at $T=0$, $(E_s-E_b)/|E_b|$, decreases from 0.29 for LJ(111) (Ref. 6) to 0.20 (ideal unrelaxed) to 0.17 (relaxed unreconstructed) to 0.13 [relaxed and reconstructed (111) surface]. As a consequence, all entropy-related quantities, such as thermal vibration and expansion, defect concentration, etc., are expected to rise much higher near T_M in a two-body system than in a many-body system. This particular point finds a direct confirmation by comparison of, e.g., the LJ(111) results of Broughton and Gilmer[6] with our Au(111) results. While the LJ surface is wobbly and full of defects already 5% below T_M, our reconstructed surface is still very much bulklike even *at* T_M. This point is particularly evident from the mean-square-vibration amplitudes of our reconstructed surface, shown in Fig. 4. Even as close to T_M as 5 K, the surface is clearly still vibrationally stable, in contrast with the LJ case.[6] The mean-square first-layer vibration amplitude relative to the bulk for our reconstructed Au(111) surface is only 1.5 just below $T_M (T=1350$ K). The corresponding value for LJ(111) is already as high as 2 at $T/T_M=0.5$.[5]

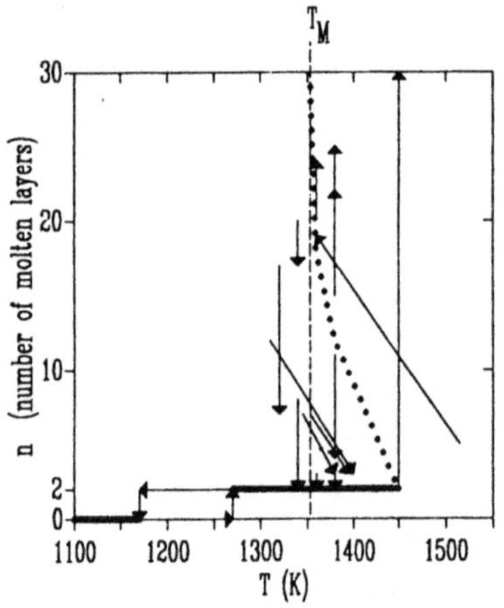

FIG. 3. Summary of the runs for the unreconstructed Au(111) surface. The dotted line represents the curve of instability $\nu(T)$.

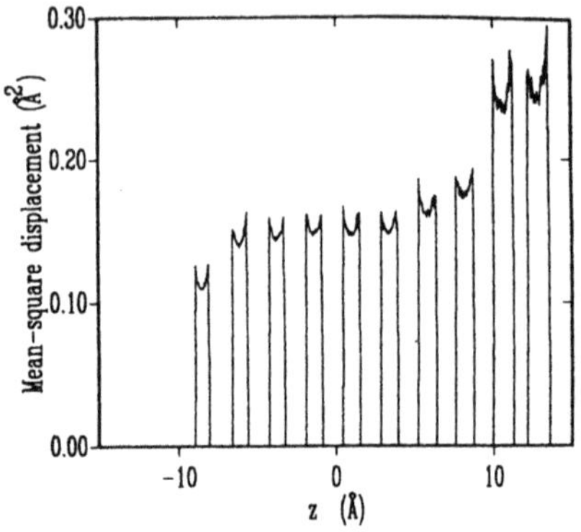

FIG. 4. Average-mean-square displacement $\langle u^2 \rangle$ (sum of the three components) as a function of z for a reconstructed sample at $T=1350$ K, just below T_M. These data refer to a case with 12 layers, of which two are rigid. The decrease in $\langle u^2 \rangle$ for the leftmost layers is due to the contact with the rigid layers. The average $\langle u^2 \rangle$ for the surface (rightmost) layer is only ~ 1.5 times larger than in the bulk (see text).

Similar considerations also apply with respect to single models. In the Pietronero-Tosatti model the surface instability is caused precisely by the abnormal entropy-driven growth of surface thermal vibration and expansion. As shown by Jayanthi, Tosatti, and Pietronero,[7] even a small energetic strengthening—such as that caused by inwards relaxation—is very efficient in raising the vibrational surface instability and possibly eliminating surface melting. The conclusion to be drawn from these considerations is that the improved surface energetics of our metal as compared with, say a LJ crystal, implies a better surface stability against melting. The case of Au(111) is perhaps extreme, and surface stability is so strong as to completely prevent microscopic melting, allowing even the surface to be overheated, at least in the absence of surface steps.

In summary, we have shown the following. (a) A warm metal surface such as Au(111) behaves very differently from a LJ surface. (b) Microscopic surface melting does not occur (at least within a time scale of 10^{-9} s and with a temperature uncertainty of 5 K) on the well-packed reconstructed surface. (c) Melting of the first two layers does take place on a poorly packed surface, such as the unreconstructed Au(111); however, no more than two layers melt. (d) The "nonmolten" surfaces can be overheated for as much as 100 to 150 K above bulk melting. (e) The contrasting behavior of Au(111) and LJ(111) is plausibly related to the improved surface stability of the former, brought about by many-body forces.

[1]A. R. Ubbelohde, *The Molten State of Matter* (Wiley, New York, 1978), Chap. 12, and references therein.

[2]D. Nenow, in *Progress in Crystal Growth and Characterization*, edited by B. R. Pamplin (Pergamon, Oxford, 1984), Vol. 9, p. 185, and references therein.

[3]L. Pietronero and E. Tosatti, Solid State Commun. **32**, 255 (1979).

[4]J. W. M. Frenken and J. F. van der Veen, Phys. Rev. Lett. **54**, 134 (1985); J. W. M. Frenken, P. M. J. Marée, and J. F. van der Veen, Phys. Rev. B **34**, 7506 (1986).

[5]R. E. Allen, F. W. De Wette, and A. Rahman, Phys. Rev. **179**, 887 (1969).

[6]J. Q. Broughton and L. V. Woodcock, J. Phys. C **11**, 2743 (1978); J. Q. Broughton and G. H. Gilmer, J. Chem. Phys. **79**, 5105 (1983); **79**, 5119 (1983).

[7]C. S. Jayanthi, E. Tosatti, and L. Pietronero, Phys. Rev. B **31**, 3456 (1985).

[8]D. M. Zhu and J. G. Dash, Phys. Rev. Lett. **57**, 2959 (1986).

[9]J. W. M. Frenken, F. Huussen, and J. F. van der Veen, Phys. Rev. Lett. **58**, 401 (1987).

[10]However, one can prove that once surface melting has begun, roughening will necessarily ensue below the triple point; see A. C. Levi and E. Tosatti, Surf. Sci. **178**, 425 (1986).

[11]F. Ercolessi, E. Tosatti, and M. Parrinello, Phys. Rev. Lett. **57**, 719 (1986).

[12]F. Ercolessi, M. Parrinello, and E. Tosatti, Surf. Sci. **177**, 314 (1986).

[13]F. Ercolessi, M. Parinello, and E. Tosatti, Philos. Mag. (to be published).

[14]A. Bartolini, F. Ercolessi, and E. Tosatti, Surf. Sci. (to be published).

[15]U. Harten, A. M. Lahee, J. P. Toennies, and Ch. Wöll, Phys. Rev. Lett. **54**, 2619 (1985).

[16]This reasoning neglects asymmetry between surface vacancies and adatoms, which, if important, might cause a gradual decrease of surface density with *T*. We do not expect this effect to be relevant in our case.

Pulsed Melting of Silicon (111) and (100) Surfaces Simulated by Molecular Dynamics

Farid F. Abraham
IBM Research Laboratory, San Jose, California 95193

and

Jeremy Q. Broughton
Materials Science Department, State University of New York at Stony Brook, Stony Brook, New York 11794
(Received 28 October 1985)

The pulsed heating of Si (100) and (111) surfaces has been simulated by molecular dynamics. The (111) crystal-melt interface propagates by layer-by-layer growth whereas the (100) interface grows in a continuous fashion. The equilibrium crystal-melt interface is sharp for the (111) orientation and broad for the (100) orientation. These simulations are the first use of nonpairwise potentials to study interfaces between condensed phases, and the results support models of interfaces which heretofore had to be deduced from indirect experimental information.

PACS numbers: 68.45.−v, 64.70.Dv

The purpose of statistical mechanical simulation is not only to describe materials properties at an atomistic level but also to predict behavior under conditions not experimentally accessible or not yet tried. For the simplest systems such as the inert gases and ionic materials, the state-of-the art has probably achieved many of these objectives. This is not presently true for more complex covalent and/or metallically bonded systems. The problem here is not only a lack of knowledge of good potential functions to describe the interaction between atoms but also the lack of computer power that such complex potentials require for their configuration-space exploration.

Silicon is a case in point. Although a harmonic potential for low-temperature crystalline Si has been known for some time,[1] it is only recently that potentials capable of approximately describing its high-temperature states have been developed.[2–6] The calculation of the force on each atom takes ~ 3 times longer to evaluate than for Lennard-Jones atoms at normal densities. This paper describes an intercomparative study by molecular dynamics of the pulsed heating of the (100) and (111) surfaces using the Stillinger-Weber (SW) potential[3] (described below) with emphasis on the growth characteristics of the melt and the subsequently achieved equilibrium crystal-melt interface. Since this potential was optimized along an isochore for the bulk crystal and liquid phases, we expect it to describe the interface between the two phases rather accurately. Our isobaric calculations of the triple-point properties further illustrate the quality of the SW potential. The simulation results support expectations that until now have only been deduced from indirect experimental observation, the details of which we now describe.

At room temperature the (111) crystal-vapor surface stabilizes in the well-known 7×7 structure.[7] A universally accepted atomistic model of this system is still lacking. At higher temperatures, a 1×1 pattern is observed which has been interpreted by some as due to an amorphous (disordered) overlayer and by others as single termination of the bulk.[8] The (100) surface forms a 2×1 structure which has been interpreted as being due to the formation of Si-Si dimers in the top layer.[9,10] We do not expect the SW potential to describe the 7×7 structure properly since it is generally believed that rehybridization of surface orbitals is a cause of this reconstruction.[7] The SW potential favors tetrahedral bonding. On the other hand, the driving force for (100) dimer formation is thought to be caused by silicon's attempt to achieve tetrahedrality.[9,10] Preliminary results using the SW potential do show in-plane dimer formation.[11,12] In the Czochralski growth of Si from the melt it is known that a slight undercooling exists at the (111) interface but not at any of the other faces.[13] This implies that the (111) interface is just below its roughening transition and grows by a layerwise mechanism. The other faces are above roughening and grow without nucleation by a continuous process. The (111) interface is expected to be smooth and sharp; the (100) rough and broad.

The SW potential energy (PE) of the system is given as the sum over all pairs of atoms of a Lennard-Jones-type term of depth ϵ which smoothly goes to zero at a distance a (approximately the second-neighbor distance) plus the sum over all triplets of a three-body term of the form

$$\phi_3(r_{ij},r_{ik},\theta_{jik}) = \lambda \exp[\gamma(r_{ij}-a)^{-1}+\gamma(r_{ik}-a)^{-1}] \times(\cos\theta_{jik}+\tfrac{1}{3})^2.$$

This term vanishes if either r_{ij} or r_{ik} is greater than a. The angular term is zero at the ideal tetrahedral angle and positive otherwise.

Each of the (111) and (100) systems were periodically connected in the x, y, and z directions with sufficient extent in the z to allow for vapor phase. The standard techniques of molecular dynamics were applied. Properties are reported in the reduced units of the SW paper; that is, the unit of length is 0.209 51 nm, the unit of energy is 3.4723×10^{-19} J, and the unit of mass (^{28}Si) is 4.6459×10^{-26} kg. The systems each comprise approximately 1800 atoms and were equilibrated for $20000\,\Delta t$ ($\Delta t = 3.8\times10^{-16}$ sec) at a reduced temperature of 0.070 (1760 K), our best guess for the triple point. The x-y period was set by use of the density from a zero-pressure bulk-crystal calculation at this temperature. Neither the (111) nor (100) systems exhibited interface melting in contrast with the same faces of a fcc Lennard-Jones system at its triple point.[14] The (111) surface, in accord with the conclusion of high-temperature LEED experiments,[8] exhibits a structure which is simple termination of the bulk. The kinetic energy of the top four layers was then instantaneously raised on one side of the slab a sufficient amount to cause approximately half the system to eventually melt. Whereas we were able to guess the amount of energy to be added to do this for the (111) surface, our first attempt on the (100) produced only a limited amount of melt. After the latter equilibrated, further kinetic energy was added and it is the results of this experiment which are reported below. The history of the constant-energy systems was followed until equilibrium was again achieved.

After the heat pulse (time zero), it took $\sim 60000\,\Delta t$ for the system to reach equilibrium, that is, for the total system temperature to be invariant with time and for the temperature profile through the system to be flat. Figure 1 gives trajectory plots in a thin x-z section for the (111) and (100) interfaces as a function of time. The (111) melt front clearly progresses a layer at a time which we expect by microscopic reversibility from Czochralski-crystal-growth experiments. Trajectory plots in the x-y plane and for other x-z sections show the front to be flat in the x-y plane. In contrast, the (100) interface grows over several layers simultaneously. Figure 2 shows x-y plots over the same time subinterval during the growth of four neighboring layers. The region in the top left corner of the computational box shows particles vibrating on lattice sites over four layers while the lower right indicates substantial disorder in those same layers. This behavior is also observed at "equilibrium" and is symptomatic of a rough interface. Irregular hill-and-valley structures at the interface form and unform dynamically as the system fluctuates around equilibrium. A similar trajectory analysis for the equilibrium (111) interface shows a transition from crystal to a highly mobile region to occur over one layer. Such a smooth interface is in keeping with the experimental observation that Si(111) is above its roughening temperature T_r and the spatial width of the transition region is similar to that observed in Lennard-Jones (100) and (111) molecular-dynamics simulations.[15] We cannot determine from our small finite-size systems whether these interfaces exhibit roughening behavior, but what is striking is the very dissimilar behavior of Si(100). T_r for Si(100) is so low that at the melting temperature the length scale for roughening fluctuations is apparently shorter than the dimensions of the computational cell. To our knowledge, this is the first three-dimensional molecular-dynamics simulation of crys-

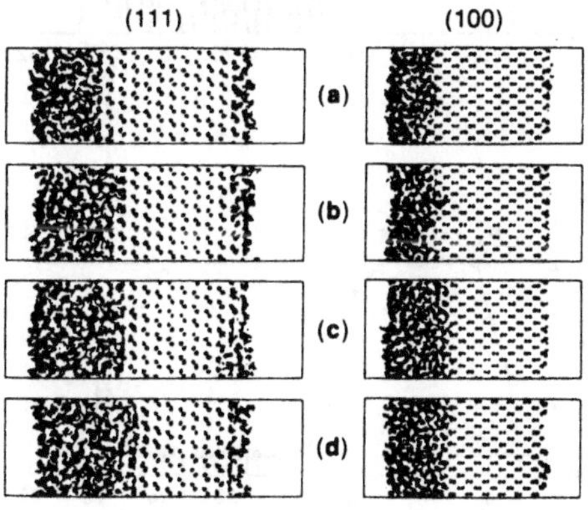

FIG. 1. Trajectory plots of thin x-z slice through (111) and (100) systems covering elapsed time ranges of (a) $(15-20)\times10^3\,\Delta t$, (b) $(25-30)\times10^3\,\Delta t$, (c) $(45-50)\times10^3\,\Delta t$, and (d) $(95-100)\times10^3\,\Delta t$.

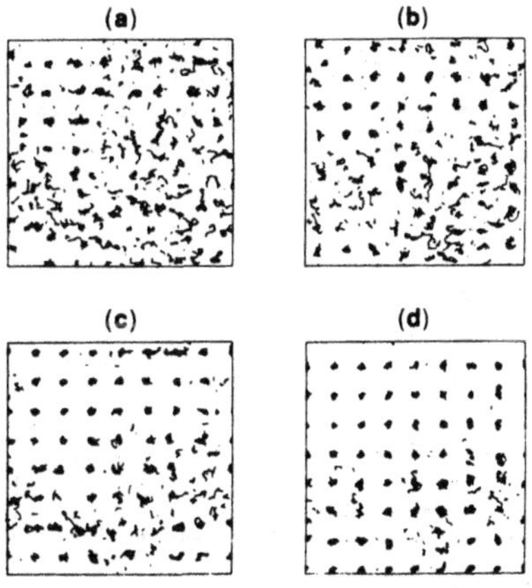

FIG. 2. Trajectory plots in x-y plane of adjacent layers covering elapsed time between $45000\,\Delta t$ and $50000\,\Delta t$. The depth range equals (a) 10.30–10.95, (b) 10.95–11.60, (c) 11.60–12.25, and (d) 12.25–12.90.

tal-fluid systems to indicate interface fluctuations characteristic of roughening. By way of rationalization of this behavior, we note that the nearest-neighbor bond model for Si predicts $T_r = 0$ for the (100) face because there is no in-plane fully connected bond network. Consequently, it costs zero energy to insert a step at the interface.[16]

Note the growth of a quasiliquid layer on the opposite side of the crystalline film during the (111) simulation. The quasiliquid layer remains after the system equilibrates. The shock wave, generated by the heat pulse, is able to overcome the nucleation barrier to quasiliquid-layer formation. The barrier is apparently more significant than that for the Lennard-Jones (111) or (100) faces since these surfaces melt both rapidly and with facility.[14] We discuss this point more fully elsewhere.[17] A similar mobile surface region is created at the Si(100) surface after the heat pulse but is less evident perhaps because the pulse was of lower magnitude (see earlier discussion). The evidence for a high-temperature quasiliquid layer on Si(111) is an attractive explanation for the 1×1 LEED structures and the amorphous overlayer suggested by ion scattering. A further temperature-dependent study of the Si(111) crystal-vapor interface is required to pursue this point further.

Figures 3 and 4 give the number-density $\rho(z)$ and three-body energy-density $E_3(z)$ profiles through the equilibrated systems. The (111) interface is sharp, the transition occurring over ~ 3 reduced units of distance (0.6 nm). The three-body profile shows a large positive contribution in the liquid, where the coordination is near 8 and well away from tetrahedrality, and a small contribution in the crystal caused by the finite mean-square displacement of atoms away from lattice sites. Again, the quasiliquid region on the right-hand end of the crystal is evident. The (100) interface, in contrast, is broader being approximately 4 reduced units wide (0.8 nm). The ripples on the three-body profile on traversing the crystal-liquid interface nicely demonstrate the occurrence of both ordered and disordered regions within the same interfacial layers. Trajectory plots indicate significant diffusion for all z values less than 13.5. Lastly, note that both the (111) and (100) systems have a shoulder in $\rho(z)$ at the liquid-vapor interface. This indicates a partially ordered structure at the interface which is not observed at simple liquid-vapor Lennard-Jones interfaces,[18] and which will be discussed in detail in a future paper.[17]

Table I gives the triple-point properties of the two

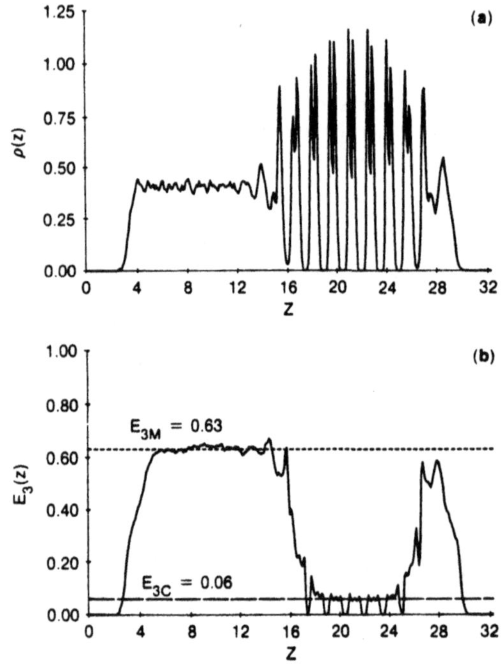

FIG. 3. (a) Equilibrium number density and (b) three-body potential energy density profiles for (111) triple-point system.

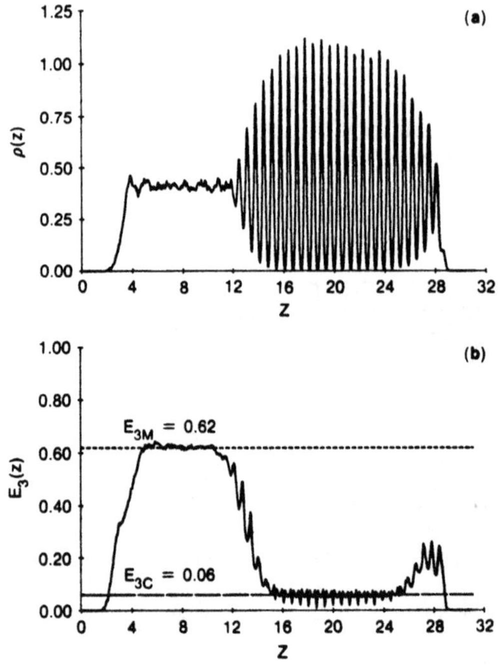

FIG. 4. (a) Equilibrium number density and (b) three-body potential energy density profiles for (100) triple-point system.

TABLE I. Triple-point properties in reduced units. Potential energies given per particle.

Surface	T	E_C	E_M	E_{2C}	E_{2M}	E_{3C}	E_{3M}	L	ρ_C	ρ_M
(111)	0.069(4)	−1.86	−1.72	−1.92	−2.35	0.06	0.63	0.14	0.45	0.49
(100)	0.070(2)	−1.86	−1.71	−1.92	−2.33	0.06	0.62	0.15	0.45	0.49

systems. They are in good agreement with one another. The melting point is ~ 1760 K which is to be compared with the experimental value of 1683 K. The two-body energy of the liquid is actually lower than that of the crystal. This is consistent with the liquid having the higher coordination number. The three-body energy difference more than compensates for this, however, and a positive heat of fusion (L) results. The calculated value of L of 30.3 kJ/mole[19] agrees poorly with an experimental value of 50.7 kJ/mole, and implies a sizable electronic contribution to the entropy of fusion. Notice that, in agreement with experiment, there is a density increase upon melting and the calculated values of 2.28 and 2.45 compare favorably with the experimental values of 2.30 and 2.53 g/cm^3.[20] Lastly, the (essentially) zero concentration of particles in the vapor phase of the simulation is compatible with the low experimental triple-point pressure of $\sim 10^{-1}$ Pa.[21] This latter quantity, assuming ideality, is equivalent to $\sim 10^{-8}$ atom in the vapor volume of this simulation.

In conclusion, our simulations of silicon have shown for the first time the ability of the SW potential to describe satisfactorily a variety of triple-point properties and to explain growth characteristics which heretofore have been inferred from macroscopic experiment.

The work of one of us (J.Q.B.) was supported in part by an Olin Corporation Charitable Trust grant from the Research Corporation and by a contract from the U.S. Department of Energy (No. DE-FG02-85ER45218).

[1]P. N. Keating, Phys. Rev. **145**, 637 (1966).

[2]D. A. Smith, Phys. Rev. Lett. **42**, 729 (1979).

[3]F. H. Stillinger and T. A. Weber, Phys. Rev. B **31**, 5262 (1985).

[4]R. Biswas and D. R. Hamman, to be published.

[5]E. Pearson, T. Takai, T. Halicioglu, and W. A. Tiller, J. Cryst. Growth **70**, 33 (1984).

[6]B. W. Dodson and P. A. Taylor, Conference Abstracts, Materials Research Society Meeting, Boston, 1985 (unpublished).

[7]P. M. Petroff and R. J. Wilson, Phys. Rev. Lett. **51**, 199 (1983), and references therein.

[8]W. S. Yang and F. Jona, Phys. Rev. B **28**, 1178 (1983), and references therein.

[9]D. J. Chadi, J. Vac. Sci. Technol. **16**, 1290 (1979).

[10]M. T. Yin and M. L. Cohen, Phys. Rev. B **24**, 2303 (1981).

[11]F. F. Abraham and I. P. Batra, Surf. Sci. (to be published).

[12]T. Weber, private communication.

[13]T. F. Ciszek, J. Cryst. Growth **10**, 263 (1971).

[14]J. Q. Broughton and G. H. Gilmer, Acta Metall. **31**, 845 (1983).

[15]J. Q. Broughton, A. Bonissent, and F. F. Abraham, J. Chem. Phys. **74**, 4029 (1981).

[16]G. H. Gilmer, Mater. Sci. Enger. **65**, 15 (1984).

[17]F. F. Abraham and J. Q. Broughton, to be published.

[18]F. F. Abraham, in *Surface Science: Recent Progress and Perspectives*, edited by R. Vanselow (CRC Press, Cleveland, 1980).

[19]This value of L agrees favorably with the constant-density value extracted from Fig. 3 of SW's paper. The value quoted in their text is in error. F. H. Stillinger, private communication.

[20]V. M. Glazov, S. N. Chizhevskaya, and N. N. Glagoleva, *Liquid Semiconductors* (Plenum, New York, 1969).

[21]F. Rosebury, *Handbook of Electron Tube and Vacuum Techniques* (Addison-Wesley, Reading, Mass., 1965).

Faceting at the Silicon (100) Crystal-Melt Interface: Theory and Experiment

Uzi Landman, W. D. Luedtke, R. N. Barnett, C. L. Cleveland, and M. W. Ribarsky
School of Physics, Georgia Institute of Technology, Atlanta, Georgia 30332

and

Emil Arnold, S. Ramesh, H. Baumgart, A. Martinez, and B. Khan
Philips Laboratories, Briarcliff Manor, New York 10510
(Received 30 September 1985)

Molecular-dynamics simulations and *in situ* experimental observations of the melting and equilibrium structure of the crystalline Si(100)-melt interface are described. The equilibrium interface is structured, exhibiting facets established on (111) planes.

PACS numbers: 68.55.Rt, 64.70.Dv, 68.45.Kg

The study of the solid-vapor and solid-liquid interfaces has attracted a recent surge of interest because of improved experimental and theoretical techniques for probing phenomena such as surface melting, surface roughening, and interface morphology. In particular, it has been suggested that the structure of the solid-melt interface during zone-melting recrystallization of silicon critically determines the generation of the observed networks of low-angle grain boundaries.[1] Ample evidence exists that surfaces of crystalline silicon become faceted upon melting, and the solid-melt interface of a growing Si crystal establishes itself on the (111) crystal planes.[2] Several models have been proposed to understand the solid-liquid interface morphology at equilibrium, notably the simple two-layer model of Jackson.[3] Silicon marginally satisfies the Jackson criterion for facet formation. It could be expected that under appropriate conditions such material would exhibit facets both on solidification and in melting. In previous experimental[4,5] and theoretical[1] studies the silicon crystal-melt interface was investigated under nonequilibrium conditions during growth by laser-induced zone melting of silicon films on SiO_2. In this Letter we report on the first theoretical microscopic simulations and experimental *in situ* observations of the melting and *equilibrium* structure of that interface, which provide evidence for a (111)-faceting transformation at the interface.

In our theoretical studies we have used the molecular dynamics method,[6] which consists of a numerical solution of the equations of motion of a large ensemble of interacting particles on refined spatial and temporal scales. The structural, dynamical, and stability properties of homogenous phases and equilibrium interphase interfaces, as well as the kinetics and dynamics of phase transformations (such as melting and solidification), are governed by the various contributions to the total energy (or free energy, at finite temperatures).[6b] The potential energy of an interacting system of particles can be written in general as a sum of contributions of varying order in the number of particles (one-body, two-body, . . . terms). Because of the directional covalent bonding, characteristic of tetrahedral semiconductors, a model of the potential for these materials must go beyond the often-used pair interactions, via the inclusion of nonadditive, angle-dependent contributions (three-body and higher order). In our simulations we have employed optimized two- and three-body potentials, V_2 and V_3, respectively, which have yielded a rather adequate description of the structural properties of crystalline and liquid silicon[7]:

$$V_2(r_{ij}) = A[Br_{ij}^{-P} - 1]g_\beta(r_{ij}), \quad (1a)$$

$$V_3(r_i, r_j, r_k) = v_{jik} + v_{ijk} + v_{ikj}, \quad (1b)$$

$$v_{jik} = \lambda g_\gamma(r_{ij}) g_\gamma(r_{ik}) [\cos\theta_{jik} + \tfrac{1}{3}]^2, \quad (1c)$$

where r_{ij} is the distance between atoms i and j, and $g_\gamma(r) = \exp[\gamma/(r-a)]$ for $r < a$ and vanishes for $r \geqslant a$. In Eqs. (1) r is expressed in units of $\sigma = 0.209\,51$ nm, the unit of energy is $\epsilon = 50$ kcal/mole, and that of temperature is $T = \epsilon/k_B$ (to convert to T in Kelvins multiply the reduced temperature by 2.5173×10^4); $A = 7.044\,556\,277$, $B = 0.602\,224\,558\,4$, $P = 4$, $a = 1.8$, $\lambda = 21$, $\beta = 1$, and $\gamma = 1.2$. The time unit, t.u., is $\sigma(m/\epsilon)^{1/2} = 7.6634 \times 10^{-14}$ s and the integration time step Δt, with use of a fifth-order predictor-corrector algorithm, is taken to be 1.5×10^{-2} t.u.; with this choice and a frequent updating of the interaction lists, the total energy is conserved to at least six significant figures. As seen from Eq. (1c), the three-body contribution to the potential energy vanishes for the perfect tetrahedral angle. Therefore, the liquid is characterized by a higher magnitude of V_3 than the solid. While improvements to the potential functions are desirable, their general form and the degree of agreement with observed data which they provide[7] warrant their use in our study of the interface between condensed phases, i.e., at the solid-melt interface (conduction-electron-density–dependent potential-energy terms could be incorporated in future studies in a manner similar to that proposed by us recently).[8] Additionally, the molecular dynamics technique which we use allows for dynamical variations in particle density and structural changes via the *Ansatz* Lagrangean of Parrinello and Rahman,[9] extended to include three-body interactions and planar 2D periodic boundary conditions.

Since zone-melted Si films on SiO_2 tend to recrystallize with (100) texture,[4,5] we start with a silicon crystal consisting of N_L dynamic layers, with N_P particles per layer, exposing the (001) face. The z axis is taken parallel to the [001] direction, and the 2D cell is defined by the [110] and [$\bar{1}$10] directions. The bottom layer of the crystal (layer number 1) is positioned in contact with a static silicon substrate. Simulations for two systems were performed: (i) $N_L=28$ and $N_P=36$, for which results are shown, and (ii) $N_L=24$ and $N_P=144$, yielding similar results. The N_L values were chosen so as to minimize the static substrate effects at the interface.

Following equilibration of the total system at a temperature $T=0.064$ and zero external pressure, a portion of the system (about $N_L/2$ from the top, free surface) was heated via scaling of particle velocities. During the subsequent dynamical evolution towards equilibrium, melting initiated, with the melting front propagating from the exposed surface towards the bulk, exhibiting a tendency for facet formation upon melting. After a prolonged period (for the data presented below, runs in excess of $1.5\times10^5\Delta t$) an equilibrium crystal-melt coexistence was established, at an average kinetic temperature $T_m=0.0662\pm0.0016$, uniform throughout the system. This value may be compared to the experimental melting temperature of silicon, $T_m=0.0669\equiv1683$ K. The equilibrium crystal-melt interface exhibits a pronounced structure, demonstrated by the sample particle trajectories, shown in Fig. 1(a). These are recorded for $2000\Delta t$ and viewed along the [$\bar{1}$10] direction (denoted by a circle at the bottom left), where the breakup into alternating (111) and ($\bar{1}\bar{1}$1) crystalline planes is indicated. The melt region in the vicinity of the solid (facet) planes exhibits a certain degree of ordering due to the crystalline potential, resulting in a diffuseness of the interface at that region.[6b] To complement the picture, we show in Fig. 1(b) particle trajectories in the region of the seventh layer ($l=7$), projected onto the 2D plane, exhibiting solid and partial-liquid characteristics. In extended runs we observed that the morphology of the interface fluctuates (on a time scale of $\sim5\times10^3\Delta t$) between equivalent facet configurations, the one shown in Fig. 1(a) where the facet runs along the ($\bar{1}$10) direction and the other one where the facet runs along the (110) direction.

Further insights are provided by the equilibrium particle density profile [Fig. 2(a)] and per-particle potential energies [Figs. 2(b)–2(d)], which show both the crystal-melt and melt-vacuum interfaces. Focusing on

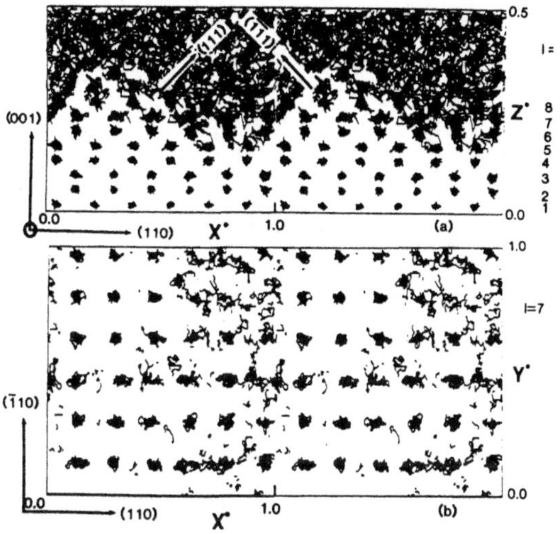

FIG. 1. Real-space particle trajectories at the interface region recorded at equilibrium. (a) Viewed along the [$\bar{1}$10] direction; (b) trajectories for particles at the region of the seventh layer ($l=7$) viewed from the [001] direction. $Z^*=1\equiv18.14\sigma$ and $X^*=Y^*=1\equiv10.9\sigma$. The 2D computational cell ($0\leq X^*,Y^*\leq1$) is replicated along the X^*[110] direction to aid visualization.

the former interface, we observe the opposing trends in the behavior of the two- and three-body potentials, V_2 and V_3, which, when added, yield the result shown in Fig. 2(b). The pronounced minima correspond to the locations of crystalline or partially crystalline layers. We observe that the variation in the total potential energy, V_2+V_3, upon transformation from the solid to the melted region, is smaller than the variation in the individual contributions V_2 and V_3. The behavior of the three-body potential can be used to distinguish the solid from the liquid and thus provide a vivid visualization of the solid component of the interface. In Fig. 2(e) contours of V_3 for $V_3\leq0.45$ are shown. By restriction of the value of V_3, only solidlike regions are captured (the appearance of the V_3 contour map does not change significantly with small changes in the cutoff value). Figure 2(e), in conjunction with the real-space trajectories shown in Fig. 1, complements and corroborates our picture of the structure of the interface.

To affirm further our prediction of the (111) faceting of the equilibrium crystalline Si(100)-melt interface, we present in Fig. 3 results for the Si(111)-melt interface, simulated in a manner similar to that described above, with $N_L=20$ and $N_P=49$. The particle trajectories viewed along the (110) direction [Fig. 3(a)] and in-layer trajectories for layers 10,11 and 8,9, shown in Figs. 3(b) and 3(c), respectively, along with the density and per-particle potential energy profiles in the (111) direction [Figs. 3(d) and 3(e), respectively],

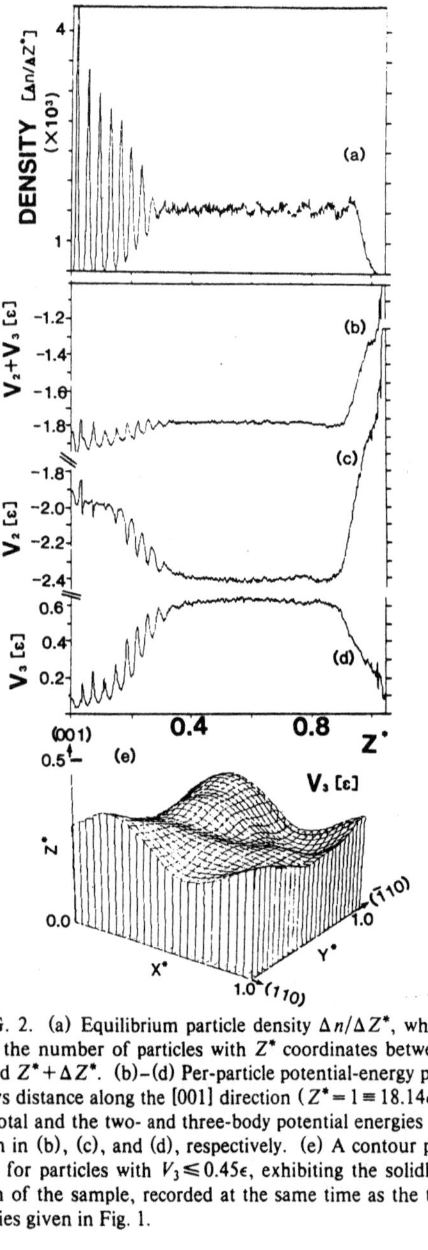

FIG. 2. (a) Equilibrium particle density $\Delta n/\Delta Z^*$, where Δn is the number of particles with Z^* coordinates between Z^* and $Z^*+\Delta Z^*$. (b)–(d) Per-particle potential-energy profiles vs distance along the [001] direction ($Z^*=1\equiv 18.14\sigma$). The total and the two- and three-body potential energies are shown in (b), (c), and (d), respectively. (e) A contour plot of V_3 for particles with $V_3 \leq 0.45\epsilon$, exhibiting the solidlike region of the sample, recorded at the same time as the trajectories given in Fig. 1.

provide clear evidence for a sharp, abrupt, and flat Si(111)-melt interface. Additionally, the melt region adjacent to the interface ($L=10$ and 11) possesses a residual degree of in-planar order.

To compare the predictions of the molecular-dynamics simulations with an experimental system, we have observed the solid-melt interface *in situ*, during melting with a shaped beam of a cw argon laser. The starting material was a 600-nm-thick film of polycrystalline silicon deposited on a quartz wafer by chemical vapor deposition and capped by a layer of silicon dioxide. The sample was mounted on a translation stage positioned upon a vibration-isolated table, and the liquid-solid interface was observed by means of a microscope equipped with a visible and near-infrared video camera. Two laser beams, shaped into elliptical cross sections, and positioned in close spatial proximity, were used to maintain a closely controlled temperature gradient across the molten zone. The size of the molten zone was approximately $25\times 100\ \mu m^2$. After an initial scan to convert the polycrystalline material into a monocrystalline ribbon, the scanning was stopped to establish quasiequilibrium temperature distribution across the molten zone. A view of the solid-melt interface is shown in Fig. 4. The recrystallized film has predominantly (100) texture, as has been confirmed by subsequent x-ray diffraction. In the direction normal to the original scan, small periodic deviations from the ⟨100⟩ axis occur, with a characteristic length of approximately 10 μm. The various sets of facets defined by alternating (111) and $(\bar{1}\bar{1}1)$ planes seen in Fig. 4 appear slightly tilted with respect to each other so as to preserve the relative alignment with the bulk texture. The structure shown in Fig. 4 is typical of many faceted configurations that were observed. The shape of the solid-melt interface fluctuates, as certain facets spontaneously disappear and new facets form. Individual facets typically persist for several video-frame times ($\frac{1}{30}$ s).

In comparison of the experimental and simulation results, the similarity in appearance is rather striking, particularly when one notices that the two differ vastly in spatial and temporal scales. On the foundation of the premise that the origins of all physical phenomena are microscopic in nature, we venture the hypothesis that above a certain size the energetics that govern the structure of the solid-melt interface in this system operate in a similar manner in both the microscopic (theory) and macroscopic (experiment) regimes, and that the dimensions of the morphological characteristics scale with system size. Clearly, the dimensions of the computational cell limit the size of facets that form. For a small facet, the relative number of edge and corner atoms to those located on the facet plane is high, thus affecting its stability. While analysis of the results obtained for the two system sizes which we simulated provides some support to this hypothesis, more theoretical and experimental work is needed. Such theoretical efforts would involve further, extended simulations and the development of improved microscopic theories of interfacial energetics, dynamics, and morphological stability.[10] On the experimental side, it would be desirable to observe the solid-melt interface on much smaller spatial and temporal scales to record the process of facet formation in greater detail.

The structure of the interface is governed by the en-

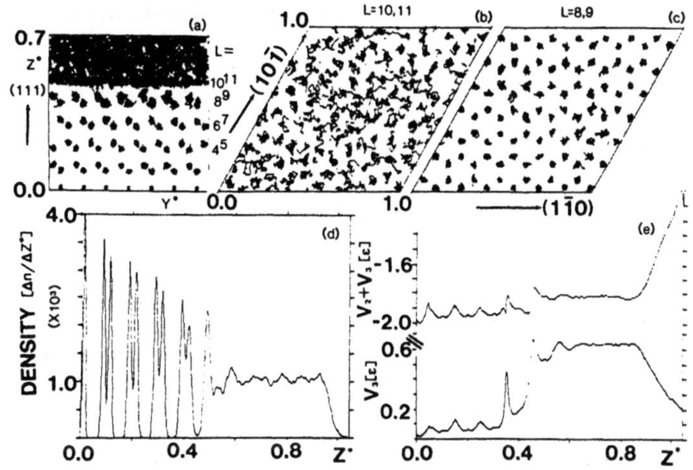

FIG. 3. (a) Particle trajectories recorded at equilibrium, viewed along the $[1\bar{1}0]$ direction. (b)–(c) In-plane trajectories, exhibiting solid and melt characteristics, in layers 10,11 and 8,9, viewed along the [111], Z^*, direction. $Z^* = 1 \equiv 14.96\sigma$ and the unit of length in the $[1\bar{1}0]$ and $[10\bar{1}]$ directions is 12.85σ. (d) Equilibrium particle density, $\Delta n/\Delta Z^*$, vs Z^*. (e) Per-particle total and three-body potential energies, $V_2 + V_3$ and V_3, vs Z^*.

ergetics, particularly the interplay between the two- and three-body contributions to the potential energy, and by entropic factors. Our observation that faceting initiates upon melting and then further refines upon achieving equilibrium is most likely a result of the fact that the closest-packed (111) planes of silicon are the directions of slowest growth.[11] In a dynamic equilibrium situation these faces are also the slowest to melt. An important issue in investigations of the recrystallization kinetics of zone-melted silicon is the role of the interface morphology in the generation of low-angle grain boundary defects, whose branching behavior was rather well described by a recent kinetic model,[1] with the assumptions of a faceted structure of the growth interface and that the growth rate is limited only by the nucleation rate of new monolayers. Having established the faceted equilibrium structure of the interface, including the partial ordering in the nearby melt region, we expect that further studies will allow us to elucidate the microscopic kinetics and dynamics of growth and subboundary formation.

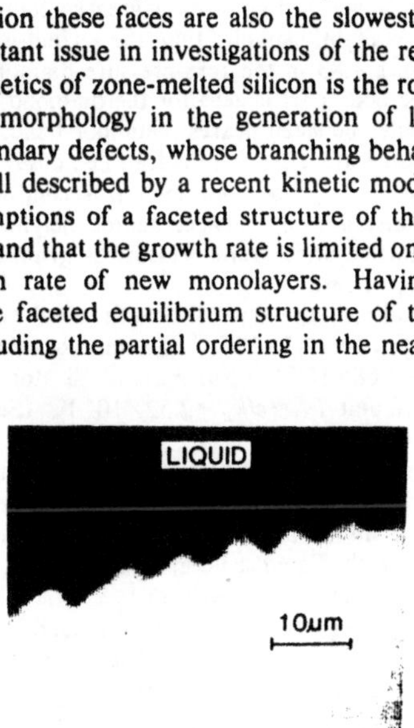

FIG. 4. *In situ* view (on a video monitor) of the solid-melt interface, in quasiequilibrium, exhibiting (111)-faceted structure.

The theoretical calculations were supported, in part, by U. S. Department of Energy Contract No. EG-S-05-5489 and were performed on the Cray-XMP at the the National Magnetic Fusion Energy Computer Center, Livermore, California.

[1]L. Pfeiffer, S. Paine, G. H. Gilmer, W. van Saarloos, and K. W. West, Phys. Rev. Lett. **54**, 1944 (1985).

[2]G. K. Celler, K. A. Jackson, L. E. Trimble, Mc. D. Robinson, and D. J. Lischner, in *Energy Beam–Solid Interactions and Transient Thermal Processing,* edited by J. C. C. Fan and N. M. Johnson (North-Holland, New York, 1984), p. 409.

[3]D. P. Woodruff, *The Solid–Liquid Interface,* (Cambridge Univ. Press, Cambridge, England, 1973), Chap. 3.

[4]M. W. Geis, H. J. Smith, B.-Y. Tsaur, J. C. C. Fan, D. J. Silversmith, and R. Mountain, J. Electrochem. Soc. **129**, 2813 (1982).

[5]K. F. Lee, T. J. Stultz, and J. F. Gibbons, in *Semiconductors and Semimetals,* edited by R. K. Willardson and A. C. Beer (Academic, New York, 1984), Vol. 17, p. 227.

[6(a)]F. F. Abraham, J. Vac. Sci. Technol. B **2**, 534 (1984).

[6(b)]U. Landman, R. N. Barnett, C. L. Cleveland, and R. H. Rast, J. Vac. Sci. Technol. A **3**, 1574 (1985).

[7]F. H. Stillinger and T. A. Weber, Phys. Rev. B **31**, 5262 (1985).

[8]R. N. Barnett, C. L. Cleveland, and U. Landman, Phys. Rev. Lett. **54** 1679 (1985).

[9]M. Parinello and A. Rahman, Phys. Rev. Lett. **45**, 1196 (1980).

[10]J. S. Langer, Rev. Mod. Phys. **52**, 1 (1980); R. F. Sekerka, Physica (Amsterdam) **12D**, 212 (1984).

[11]G. H. Gilmer, Mat. Res. Soc. Proc. **13**, 249 (1983); see also Ref. 6b.

PHYSICAL REVIEW B VOLUME 36, NUMBER 2 15 JULY 1987-I

Epitaxial growth of silicon: A molecular-dynamics simulation

M. Schneider and Ivan K. Schuller
Materials Science Division, Argonne National Laboratory, Argonne, Illinois 60439

A. Rahman*
Supercomputer Institute, School of Physics and Astronomy, University of Minnesota, Minneapolis, Minnesota 55455

(Received 5 December 1986)

We have studied the epitaxial growth of silicon using molecular-dynamics techniques. The model consists of a temperature-controlled Si(111) substrate, with the Si atoms projected towards the substrate as is done in the laboratory. The atoms interact via a potential developed by Stillinger and Weber to simulate the bulk properties of Si. We find that at low substrate temperatures the growth is not well ordered; this is in accordance with experimental observation. It is precisely the opposite of what occurs in spherically symmetric potentials that were used to simulate the growth of metallic films. At higher substrate temperatures the growth is into properly stacked, crystalline Si layers. In contrast to the growth of metals (spherically symmetric potentials), the atomic mobility on the growing surface and the thermal conductivity of the system are much lower for Si; the results of this simulation and those of our previous work are in agreement with experimental observations showing, as expected, that a major determining factor in epitaxial growth of films is the nature of the interaction potential.

Epitaxial growth from the vapor phase is a subject of much current interest.[1] This is motivated by the unique physics and materials which can be studied using this technique, and by important applications in the fields of semiconductors, magnetism, and optics. The theoretical studies to date have mostly been based on phenomenological, thermodynamic models which rely on a variety of parameters whose significance, quantitative and qualitative, is unknown *a priori*.[2] Recent advances in computer technology have opened up the possibility for realistic simulations of epitaxial processes.[3-5] These simulations are useful in understanding the role of the various parameters important for epitaxial growth. We present here the first molecular-dynamics (MD) simulation of epitaxial growth of silicon on a Si(111) substrate. At very low substrate temperatures the growth is found to be in a disordered structure. Our results show that there is an optimum range of temperatures for which epitaxial growth occurs, and that dynamical relaxation is an essential factor in this process. However, the surface mobility for the case of fourfold-coordinated elements such as silicon is much smaller than the one found for spherically symmetric potentials. We speculate that in the laboratory this is the reason for the easier epitaxial growth of metals as compared to semiconductors and that the interfaces might be sharper in semiconducting than in metallic superlattices.

In the calculation presented here the atoms interact via a potential developed by Stillinger and Weber[6] (SW) to simulate the properties of liquid and solid silicon. The potential comprises both two-body and three-body contributions. The two-body part has a Lennard-Jones–type form

$$f_2(r_{ij}) = A(Br_{ij}^{-p} - 1)\exp[(r_{ij} - a)^{-1}], \quad (1)$$

with $A = 7.049556$, $B = 0.6022246$, $p = 4$, $a = 1.80$. The minimum of f_2 is at $r_{\min} = 2^{1/6}$ with a depth $f_2(r_{\min}) = -1$. The three-body part of the potential energy consists of a sum of the following term over all triplets:

$$f_3(r_{ij}, r_{ik}, \theta_{jik}) = \lambda \exp[\gamma(r_{ij} - a)^{-1} + \gamma(r_{ik} - a)^{-1}] \times (\cos\theta_{jik} + \tfrac{1}{3})^2, \quad (2)$$

with $\lambda = 21.0$, $\gamma = 1.20$.

Both f_2 and f_3 vanish if r_{ij} or r_{ik} is greater than a, this being a distance slightly smaller than the second-nearest-neighbor distance 1.83 in the zero-pressure diamond lattice. The three-body part is zero for the diamond lattice because all angles between nearest-neighbor bonds starting from the same atom are tetrahedral, i.e., $\cos\theta = -\frac{1}{3}$. Since the range of f_2 does not include second-nearest neighbors, the distance r_{NN} between nearest neighbors for the diamond lattice under zero pressure is given by the minimum of f_2, $r_{NN} = 2^{1/6}$.

Applied to silicon, the unit of length $\sigma = 0.210$ nm, unit of energy $\varepsilon = 3.47 \times 10^{-19}$ J, unit of time $t^* = \sigma(m/\varepsilon)^{1/2} = 7.66 \times 10^{-14}$ s (m mass of Si atom), and the temperature unit $T^* = \varepsilon/k_B = 2.52 \times 10^4$ K. (See Ref. 6 for full details.)

The parameters in the SW potential were chosen so as to make the diamond structure the most stable periodic arrangement of atoms at low pressure. Moreover, the melting point and the liquid structure were sought to be in reasonable agreement with experiment.[6]

The (111) stacking sequence for the diamond lattice is *AaBbCcAa*.... Within each plane the atoms are arranged in a triangular lattice with a lattice constant equal to the second-nearest-neighbor distance $\frac{2}{3}\sqrt{6}r_{NN} = 1.83$. The distance between planes A-a, B-b, etc. is equal to the nearest-neighbor distance r_{NN}, and the atoms are stacked directly above each other. The distance between planes a-B, b-C, etc. is $r_{NN}/3$, atoms of plane B, C... being stacked above the center of the triangles in the lower plane $a,b,\ldots$.

It should be mentioned that the SW potential, because of its short-ranged nature, does not distinguish between the diamond and the wurtzite structure (*AaBbAaBb*...). The difference in energies between these two structures, however, has been calculated to be much smaller than the difference between diamond and other structures (sc, β-tin, hcp, fcc, and bcc).[7] Also, one does not expect the SW potential to reproduce the rather complicated 7×7 reconstruction of the Si(111) surface. However, the SW potential properly describes the magic numbers and the topology of ground-state structures of small Si clusters found from quantum-chemistry calculations.[8-10]

Our molecular-dynamics model consists of a Si(111) substrate with three atomic planes at $z=-r_{NN},0,\frac{1}{3}r_{NN}$ (stacking sequence *AaB*). The atoms in the first two layers are fixed at their ideal lattice sites, whereas the atoms in the third layer *B* are allowed to move as part of the dynamical system. The system is open along the positive z axis, and periodic boundary conditions are applied in the x-y plane which contains the substrate. The x and y dimensions of the simulation cell are adapted to a 14×16 triangular array of atoms. In order to include thermal expansion we make the lattice constant, and by that the x-y cell dimensions, dependent on temperature. For the simulation at higher temperatures we take a lattice constant which is 0.6% larger than the lattice constant at $T=0$. This relatively small value for the thermal expansion agrees with zero-pressure simulations of a SW crystal at higher temperatures.[11]

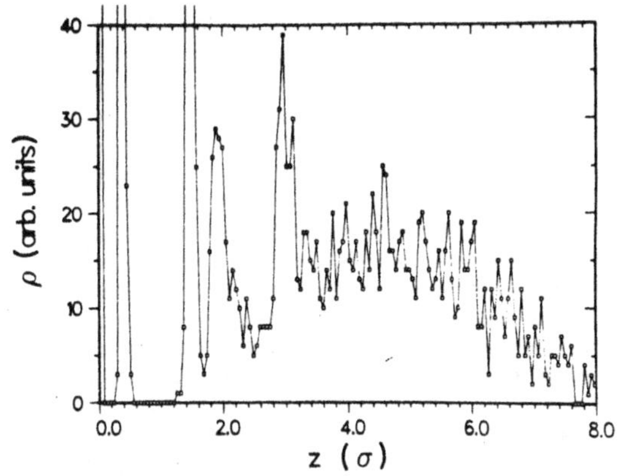

FIG. 1. Density of atoms in the perpendicular z direction to the substrate after the deposition of 1828 atoms at $T_s=0$.

To simulate the growth process, an atom is introduced every $90\Delta t$ ($\Delta t=0.02t^*$), moving, at the moment of introduction, perpendicularly towards the substrate. The beam temperature is 40% higher than the melting temperature T_m of the SW model (T_m was determined to be about 0.07).[12,13] Due to computational limitations, we have to choose a deposition rate which is orders of magnitude higher than the deposition rates usually found in real experiment. Yet, the time between the introduction of atoms in our simulations is comparable to the vibrational

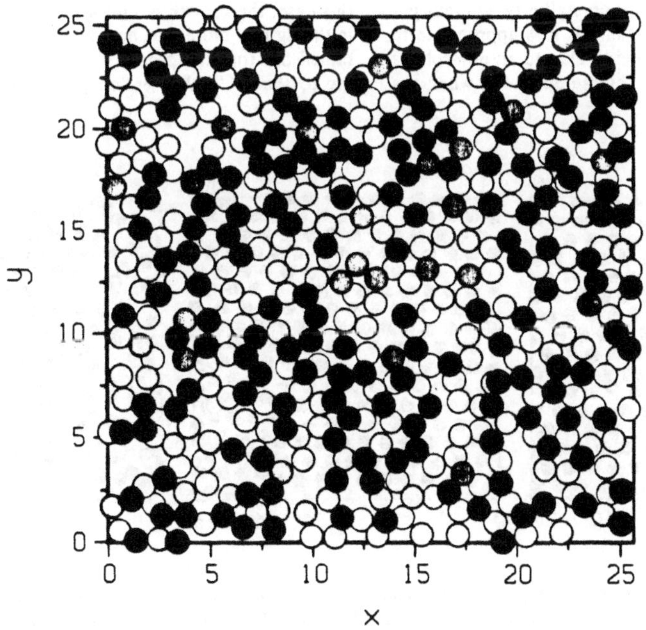

FIG. 2. Atomic arrangement in a horizontal slice of thickness $\Delta z=1.5$. Different gray values correspond to different z coordinates of atoms.

time of atoms, and most of the atoms approach the adsorbate individually. We think that important short-time relaxations within the system are included in our simulations. Of course, relaxations on a macroscopic time scale cannot be included in such calculations. Since the simulations are rather costly, the influence of the deposition rate on the growth behavior has not been investigated.

The temperature of the adsorbate is controlled via the substrate temperature T_s which in turn is adjusted by periodically scaling the velocities of the atoms in the movable substrate layer. This is an effective procedure to cool the adsorbate to the desired temperature. For the low-temperature simulation with $T_s=0$ the temperature within the adsorbate varies as a function of height z from $T_1=0.004$ in the bottom layers to $T_2=0.012$ in the top layers of the film. In the case of growth at intermediate temperatures the temperature profile ranges from $T_1=0.035$ in the bottom layers to $T_2=0.042$ in the top layers. This temperature of the film surface ($T_2=0.042$) is well below the melting temperature T_m (this can also be seen from the relatively low mobility of the surface atoms in Fig. 5).

The equations of motion are solved numerically with an integration step Δt, and the trajectories of all atoms are followed throughout the simulation as in all standard molecular-dynamics calculations. On a Cray-XMP computer about 68 h CPU time is required to deposit 2500 atoms.

Figure 1 shows the particle density in the z direction, i.e., the direction perpendicular to the substrate, after the deposition of 1828 particles at a temperature $T_s=0$. After the first three layers, i.e., beyond $z=3.0$ in Fig. 1, the atomic distribution is random without any evidence for layered growth. The particle arrangement in the plane of the film shows no evidence for crystallinity; without going into a detailed structural analysis it is appropriate to characterize the structure as being akin to an amorphous material, as shown in Fig. 2. This is opposite to what was

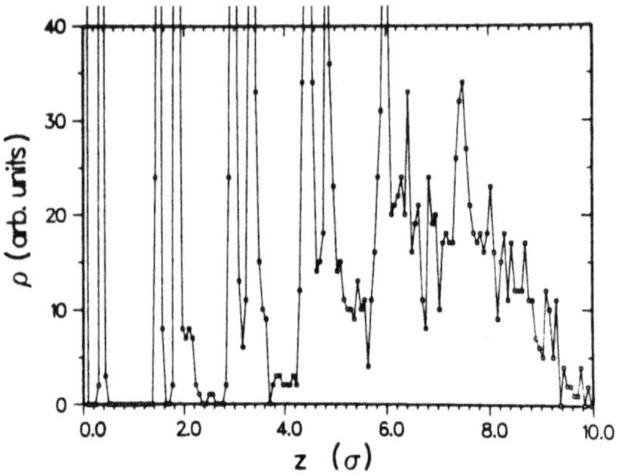

FIG. 3. Atomic density after the deposition of 2492 atoms for a system grown at intermediate temperatures. In order to extract the thermal motion, the system has been cooled to a low temperature without further deposition.

found earlier[5] using a Lennard-Jones potential where even at the lowest substrate temperatures the growth was into a layered, crystalline structure. Our finding of an amorphous growth for Si at low substrate temperature is in agreement with experimental results which invariably produce an amorphous type of structure.[14] Note the large number of 5–7-membered rings observable in Fig. 2. Long time ($30000\Delta t$) annealing of this structure (at low temperature) did not change these results.

The growth at an intermediate temperature is quite different, as illustrated in Fig. 3 after the deposition of 2492 particles at an adsorbate temperature $T=0.04$. In this case, the growth is into a layered structure as shown in the figure. Note that not only is a layered structure obtained but that the proper stacking distances corresponding to the Si(111) direction persists over 9 layers. The

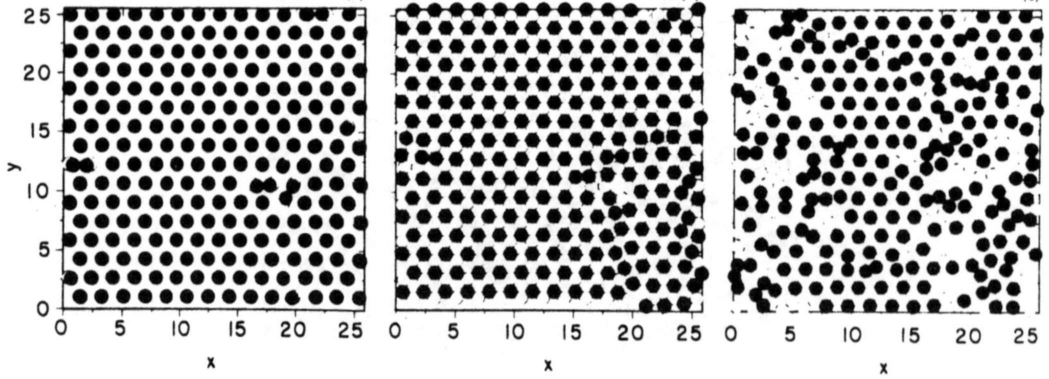

FIG. 4. Atomic arrangements at various heights for the system in Fig. 3. (a) First deposited layer on top of substrate layer (stacking *Bb*). (b) Second deposited layer on top of first layer (stacking *bC* and *bA*). (c) Sixth deposited layer on top of the fifth layer. Figure 4(b) exhibits a grain boundary between two differently stacked regions in the second deposited layer (lower right-hand corner). Such grain boundaries were found to heal at comparable temperatures for the LJ potential.

in-plane structure, however, exhibits some disorder. This is illustrated by an in-plane plot of the atomic positions of two adjacent layers [Fig. 4(a)]. In the perfect single-crystal stacking the atomic positions should be just above each other. The presence of defects is indicated by the fact that in certain areas of the sample, atoms are not perfectly overlapping. At higher levels in the sample more disorder is found, as shown in Figs. 4(b) and 4(c), although large parts of the sample are always found to be well crystallized and are stacked at a proper height. We have observed that the presence of atoms which form overlayers tends to improve the crystallinity of the lower layers. Intuitively this might be expected from a very directional potential such as the one used here.

We recall that in the case of a system of Lennard-Jones particles[5] a triangular, locally ordered structure was obtained during deposition and the presence of overlayers, as observed here, did not seem necessary to obtain a well-stacked crystal structure.

In order to understand the origin of the differences between the two kinds of simulation, i.e., metallic versus silicon, we have followed the atomic trajectories of particles for $5000\Delta t$ in a vertical slice of the sample (Fig. 5). The figure shows that the atomic mobility in the growing front is somewhat higher than in the bottom layers. A comparison of similar trajectory plots for the Lennard-Jones (LJ) (Fig. 4 of Ref. 5) and the Stillinger-Weber crystals shows that the mobility is much higher in the former (LJ) than in the latter (SW) case. This is probably the reason for the very high order encountered in the LJ growth even without the existence of an overlayer. The physical reason for these differences is perhaps quite easy to understand. A fourfold-coordinated structure is more "rigid" (because of the bond-bending part of the potential) than the structures arising out of the LJ potential, and, therefore, longer times are necessary for the atoms to find their most favorable equilibrium positions.

A comparison with other theoretical and experimental results is quite revealing. Experimental low-energy electron diffraction (LEED) results prove that at low temperatures ($\lesssim 350\,^{\circ}\mathrm{C}$) the growth of Si is into an amorphous structure, whereas at intermediate temperatures (600–1100 °C) the growth is into epitaxial films.[14] The present calculation is at odds, however, with a Monte Carlo (MC) calculation[15] which claims that the SW potential does not produce growth beyond $\frac{1}{3}$ of a monolayer. Even a

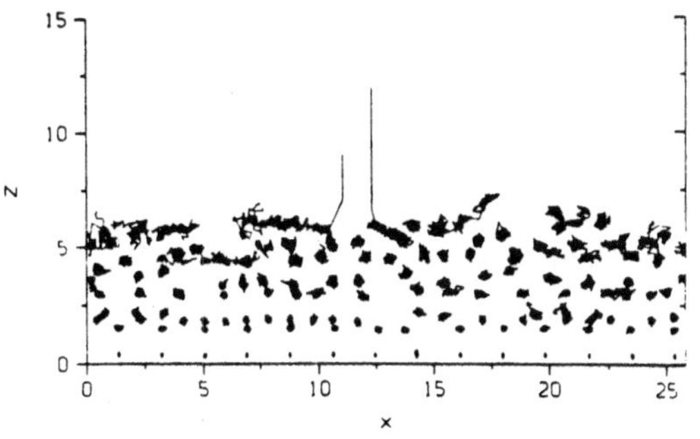

FIG. 5. Trajectory plot at intermediate temperature. A comparison of this figure with Fig. 4 of Ref. 5 clearly shows that the high mobility observed for a LJ potential is not present here.

modification of the potential parameters was suggested as a remedy. A successful crystallization of a disordered SW silicon system has not yet been reported in the literature. In this context one can, in addition, raise questions concerning the efficacy with which escape out of metastable pockets is handled by MC or MD methods, especially when large collective motions in configuration space are needed to obtain the lower-energy minimum. Obviously, the calculations we have presented here cannot give an answer to this difficult question.

In summary, we have studied the vapor-phase growth of silicon using molecular-dynamics techniques and a model potential developed by Stillinger and Weber. At low temperature the system does not show proper stacking and since we have not made a thorough analysis of pair correlations and other properties for a complete characterization we refer to the result as an "amorphouslike" structure; on the other hand, at intermediate temperatures the growth is into well-stacked and properly crystallized structures. The results are in good agreement with experimental results.

This work was sponsored by the U. S. Department of Energy, Basic Energy Sciences Program, Materials Sciences, under Contract No. W-31-109-ENG-38. The calculations were performed on the ER-Cray.

*Deceased.

[1]For a review, see, for instance, *Epitaxial Growth*, edited by J. W. Matthews (Academic, New York, 1975).

[2]See, for instance, J. A. Venables, Vacuum **33**, 701 (1983), and references therein.

[3]H. J. Leamy, G. H. Gilmer, and A. C. Dirks, in *Current Topics in Materials Science*, edited by E. Kaldis (North-Holland, Amsterdam, 1980), Vol. 6, p. 309.

[4]J. Singh and A. Madhukar, J. Vac. Sci. Technol. B **1**, 305 (1983).

[5]M. Schneider, A. Rahman, and I. K. Schuller, Phys. Rev. Lett. **55**, 604 (1985).

[6]F. H. Stillinger and T. A. Weber, Phys. Rev. B **31**, 5262 (1985).

[7]M. T. Yin and M. L. Cohen, Phys. Rev. B **26**, 5668 (1982).

[8]E. Blaisten-Barojas and D. Levesque, Phys. Rev. B **34**, 3910 (1986).

[9]B. P. Feuston, R. Kalia, and P. Vashishta, Phys. Rev. B **35**, 6222 (1987).

[10]B. P. Feuston and R. Kalia (private communication).

[11]J. R. Ray (private communication).

[12]F. F. Abraham and J. Q. Broughton, Phys. Rev. Lett. **56**, 734 (1986).

[13]U. Landman *et al.*, Phys. Rev. Lett. **56**, 155 (1986).

[14]See, for instance, R. N. Thomas and M. H. Francombe, Appl. Phys. Lett. **11**, 108 (1967).

[15]B. W. Dodson, Phys. Rev. B **33**, 7361 (1986).

MOLECULAR-DYNAMICS SIMULATION OF THIN-FILM GROWTH

MATTHIAS SCHNEIDER,* IVAN K. SCHULLER,* AND A. RAHMAN**

*Materials Science Division, Argonne National Laboratory, Argonne, IL 60439
**Supercomputer Institute, School of Physics and Astronomy, University of Minnesota, Minneapolis, MN 55455

ABSTRACT

The epitaxial growth of thin films has been studied by molecular-dynamics computer simulation. In these simulations atoms are projected towards a temperature-controlled substrate, and the equations of motion of all atoms are solved for a given interaction potential. The calculations give insight into the microscopic structure of thin films, the dynamics of the adsorption process, and they help answer the way in which substrate temperature, form of the substrate, flux of impinging atoms, and form of the interaction potential, affect epitaxial growth. Simulations were performed for monatomic and binary systems with spherically symmetric atomic interactions, and for systems in which the atoms are interacting via a three-body potential to simulate the epitaxial growth of silicon.

INTRODUCTION

Epitaxial growth from the vapor phase is a subject of much experimental and theoretical interest [1]. The understanding of epitaxial thin-film growth is of importance for the preparation of new materials and novel devices that exhibit unusual physical phenomena; in addition, thin-film growth is already of technological importance for semiconductor and superconducting devices, solar cells, magnetic recording, etc.

The theoretical studies to date have mostly been based on phenomenological, thermodynamic models which rely on a variety of parameters whose significance, quantitative and qualitative, is unknown a-priori [2]. With the advent of supercomputers, it has now become practicable to apply the method of computer simulation to problems of epitaxial growth through the use of realistic models with a sufficiently large number of atoms [3]. We have done what we believe are the first full molecular-dynamics (MD) studies of epitaxial growth; in other words, once the interatomic potential and the procedure for adjusting the substrate temperature are given, the classical equations of motion for all atoms are solved in the usual way without any further approximations.

SIMULATION MODEL

Our molecular-dynamics models consist of an atomic substrate which is placed in the plane z=0. The substrate is built of two or three atomic planes stacked properly above each other. Each of the substrate planes contains 224 atoms in a triangular arrangement, thus simulating the (111) surface of the systems studied by us. The atoms in the bottom substrate layers are fixed at their ideal lattice sites, whereas the atoms in the uppermost substrate layer are allowed to move as part of the dynamical system. The rectangular simulation cell is open along the positive z axis, and periodic boundary conditions are applied in the x-y plane which contains the substrate. To simulate the deposition process, atoms with a Gaussian

Mat. Res. Soc. Symp. Proc. Vol. 77. © 1987 Materials Research Society velocity distribution are introduced periodically at a certain height from the substrate, and the atoms are moving, at the moment of introduction, perpendicularly towards the substrate. The beam temperature is 30-40% higher than the bulk melting temperature of the system being simulated. The temperature of the adsorbate is controlled via the substrate temperature T_S which in turn is adjusted by periodically scaling the velocities of the atoms in the movable substrate layer. The equations of motion are solved numerically with an integration step Δt, and the trajectories of all atoms are followed throughout the simulation as in all standard MD calculations.

RESULTS

In this paragraph, results of growth simulations will be presented for two different types of systems: monatomic systems with spherically symmetric atomic interaction (metal-like), and binary metal-like systems.

Metal-like Systems

In these systems [4] the atoms interact with each other through the Lennard-Jones potential

$$v(r) = 4\varepsilon\,[(\sigma/r)^{12} - (\sigma/r)^{6}]\ . \qquad (1)$$

The units of length and of energy are, as usual, taken to be σ and ε, respectively. The substrate consists of two close-packed planes representing the (111) surface of an fcc crystal (for the Lennard-Jones potential a close-packed lattice is the stable configuration of atoms).

The calculations show that for spherically symmetric pair potentials, homoepitaxial growth yields well-formed crystallites at all substrate temperatures T_S, including absolute zero. Moreover, the dynamics plays a key role in the growth of these crystals even at the lowest temperatures.

For the case T_S=0, the atom density along the z axis (perpendicular to the substrate) after the deposition of 2052 atoms is shown in Fig. 1. It can clearly be seen that the adsorbate consists of quite distinct layers

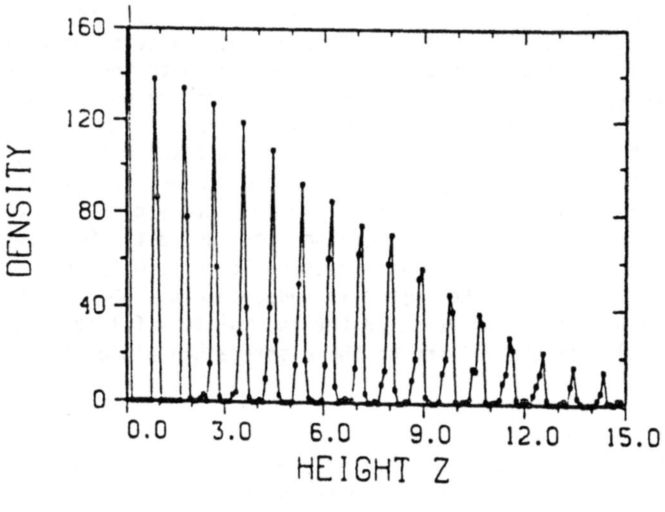

Fig. 1. Histogram of the number of atoms as a function of height z.

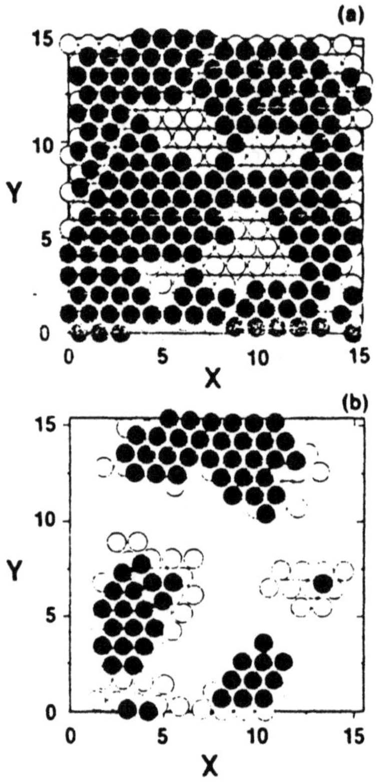

Fig. 2. Arrangements of atoms after the deposition of 680 atoms at $T_s=0$. (a) Atoms in the first deposited layer on top of the movable substrate layer. (b) Atoms in the fifth deposited layer on top of the fourth layer.

with no appreciable disorder in the z direction. Pictures of the arrangement of atoms within the layers (see Fig. 2) show the existence of grain boundaries and voids. The size of the voids increases with the distance from the substrate (compare Fig. 2a and 2b). The atoms are well arranged in triangular patterns with no evidence of in-plane disorder.

At intermediate substrate temperatures (about half the melting temperature), the layers become completely populated. The process of relaxation by which the layers are filled is illustrated by a trajectory plot in Fig. 3. This figure shows the trajectories of the atoms for the last 1500 integration steps of the simulation at a substrate temperature $T_s=0.4$, where $T_m=0.7$ is the melting temperature for the system in reduced units. During the deposition, the high mobility of atoms in the incompletely filled upper layers is responsible for the filling of the close-packed planes further below.

Binary Metal-like Systems

The purpose of these simulations [5] was to study the growth of a mixture of two differently sized Lennard-Jones particles as a function of relative atomic sizes. The results indicate that the ratio of atomic radii is the determining factor for amorphous growth.

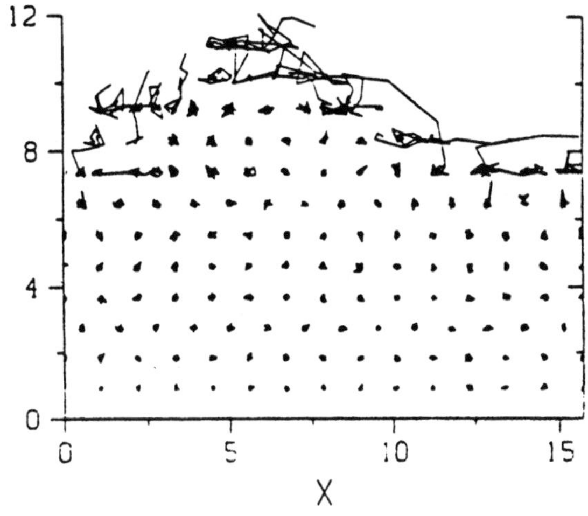

Fig. 3. Particle trajectories at T_s=0.4 (vertical cross-section).

There are two types of particles A and B interacting via standard Lennard-Jones potentials

$$v_i(r) = 4\varepsilon\,[(\sigma_i/r)^{12} - (\sigma_i/r)^6]\ (i = AA,\ BB,\ AB)\ . \qquad (2)$$

The unit of length is taken to be σ_{AA}. Here i=AA,BB,AB refer to the interaction between like particles of type A or B or unlike particles A and B. The length parameter σ_{AB} for the unlike particles is given by

$$\sigma_{AB} = (\sigma_{AA} + \sigma_{BB})/2\ . \qquad (3)$$

This is the natural choice if σ is interpreted as usual as the atomic diameter. The substrate consists of two close-packed planes; each plane contains 224 Lennard-Jones atoms of type A. During deposition the two types of particles are introduced alternately, i.e., in equal numbers. Figure 4 shows the density of particles in the z direction for σ_{BB}/σ_{AA}=0.900 after the deposition of 3052 atoms at T_s=0.4. (In the preceding paragraph we have shown that T_s=0.4 is the optimum substrate temperature for layer-by-layer epitaxial growth in the monatomic case.) The figure shows that for this ratio of atomic sizes the growth is into well-formed distinct layers with no evidence for amorphous growth. Figure 5 shows the system in a vertical cross-section. The particles are mainly arranged in a lattice corresponding to the bigger atoms.

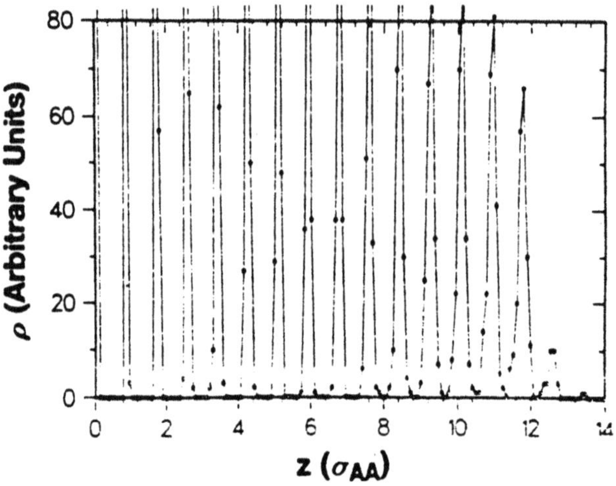

Fig. 4. Particle density versus height z for T_s=0.4 and σ_{BB}/σ_{AA}=0.900.

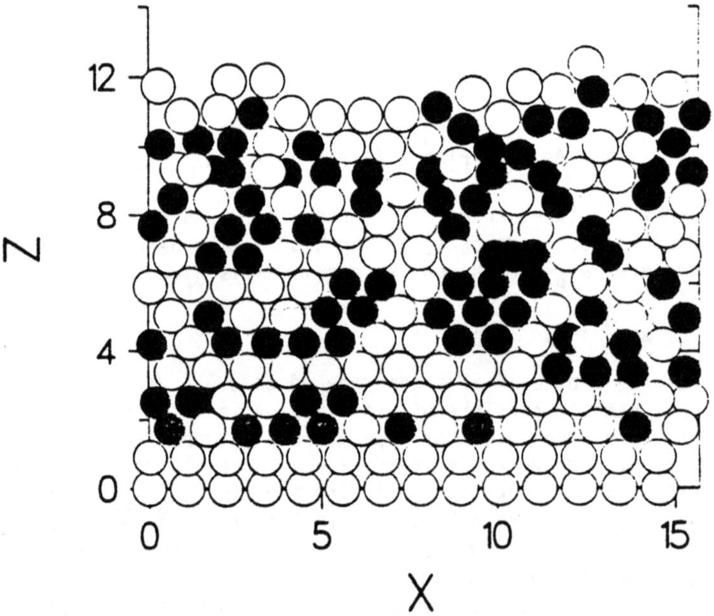

Fig. 5. Arrangement of atoms in a vertical slice, with σ_{BB}/σ_{AA}=0.900. Particles of type A are drawn with open circles, those of the smaller type B with solid circles.

As a function of σ_{BB}/σ_{AA}, a transition from crystalline-layered to a disordered growth occurs at a value of σ_{BB}/σ_{AA}, which is slightly lower than the above value σ_{BB}/σ_{AA}=0.900. Figure 6 shows the atomic density as a function of height z from the substrate for σ_{BB}/σ_{AA}=0.875. After about five atomic layers from the substrate the structure is not layered. Pictures of atomic arrangements show that the structure is quite disordered when compared to the σ_{BB}/σ_{AA}=0.900 case.

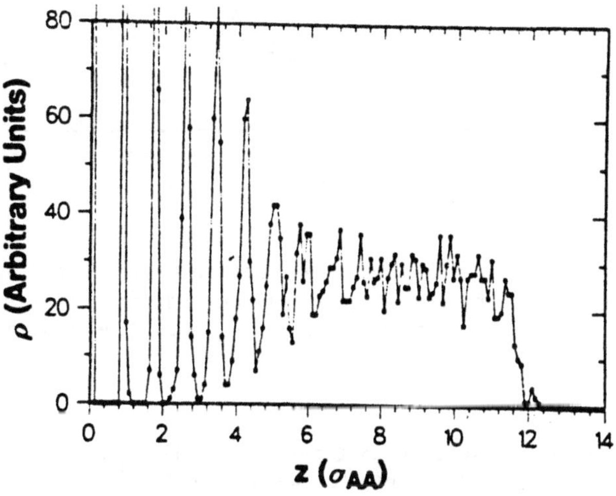

Fig. 6. Particle density versus height z for T_s=0.4 and σ_{BB}/σ_{AA}=0.875.

These calculations together with additional calculations for different values of σ_{BB}/σ_{AA} show that a rather abrupt change occurs in the growth mode at a value of σ_{BB}/σ_{AA}=0.89 ± 0.01. Figure 7 shows the number of distinct layers as a function of σ_{BB}/σ_{AA} ratio. It is interesting that the transition is quite abrupt, almost like in a phase transition. Above σ_{BB}/σ_{AA}=0.89, the growth is into a layered crystalline structure with the lattice parameter determined by the largest of the two components. Below this value the growth is into an unlayered disordered structure after a few monolayers from the substrate. The change in the growth mode is rather abrupt contrary to what one might expect naively.

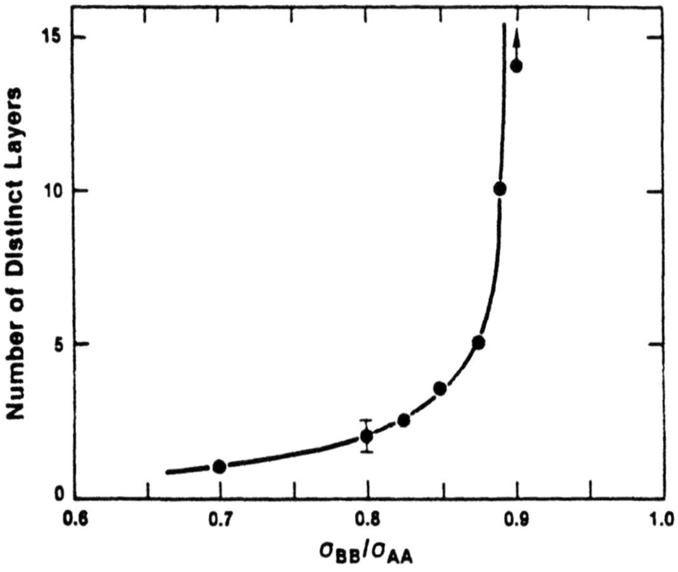

Fig. 7. Number of distinct layers as a function of the ratio σ_{BB}/σ_{AA} (substrate layers are not counted). The arrow indicates that even more layers are expected to form in the case σ_{BB}/σ_{AA}=0.9 (see Fig. 4).

CONCLUSIONS

A comparison of our results with experimental data is quite revealing [1]. It is well known that monatomic metals cannot be grown as amorphous films by using vapor-growth techniques. Metallic elements generally form small crystallites. However, elements that show directional bonding, e.g., Si, Bi, or Ga, can be grown as amorphous films with relative ease. Our results on the monatomic Lennard-Jones system are an indication of the former type of behavior. Since the interatomic force in our case is spherically symmetric, the growth is invariably into well-layered, close-packed structures, similar to what is observed for most metals.

We show that the stabilization of amorphous structures occurs if there are particles of differing sizes in the system. The results on the binary Lennard-Jones system are in good agreement with experimental observations. It is a well known empirical rule that the atomic radii of two metallic elements must differ by more than 10% in order for their binary alloys to form a glass [6]. The value of 10% is in accordance with our critical value for the ratio of atomic radii σ_{BB}/σ_{AA}=0.89.

An interesting and exciting recent development is the growth of silicon using phenomenological models [7]. Our full molecular-dynamics simulations confirm that amorphous structures can be obtained if the interatomic potential exhibits some anisotropy due to directional bonding [8]. In these simulations the atoms interact via a three-body potential developed by Stillinger and Weber [9] to simulate the properties of covalently bonded silicon. At intermediate temperatures the growth is into well-stacked and properly crystallized Si structures. At low temperatures the films grow in an "amorphous-like" structure. This is in accordance with experimental observation [10].

We thank the U.S. Department of Energy for computing time made available on the ER-Cray. Work supported by the U.S. Department of Energy, BES-Materials Sciences, under Contract #W-31-109-ENG-38.

REFERENCES

[1] For a review see, Epitaxial Growth, edited by J. W. Matthews (Academic Press, New York, 1975).
[2] See for instance, J. A. Venables, Vacuum 33, 701 (1983), and references cited therein.
[3] For a review see, F. F. Abraham, Adv. Phys. 35, 1 (1986).
[4] M. Schneider, A. Rahman and I. K. Schuller, Phys. Rev. Lett. 55, 604 (1985).
[5] M. Schneider, A. Rahman and I. K. Schuller, Phys. Rev. B 34, 1802 (1986).
[6] B. C. Giessen, Proc. 4th Int. Conf. on Rapidly Quenched Metals, eds., T. Masumoto and K. Suzuki, Vol. 1 (Japan Inst. Metals, Sendai, 1982), p. 213; F. Sommer, M. Fripan, B. Predel, ibid., p. 209.
[7] B. W. Dodson, Phys. Rev. B 33, 7361 (1986).
[8] M. Schneider, I. K. Schuller and A. Rahman (to be published).
[9] F. H. Stillinger and T. A. Weber, Phys. Rev. B 31, 5262 (1985).
[10] See for instance, R. N. Thomas and M. H. Francombe, Appl. Phys. Lett. 11, 108 (1967).

Structural, Dymanical, and Electronic Properties of Amorphous Silicon: An *Ab Initio* Molecular-Dynamics Study

R. Car and M. Parrinello

Scuola Internazionale Superiore di Studi Avanzati, Trieste 34014, Italy

(Received 31 August 1987)

An amorphous silicon structure is obtained with a computer simulation based on a new molecular-dynamics technique in which the interatomic potential is derived from a parameter-free quantum-mechanical method. Our results for the atomic structure, the phonon spectrum, and the electronic properties are in excellent agreement with experiment. In addition we study details of the microscopic dynamics which are not directly accessible to experiment. We find in particular that structural defects are associated with weak bonds. These may give rise to low-frequency vibrational modes.

PACS numbers: 61.40.+b, 63.50.+x, 71.25.Mg

A very important problem in the physics of amorphous semiconductors is the understanding of the way short-range order (SRO) is incorporated in the disordered network.[1] Many of the characteristic properties of amorphous semiconductors depend crucially on SRO. However, despite the very substantial progress made in the field over the last twenty years, our understanding of SRO and of how it affects structural and electronic properties is still largely incomplete. Experimentally, the presence of disorder means that one can measure only averaged properties and an accurate determination of the individual atomic coordinates is impossible. Yet a quantitative characterization of the atomistic disorder, i.e., an accurate microscopic model of the amorphous network, is a prerequisite of any quantitative theoretical approach to the properties of amorphous semiconductors.

Not surprisingly, a large effort has been devoted to the attempt to construct structural models of disordered semiconductors. Some of the more systematic approaches in this area make use of interatomic potentials in conjunction with molecular dynamics (MD)[2] or Monte Carlo[3] techniques that allow simulation of thermal treatments similar to those used in laboratories to prepare glasses. The main difficulty relies on the need for realistic interatomic potentials. So far these have been constructed empirically,[4] trying to model the complex many-body interatomic interactions by means of a few simple potential functions whose parameters are fitted to experimental data. Although very useful, such an approach suffers some basic limitations since (i) it is difficult to assess the range of validity of the empirical potentials, and (ii) the approach does not explicitly display the dependence existing in nature between bonding electronic properties and atomic dynamics. This may be particularly important in covalent semiconductors.

These limitations are overcome in a first-principles MD scheme that we have recently proposed.[5] In this approach the interatomic potential is constructea directly from the electronic ground state, and this is treated with accurate density-functional techniques such as those used in electronic-structure calculations. We have already shown in a preliminary report[6] that both liquid and amorphous silicon can be simulated in this way.

Here we focus on the numerical simulation of amorphous silicon and present results for a number of properties ranging from atomic to electronic structure. Our calculated radial distribution function (RDF) and phonon spectrum are in excellent agreement with experimental neutron-diffraction studies, showing that not only static but also dynamic properties of disordered systems may be obtained in this way. Quite remarkably our calculated electronic density of states reproduces both the metallic character of the liquid and the presence of a gap in the amorphous state. In addition we study coordination defects and topological properties of the computer-generated disordered network. Our findings have a number of consequences concerning both the nature of defects and the low-T and vibrational properties of amorphous silicon.

The theoretical method and the computational details have already been discussed elsewhere[5,6] and will not be further elaborated here. Basically we perform a constant-volume MD simulation for both the particles and the electronic orbitals belonging to a periodically repeated unit cell. As in standard MD simulations of condensed matter systems, this is justified if the unit cell is large compared with the longest relevant length scale of the problem, either electronic or atomic.

The present simulation was started from the liquid silicon data at an average temperature of ≈ 2200 K already described in our earlier report.[6] These were in fair agreement with experiment, even though the theoretical RDF was more structured than the experimental one and, correspondingly, the average coordination number was 5.4 instead of 6.4. The origin of such discrepancies is not yet completely understood. The liquid data were obtained with use of an fcc cell containing 54 atoms in a volume appropriate to liquid silicon at the melting point ($\Omega = 1033$ Å^3). We used an accurate nonlocal pseudopotential[7] and adopted for exchange-correlation effects the parametrized local-density form of Perdew and Zunger.[8] We used a plane-wave expansion of the electronic states at $k=0$ with an energy cutoff of 5.5 Ry. Surface[9] and crystal[10] calculations have shown that such a cutoff, albeit not enough for perfect convergence, is sufficient for a reasonably good description of the bonding properties in silicon. The integration of the generalized MD equations[5] was performed with the Verlet algorithm and a time step of 1.7×10^{-16} s, while the "mass" parameter for the electronic degrees of freedom was set equal to 300 a.u.

An amorphous structure was generated by our quenching from the melt, as in Ref. 6, but starting from a different liquid configuration. This was obtained by our letting the liquid evolve for $\approx 0.7\times10^{-12}$ s after the configuration used in Ref. 6. Each atom in the box moved on average by two bond lengths during such a time interval. The thermal treatment consisted of a quench from ≈ 2200 to ≈ 300 K at a cooling rate $\approx 2\times10^{+15}$ K/s, followed by an annealing cycle in which the temperature was raised to ≈ 1000 K before performance of a final equilibration at a temperature of ≈ 300 K. During the initial quenching, the volume of the cell was gradually increased to the value $\Omega = 1080$ Å^3, which is more appropriate to amorphous silicon.[11] The total thermal treatment took $\approx 2\times10^{-12}$ s during which each atom in the box moved on average by more than one bond length, a substantial distance. The structure so obtained is different from the inherent structure of the liquid.[12] Compared to Ref. 6 this new amorphous structure has been obtained with a more careful annealing and the volume variation in the process of amorphization has also been considered. Even so, the newly generated amorphous structure is remarkably similar to that of our earlier report. The most noticeable difference is a substantial reduction in the number of coordination defects in the new structure as a consequence of the additional relaxation and of the more careful annealing. Accurate structural data on tetravalent amorphous semiconductors are available in real space only for a-Ge.[13] Since no RDF of corresponding quality has to our knowledge been reported for a-Si, we suitably scale our simulation data and compare them with the experimental a-Ge data in Fig. 1. The agreement between theory and experiment is very good. Both data sets reveal the disappearance of the third crystalline coordination shell. The area under the first peak of the RDF gives a local coordination of four in both cases. The main difference between theory and experiment is a slightly broader second peak in the theoretical data, which reflects a broader distribution of bond angles around the tetrahedral angle with an rms deviation of $\approx 14°$ whereas experiment gives an rms deviation of $\approx 10°$. This may be a consequence of the preparation and/or the small size of our unit cell. Because of the size of our MD cell, we cannot study distances that are larger than ≈ 6 Å. Since the experimental RDF suggests that SRO effects are present up to distances of ≈ 10 Å, it would be very interesting to perform simulations with larger cells.

Recently the vibrational spectrum of a-Si has been measured by inelastic neutron scattering.[14] In our simulation it is relatively easy to obtain the phonon density of states (PDOS) by taking the Fourier transform of the velocity autocorrelation function: $\langle v(0)v(t)\rangle$. The results are displayed in Fig. 2. Again the agreement between theory and experiment is remarkable. The main difference is an almost rigid shift by ≈ 50 cm^{-1} of the theoretical TA peak toward higher frequencies. This is likely to be a consequence of the small cutoff used for the plane-wave expansion, as suggested by the perfect-

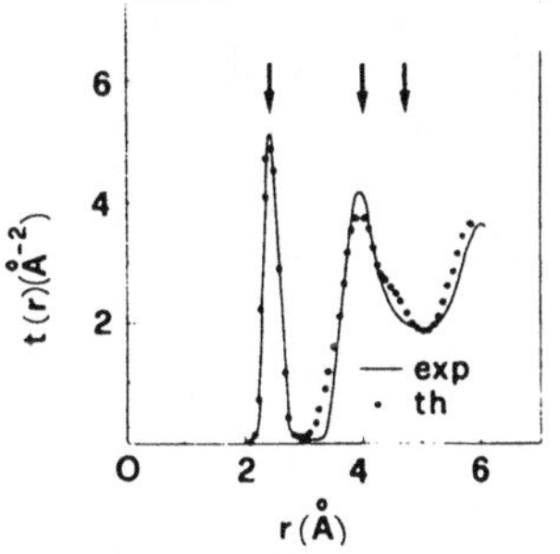

FIG. 1. Plot of $t(r)=4\pi\rho r g(r)$, where ρ is the average density and $g(r)$ the pair correlation function. Experimental data are for a-Ge (see text) (Ref. 13) and the theoretical a-Si data have been properly scaled. The theoretical curve has been convoluted with the experimental resolution function (Ref. 13). Arrows indicate peak positions in the perfect crystalline structure.

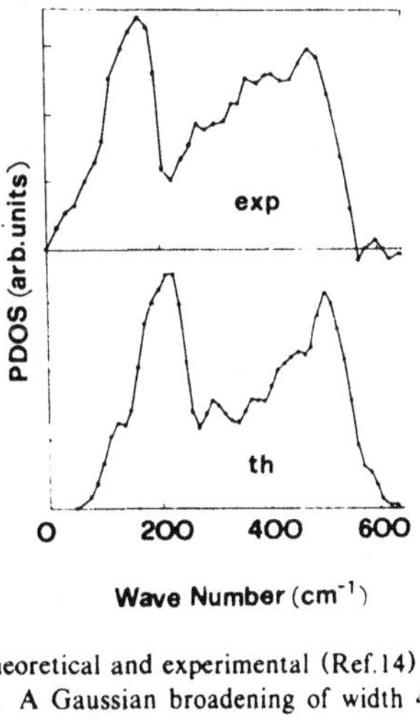

FIG. 2. Theoretical and experimental (Ref. 14) phonon density of states. A Gaussian broadening of width 40 cm^{-1} has been used to smooth the theoretical density of states.

205

crystal phonon calculations by Yin and Cohen.[10] Note that the theoretical PDOS decays to zero more rapidly than the experimental one at low frequencies, because of the low-frequency cutoff set by the finite MD observation time for the velocity autocorrelation function. We observe that the relative strength of the TA and the TO peaks is correctly reproduced in our data, whereas all model calculations assign a larger amplitude to the TO peak.[14,15] It is tempting to relate this to structural defects present both in our simulation and in experiment and which are instead absent in conventional models. These defects can give rise to low-frequency modes which result in a transfer of oscillator strength from higher to lower frequencies. This is indirectly confirmed by the low-frequency shoulder in the TA peak, which is quite prominent in our data and is also present in experiment. An increase in the density of low-frequency vibrational modes with respect to the crystalline case is a common feature of amorphous systems and can be seen, for instance, in low-T measurements of the specific heat.[1]

Trying to discover the structural defects that might be responsible for the above behavior, we have identified two weak bonds which are well separated in space and are associated with coordination defects. The time variation of these two bond lengths r_b is plotted in Fig. 3, and it is seen to undergo large variations on a rather long time scale. If we assign to the first coordination shell all bond distances that are smaller than r_m, where $r_m = 2.75$ Å is the first minimum position in the RDF, we find that in these two stretched bonds the partner atom can have coordination 3 and 4, or 4 and 5, for $r_b > r_m$ or $r_b < r_m$, respectively. Therefore, the ample variations observed in r_b lead to a continuous switching from a threefold coordinated atom (T_3) to a fivefold one (T_5) and vice versa. A recent suggestion[16] that T_5 defects might exist and play an important role in a-Si[16] has attracted considerable interest. Our data confirm that T_3 and T_5 should have comparable formation energies and may be converted into one another via network distortion.[16] However, we find that T_3 dominates on the average, and that the average length of our weak bonds ($\approx 2.8-2.9$ Å is considerably smaller than that expected for an ideal isolated dangling bond pointing into a microvoid or residing on an internal surface. Consequently the wave function associated with such defects will be more delocalized than for a dangling bond. This is quite consistent with the discussion on the localization of the D-center wave function in Ref. 16. However, when comparing our findings with experiment one must be careful. It is very likely that the type and concentration of defects depend on the size of the system and the annealing procedure. Clear evidence is that in the simulation we find an unrealistically high concentration of defects. In addition, intermediate-range order effects, like, e.g., voids occurring on a 100-Å scale, may play an important role in real amorphous systems.[1] Nonetheless, our calculation is highly suggestive of the nature of defects that may be found in real-life a-Si.

In our Md scheme one generates not only a sequence of atomic configurations but also the corresponding electronic self-consistent potentials. These may then be used to study electronic properties. A striking difference between liquid and amorphous silicon is that the former is metallic while the latter has semiconducting character. We were able to reproduce this feature in our simulation. In Fig. 4 the electronic density of states (EDOS) of liquid Si is displayed together with that of our equilibrated a-Si. These EDOS have been calculated as an average over several atomic configurations for both liquid and amorphous silicon. The change in electronic structure that accompanies the change in atomic arrangement as the system becomes amorphous is remarkable, since in

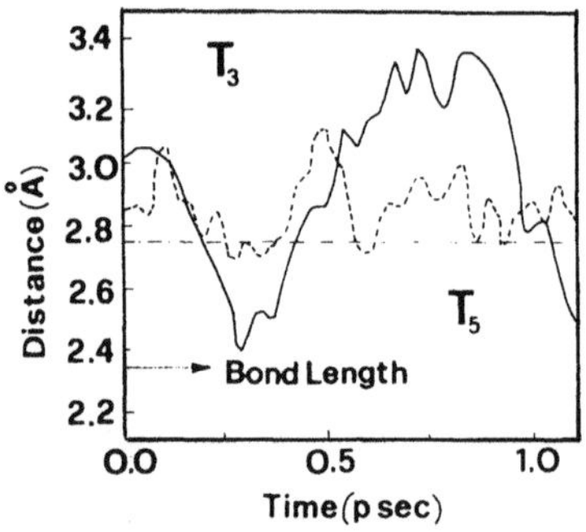

FIG. 3. Weak-bond length as a function of time for the two defects found in the computer simulation. The dash-dotted line is the conventional divide at 2.75 Å between T_3 and T_5 defects. The bond-length distance is also indicated in the picture.

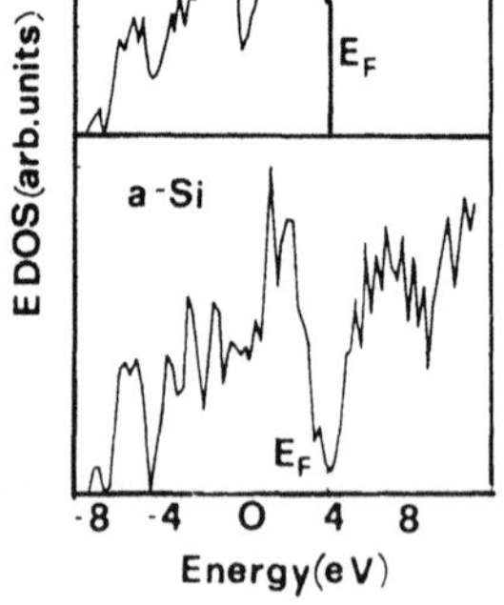

FIG. 4. Electronic density of states for liquid (l-Si) and amorphous silicon. The energy resolution in the histogram is 0.27 eV. The averages have been calculated over 12 and 32 different configurations for l-Si and a-Si, respectively.

a-Si a gap opens at the Fermi level while liquid-Si EDOS exhibits metal-like behavior. A rather similar EDOS for liquid Si has also been calculated by Broughton and Allen[17] using a more conventional approach in which the atomic positions were generated in an MD run that used effective potentials and the electronic levels were subsequently determined with an empirical tight-binding method. Their results are similar to ours, but it must be stressed that in our approach the atomic positions and the electronic states are the result of a single and self-consistent procedure. The formation of a gap in *a*-Si is accompanied by a marked increase in the localization of the wave functions belonging to gap states. This was checked by measurement of the participation ratio.[18] The localized states are absent in the liquid, suggesting that the metallic behavior should be associated with the presence of diffusive motions and the resulting overcoordination. The gap states in our amorphous structure are due to localized defect states induced by the weak bonds discussed above. The rather large concentration of gap states is simply a consequence of the large concentration of defects in our simulated amorphous structure. The poor energy resolution in the EDOS is a consequence of our small unit cell and limited statistics. This does not allow us to say anything conclusive concerning the shape of the EDOS in the lower valence-band region. It has been suggested[19] that experimental x-ray photoemission spectra in that region might provide evidence for the existence of odd-membered rings in the amorphous network in addition to the even-membered rings typical of the diamond structure. Even if we cannot say how odd-membered rings correlate with the EDOS, an analysis of the ring statistics of our amorphous structure has revealed a significant fraction of odd-membered rings with a dominance of fivefold rings.[12]

In conclusion, we have demonstrated that first-principles studies of disordered systems are feasible and may yield valuable information. A distinctive feature of our approach is that it treats simultaneously and in a consistent way both atomic and electronic properties: The agreement with experiment found consistently for several properties either atomic and electronic is therefore particularly encouraging. An analysis of the data obtained with the computer simulation has allowed a first realistic microscopic identification of structural defects that may be responsible for several observed properties of amorphous semiconductors. Extensions to larger cells and to different systems and processes are now underway.

This work has been supported by the Scuola Internazionale Superiore di Studi Avanzati–Centro di Calcolo Elettronico Interuniversitario dell'Italia Nord-Orientale collaborative project, under the sponsorship of the Italian Ministry for Public Education.

[1]See, e.g., N. F. Mott and E. A. Davis, *Electronic Processes in Noncrystalline Materials* (Clarendon, Oxford, 1979), and review articles, e.g., in *Amorphous Semiconductors*, edited by M. H. Brodsky (Springer-Verlag, Berlin, 1985).

[2]K. Ding and H. C. Andersen, Phys. Rev. B **34**, 6987 (1986).

[3]F. Wooten, K. Winer, and D. Weaire, Phys. Rev. Lett. **54**, 1392 (1985).

[4]Empirical potentials for silicon have been recently constructed by F. H. Stillinger and T. A. Weber, Phys. Rev. B **31**, 5262 (1985); R. Biswas and D. R. Hamann, Phys. Rev. Lett. **55**, 2001 (1985); J. Tersoff, Phys. Rev. Lett. **56**, 632 (1986).

[5]R. Car and M. Parrinello, Phys. Rev. Lett. **55**, 2471 (1985).

[6]R. Car and M. Parrinello, in *Proceedings of the Eighteenth International Conference on the Physics of Semiconductors, Stockholm, 1986*, edited by O. Engstrom (World Scientific, Singapore, 1987), pp. 1165–1172.

[7]D. R. Hamann, M. Schlueter, and C. Chiang, Phys. Rev. Lett. **43**, 1494 (1979).

[8]J. P. Perdew and A. Zunger, Phys. Rev. B **23**, 5048 (1981).

[9]M. T. Yin and M. L. Cohen, Phys. Rev. B **24**, 2303 (1981).

[10]M. T. Yin and M. L. Cohen, Phys. Rev. B **26**, 3259 (1982).

[11]This is the volume appropriate to crystalline silicon. In real amorphous silicon the density is lower because of the presence of microvoids (Ref. 1).

[12]I. Stich, R. Car, and M. Parrinello, to be published.

[13]J. H. Etherington, A. C. Wright, J. T. Wenzel, J. C. Dore, J. H. Clarke, and R. N. Sinclair, J. Non-Cryst. Solids **48**, 265 (1982).

[14]W. A. Kamitakahara, H. R. Shanks, J. F. McClelland, U. Buchenau, F. Gompf, and L. Pintchovius, Phys. Rev. Lett. **52**, 644 (1984).

[15]K. Winer, Phys. Rev. B **35**, 2366 (1987).

[16]S. T. Pantelides, Phys. Rev. Lett. **57**, 2979 (1986).

[17]J. Q. Broughton and P. B. Allen, in *Computer-Based Microscopic Description of the Structure and Properties of Materials*, edited by J. Q. Broughton, W. Krakow, and S. T. Pantalides, MRS Symposium Proceedings Vol. 63 (Materials Research Society, Pittsburgh, 1986).

[18]R. J. Bell and P. Dean, Discuss. Faraday Soc. **50**, 55 (1970).

[19]J. D. Joannopoulos and M. L. Cohen, Phys. Rev. B **7**, 2644 (1973); for a review, see J. D. Joannopoulos and M. L. Cohen in *Solid State Physics*, edited by H. Ehrenreich, F. Seitz, and D. Turnbull (Academic, New York, 1976), Vol. 31, pp. 71–148.

Equilibrium Structures and Finite Temperature Properties of Silicon Microclusters from *Ab Initio* Molecular-Dynamics Calculations

Pietro Ballone and Wanda Andreoni
Zurich Research Laboratory, IBM Research Division, 8803 Rüschlikon, Switzerland

and

Roberto Car and Michele Parrinello
International School for Advanced Studies, 34014 Trieste, Italy
(Received 20 October 1987)

We show that new aspects of the physics of microclusters can be investigated accurately with *ab initio* molecular dynamics. We present results on a number of properties of Si_N aggregates (N up to 10) at both zero and finite temperatures. The results of dynamical simulated annealing for the ground state point to a complex growth sequence. Simulations at finite temperatures show the existence of two regimes, solidlike and liquidlike, with substantially different electronic and structural properties.

PACS numbers: 36.40.+d, 71.45.Nt

Understanding the behavior of covalent materials in different states of aggregation under different physical conditions is a central issue in the physics of semiconductors. Microclusters are particularly challenging since their bonding properties are expected to be rather different from those of the crystal and crucially dependent on the specific size. In fact, experiments are by now available on microclusters of Si, Ge, and GaAs,[1] which illustrate the strong size dependence of both stability and reactivity. Very recently Cheshnovsky *et al.*[2] have introduced ultraviolet photoelectron spectroscopy for mass-selected negatively charged small clusters to probe the electronic structure. These measurements have once more raised questions about the equilibrium configurations and their link to the electronic structure, as well as about the influence of temperature. So far, these questions have been investigated only partially. This was due in part to the limitations of standard theoretical approaches, i.e., *ab initio* methods of various degrees of sophistication [density-functional theory (DFT) in the local-density approximation, Hartree-Fock, etc.][3,4] and computer simulation [molecular dynamics (MD)][5] with empirical potentials for the description of the atomic interactions.[6] The former have been applied to search for the state of lowest energy by exploration of a severely limited number of configurations. The latter is more efficient in the search for the ground state and suitable for the study of the thermal behavior of the cluster since it exploits the techniques of statistical mechanics to explore large fractions of configuration space. However, it misses the link between electronic and structural properties, and the predictive capability of the potentials (usually fitted to the solid) in microclusters has still to be tested against quantum-mechanical calculations.

To achieve a reliable as well as comprehensive description of the physics of semiconductor microclusters, it is crucial to combine the accuracy of *ab initio* calculations with the advantages of statistical-mechanics approaches. This can be achieved by use of the unified approach for MD and DFT, which has been recently introduced by two of us.[7] In this scheme, a fictitious classical Lagrangean is introduced whose dynamical variables are the ionic positions and the Kohn-Sham orbitals. The dynamical trajectories generated by the Lagrangean can be used in several ways.[8,9] (i) to relax ions and electrons simultaneously to the closest local minimum; (ii) to search for the absolute minimum with the dynamical simulated annealing (DSA) strategy; (iii) to perform MD simulations, with forces calculated with DFT-local-density approximation accuracy.

Here we present results for Si_N clusters ($N=7-10$) on the equilibrium structures and electronic properties at $T=0$, as well as on static and dynamical properties at finite temperatures. The study at low temperatures reveals the existence of unforeseen equilibrium structures, and may lead one to modify some of the existing views on the growth of Si microclusters. The extension to higher temperatures offers, for the first time, the possibility of investigating *ab initio* the effect of thermal motion on the properties of covalent clusters. Preliminary results have been published elsewhere.[10] Calculations on Si_3, Si_4, and Si_5 were used as a successful test[8] of the ability of the DSA to reproduce their known equilibrium structures.[4]

We used a norm-conserving nonlocal pseudopotential to describe the electron-ion interaction.[11] The Kohn-Sham orbitals at Γ were expanded in plane waves with an energy cutoff of 6 Ry.[12] We assumed a fcc supercell geometry with edge $a=35$ a.u. In the MD runs, an integration time step $\Delta t=1.2\times10^{-16}$ sec was used. The fictitious mass μ of Ref. 7 was set at 300 a.u. This led to trajectories which lay on the Born-Oppenheimer surface to within 10^{-5} a.u.

In order to obtain the most stable $T=0$ configuration, we have applied the DSA strategy. In a typical run, we heated the system to 3300 K for about $2000\Delta t$, and then cooled it down at a rate of 10^{14} K/sec. Obviously, the final result is not guaranteed to coincide with the global minimum in the potential energy surface. To enhance confidence in our findings, we also considered configurations generated by faster cooling rates and relaxation around preassigned geometries. In all cases, the stability of the resulting structures was tested by heating the clusters up to about 700 K, and monitoring whether a canonical drift was observed.

The structures of minimum energy are invariably the result of DSA and are displayed in Fig. 1. Si_7 is a pentagonal bipyramid,[13] Si_8 is a bicapped octahedron,[3] Si_9 can be viewed as a strongly reconstructed tricapped octahedron, and Si_{10} is a distorted tetracapped triangular prism (TTP). Some of these structures, most noticeably the TTP, have so far escaped the attention of other investigators. It would be very interesting to have fully *ab initio* calculations for these new geometries in order to test the accuracy of our predictions. Within our model, the growth pattern in this size range appears to be more complex than previously speculated.[3]

As expected, the number of stable isomers increases rapidly with cluster size. In the case of Si_{10}, in order of increasing energy from the TTP state we find, among others, (i) the tetracapped octahedron proposed as ground state in Ref. 3, $E \lesssim 0.01$ eV/atom (i.e., essentially degenerate with the TTP within the accuracy of the calculations); (ii) a bicapped tetragonal antiprism, $E=0.05$ eV/atom; and (iii) two parallel but shifted planar zigzag chains of five atoms, $E=0.07$ eV/atom. Relaxing these configurations in a larger cell ($a=40$ a.u.) and using an energy cutoff of 8 Ry preserves the energy ordering. Moreover, the structural energy differences increase slightly ($E=0.03$, 0.06, and 0.09 eV/atom, respectively).

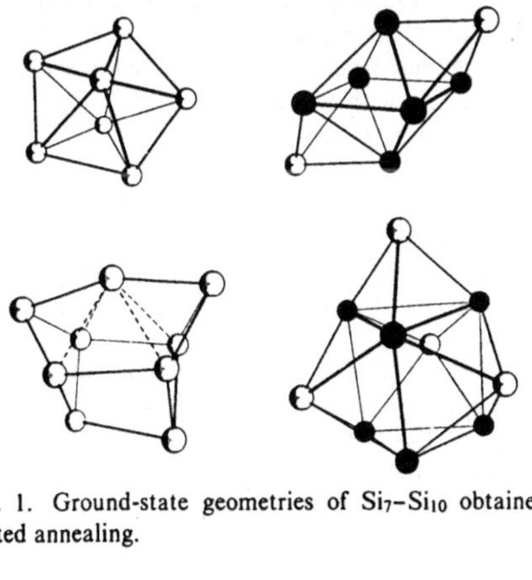

FIG. 1. Ground-state geometries of Si_7–Si_{10} obtained by simulated annealing.

In agreement with previous studies, all low-energy structures are quite different from crystalline fragments which have open configurations. Instead, they are very compact, with the majority of bond angles in the range of 60°, an average coordination number higher than four, and the atomic positions confined within a narrow shell at $\simeq 4.5$ a.u. from the center of mass of the cluster. The spherical average $\bar{\rho}_e(r)$ of the electronic density for these structures looks very similar, i.e., peaking around the ionic shell with a tail mostly due to the highest-energy orbitals.

It would be interesting to compare the electronic structure of the different low-energy geometries with the ultraviolet photoelectron spectroscopy data.[2] This re-

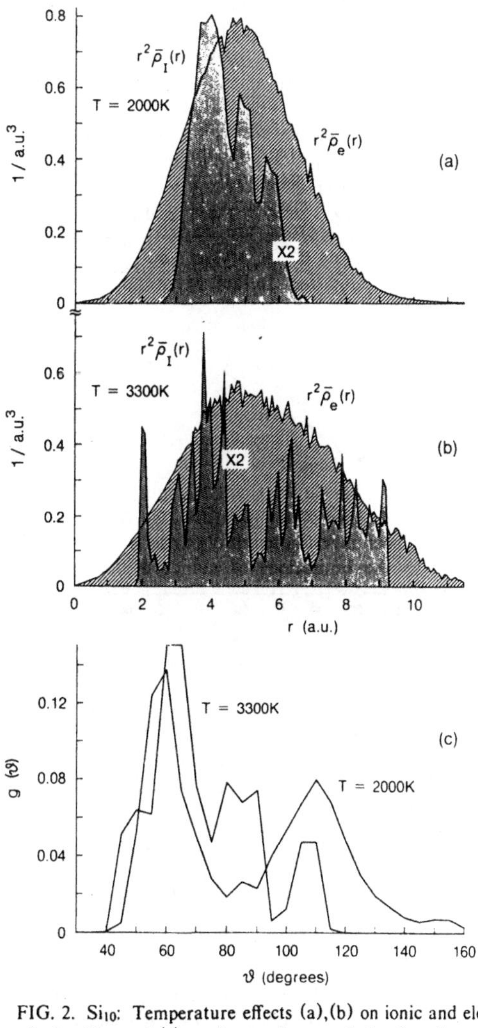

FIG. 2. Si_{10}: Temperature effects (a),(b) on ionic and electronic densities and (c) on the angular correlation function.

quires the computation of transition probabilities between ground and excited states of Si_N and Si_N^-. Here we only look at the energy distribution of the final states, i.e., neutral configurations corresponding to the removal of one electron from the occupied states of the anion. We have calculated these energies for Si_{10}, keeping the ionic positions fixed at the two lowest minima of the neutral cluster (TTP and tetracapped octahedron) but allowing for full electron relaxation. This amounts to the assumption that the extra electron does not substantially modify the geometry. Both structures exhibit a large lowest unoccupied–highest occupied molecular orbital energy gap of $\simeq 2$ eV, to be compared with the experimental value of $\simeq 1.7$ eV for the distance between the two highest peaks. The remaining part of the quasiparticle spectra is rather different for the two structures, with the TTP resembling the experimental results more closely. However, an unambiguous identification of the structure from the experimental data requires higher experimental resolution and more elaborate theoretical analysis.

We have studied the changes with temperature of several properties of Si_{10}. The averages were typically taken over a time interval of 1 ps, after equilibration runs of 0.2 ps. In Fig. 2, the spherically averaged electronic and ionic densities $\bar{\rho}_e(r)$ and $\bar{\rho}_I(r)$ are plotted versus the distance from the center of mass, (a) at

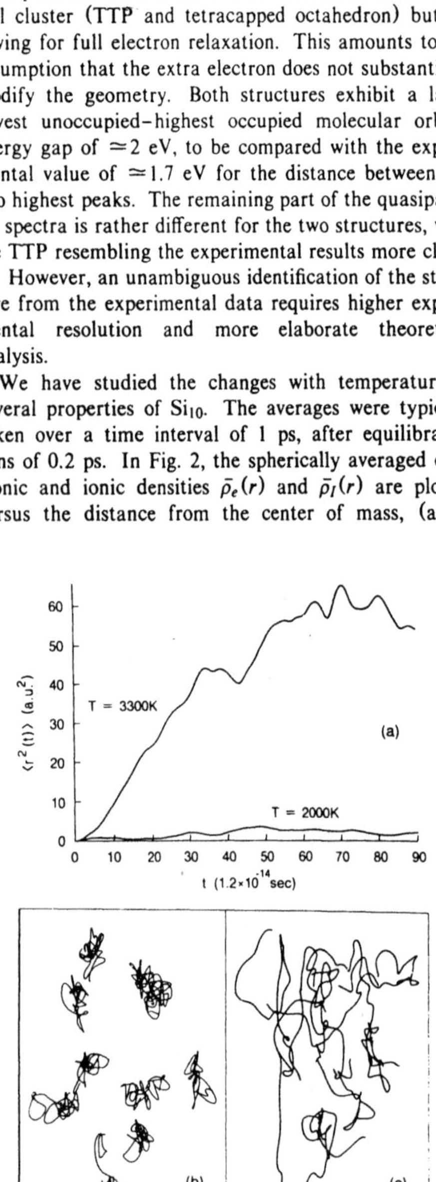

FIG. 3. Si_{10}: (a) Time dependence of mean square displacement and (b),(c) ionic trajectories at $T = 2000$ and 3300 K.

$T_1 = 2000$ K and (b) at $T_2 = 3300$ K. At T_1, both densities are only slightly broadened relative to low temperatures. In contrast, at T_2 more pronounced effects are observed: $\bar{\rho}_e(r)$ becomes much smoother and $\bar{\rho}_I(r)$ shows an approximate ionic shell structure with three atoms moving from the main shell, one inwards and two outwards.

In this temperature range, the two-body ionic correlation function shows only some thermal broadening. More striking structural changes are revealed by the bond-angle distribution [Fig. 2(c)], which shows that angles in the range 80°–90° become more likely at higher temperatures. Significant variations are also observed in the density of the valence-electron states, which extends invariably over $\simeq 12$ eV. At T_1 it shows four well-separated peaks, whereas at T_2 only the high-energy peak can be distinguished.

New information comes from the study of the dynamical properties of the clusters. We have calculated the atomic mean square displacement for Si_{10} as a function of time different temperatures. This shows a diffusive behavior at T_2, while at T_1 only sporadic jumplike diffusion processes were observed (Fig. 3). Since at T_2 the ionic shell structure is still maintained over relatively long times, we could analyze the mean square displacement for atoms in different shells. We find (i) marked increased in the ionic mobility on passing from the inner to the outer shell, (ii) predominance of tangential over radial motion, and (iii) strong correlation of the two outermost atoms, which tend to sit on opposite sides of the cluster.

When one is interested only in ionic properties, empirical potentials offer an efficient scheme for computer simulations, especially if one wants to extend our studies to large clusters. By using a DSA strategy similar to that described above, we have checked the accuracy of a recently improved potential.[14] The results found are at variance with ours. For instance, in the case of Si_{10}, the bicapped tetragonal antiprism is the ground state, and the tetracapped octahedron is much too high (by 0.6 eV/atom). The results published for other potentials[5] point to even larger discrepancies.

In conclusion, new results can be obtained in the physics of microclusters with the unified MD-DFT approach. On the basis of our present calculations, we can make two remarks:

(i) The success of DSA in generating ordered configurations suggests that the experimental clusters, which are obtained after cooling on a time scale presumably longer by several orders of magnitude, are likely to be in the ground state.

(ii) The study of dissociation processes requires simulation runs longer than the ones used in this work, but is feasible. On the basis of our experience at high temperatures, we expect that the dissociation products are not simply related to the $T = 0$ structures, as already suggested by classical MD calculations by Feuston, Kalia, and Vashishta.[5]

We are grateful to F. Ercolessi and P. Giannozzi for kind cooperation. One of us (W.A.) thanks J. Friedel, J. C. Phillips, K. Raghavachari, M. Schlüter, O. Cheshnovsky, C. L. Pettiette, and R. E. Smalley for useful discussions.

[1] L. A. Bloomfield, R. R. Freeman, and W. L. Brown, Phys. Rev. Lett. **54**, 2246 (1985); T. P. Martin and H. Schaber, J. Chem. Phys. **83**, 855 (1985); W. Begemann, K. H. Meiwes-Broer, and H. O. Lutz, Phys. Rev. Lett. **56**, 2248 (1986); Y. Liu *et al.*, J. Chem. Phys. **85**, 7434 (1986); M. L. Mandich, V. E. Bondybey, and W. D. Reents, Jr., J. Chem. Phys. **86**, 4245 (1987); J. L. Elkind *et al.*, J. Chem. Phys. **87**, 2393 (1987).

[2] O. Cheshnovsky *et al.*, Chem. Phys. Lett. **138**, 119 (1987).

[3] K. Raghavachari, J. Chem. Phys. **84**, 5672 (1986); D. Tománek and M. A. Schlüter, Phys. Rev. B **36**, 1208 (1987).

[4] G. Pacchioni and J. Koutecký, J. Chem. Phys. **84**, 3301 (1986).

[5] R. Biswas and D. R. Hamann, Phys. Rev. B **34**, 895 (1986); E. Blaisten-Barojas and D. Levesque, Phys. Rev. B **34**, 3910 (1986); B. P. Feuston, R. K. Kalia, and P. Vashishta, Phys. Rev. B **35**, 6222 (1987).

[6] F. H. Stillinger and T. A. Weber, Phys. Rev. B **31**, 5262 (1985); R. Biswas and D. R. Hamann, Phys. Rev. Lett. **55**, 2001 (1985); J. Tersoff, Phys. Rev. Lett. **56**, 632 (1986).

[7] R. Car and M. Parrinello, Phys. Rev. Lett. **55**, 2471 (1985).

[8] R. Car, M. Parrinello, and W. Andreoni, in *Microclusters*, edited by S. Sugano, Y. Nishina, and S. Ohnishi, Springer Series in Materials Science Vol. 4 (Springer-Verlag, Berlin, 1987), p. 134.

[9] R. Car and M. Parrinello, in *The Eighteenth International Conference on the Physics of Semiconductors*, edited by O. Engström (World Scientific, Singapore, 1987), p. 1165; M. C. Payne, P. D. Bristowe, and J. D. Joannopoulos, Phys. Rev. Lett. **58**, 1348 (1987); M. Needels, M. C. Payne, and J. D. Joannopoulos, Phys. Rev. Lett. **58**, 1765 (1987); D. Hohl, R. O. Jones, R. Car, and M. Parrinello, Chem. Phys. Lett. **139**, 540 (1987); D. C. Allan and M. P. Teter, Phys. Rev. Lett. **59**, 1136 (1987).

[10] W. Andreoni and P. Ballone, Phys. Scr. (to be published).

[11] D. R. Hamann, M. Schlüter, and C. Chiang, Phys. Rev. Lett. **43**, 1494 (1979). For the exchange-correlation functional see J. P. Perdew and A. Zunger, Phys. Rev. B **23**, 5048 (1981).

[12] In the case of Si_{10} the dispersion turns out to be at most 0.5 eV for the levels of highest energy.

[13] This is in agreement with recent configuration-interaction calculations by K. Raghavachari (private communication) and with the proposal in Ref. 4.

[14] J. Tersoff, private communication.

FREE ENERGY OF FORMATION OF LATTICE VACANCIES IN SILICON

G.B. Bachelet and G. Jacucci
Centro Studi C.N.R. and Dipartimento di Fisica
Universita' di Trento, I-38050 Povo, Italy

R. Car and M. Parrinello
International School for Advanced Studies and Dipartimento di Fisica Teorica dell'Universita' di Trieste, I-34100 Trieste, Italy

We propose an accurate method to compute finite-temperature properties of defects in semiconductors by combining quasiharmonic lattice dynamics with state-of-art electronic structure calculations. We present a preliminary application of our scheme to the study of thermal properties of the doubly positively charged lattice vacancy (V^{++}) in silicon. By using an 8-atom supercell we find an entropy of formation of about 3 k_B at constant lattice parameter. Our analysis of phonon modes around the vacancy suggests a very low barrier along the split-vacancy migration path.

I. INTRODUCTION

Formation and migration energies of intrinsic defects in semiconductors play a crucial role in understanding important processes like atomic diffusion. Since it is difficult to obtain direct information on defect energetics from experiment, there has been recently large interest in the numerical evaluation of defect total energies using reliable first-principle approaches[1]. Such calculations have been limited to the study of T=0 properties, whereas atomic diffusion is mostly relevant at high temperature. We propose a scheme which allows to extend such calculations at finite temperature. It combines quasiharmonic (QH) Lattice Dynamics, extensively tested for Lennard-Jones systems[2], with electronic structure calculations in the Local Density Approximation (LDA)[3]. We note that a fully anharmonic molecular dynamics study, also possible[4], may become a valid alternative to the scheme proposed here only if refined subtraction techniques are introduced: otherwise point defect properties, contributing only terms 1/N to total crystal properties, are buried in the statistical noise[5].

II. METHOD

In the harmonic approximation the Helmholtz free energy of an N-atom system is given by:

$$F = E_0 + 1/2 \sum_n h\nu_n + k_B T \sum_n \ln [1 - \exp(-h\nu_n/k_B T)] \qquad (1)$$

where E_0 and v_n are the total energy and the phonon frequencies, respectively. Other thermodynamic quantities can be obtained from standard derivations. In the QH approximation anharmonicity is partially incorporated into eq.1 through the volume dependence of E_0 and v_n[5]. The free energy of formation of a lattice vacancy is:

$$f = F_{vac}(V_1,T) - [(N-1)/N]\, F_{bulk}(V_2,T) \qquad (2)$$

Here V_2 equals $V_{bulk}(p,T)$, the bulk equilibrium volume at p,T ; depending on the choice of thermodynamic conditions we can choose V_1 equal to V_2 (constant lattice parameter,CL) or to $[(N-1)/N]\ V_2$ (constant atomic volume,CV); with $V_1=V_{vac}(p,T)$ eq.2 gives the Gibbs free energy of formation g (constant pressure,CP) at p=0; at p=0, g can also be roughly estimated from CL results as:

$$g_{CP} = f_{CL} + [p_{vac}(V_2,T)]^2\ (dV_2/dP)_T \qquad (3)$$

In the present work self-consisent LDA calculations are used to evaluate E_0, v_n and their volume dependence. We use a small (N=8 atoms) supercell with periodic boundary conditions and evaluate the free energy of formation at CL. We check CP using eq.3. Larger supercells are currently uder study[6]. For the electronic part we use norm-conserving pseudopotentials[7], one Baldereschi point[8] and about 400 plane waves per cell; we add to total energies a volume dependent correction which balances the lack of completeness[6] so that in the end our bulk modulus and equilibrium lattice constant are in agreement with converged (T=0) LDA results and thus very close to experiments[3].To implement eqs.1-3 we need an accurate evaluation of the dynamical matrix (involving the electronic structure of many independent ionic configurations) for the bulk and for the relaxed silicon vacancy at many different volumes. The large number of electronic calculations needed have been performed by using the method of Ref.4.

III. BULK SILICON

When replacing an infinite crystal with a periodic box containing N atoms (eight in our case) a discrete spectrum approximates the true phonon density of states. To examine this effect we test the method and our approximations on the bulk thermal expansion coefficient. The 8-atom supercell allows only bulk phonons at **Γ** and **X**, so we can directly test (experiment vs. experiment) how well these phonons alone approximate the full phonon density of states. In Table I we compare the linear thermal expansion coefficient obtained from experiment[9] (first column), from QH formulas using experimental phonons at **Γ** and **X** (second column), and from QH formulas using our theoretical phonons (third column). Note that the differences between column 2 and 3 are small and due to small inaccuracies in our theoretical phonons and Grüneisen parameters with respect to experiment. Assuming the validity of the QH approximation, it appears that the largest error, which shows up in the differences between column 1 and the other two, is due to the small box size. However it is apparent from Tab.I that an 8-atom cell gives results in semiquantitative agreement with experiment.

Table I

T(K)	$1/L(dL/dT) \times 10^7$		
0.01	0.0	0.0	0.0
50.00	-2.8	-6.2	-5.4
100.00	-3.4	-13.2	-13.0
150.00	+5.3	-7.6	-6.7
300.00	25.6	11.4	15.3
400.00	32.0	17.2	22.1
500.00	35.9	20.3	25.8
600.00	38.6	22.3	28.1
700.00	39.8	23.5	29.6

Si : linear thermal expansion coefficient vs. T (see text)

IV. SILICON VACANCY

In Table II we give results for the entropy and free energy of formation (s and f) of a double-plus vacancy at constant lattice parameter. With our small cell this is a convenient choice, since the other two thermodynamic conditions, CV and CP, involve relative volume changes of order (N-1)/N, which are unrealistically large with N=8. Thus to estimate g (third column) we use eq.3, based on results at constant lattice parameter. Above RT the entropy of formation ranges from 3.4 to 2.7 k_B, which is significantly higher

Table II

T(K)	s/k_B	f(eV)	g(eV)
300	3.35	4.38	4.03
450	3.25	4.34	3.97
600	3.14	4.30	3.92
900	2.94	4.24	3.81
1200	2.78	4.19	3.72
1500	2.65	4.14	3.63

$Si:V^{++}$ entropy and free energy of formation vs.T

than what is found for vacancies in metals yet insufficient to give a strong temperature dependence to the free energy of formation; thus our new results at high T are not dramatically different from previous theoretical results at T=0[1]. Our entropy is in good agreement with the valence force model of Lannoo and Allan[10] and compatible with Dannefaer's[11] experimental estimate for V^0, for which a larger entropy than V^{++} is expected (Jahn-Teller relaxation). Comparison with diffusion data is difficult because (1) we have not calculated the free energy of migration, and (2) the specific contribution of V^{++} to atomic diffusion is unknown.

From the volume dependence of phonon frequencies (not shown) we gain additional information: bulk frequencies are well represented by a linear fit over a wide range of volumes, but some frequencies of the vacancy show a quadratic behavior even for 1-2% volume changes around the experimental lattice constant of bulk Si. In particular the lowest frequency, corresponding to a split-vacancy migration coordinate, decreases very rapidly as the volume is decreased. The actual existence of a soft mode (i.e., an instability under pressure of the T_d relaxed geometry w.r.t. the split-vacancy geometry for the double-plus state) could be an artifact of the small cell, yet the finding suggests that the migration barrier along the split-vacancy coordinate is indeed very small.

In conclusion, the results of the eight-atom model provide a good test for our method and interesting new suggestions. A systematic study of calculated thermodynamic properties as a function of supercell size is needed to confirm our findings.

REFERENCES

1. R.Car,P.J.Kelly,A.Oshiyama and S.T.Pantelides, Proceedings of 17th ICPS, p.713, Springer 1985; Y.Bar-Yam and J.D.Joannopoulos, ibid. p.720; G.A.Baraff and M.Schlüter, ibid. p.725
2. G.DeLorenzi and G.Jacucci,Phys.Rev.B 33,1993 (1985)
3. M.T.Yin, Ref.1, p. 927
4. R.Car and M.Parrinello,Phys.Rev.Lett. 55,2471 (1985)
5. G.Jacucci,Diffusion in Crystalline Solids,p.429, Academic 1985
6. G.B.Bachelet, G.Jacucci, R.Car and M.Parrinello, to appear
7. D.Hamann,M.Schlüter and C.Chiang,Phys.Rev.Lett. 43,1494 (1979)
8. A.Baldereschi,Phys.Rev.B 7,5212 (1973)
9. H.Ibach, phys.stat.sol. 31,625 (1969)
10. M.Lannoo and G.Allan,Phys.Rev.B 25,4089 (1982);33,8789 (1986)
11. S.Dannefaer,P.Mascher and D.Kerr,Phys.Rev.Lett. 56,2195 (1986)

PHYSICAL REVIEW B VOLUME 36, NUMBER 4 1 AUGUST 1987

Ground state of solid hydrogen at high pressures

D. M. Ceperley and B. J. Alder
Lawrence Livermore National Laboratory, University of California, Livermore, California 94550
(Received 13 March 1987)

Quantum Monte Carlo calculations of the properties of bulk hydrogen at zero temperature have been performed. The only approximations involved in these calculations are the restriction to finite systems (64 to 432 atoms), the use of the fixed-node approximation to treat Fermi statistics, and the finite length of the Monte Carlo runs. The Born-Oppenheimer approximation was avoided by solving the quantum many-body problem simultaneously both for the electron and proton degrees of freedom. Using different trial functions and several different crystal structures the transition between the explored molecular and atomic phases was determined to occur at 3.0±0.4 Mbar. The transition to a rotationally ordered molecular phase occurred at about 1.0 Mbar. A lower bound to the static dielectric constant, given in terms of the static structure factor, was found to lie close to experimental values and became large for pressures greater than 500 kbar.

I. INTRODUCTION

The properties of bulk hydrogen have yet to be calculated from first principles, even though it is the simplest of the elements. There is a long tradition of calculations of the structure and properties of hydrogen dating back at least to the pioneering work of Wigner and Huntington[1] in 1935. They predicted that hydrogen will undergo a molecular to atomic transition as the density is increased and estimated that this transition occurs above a pressure of 0.25 Mbar. While the sophistication and accuracy of the calculations of the atomic phase have increased over the years, the knowledge of the equation of state of the molecular phase still comes primarily from experiment. Recent diamond-anvil measurements[2,3] have increased this knowledge to pressures close to 1 Mbar, and there are good prospects for experiments to still higher pressures. In addition, shock-wave experiments[4] have determined some properties of hydrogen at Mbar pressures, but at much higher temperatures (10^4 K). To date, there has been no reproducible observation of the transition of hydrogen into the atomic phase. One of the goals of this paper is to calculate from first principles this transition density and pressure.

For pressure of less than 100 kbar, the molecules of hydrogen in the molecular crystal are relatively undistorted from their state in the vacuum.[5] Consequently, up to these pressures the angular momentum of a single molecule is almost a good quantum number and the molecules are almost freely rotating at zero temperature so that the rotation of nearby molecules are mutually independent. However, at higher pressures it becomes energetically favorable for the molecular axes to align relative to each other. This transition has been observed[6] to occur in deuterium at a pressure of 280 kbar but it has not yet been observed in hydrogen. Because this transition involves a small energy change, it is possible to calculate the transition density only crudely.

Other possible transitions in hydrogen at zero temperature have been proposed. It is possible, through a mechanism known as band crossing, for molecular hydrogen to become metallic before the transition to an atomic metal. Such a transition occurs in iodine. Calculations based on density-functional theory have predicted[7] that such band crossing will occur in hydrogen at 9-fold compression ($r_s = 1.48$). In these calculations, the distortions of the molecules due to pressure have not been included, so their reliability is unclear. Although the present method avoids these approximations, the investigation of this particular transition is postponed to a future publication.

Another transition that has been proposed[8] is that of melting of the atomic solid to an atomic liquid. Such a transition is inevitable at enormously high pressures, but it has been argued that it could occur at relatively low pressures due to electron screening of the proton-proton interaction. In support of this mechanism, band-structure calculations[9] of metallic hydrogen at relatively low densities (which, however, neglect the zero-point motion of the protons) find a highly distorted, low-symmetry lattice as giving the minimum-energy structure. However, these distorted lattices have higher energy when the proton zero-point energy is added.

To resolve these questions an accurate computational method is required, that can determine the properties and energy of hydrogen in these various phases. The quantum Monte Carlo method is such a promising technique and has been successfully employed to determine the properties of such diverse systems as liquid and solid helium,[10] small chemical molecules,[11] and the electron gas,[12] including the melting of the latter. The simulation by Monte Carlo calculations of hydrogen is a natural extension of these studies. In fact, hydrogen is simulated simply as a two-component system of charged particles with unequal masses. Such a simulation in the experimentally accessible regime allows an unambiguous comparison of experiment with theory, perhaps for the first time on a many-body system. Previous comparisons of such simulation results with experiments have involved empirically determined interatomic potentials or other simplified Hamiltonians.

One advantage of the Monte Carlo method is that the zero-point motion of the protons can be treated exactly in both the molecular and atomic phases. Another is that the accuracy of the results can be determined by statistical means within the calculation itself. To make the error bars small requires a great deal of computer time. However, the human time needed to set up the calculation is probably less than for the other types of methods; particularly since the same program works for all phases and, in principle, for any combination of elements. In practice, the method as so far developed is only practical for low-Z elements. These simulations became only practical when computers of the speed of the Cray Research Cray-I computer were available. The amount of computer time needed (a few hours of central-processing unit time for each density) will not be as large in the near future, particularly as cheap and fast parallel processors become widespread. It will be possible to do much more refined calculations than presented here as computers and the methodology improve.

The main difficulties and limitations of the quantum Monte Carlo method are its restriction to a small number of atoms (less than a few hundred), the difficulty of determining the most stable crystal structure at a fixed density, the slow motion of the protons relative to the electrons, the upper-bound aspect of the fixed-node approximation, and the transient nature of the release-node method for fermions.

In Sec. II the Monte Carlo methods used in this paper will be briefly discussed. Section III presents the crucial choice of the trial function used in the various phases of hydrogen. Section IV contains the results of the simulations in the molecular phase, Sec. V in the metallic phase, Sec. VI introduces the conductivity, and Sec. VII presents a summary.

Throughout this paper energies are in units of Rydbergs/hydrogen atom and densities are in r_s units, where $4\pi a^3/3=v$ and $r_s=a/a_0$, where a_0 is the Bohr radius and v is the volume of an hydrogen atom, that is $r_s^3=1.338v_m$, where v_m is in units of cm^3/mole H_2.

II. NUMERICAL METHOD

Since the quantum Monte Carlo methods used here have been already discussed in detail elsewhere, we will simply define the three types of methods used on this problem, namely the following.

(i) *Variational Monte Carlo (VMC):* Let $\Psi(R)$ be a known trial function, where R refers to the full $3N$ set of particle coordinates and N is the total number of particles. Then VMC uses the Metropolis algorithm[13–15] to sample $|\Psi|^2$ and thus any expectation value with respect to this trial function can be computed, the most important of which is the variational energy of the trial function. The VMC method is used to select a good trial function, by minimizing the variational energy with respect to parameters in the trial function, and to initialize the ensemble for methods (ii) and (iii). VMC has the advantage in that it is considerably faster than the other two methods, a factor of 10 is typical, and there are no difficulties with fermions. However, it only yields an upper bound, not the true ground-state energy.

(ii) *Fixed-node diffusion Monte Carlo (DMC).* The Schrödinger equation in imaginary time, and transformed by the trial function is

$$\frac{\partial f(R,t)}{\partial t}=\sum_{j=1}^{N}\frac{\hbar^2}{2m_j}\nabla_j(\nabla_j f-f\nabla_j \ln\Psi^2)$$
$$-(\Psi^{-1}H\Psi-E_T)f\ , \qquad (1)$$

where E_T is the trial energy. The DMC algorithm[11,12] interprets $f(R,t)=\Psi(R)\phi(R,t)$ as a probability distribution in configuration space. The function $\phi(R,t)$ tends at large "time" t to the ground-state wave function. An initial ensemble of several hundred points with density $f(R,0)=\Psi(R)^2$, is evolved forward in "time." The three terms on the right-hand side then correspond to diffusion (with a diffusion constant equal to $\hbar^2/2m_j$), a drift derived from the trial function, and branching. For fermions, to interpret f as a probability, one also assumes that the nodes of the ground-state wave function are identical to the nodes of the trial function, so that their product, f, is always positive. This converts the fermion system to a distinguishable particle system. With this restriction, it can be shown[16] that the calculated energy is an upper bound to the exact ground-state energy.

In DMC, an additional approximation is made in "solving" Eq. (1). It is assumed that the drift and the local energy are constant in the region about the current position. This is only valid if the time step is small enough, so in practice, simulations for several different time steps need to be performed to test the accuracy of the simulations. More details of this algorithm are given in Ref. 11.

After convergence in time is reached, that is t is large enough so that the steady-state solution of Eq. (1) is obtained, the probability distribution of points in the ensemble is $\phi(R)\Psi(R)$. This is called the *mixed distribution* since it contains information about both the exact ground state, $\phi(R)$ (with the fixed-node restriction) and the trial function. The ground-state energy is the average value of $H\Psi/\Psi$, averaged over this mixed distribution, and has the *zero variance property* of quantum Monte Carlo; since as $\Psi(R)$ approaches an exact eigenfunction, the variance of the MC estimate of the energy approaches zero. Thus, energies in QMC can be calculated more accurately than for classical systems since the accuracy depends only on the accuracy of the trial function.

If the radial distribution function, $g(r)$, or any other average over the wave function other than the total energy is calculated using the mixed distribution, the result is somewhere between the variational $g(r)$ and the exact $g(r)$. In fact, if the trial function is sufficiently accurate the mixed pair correlation function should be halfway in between. Linear extrapolation[17] is the simplest way of calculating averaged quantities other than the energy, that is:

$$g(r)=2g_{\rm mix}(r)-g_{\rm var}(r)\ . \qquad (2)$$

(iii) *Green's-function Monte Carlo (GFMC).* This combined with the release-node method for treating Fermi statistics is an exact procedure; that is it completely removes the two approximations discussed above.[17,18] The method is practically only convergent, however, for systems where the Fermi energy is sufficiently close to the Bose energy and the trial function has reasonably accurate nodes. Both of these assumptions are satisfied for hydrogen up to quite high densities (a few Mbar) as checked here by Monte Carlo calculations. Because the GFMC method is the slowest of the three methods, it was not possible to use it for all of the calculations, but only to benchmark the DMC runs.

There are additional complications to be considered in these calculations which will be mentioned here and discussed later. First, the results for finite systems must be extrapolated to the bulk limit. To minimize the finite-size effects, the calculations are always performed using periodic boundary conditions. Because the Coulomb interaction as well as the many particle correlations in the trial function are long ranged, the Ewald image potential summation[19] must be used. Nevertheless, whenever the electrons are delocalized, there still remains an appreciable dependence of the properties on the size of the system, which can be traced to the discontinuity in the momentum distribution at the Fermi surface. This dependence is removed by an extrapolation based on Fermi-liquid theory.

Secondly, because the proton is 1836 times more massive than the electron, its diffusion with the DMC or GFMC algorithm (but not VMC) is that much slower, and its root-mean-square displacement per step is hence 42 times smaller than that of the electron. While the electronic distribution converges rapidly to its ground state, it is easy to find situations where the protonic distribution does not equilibrate in a reasonable amount of computer time. In principle, the simulations for the electron-proton system should be thousands of times longer than for a one-component system, and that is not practical. It is only possible with present computers to make one order of magnitude longer runs than that for the electrons alone. Several things can be done to improve the rate of convergence and establish a test for it, short of finding a general procedure to remove the disparate time scale problem. The initial ensemble should be thoroughly equilibrated by VMC so that for accurate trial functions the proton distribution will, in fact, be close to its final equilibrium value. In crystal phases the motion of the protons is severely limited in any case, so that the relevant time for equilibration is much shorter, namely the inverse of the Debye temperature. Some care has been taken to make the proton trial function accurate. For example, at low density an accurate approximation to the exact proton-proton molecular wave function has been used. To test for convergence, very long runs have been made on small systems of eight atoms.

Convergence to the state of lowest energy is also inhibited by the initial conditions, that is, for example, by the assumed crystal structure and by the nature of the trial function. Both of these conspire to keep the points of the random walk in the region of phase space appropriate to the initially selected phase. Although a constant pressure ensemble[20] instead of the constant volume ensemble used here could, in principle, partially overcome this difficulty, it is not yet clear whether that method can in reasonable computer time make the transition to the most favorable phase. For the constant volume ensemble each crystal structure must be tested separately. The ground state at a given density is then the one with the lowest energy. By this procedure it is certainly possible that a relevant phase, particularly of crystalline molecular hydrogen at high pressure, has been missed. Furthermore, phase transitions often involve very small energy differences. Since even with long runs the error bars on our calculations are roughly 0.001 Ry/atom, phase transitions driven by energies less than this are not capable of being resolved by this direct quantum Monte Carlo method. The development of differential Monte Carlo could circumvent this difficulty.

III. THE TRIAL FUNCTION

A trial function is selected to be as good an approximation as possible to the ground state, however, the function must be quick to evaluate on the computer as well, since at each step of the random walk the trial function and its first and second derivatives must be evaluated. Since the protons are not in a periodic array because of their zero-point motion, conventional band functions are not appropriate. Pair-product or Slater-Jastrow trial functions are employed because they have been found to be quite accurate in studies of the electron gas,[19] helium[10], and chemical molecules.[11]

$$\Psi(R)=\exp\left[\sum_{i<j} u_{ij}(r_{ij})\right]\prod_{\sigma} D_{\sigma}\ , \tag{3}$$

where $u_{ij}(r)$ is the "pseudopotential" acting between particles i and j a distance r apart and D_σ is the Slater determinant of a group of particles σ. Therefore,

$$D_\sigma=\det\{\phi_k(r_j)\} \tag{4}$$

and $\phi_k(r)$ is the kth orbital function. In principle, there should be four determinants in Eq. (3); two for electrons and two for protons, each with up or down spin. However, for the phase of hydrogen considered in this paper, the protons are always localized and hence distinguishable to a high degree of accuracy, so their antisymmetric exchange can be ignored.

The trial function of Eq. (3) contains three pseudopotentials, that acting between two electrons, between two protons, and between an electron and a proton. In addition, there are orbitals for both the electrons and for the protons. It is impractical to parametrize these functions and then find by brute force the optimal such trial function, as the dimensionality of the search for all the parameters of the pseudopotentials and orbitals is too high. For simpler problems, such as the electron gas and liquid helium, the optimal pseudopotential has been determined within the hypernetted chain approximation.[21] However, the simple integral equations are not sufficiently accurate for hydrogen and they would be much more complicated to solve numerically for a mixture of electrons and protons in a crystalline phase.

For the electron gas it was found[19] that pseudopotentials derived from the random-phase approximation (RPA) gave quite reasonable trial functions, and these RPA pseudopotentials are also found to be acceptable for hydrogen, as will be demonstrated. The RPA pseudopotentials[22] are derived from an expression for the variational energy in momentum space in which all terms involving three wave vectors are dropped (the RPA approximation). The structure function is related to the Hartree-Fock structure function by perturbation theory and then the energy is minimized with respect to the pseudopotentials. These RPA functions have the exact limiting behavior both when any two particles approach each other, the cusp condition, and when they are far apart. The RPA functions contain no adjustable parameters. A derivation will be published elsewhere.

The electron and proton orbitals that go into the Slater determinants define the phase of the system. Electron orbitals in crystals can either be described as delocalized band functions or by Wannier functions and these two descriptions are equivalent. However, there may be significant numerical advantages to one description or the other. In the atomic crystal the electrons are delocalized with a nearly spherical Fermi surface. Under these circumstances the s-wave scattering of the electrons by the protons can be accounted for by a pseudopotential so that the orbitals will be very close to that of the homogeneous electron gas, namely plane waves. The equivalent Wannier functions are less appropriate since these functions decay slowly in real space, as r^{-3}, and are difficult to construct for finite systems because of boundary effects. On the other hand, in the molecular phase, Wannier functions are a much more compact representation of the wave function but a plane-wave expansion of a molecular wave function would involve many plane waves. It has been shown[23] in one-electron theory that Wannier functions for systems with band gaps are exponentially localized. However, an even simpler compact form has been found[11] to describe the electronic hydrogen wave function well, namely a single Gaussian centered at the middle of the bond. Gaussians were hence used for the molecular phase.

$$\phi_k(r) = \exp[-C_e(r-Z_k)^2] , \tag{5}$$

where the lattice sites, Z_k, were chosen appropriate to several crystal structures. These orbitals are not orthogonal, orthogonality is not essential with the Monte Carlo approach. The variational parameter C_e was chosen by minimizing the energy in a variational calculation.

For the proton orbitals, Gaussians were also used with one proton per lattice site in the atomic phase and two protons per lattice site in the molecular phase. The protons were treated as distinguishable particles, so the assignment to lattice sites was fixed at the beginning of the calculation. The electrons are of course free to exchange between different molecules. As for the pseudopotential between protons it was found in the molecular phase that the RPA approximation between protons within the same molecule did not work very well. The correlation energy represented by the proton-proton pseudopotential is quite small (about 0.01 Ry) but it is highly desirable to start with a quite accurate proton trial function so that the random walk will converge quickly. A better approximation is the Born-Oppenheimer wave function for an isolated hydrogen molecule, the proton-proton part of which can be well approximated by

$$\phi(r) = \exp\{-[b_1/(b_2+r)]^{b_3} - b_4 r\}/r \tag{6}$$

with the values of the parameters b_1, b_2, b_3, and b_4 equal, respectively, to 8.046, 3.0, 4.456, and 14.56 and the units of r and the b's are in bohr radii. The first term in the exponent corresponds to the repulsion of the protons and the second to the covalent binding. This simple function gives about 95% of the proton correlation energy in an isolated molecule. The variational energy of an isolated molecule, using Eq. (6) for the proton orbital and Eq. (5) for the electron orbital, is -1.157 a.u.; the exact value is -1.164 a.u.

When a molecule is incorporated into a solid, some changes have to be made. First of all, a repulsion acts between protons of different molecules. The first term in the exponent of Eq. (6) is used for that. Furthermore, since two protons are bound, via Eq. (5), onto one lattice site, a term is added to the pseudopotential to cancel out this extra proton-proton binding. This is achieved by writing the wave function for the two protons in a molecule in relative and center-of-mass coordinates. The pseudopotential depends only on the relative coordinates, while fixing of the molecule to the lattice site only depends on center of mass coordinates. Finally, a term is added to the trial function which orients the molecule in a crystal direction. The simplest function with the appropriate symmetry is $\cos^2(\theta)$, where θ is the angle between the molecular axis and a crystal direction. For the fcc lattice these crystal axes are body diagonals $(\pm 1, \pm 1, \pm 1)$. There are four possible choices for these directions and when each of those possibilities is assigned to one of the four lattice sites in the cubic unit cell one arrives at the $Pa3$ structure. The pseudopotential between any two protons is

$$u(r) = [b_1/(b_2+r)]^{b_3} \tag{7a}$$

while two protons on the same molecule have the additional term:

$$u(r) = \ln(r) + b_4 r - C_p r^2 + G\cos^2(\theta) . \tag{7b}$$

The various variational parameters have been determined using two different criterion: namely, minimum variational energy and maximum overlap with the exact ground state as generated with DMC. Since this second criterion is less well known it will be briefly described. The overlap between ϕ and Ψ is defined as

$$0 = \int \phi\Psi \Big/ \left[\int \Psi^2\right]^{1/2} . \tag{8}$$

If the logarithm of the trial function (the pseudopotential) is expanded in a linear basis, $\ln(\Psi) = \sum_\beta u_\beta f_\beta(R)$, where u_β are unknown parameters and $f_\beta(R)$ are known functions, then at the maximum of 0 with respect to variations in u_β the following equation holds:

$$\langle f_\beta \rangle_{\text{mix}} = \langle f_\beta \rangle_{\text{var}} . \tag{9}$$

This condition requires[24] that for the radial part of the pseudopotential between pairs i and j to be optimal, the radial distribution functions, as computed by VMC and

TABLE I. Trial function parameters as used in Eqs. 5–7 in units of the Wigner-Seitz radius a in both the atomic and molecular phase. The other parameter, b_3, was always equal to 4.456. The G parameter which controls orientation with the crystal axis was zero for the isotropic fcc phase, but has the tabulated values in the oriented phase.

Density	Atomic	Molecular					
r_s	C_p	C_e	C_p	b_1	b_2	b_4	G
1.13	5.0						
1.31	5.0	0.48	51	6.14	2.25	19.0	−0.33
1.45	5.0	0.55	40	5.55	2.06	10.0	−0.40
1.61	5.5	0.50	42	5.00	1.86	23.5	−0.38
1.77	4.0	0.60	25	4.55	1.69	25.8	
2.00		0.85	13	4.02	1.5	29.1	
2.20		0.75	14	3.66	1.36	32.0	
3.0		1.00	15	2.68	1.00	43.7	

DMC, be equal at all r. Similarly, if the squared displacement of particles from their lattice sites is equal in the VMC and DMC calculation then the C parameters have been correctly chosen. Finally, if the expectation value of $\cos^2(\theta)$ is identical in DMC and VMC, then the orientational parameter G is correctly chosen. Maximum overlap is in some respects an easier criterion to apply than that of minimum variational energy since one knows after doing a VMC and DMC run, which specific parameters need adjusting and in which direction. In the case of orientated molecular hydrogen one can use this criterion to help determine what oriented phase is stable. For example, if the only solution to the equation $\langle \cos^2(\theta)\rangle_{var} = \langle \cos^2(\theta)\rangle_{mix}$ is $G=0$, it seems unlikely that the oriented phase, in fact, exists at that density. The disadvantage of the maximum overlap method is that it requires a well converged DMC run for each iteration of the variational parameters. The best parameters obtained for the various phases of hydrogen are given in Table I. It should be emphasized that converged DMC results are independent of the value of these parameters, it is only the error bars for a given amount of computer time which depends on the trial function.

IV. MOLECULAR PHASE

Using the trial wave function described in Sec. III, VMC, DMC, and GFMC simulations have been performed for four different crystal structures in the molecular phase at densities ranging from 20 cc/mol to 2 cc/mol. The results are summarized in Table II.

A. Crystal structure

The four different crystal structures examined were among those from the high-pressure phases of nitrogen.

(i) fcc isotropic phase. The trial function parameter G

TABLE II. Results of Monte Carlo calculation in the molecular phase. I, O, B, and G refer to the four crystal structures studied [see Sec. IV under (i), (ii), (iii), and (iv), respectively] and the number thereafter to the number of atoms. T is the length of the run (time step times the total number of steps). The energies are in Ry/atom with number in parenthesis being the error in the last digit. The bond length is in bohrs.

r_s	Method	Crystal	T ($\times 10^{-3}$)	$-E_{var}$	$-E^N_{mix}$	P (Mbar)	Bond length
3.0	GFMC	$I64$	8.4	1.125(1)	1.162(1)	0.004	1.42
2.2	DMC	$I64$	2.7	1.114(1)	1.160(1)	0.10	
2.0	DMC	$I64$	6.4	1.117(1)	1.156(1)	0.25	1.40
1.77	DMC	$I64$	3.5	1.080(1)	1.134(1)	0.66	
1.61	GFMC	$I64$	4.3	1.056(1)	1.085(1)		1.26
1.61	DMC	$I216$		1.058(1)			
1.61	DMC	$I64$	1.8	1.058(1)	1.089(1)	1.4	1.24
1.61	DMC	$B54$	1.0	1.071(1)	1.099(3)		1.23
1.61	DMC	$B128$		1.034(1)			
1.61	DMC	$B250$		1.037(1)			
1.45	DMC	$I64$	3.4	1.002(2)	1.032(1)	2.3	1.47
1.45	DMC	$O64$	3.8	1.005(1)	1.035(1)	2.3	1.59
1.45	DMC	$O216$		1.013(1)			
1.45	DMC	$G108$	2.7	0.996(1)	1.028(1)	2.2	1.57
1.31	DMC	$I64$	1.3	0.890(2)	0.939(1)	4.9	
1.31	DMC	$O64$	1.3	0.935(1)	0.960(2)	4.9	1.13
1.31	DMC	$O216$		0.942(1)			
1.31	DMC	$G108$	0.7	0.942(2)	0.962(1)	5.0	1.16
1.31	DMC	$G256$		0.942(1)			

is set to zero which results in the molecule being almost freely rotating about an fcc site. This is the experimentally observed low-pressure phase for hydrogen.

(ii) fcc-oriented phase. The G parameter is nonzero and this means the molecules are oriented relative to a crystal axis in the way described in the preceding section. This is the $Pa3$ (T_h^b) or α-nitrogen structure.

(iii) bccp phase. This is a cubic structure with the molecules aligned along the [111] direction. This arrangement has an energy 0.024 Ry/atom higher than the fcc-oriented structure at $r_s=1.61$.

(iv) γ-nitrogen phase. This structure is constructed by putting each molecule on a bcc lattice site and orienting the bond directions of the molecules in one of the simple-cubic sublattices in the [110] direction and the other molecules in the $[\bar{1}10]$ direction. It has an energy 0.015 Ry/atom higher than the fcc structure at $r_s=1.45$ and only 0.003 Ry/atom higher at $r_s=1.31$ but at the latter density the atomic phase has an even lower energy than any of the molecular structures examined.

The results indicate that the oriented $Pa3$ phase is preferred at $r_s=1.45$, although the energy difference between that and the isotropic phase is small. At higher density the oriented phase has lowest energy. This orientational ordering transition can be placed near 1 Mbar as shown in Fig. 1. It should be noted that the finite time step error, the fixed-node, and the size-dependent effects will tend to cancel in determining the energy difference between these phases so that these limitations should have very little systematic effects. Among the large set of other possible crystal structures only two other phases were examined, the bccp and the γ-nitrogen phase both have definitely higher energies until the atomic phase becomes more stable at $r_s=1.39$.

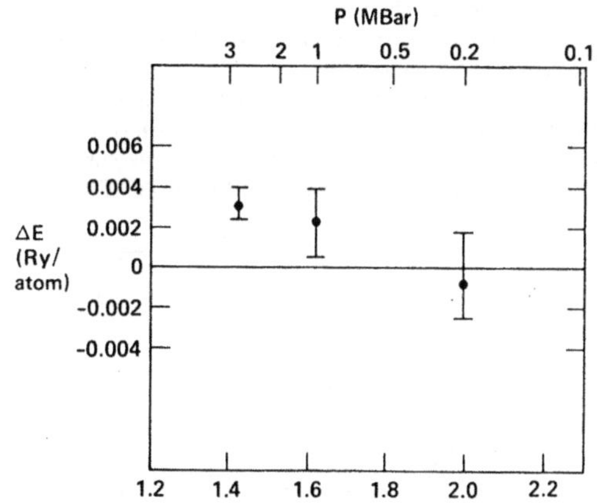

FIG. 1. The difference in energy between the isotropic fcc phase and the oriented $Pa3$ structure as a function of density. The pressure scale is indicated at the top.

B. Approximations

The size effects have been studied only at the variational level because of constraints on computer time but it is expected that these pair-product trial functions will have roughly the same size dependence as would the exact wave functions. At the density of $r_s=1.61$ there were no observable size effects in the ground-state fcc structure but larger ones in the bccp phase. Both at 2.3 Mbar ($r_s=1.45$) and at 4.9 Mbars ($r_s=1.31$) the difference between the 64 and 216 atom systems was 0.008 Ry. The effects in the metallic phase are much greater. In an insulating phase one expects size effects to be smaller because the electrons are localized and at low pressure the molecules interact with each other essentially by weak dipolar forces.

The integration time step used in these calculations varied from 0.01 to 0.02 Ry^{-1}. With these time steps the acceptance rate for moves was greater than 99%. Based on calculations for an isolated hydrogen molecule[11] the finite time-step correction for the ground-state energy will be on the order of 0.001 Ry/atom. To verify this estimate GFMC calculations have been performed at two densities. At the equilibrium density of 20.2 cc/mol ($r_s=3$), near zero pressure, an energy of -1.162 ± 0.001 Ry/atom was obtained compared with the experimental energy of -1.1645 Ry/atom. At the higher density $r_s=1.61$ ($v=3.12$ cc/mol) GFMC gives an energy of -1.085 ± 0.001 Ry/atom compared to DMC of -1.089 ± 0.001 Ry/atom. It is not clear whether this large difference is to be ascribed to time-step error, underestimation of the error bars, statistical fluctuation or the way the fixed-node approximation is treated in GFMC.

The error of the variational wave function ranges from 0.02 to 0.04 Ry and seems to be roughly independent of density. The optimization of the parameters was not carried out uniformly for all runs, so some variation in the quality of the trial function can be expected. The lengths of the computer runs are given in the table in terms of T defined as the time step times the total number of moves. The error of the ground-state energy agrees roughly (within a factor of 2) with the estimate[25] $[2(E_{\text{var}}-E_{\text{mix}})/T]^{1/2}$. Improvements in the wave function will reduce the statistical error.

C. Ground-state energy

The ground-state energies given in Table II for the molecular phases considered and values of r_s between 2.0 and 1.31 were corrected for finite-size effects by adding to them the difference in energy between the 216 and 64 particle variational energy. The results were then fitted by a cubic polynomial in r_s. The fitting coefficients in increasing powers of r_s are equal to 1.1200, -2.9017, 1.2208, and -0.1695. The corrected Monte Carlo points are shown in Fig. 2 and compared with the experimental equation[2] and its empirical extrapolation. The extrapolation is in near agreement with a lattice dynamics calculation[26] with an empirical intermolecular potential derived from shock wave data. The agreement is quite good where experiment is available but in the region $r_s\sim1.6$

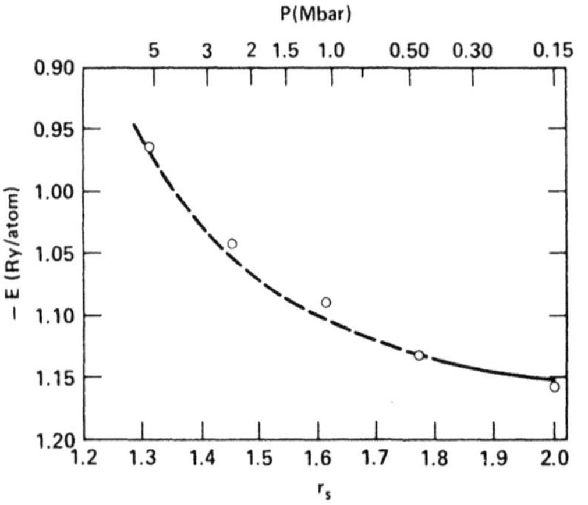

FIG. 2. The energy in the molecule phase (in Ry/atom) vs density (in r_s units). The circles are the Monte Carlo results. The solid line is the diamond-anvil results for pressures less than 400 kbar. The dashed lines are the results (Ref. 6) of lattice dynamics calculation (Ref. 26) with the pair potential fitted to shock-wave data at about 1 Mbar. The pressure scale at the top is based on this pair potential. The dashed line also agrees with an extrapolation of the experimental diamond-anvil measurements (Ref. 6).

(1.2 Mbar) the MC results are 0.010 Ry higher. Tests that have been performed suggest that the MC energies in the molecular phase are accurate to better than 0.005 Ry/atom. Thus, a discrepancy exists at intermediate pressures.

The curve of energy versus density is suggestive of a phase transition at a pressure between 1 and 2 Mbar. Of course there are not nearly enough calculations at different densities to even roughly establish that a transition occurs. The energies involved in the orientational ordering transition are too small to explain the "bump" as can be seen from Fig. 1. Given that the crystal phase is constrained by the trial wave function and the boundary conditions, the only other likely transition would involve a change in the bond length distribution, a contraction of the molecules, possibly coupled with the orientational transition. Some evidence for this is presented in Sec. IV D below. However, since the Monte Carlo run at the density $r_s = 1.61$ is not particularly long, it is possible it has not fully converged to the ground state; the proton degrees of freedom are the slowest to converge and in addition if there is a phase transition there will be a slowing down of the rate of convergence.

The pressure can be computed either by differentiating the energy fit, i.e., $P = 11.71\ \mathrm{Mbar\,Ry}^{-1}\ r_s^{-2} dE(r_s)/dr_s$ or by use of the virial theorem. The virial pressure is shown in Table II and a comparison of the two in Fig. 3.

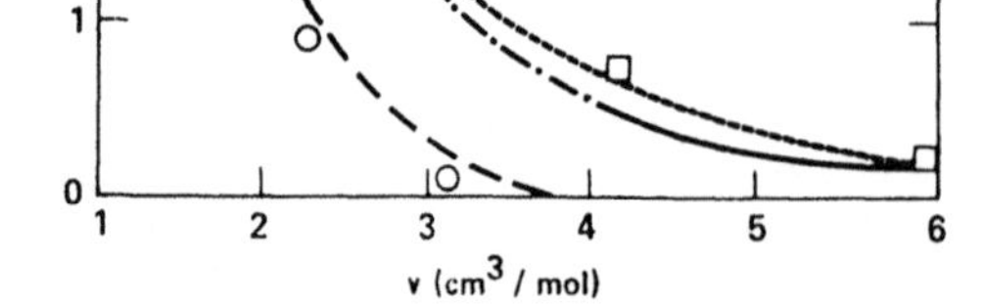

FIG. 3. The pressure vs volume. The open circles are the MC results from the virial theorem in the atomic phase, the lower dashed curve is the derivative of the fit to the atomic energies [Eq. (11)]. Similarly, the open squares and upper dashed curves are for the molecular phase. The solid line shows the transition. The solid curve is the measured (Ref. 2) EOS and its extrapolation (dashed-dotted curve) (Refs. 2 and 26).

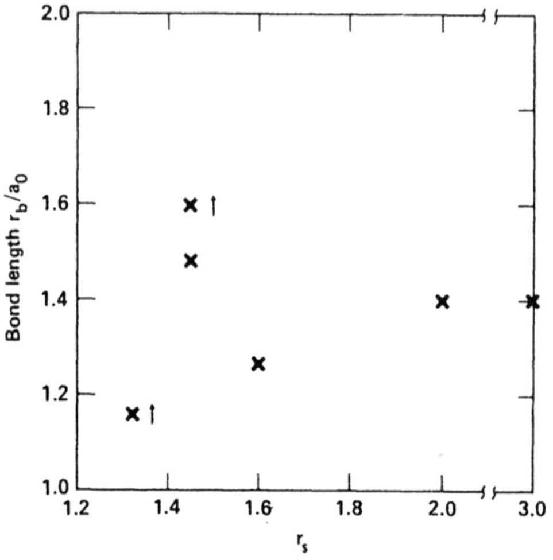

FIG. 4. The bond lengths in bohr radii with the arrows indicating the oriented phase.

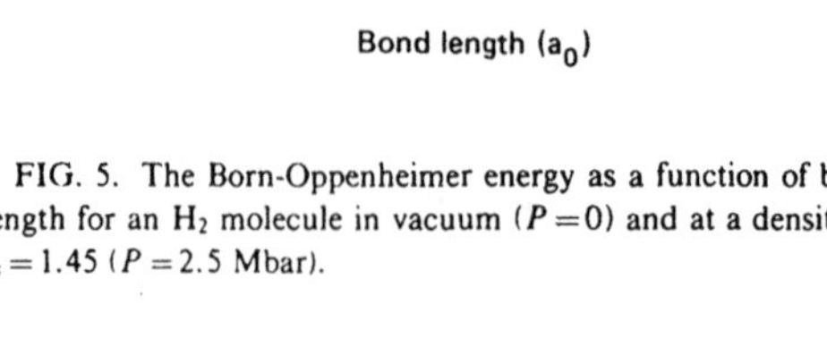

FIG. 5. The Born-Oppenheimer energy as a function of bond length for an H_2 molecule in vacuum ($P = 0$) and at a density of $r_s = 1.45$ ($P = 2.5$ Mbar).

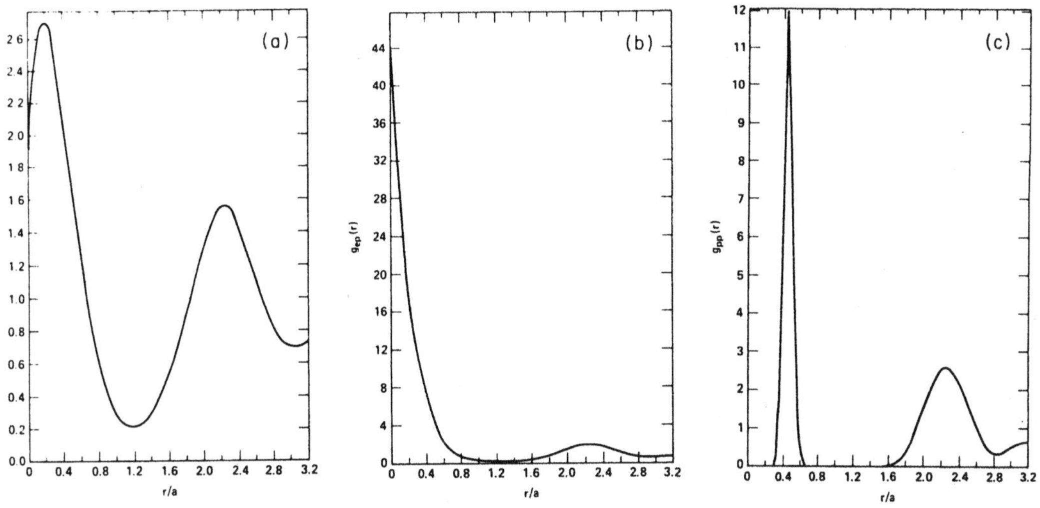

FIG. 6. The pair correlation functions in the molecular ($Pa3$) phase at zero pressure ($r_s = 3.0$). They are sphericalized and normalized to go to unity at large r.

Differentiating the energy is more accurate because it uses the zero variance property of the QMC to reduce fluctuations. The two pressures are in agreement where the polynomial fit to the calculated energies is good.

D. Bond length

The rms value of the molecular bond length is also given in Table II and shown in Fig. 4. Note that these averages are obtained by the extrapolation technique from the variational and mixed averages as explained in the previous section. In general, the bond lengths decrease with increasing density, however, the results at $r_s = 1.45$ do not fit this pattern. The statistical error bars are small for these bond lengths, however, there could well be significant systematic errors, due to either the failure of the extrapolation technique or the slow convergence of the proton degrees of freedom. To avoid the latter problem some runs were carried out at the density $r_s = 1.45$ with the Born-Oppenheimer approximation, that is by fixing the protons in the oriented $Pa3$ structure with a given bond length and setting their mass to infinity. The energy as a function of bond length is shown in Fig. 5. As a comparison, the Born-Oppenheimer energy of an isolated molecule is shown. It is seen that the effect of pressure on the molecule is to shrink the bond, but the restoring force also diminishes, leading to a much more anharmonic bonding force. These Born-Oppenheimer simulations are naturally much less expensive since they converge to the ground state within a few inverse Rydbergs, corresponding to about 50 time steps, rather than thousands of iterations, if the protons are moving with their correct mass. The unusual dependence of bond length on density could be related to the excess energy at intermediate density discussed above in Sec. IV C.

E. The pair correlation function

Shown in Fig. 6 are the three types of pair correlation functions (that is *e-e*, *e-p*, and *p-p*) for molecular hydrogen at zero pressure. The Fourier transform of these functions, the structure function will be discussed later in relation to the dielectric susceptibility. The peak at small r represents correlations between particles within the same molecule while that for distances between $2.0a$ and $2.6a$ represents correlations between nearest-neighbor molecules.

V. THE ATOMIC PHASE

In the atomic phase the RPA pseudopotentials[22] were used between all pairs of particles. Gaussians were used for the protonic orbitals while plane waves, filled up to the Fermi level, were used for the electronic orbitals. Table III contains the variational energies and the DMC results.

A. Finite-size scaling

In the metallic phase the extrapolation to the thermodynamic limit is much larger than in the molecular phase and must be done carefully if the results are to be accurate. Size effects are bigger in this phase simply because the electrons are delocalized and thus sensitive to boundary effects.

The extrapolation method is based on the picture provided by Fermi-liquid theory. According to Landau,[27] the energy of a system can be written as an energy functional of occupation numbers of quasiparticles, which behave just like an ideal Fermi gas. For small excitations from the ground state the functional can be linearized and is characterized by a few Fermi-liquid parameters. Consider how the energy of such an ideal Fermi-liquid

TABLE III. Monte Carlo results in the atomic phase. N is the number of atoms, T is the total length of the run [(time step)× (ensemble size)×(number of steps)]. E^N_{mix} is the fixed-node energy for the finite system, E^{∞}_{mix} is extrapolated to the bulk limit using Eq. (10). P is the virial pressure. Lindemann's ratio is the rms value of the proton displacement from the lattice site divided by the nearest-neighbor distance.

r_s	Crystal	N	T ($\times 10^{-3}$)	$-E_{var}$	$-E^N_{mix}$	$-E^{\infty}_{mix}$	P (Mbar)	Lindemann's ratio
				Static lattice of protons				
1.0	fcc	108	1.0	0.635(2)	0.670(1)	0.727	19.5	
1.13	fcc	108	1.6	0.816(1)	0.849(1)	0.893	8.4	
1.13	bcc	54	1.2	0.949(2)	0.983(3)	0.891	8.5	
1.13	bcc	128	1.6	0.835(2)	0.876(3)	0.897	8.3	
1.30	fcc	108	2.0	0.935(1)	0.970(2)	1.002	2.7	
1.31	bcc	128	1.2	0.957(1)	0.988(1)	1.002	2.6	
1.31	fcc	256	0.8	0.957(1)	0.992(1)	1.002	2.6	
1.45	bcc	54	3.4	1.065(1)	1.094(1)	1.035	0.88	
1.45	fcc	108	1.5	0.982(1)	1.011(1)	1.039	0.85	
1.61	fcc	108	2.8	0.999(1)	1.033(1)	1.052	0.07	
1.77	fcc	108	3.2	0.998(1)	1.033(1)	1.048	−0.18	
2.00	fcc	108	4.0	0.980(1)	1.021(1)	1.033	−0.29	
				Dynamic lattice of protons				
1.13	bcc	54	0.6	0.782(1)	0.813(1)	0.856	9.13	0.15
1.31	sc	64	5.0	0.967(1)	0.985(1)	0.962	2.84	0.16
1.31	fcc	108	0.8	0.913(1)	0.942(1)	0.973	3.05	0.13
1.45	bcc	54	2.3	1.050(1)	1.071(1)	1.012	1.17	0.16
1.45	fcc	108	1.2	0.960(1)	0.990(1)	1.016	1.18	0.15
1.61	bcc	54	1.2	1.057(1)	1.081(1)	1.032	0.25	0.15
1.77	fcc	108	0.8	0.985(1)	1.019(1)	1.035	−0.03	0.16

changes when one goes from a finite system in periodic boundary conditions to an infinite system. For a finite system the allowed values of momentum lie on a lattice reciprocal to that of the simulation cell and the ground state is obtained by filling successive "shells" of these lattice points. (A shell consists of all lattice points related to each other by symmetry.) As long as the shells are filled in a symmetric fashion then the only Fermi-liquid parameter that comes in should be the effective mass. This implies that the size corrections of the interacting system should be simply proportional to the size correction of the noninteracting system, at least for large enough systems. The difference in energy per particle between an infinite and a finite ideal Fermi gas of N particles in periodic boundary conditions is of order $1/N$ with a coefficient[19] which varies between ± 1 as N changes.

For charged systems, in addition to this number dependent effect on the kinetic energy, there is also an effect in the potential energy,[19] since the potential is long ranged. In simulations with periodic boundary conditions, the coulomb interaction is replaced by the Ewald image potential for finite systems. To have charge neutrality a particle must interact with its own image. Thus, in calculating the potential energy one term out of N is appropriate to a perfect lattice, not to a Fermi liquid, as it should be. This intuitive result is supported by both Hartree-Fock calculations, valid at small r_s, and harmonic lattice calculations, valid at large r_s, which show that the size dependence of the potential energy of the homogeneous electron gas is proportional to $1/N$.

Adding together these kinetic and potential-energy size corrections, implies that the energy per particle for a finite system is related to the bulk energy per particle by

$$E_N = E^{\infty} + c_1(r_s)(T_N - T_{\infty})/r_s^2 + c_2(r_s)/(N r_s)\,, \quad (10)$$

where c_1 and c_2 are functions of the density to be determined from the simulations and T_N is the kinetic energy of the ideal gas, at $r_s = 1$. We have multiplied c_1 and c_2 by r_s^2 and r_s, respectively, so that c_1 and c_2 will be roughly independent of r_s in the high-density limit.

VMC calculations have been performed with many different values of N to test the accuracy of Eq. (10) and to determine the unknown parameters E^{∞}, c_1, and c_2. VMC was used rather than DMC since it is much less time consuming, particularly for large systems. Because the pair product trial function has the correct long-wavelength properties it should give the same size dependence as an exact calculation. One can observe by examining Table III that the size dependence is the same for the VMC and DMC calculation to within the statistical error of 0.002 Ry/atom even though the variational energy changes by 0.101 Ry/atom in going from 54 to 108 atoms. The size-dependence correction is itself very large. Table IV contains the results of variational calculations for N ranging from 32 to 432, for $r_s = 1.31$ and Fig. 7(a) illustrates the correlation between the variational kinetic energy versus that kinetic energy of an ideal Fermi gas with the same number of particles. Figure 7(b) shows the strong correlation between the variational potential energy and $1/N$. In fact, Eq. (10) fits the total variational energies better than the variational and potential energy separately. For $r_s = 1.31$ the fitted values of c_1 and c_2 are 1.118 and −1.146. The corrected infinite system energies are also displayed in Table IV. It is seen that the maximum variation not including that of the sc phase is only 0.0033 Ry. Thus the uncertainty due to the finite system

TABLE IV. Results of variational calculations in the atomic phase at $r_s = 1.31$ for seven different sizes of systems. Columns labeled with ∞ show the extrapolation to the bulk limit. T_N is the ideal-gas kinetic energy.

Crystal	N	$-E^N_{\text{var}}$	$-V^N_{\text{var}}$	T_N/r_s^2	$-E^\infty_{\text{var}}$	$-V^\infty_{\text{var}}$	$-P^\infty_{\text{var}}$
fcc	32	0.9442(3)	2.446(1)	1.318	0.9501	2.430	2.76
bcc	54	1.0209(3)	2.451(1)	1.239	0.9506	2.430	2.75
sc	64	0.9664(7)	2.437(4)	1.280	0.9435		
fcc	108	0.9199(4)	2.431(2)	1.323	0.9511	2.431	2.75
bcc	128	0.9368(5)	2.432(2)	1.307	0.9516	2.430	2.74
bcc	250	0.9390(5)	2.430(2)	1.299	0.9483	2.429	2.77
fcc	256	0.9406(9)	2.430(3)	1.300	0.9513	2.429	2.74
bcc	432	0.9477(11)	2.429(4)	1.292	0.9509	2.428	2.74

size has been reduced by a factor of 30!

The reason the total energies are fit better by Eq. (10) than are the variational and kinetic energies separately is that some of the potential-energy correction gets mixed into the kinetic energy and vice versa. For a system of charges, the kinetic energy is related to the total energy by $T = -d(r_s E)/dr_s$. Thus the size dependence of the kinetic energy also follows Eq. (10) with $c_1^T = c_1 - r_s c_1$ and $c_2^T = r_s c_2$. Since the size dependence of the potential and kinetic energies are needed to find the pressure with the virial theorem, it is necessary to fit separately the kinetic and potential energies to Eq. (10), giving new coefficients c^T and c^V. The potential-energy corrections at $r_s = 1.31$ are $c^V = (0.165, -0.890)$. Variational calculations have also been carried out at a higher density $r_s = 1.13$, obtaining for the fitting coefficients $c = (1.104, -1.153)$ and $c^V = (0.050, -0.80)$ showing the assumed r_s dependence of c_1 and c_2 is approximately correct. This method provides a simple way of calculating the Fermi liquid parameters since c_1 should be equal to the effective mass, which is near unity in metallic hydrogen.

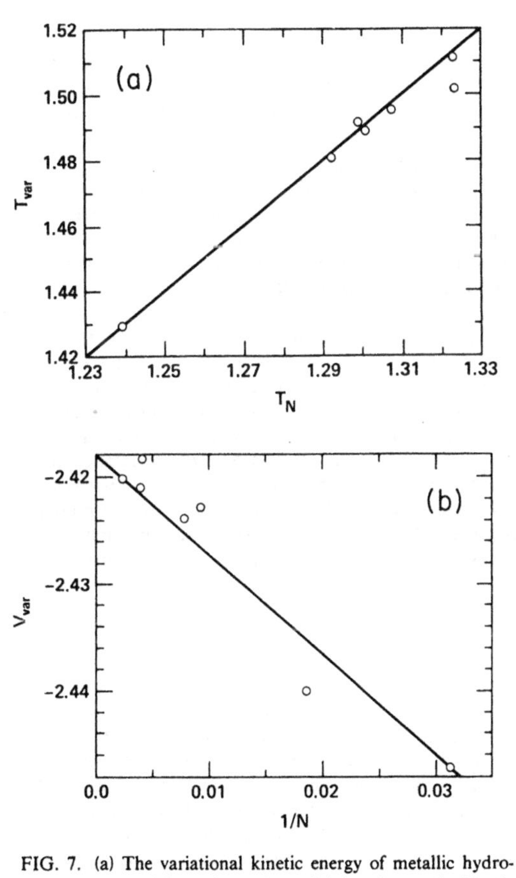

FIG. 7. (a) The variational kinetic energy of metallic hydrogen as computed with a pair product trial function vs the kinetic energy of an ideal Fermi gas with the same number of electrons at the density of $r_s = 1.31$. The line has a slope of unity. (b) The variational potential energy vs $1/N$, where N is the number of hydrogen atoms at the density of $r_s = 1.31$.

B. Lattice type

All calculations have used either fcc, bcc, or sc lattices. The simple cubic lattice has higher energy at $r_s = 1.31$ (by about 0.010 Ry/atom) but the size effects and the error bars are too large to determine whether fcc or bcc is more stable. Adding a term dependent on the lattice type to Eq. (10) does not appreciably improve the fit to the energies. Thus the energies of the two lattices are the same to the accuracy of about 0.002 Ry/atom at the density $r_s = 1.31$. For comparison, by perturbation theory[28] the static fcc lattice was found to be more stable in the range $1 < r_s < 1.6$ by 0.0012 Ry. Perturbation calculations which ignore proton zero-point motion, have found unusual planar structures that are more stable than either fcc or bcc, however, isotropic structures are favored once the protons are allowed to move. A density-functional calculation[29] found the simple cubic structure most stable at $r_s = 1.31$ by 0.003 Ry/atom. Thus, this theory cannot reliably determine even crystal structures for hydrogen.

C. Ground-state energy

Table III contains the ground-state energies corrected for finite system size, assuming $c_1 = 1.10$ and $c_2 = -1.15$ are independent of r_s. The energies for densities in the range $1.0 < r_s < 2.0$ are fit to the expression:

$$E(r_s) = 2.21 r_s^{-2} - 2.70722 r_s^{-1} + d_1 + d_2 r_s + d_3 r_s^2 , \qquad (11)$$

where the first two terms give the exact behavior[30] for a static lattice in the high-density limit. The values of

$d_1=-0.2166$, $d_2=0.0566$, and $d_3=-0.0301$ were obtained for the dynamic hydrogen lattice.

The fitted energies are shown in Fig. 8 and compared to other theoretical predictions. The results in closest agreement with the MC energies are those of perturbation theory[28] which are quite accurate in the range $1.4<r_s<1.6$ and are high by about 0.015 Ry at $r_s=1.1$. Variational correlated basis calculations[31] are too *low* by 0.02 Ry. Approximations have significantly compromised the variational principle since the variational energy with this type of trial function should be 0.03 Ry above the exact energy. Local-density-functional results lie even further below the exact energies, by 0.03 Ry/atom and the error is density dependent, getting much worse as r_s gets larger than 1, and even give the wrong crystal structure. The atomic equation of state (EOS) is shown in Fig. 3.

D. Effect of proton motion

The difference in energy between calculations on a static lattice and for the real lattice defines the proton zero-point energy and is shown in Fig. 9. The results are larger than the estimate of this energy difference using either the Debye model[32] or the self-consistent harmonic approximation.[30] Also given in Table III is the Lindemann's ratio, that is the rms displacement from the lattice divided by the nearest-neighbor distance. For hydrogen this value is about 0.15, considerably less than that of helium at melting, where the ratio is 0.26.

E. Effect of finite time step and fixed-node approximation

Simulations with different but small time steps were used to correct the energy for the timestep error. A GFMC test run at one density showed no observable

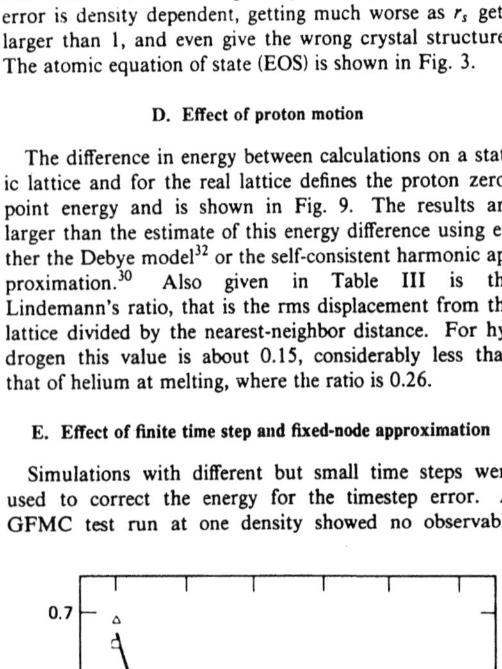

FIG. 8. The energy of metallic hydrogen in the atomic phase as a function of density. The upper line is a fit to the MC results for the finite mass proton lattice of hydrogen while the lower line is for a static lattice (protons of infinite mass) either fcc or bcc. Open triangles represent results of perturbation theory (Ref. 28); open circles the variational correlated theory (Ref. 31); open squares the density-functional theory (Ref. 29).

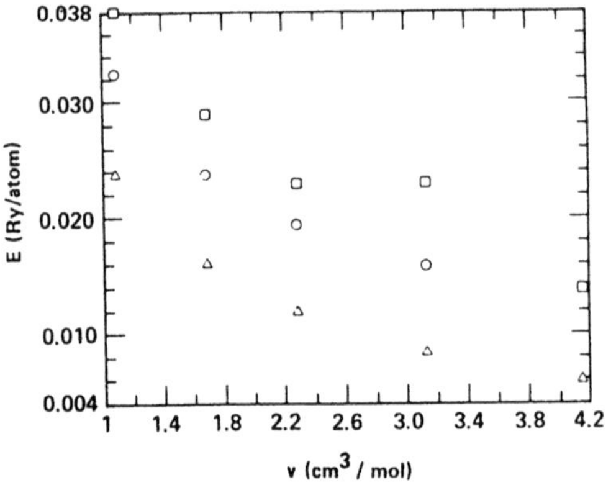

FIG. 9. The zero-point vibrational energy of atomic hydrogen (Ry/atom) as a function of density. The open squares are the MC results, the open circles the results of a Debye model (Ref. 32), and the open triangles of a lattice dynamics calculation (Ref. 30).

effect of timestep errors or fixed-node errors. This conforms to a similar nodal-release calculation for the electron gas,[12] where the energy as lowered by about 0.001 Ry/atom. The error in the fixed-node approximation should be less important in hydrogen than in the electron gas since the protons partially localize the electrons. The final accuracy in the ground-state energy is estimated to be 0.003 Ry/atom, at least twice as small as the error in the molecular phase. The calculations are more accurate in the atomic phase because the motion of the protons is much more restricted.

The amount of energy missed by the RPA pseudopotential is about 0.035 Ry/atom, 35% of the total electron-correlation energy. Thus the electronic part of the trial wave function is equally accurate in the atomic and molecular phase. However for the electron gas the RPA pseudopotential picks up 95% of the correlation energy at this density.[16] This much larger variational energy for hydrogen probably comes from electron-proton correlations. The simple pair-product trial function used here has an inadequate treatment of the band structure which raises the variational energy by about 0.03 Ry/atom.

F. Atomic-molecular transition

Using the fits to the molecular and atomic energies, it can be determined that hydrogen changes from an fcc molecular phase to a cubic atomic crystal at a pressure of 3.0 Mbar. The atomic phase is stable for $r_s<1.30$ (1.65 cc/mol) and the assumed molecular phase is stable for $r_s>1.39$ (2.01 cc/mole). Thus, the relative volume change is 20%. There have been many previous estimates of this transition density some of which are similar to the present one. It is difficult to give good error bounds to this estimate of the transition density. The errors coming solely from statistical fluctuations of the Monte Carlo runs are quite small (on the order of 0.002 Ry/atom) but

TABLE V. The energy needed to add or subtract a zero-momentum electron for $N/2$ molecules of H_2 and the resulting energy gap at $k=0$. The protons were held rigid in the $Pa3$ structure with the given bond length (r_{HH}).

r_s	N	r_{HH} (Å)	$E^- - E$ (eV)	$E^+ - E$ (eV)	E_{gap} (eV)
3.0	8	1.4	−3.0	12.6	9.6±2.0
3.0	64	1.4	1.2	13.6	14.8±1.0
1.45	64	1.3	−1.0	1.9	0.9±1.0

systematic effects can be quite a bit larger, the largest ones are (i) poor convergence of the proton degrees of freedom in the molecular phase, (ii) uncertainty about the crystal structure of the molecular phase, (iii) finite-size effects, (iv) incomplete treatment of antisymmetry (Fermi statistics), and (v) inadequate fit to the calculated energies. Attributing an uncertainty of no more than 0.004 Ry/atom to each phase, and assuming that this error is roughly independent of density in the 3-Mbar region, these factors could change the transition pressure by 15%, the molecular critical volume by 8%, and the atomic critical volume by 5%.

G. Pair correlation function

The pair correlation function in the atomic sc phase is shown in Fig. 10 at the lowest stable atomic density, $r_s=1.31$. The electron-electron function is similar to that of an electron gas at the same density.

VI. DIELECTRIC RESPONSE OF HYDROGEN

It has been suggested that molecular hydrogen could become a metal at high pressures before undergoing a transition to an atomic phase since as the density increases, its bands broaden, closing an indirect gap and the system would undergo a transition to a metallic state.[7] Such a phenomenon has been observed in iodine at high pressures.

For a system of rigid molecules in the $Pa3$ structure, the energy change in adding and in removing an electron was calculated with DMC as shown in Table V. Both the electron and hole were assumed for simplicity to be in a zero-momentum state, thus the sum of the energy changes is the band gap at $k=0$. For a system of 64 molecules at zero pressure ($r_s=3$) the band gap was found to be 15±1 eV while at 2.4 Mbar ($r_s=1.45$) the band gap was 1±1 eV. It is difficult both to assess the reliability of this calculation and to compare with experiment. The restriction to zero-momentum particle and hole states increases the gap at $r_s=3$ by 1.5 eV in a band-structure calculation[7] and 3.5 eV in a Hartree-Fock calculation,[33] the band gaps in these two methods are also very different. Band structure predicts a minimum gap of 9.2 eV while Hartree-Fock predicts a gap of 15 eV, possibly because neither of these methods can calculate reliable electron affinities of molecules. It is also difficult to determine the band gap from either the energy-loss experiments,[34] which show an edge at 10.9 eV plus other features at 15 eV or from absorption in the optical spectrum[35] which has features at 10, 12, 15, and 17 eV. At the higher pressure of 2.4

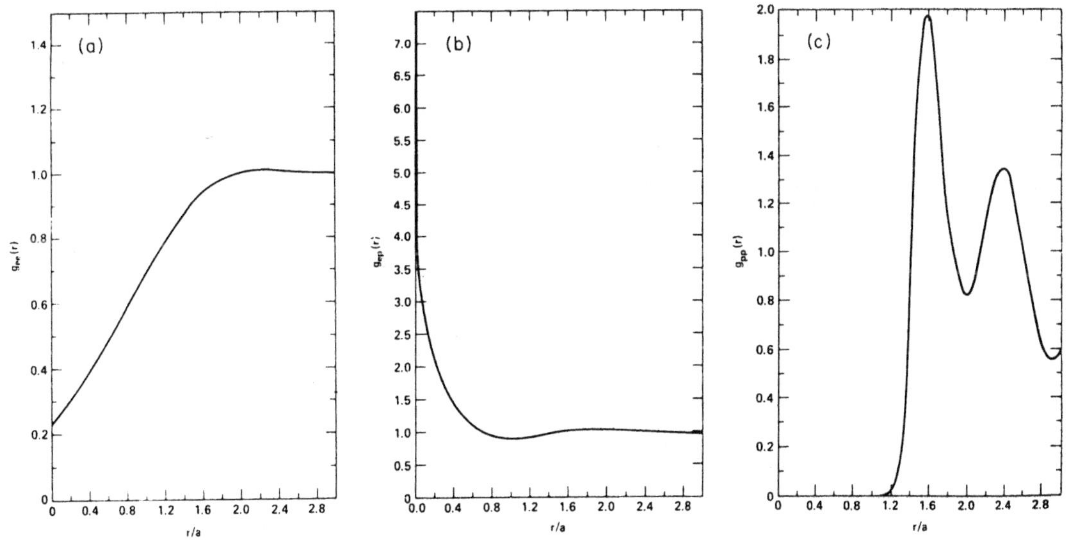

FIG. 10. The pair correlation function in the atomic sc phase at the density $r_s=1.31$. Each of the three functions is sphericalized and normalized to be unity at large r.

Mbar the gap at $k=0$ is almost closed in our calculation. In density-functional theory[7] the gap is predicted to be about 6 eV, but an indirect gap has almost closed.

Not withstanding the approximations employed in this gap calculation (small systems, rigid molecules, zero-momentum excitations, and the fixed-node approximation) the results indicate that this direct subtraction method could be useful. The most worrisome problem is the restriction to small systems as can be seen from Table V. The short run of 8 atoms shows that one must clearly go to larger systems. It would be highly desirable to explore the finite-size effects by doing the calculations on larger systems because such effects are much larger for charged systems. The present method however cannot be practically used for very large systems since the statistical error for a given amount of computer time, t, can be shown to be proportional to $(N^4/t)^{1/2}$; thus computer time requirements become excessive. A direct calculation of the energy necessary to promote an electron into the conduction band would solve both the problem of charge neutrality and the restriction to small system size and appears feasible. In addition, information about the entire band structure could be obtained by varying the momentum of the hole and electron.

Alternatively, the behavior of the dielectric constant can be used to indicate the onset of metallic conductivity. An inequality between the structure factor, which can easily be found with Monte Carlo calculations, and the static dielectric function restricts the range when the fcc molecular system is predicted to conduct at zero temperature.

The dielectric function, $\epsilon(k,\omega)$, for any system with translational invariance (valid here since the protons are free to move) can be related[36] to the dynamical structure factor, $S(k,\omega)$, by

$$1-\frac{1}{\epsilon(k,\omega)}=\lim_{\delta\to 0}\left\{[4\pi/(k^2v)]\times\int_{-\infty}^{\infty}d\omega' S(k,\omega')[(\omega'+\omega+i\delta)^{-1}+(\omega'-\omega-i\delta)^{-1}]\right\}. \quad (12)$$

The static dielectric function (i.e., when $\omega=0$) is then proportional to the ω^{-1} moment of $S(k,\omega)$; the singularity at $\omega'=0$ is treated as given by the limit in Eq. (12). The dynamic structure factor can be expanded in terms of the full set of eigenvalues and eigenfunctions:

$$S(k,\omega)=\sum_n \delta(\omega-E_n+E_0)\,|\langle n|\rho_k|0\rangle|^2, \quad (13)$$

where $\rho_k=\sum_i e_i\exp(i\mathbf{k}\cdot\mathbf{r}_i)/\sqrt{N}$ is the fourier transform of the charge density, N is the number of atoms, and e_i is the charge of the ith particle. The first two moments of $S(k,\omega)$ satisfy the two sum rules

$$S(k)=\int_{-\infty}^{\infty}d\omega\,S(k,\omega)=\langle\rho_k\rho_{-k}\rangle,$$
$$S_1(k)=\int_{-\infty}^{\infty}d\omega\,\omega S(k,\omega)=\sum_i \hbar^2k^2e_i^2/2m_iN. \quad (14)$$

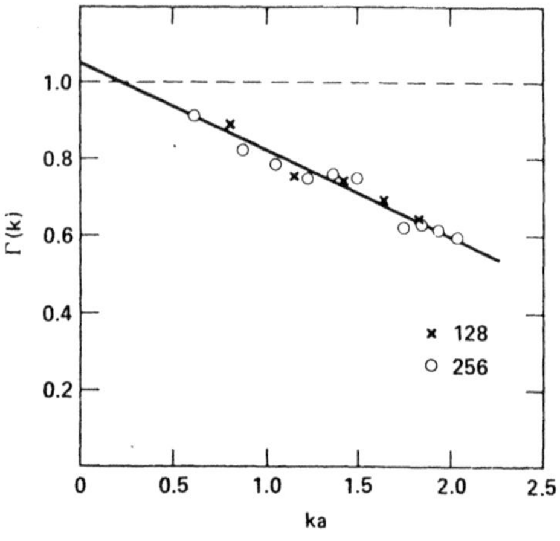

FIG. 11. The low wave-number (ka) behavior of the charged structure factor $\Gamma(k)$ [defined in Eq. (16)] for the atomic metal at $r_s=1.31$ for two different sized systems.

Since from Eq. (13), $S(k,\omega)$ is non-negative, the following inequality holds for any value of b:

$$\int_0^{\infty}d\omega\,S(k,\omega)(\omega^{-1/2}-b\omega^{1/2})^2\geq 0. \quad (15)$$

Choosing the value of b which minimizes the integral, that is, $b=S(k)/S_1(k)$, and using Eqs. (12) and (13) and the observation that $S(k,\omega)$ is zero for $\omega<0$ leads to the following inequality for any translationally invariant system of charges at zero temperature:

$$1-\frac{1}{\epsilon(k,0)}\geq 8\pi[S(k)]^2/[k^2vS_1(k)]\equiv\Gamma(k)^2. \quad (16)$$

The static structure factor thus provides a rigorous lower

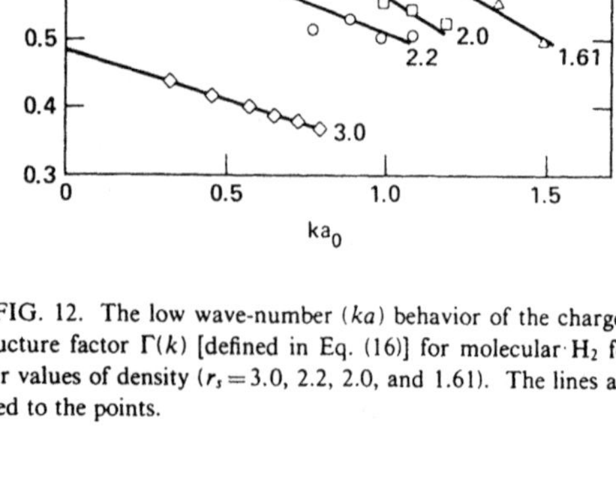

FIG. 12. The low wave-number (ka) behavior of the charged structure factor $\Gamma(k)$ [defined in Eq. (16)] for molecular H_2 for four values of density ($r_s=3.0$, 2.2, 2.0, and 1.61). The lines are fitted to the points.

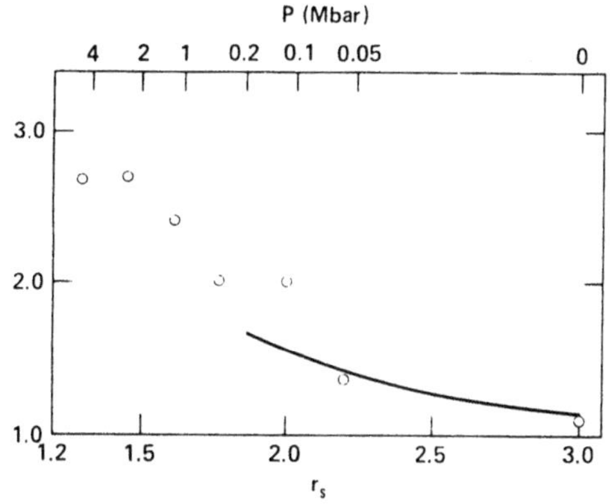

FIG. 13. The index of refraction, n, vs density for hydrogen in the molecular phase. The solid line represents experiment (Ref. 34). The open circles are the MC lower bond obtained from Eq. (16) by extrapolating as in Fig. 10. The pressure scale is the same as used in Fig. 2.

bound to the dielectric function. If at particular wavelength $S(k,\omega)$ is sharply peaked at a single frequency then it is easy to show from Eq. (15) that the inequality becomes an equality. Such is the case in a metal where plasmons are the only long-wavelength excitations. An illustration of that is given in Fig. 11 for $\Gamma(k)$ in the atomic phase at a density of $r_s = 1.31$. Linear extrapolation of $\Gamma(k)$ to zero wave number gives 1.05, which is consistent within the errors and the extrapolation to a value of Γ of 1, that is an infinite dielectric constant or, in other words, a metallic state.

Shown in Fig. 12 is $\Gamma(k)$ in the molecular phase for four densities, $r_s = 3.0$, 2.2, 2.0, and 1.61. Most of the simulations have been performed with 64 atoms, which corresponds to a minimum value of k of 0.97, and hence a relatively far extrapolation to $k = 0$. At zero pressure ($r_s = 3$) the extrapolated $\Gamma(0)$ values of 0.47 yields a lower bound to the index of refraction of 1.14. This value is to be compared with the experimental value of 1.12. In Fig. 13 a comparison of the computed lower bound to the index of refraction with experiment[37] is shown. Higher calculated index of refraction values than experiment around 100 kbar could be accounted for either by statistical fluctuations or more likely by extrapolation errors. Also it has been assumed that the structure factor is isotropic at small k. The index of refraction levels out above 500 kbar at about 2.7. A sharp transition to a metallic state is not expected to be obtained since a finite system contains so few points in reciprocal-lattice space near the Fermi surface. However, this method based on the charged structure factor gives a rough indication of the onset of the metallic state.

VII. CONCLUSION

This work represents an attempt to calculate the properties of a real material from first principles with quantum Monte Carlo. Much more work needs to be done to improve the trial functions, to improve the methods for dealing with fermions, to improve the problem faced by the separation of proton and electron time scales and to simulate much larger systems. However, the feasibility of a realistic calculation of the atomic and molecular phases and the transition between them in hydrogen without using experimental information has been demonstrated. Similar calculations can be done for other low Z elements and are now in progress for lithium. It is straightforward to simulate also mixtures of hydrogen, helium and lithium. It has been shown[25] that this method scales as $Z^{5.5}$, thus making it impractical for large Z elements unless elimination of the inner electrons by the use of pseudopotentials is employed. Finite-temperature extensions of these techniques are well developed for boson systems and could be extended as well to charged fermion systems.

ACKNOWLEDGMENTS

This work was performed under the auspices of the U.S. Department of Energy by the Lawrence Livermore National Laboratory under Contract No. W-7405-Eng-48. We wish to thank G. Sugiyama for performing the energy-gap calculations.

[1] E. Wigner and H. B. Huntington, J. Chem. Phys. **3**, 764 (1935).
[2] J. van Straaten, R. J. Wijngaarden, and I. F. Silvera, Phys. Rev. Lett **48**, 97 (1982).
[3] S. K. Sharma, H. K. Mao, and P. M. Bell, Phys. Rev. Lett. **44**, 886 (1980).
[4] W. J. Nellis *et al.*, Phys. Rev. A **27**, 608 (1983); J. Chem. Phys. **79**, 1480 (1983).
[5] I. F. Silvera, Rev. Mod. Phys. **52**, 393 (1980).
[6] I. F. Silvera and R. J. Wijngaarden, Phys. Rev. Lett. **47**, 39 (1981).
[7] C. Friedli and N. W. Ashcroft, Phys. Rev. B **16**, 662 (1977).
[8] N. W. Ashcroft, Phys. Rev. Lett. **21**, 1748 (1968).
[9] E. G. Brovman, Y. Kagan, A. Kholas, Zh. Eksp. Teor. Fiz. **61**, 2429 (1971) [Sov. Phys—JETP **34**, 1300 (1972)].
[10] M. H. Kalos, M. A. Lee, and P. A. Whitlock, Phys. Rev. B **24**, 115 (1981).
[11] P. J. Reynolds, D. M. Ceperley, B. J. Alder, and W. A. Lester, J. Chem. Phys. **77**, 5593 (1982).
[12] D. M. Ceperley and B. J. Alder, Phys. Rev. Lett. **45**, 566 (1980).
[13] N. Metropolis, A. W. Rosenbluth, M. N. Rosenbluth, A. H. Teller, and E. Teller, J. Chem. Phys. **21**, 1087 (1953).
[14] W. L. McMillan, Phys. Rev. **138**, A442 (1965).
[15] D. M. Ceperley, G. V. Chester, and M. H. Kalos, Phys. Rev. B **16**, 3081 (1977).
[16] D. M. Ceperley, in *Recent Progress in Many-Body Theories*, Vol. 142 of *Lecture Notes in Physics*, edited by J. G. Zabolitzky (Springer-Verlag, Berlin, 1981), p. 262.
[17] D. M. Ceperley and M. H. Kalos, in *Monte Carlo Methods in Statistical Physics*, edited by K. Binder (Springer-Verlag, Berlin, 1979).
[18] D. M. Ceperley and B. J. Alder, J. Chem. Phys. **81**, 5833

(1984).
[19]D. M. Ceperley, Phys. Rev. B **18**, 3126 (1978).
[20]M. Parrinello and A. Rahman, Phys. Rev. Lett **45**, 1196 (1980).
[21]F. J. Pinski and C. E. Campbell, Phys. Lett. **79B**, 23 (1978); L. J. Lantto, Phys. Rev. B **22**, 1380 (1980).
[22]D. M. Ceperley, Physica **108B**, 875 (1981).
[23]J. des Cloizeaux, Phys. Rev. **135A**, 698 (1964).
[24]L. Reatto, Phys. Rev. B **26**, 130 (1982).
[25]D. M. Ceperley, J. Stat. Phys. **43**, 815 (1986).
[26]M. Ross, F. H. Ree and D. A. Young, J. Chem. Phys. **74**, 1487 (1983).
[27]L. D. Landau and E. M. Lifshitz, *Statistical Physics*, 2nd ed. (Addison-Wesley, Reading, Mass, 1970), p. 194.
[28]J. Hammerberg and N. W. Ashcroft, Phys. Rev. B **9**, 409 (1974).
[29]B. I. Min, H. J. F. Jansen, and A. J. Freeman, Phys. Rev. B **30**, 5076 (1984).
[30]L. G. Caron, Phys. Rev. B **9**, 5025 (1974).
[31]V. T. Rajan, C.-W. Woo, Phys. Rev. B **18**, 4048 (1978).
[32]G. A Neece, F. J. Rogers, and W. G. Hoover, J. Comp. Phys. **7**, 621 (1971).
[33]P. Giannozzi and S. Baroni, Phys. Rev. B **30**, 7187 (1984).
[34]L. Schmidt, Phys. Lett. **36A**, 87 (1971).
[35]G. Baldini, Jpn. J. Appl. Phys. Suppl. **14**, 613 (1965).
[36]G. D. Mahon, *Many-Particle Physics* (Plenum, New York, 1981).
[37]H. Shimizu, E. M. Brody, H. K. Mao, and D. M. Bell, Phys. Rev. Lett. **47**, 128 (1981).

Calculation of Exchange Frequencies in bcc ^{3}He with the Path-Integral Monte Carlo Method

D. M. Ceperley[a] and G. Jacucci
Department of Physics, University of Trento, I-38050 Povo, Italy
(Received 21 July 1986)

The exchange frequency in crystal ^{3}He is calculated from first principles with a combination of the path-integral Monte Carlo method and a method used in classical statistical mechanics to determine free-energy differences. The frequency of nearest-neighbor exchange at melting density is 0.46 mK, that of triple exchange is 0.19 mK, and that of four-particle planar exchange is 0.27 mK. These exchange frequencies are within 30% of the values obtained from the empirical multiple-exchange model and agree with measurements. Many other types of high-order cyclic exchanges make significant contributions to thermodynamic properties, showing that ^{3}He is more complex than previously thought.

PACS numbers: 67.80.Mg, 02.70.+d, 05.30.Fk

Crystal ^{3}He at millikelvin temperatures is one of the simplest and cleanest examples in nature of a lattice-spin system. This simplicity arises because atomic helium is, at those temperatures, practically a hard spherical atom with very weak bonding to other helium atoms. Also, thermal phonons and vacancies are not excited. Thus the properties of the magnetic crystal result only from atomic exchange which occurs very rarely, roughly every 10^5 atomic vibrations. Originally it was anticipated that crystal ^{3}He would be described by the antiferromagnetic Heisenberg spin-$\frac{1}{2}$ model since one can show that if only pairs of nearest-neighbor atoms exchange, the system in continuous space can be mapped onto this spin model. The experimental phase diagram is totally at variance with this model: For example, the symmetry of the ground state is different from that of the Heisenberg model. To fit experimental data it is necessary to assume that two-, three-, and four-atom exchanges are approximately equally frequent. This is known as the multiple-exchange model.[1] Since exchange of atoms is a tunneling process, one would expect the barriers for these different processes to be, in general, different and to have different density dependences. It therefore seems improbable that the various frequencies have equal orders of magnitude. The calculation of the exchange frequencies from first principles is necessary for this model to be finally verified.[2]

To define the exchange frequencies and the lattice-spin model, one assumes that most of the time the atoms are close to lattice sites.[3] If there are no ground-state vacancies, there are $N!$ ways of arranging the N atoms onto N lattice sites. This degeneracy is broken by the exchange of atoms. Suppose that we allow only two ways of arranging the N atoms onto the lattice sites which we will denote as Z and PZ. Here P is a cyclic permutation of a few atoms and Z is the vector of a perfect bcc lattice $Z=\{z_1,z_2,\ldots,z_N\}$. Then the ground state is split into two states ϕ_0 and ϕ_1 with even and odd symmetry and energies E_0 and E_1. The frequency with which the system oscillates from Z to PZ is $2J_P=E_1-E_0$. This definition applies to any permutation, but it is expected that only for small cyclic exchanges will the frequency be significant. Then for temperatures below 10 mK, the system of spin-$\frac{1}{2}$ fermions is described by a lattice Hamiltonian[3] acting only on the spins of the atoms: $\sum_P J_p(-1)^P P_\sigma$, where $(-1)^P$ is the parity of the permutation and P_σ is the spin permutation operator.

The exchange frequencies are difficult to calculate since the helium atoms have large zero-point motion and correlation. An exchange of a pair of atoms necessarily involves neighboring atoms moving out of the way, and so it is really a collective process involving on the order of twenty atoms.[4] This explains qualitatively why the frequencies of three- and four-atom exchange are similar to that of two-atom exchange since those exchanges impinge on neighboring atoms much less.

Consider the many-body density matrix for distinguishable particles:

$$\rho(R,R';\beta)=\sum_n e^{-\beta E_n}\phi_n(R)\phi_n(R'), \qquad (1)$$

where $\beta=1/kT$ and R represents the $3N$ spatial coordinates of the atoms. Again let us make the restriction that the atoms must be near Z or PZ. Then for temperatures well below the Debye temperature only two states will contribute to the expansion in Eq. (1). To determine the exchange frequency, consider the density matrix taken between the perfect lattice and a permutation of the perfect lattice, normalized by the diagonal density matrix

$$F_P(\beta)\equiv\frac{\rho(Z,PZ;\beta)}{\rho(Z,Z;\beta)}=\tanh[J_P(\beta-\beta_P)], \qquad (2)$$

where the second equality follows from symmetry properties and $\beta_P=\ln[\phi_1(Z)/\phi_0(Z)]/J_P$. If $F_P(\beta)$ is calculated at two values of β, the exchange frequency can be determined.

To evaluate these density matrices, $M-1$ intermediate points $R_1, R_2, \ldots, R_{M-1}$ are inserted by use of the product property of density matrices giving

$$F_P(\beta)=\frac{\int dR_1\cdots dR_{M-1}\rho(Z,R_1;\tau)\rho(R_1,R_2;\tau)\cdots\rho(R_{M-1},PZ;\tau)}{\int dR_1\cdots dR_{M-1}\rho(Z,R_1;\tau)\rho(R_1,R_2;\tau)\cdots\rho(R_{M-1},Z;\tau)}, \qquad (3)$$

where $\tau=\beta/M$. The path-integral Monte Carlo method is based on making M large enough so that an accurate expression can be written down for $\rho(R,R';\tau)$. We have used for $\rho(R,R';\tau)$ the product of two-atom density matrices[5] and its most important correction term, and $\tau=0.025/\mathrm{K}$ where it was assumed that the helium atoms interact with a pair potential derived from theory and atom-atom scattering data.[6] This high-temperature density matrix gives the low-temperature properties of liquid and solid ^{4}He to an accuracy of 0.1 K/atom.[5]

There has been considerable development in classical statistical mechanics of methods for the computation of such ratios with the Monte Carlo method and, as suggested by one of us,[7] such methods are useful for quantum problems. First let us review the Metropolis Monte Carlo method[8] for sampling from an arbitrary probability distribution function, $\pi(s)$, where s represents the state of some system. The sampling is achieved by the construction of a Markovian random walk where at each step of the walk the state of the system is moved to a nearby state based on a transition probability $T(s\rightarrow s')$ chosen so that detailed balance applies. This guarantees that the probability of the walk visiting the state s is $\pi(s)$. Rejections, which mean that the state is not always changed at each step, are used to satisfy the detailed-balance condition.

To calculate $F_P(\beta)$ we define the state of the walk to consist not only of the path but also of a discrete variable called σ which takes two values, I or P. When $\sigma=I$ we set $R_M=Z$ and when $\sigma=P$ then $R_M=PZ$; the paths either close on themselves or are cross linked. Thus the state is $\{R_1,R_2,\ldots,R_{M-1},\sigma\}$. Now the Metropolis method is used to sample the distribution function, $\pi(s)=\rho(Z,R_1;\tau)\cdots\rho(R_{M-1},R_M;\tau)$, allowing, of course, for transitions between the two states I and P. Then the average number of steps the walk spends in the state P divided by the average number in the state I equals $F_P(\beta)$.

The success of the method clearly depends on the ability to make many transitions between the two states. Luckily this problem has been studied[5] since it arises in calculations of the lambda transition in Bose ^{4}He. Our procedure for making a transition from I to P is to choose a section of the path as likely to allow an exchange. The total path is divided into three parts. The coordinates $R_1,\ldots,R_k$ are left unchanged. The coordinates $R_{k+j},\ldots,R_M$ are reflected into $PR_{k+j},\ldots,PR_M$ which by definition changes the state from I to P. Since the Hamiltonian is symmetric this does not change $\pi(s)$. Finally, a new section of the path is sampled to connect R_k onto PR_{k+j}. The coordinates of the atoms not involved in the exchange are left unchanged while those exchanging are chosen according to the bisection algorithm where first the midpoint between R_k and PR_{k+j} is sampled, then midpoints of the two resulting segments are chosen, etc., until the entire segment is sampled. There is an optimal size, j, for this collective move of the path from state I to P. If j is too small the springs of the polymers will be stretched too much in going from R_k to PR_{k+j}. If j is too large rejections will occur because of the accumulation of small errors in the sampling procedures. The parameter β_P gives the amount of imaginary time needed for the exchange to take place and it is about 0.3/K for most of the exchanges examined. We have found that the optimal value of j is approximately $\beta_P/(2\tau)$. For an exchange of four atoms the optimal elementary Monte Carlo move from state I to P turns out to involve changing 84 coordinates.

A similar method has been optimized for classical statistical mechanics by Bennett.[9] The relative time the system spends in the two states is equal to the rate for making transitions from I to P divided by the reverse rate. Thus it is not actually necessary to make the transitions from one state to the other but only necessary to compute the rates that one could have made the transitions. This combination of classical and quantum techniques has the important property that its accuracy is independent of the actual tunneling frequency. The exchange frequency is expressed as the ratio of rates, not as the very small difference between two eigenvalues. Computer time is only weakly dependent on the number of atoms in the system and on temperature. Our computer runs have involved 16, 54, and 128 atoms. Calculations have been performed down to 0.25 K but most are for 1 K.

The exchange frequency is given by the slope of $F_P(\beta)$ with respect to β [see Eq. (2) and Fig. 1]. One can

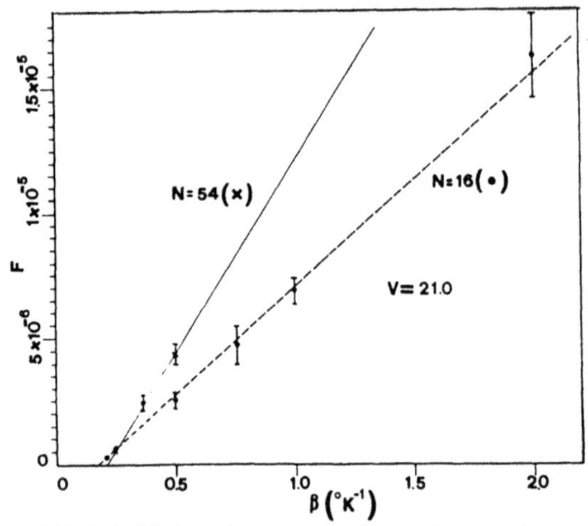

FIG. 1. $F_P(\beta)$ for pair nearest-neighbor exchange. The dots are for 16 atoms and the crosses are for 54 atoms. The solid line represents the least-squares fit of Eq. (2) through the crosses and the dashed line that through the dots. The slope is the exchange frequency J_P and the intercept is the imaginary time of the exchange β_P.

TABLE I. Exchange frequencies at two densities. The semiempirical values from the multiple-exchange model (MEM) are in the last column.[a] The type of exchange is determined (Ref. 10) by the set of $p(p-1)/2$ pair distances amongst the p atoms exchanging where 1 refers to a nearest-neighbor (nn) distance, 2 a next-nearest-neighbor (nnn), etc. The first set of numbers specify the distances of adjacent atoms on the cycle, the next set the second neighbors, etc.

Volume (cm^3/mole) p type	20.07 J_P(MC) (μK)	24.12 J_P(MC) (mK)	24.12 J_P(MEM) (mK)
2(1) nn	15 ($\pm$13%)	0.46 ($\pm$7%)	0.35
2(2) nnn		0.065 ($\pm$10%)	
3(112) triplet	4 ($\pm$20%)	0.19 ($\pm$10%)	0.14
4(1^4;23) planar	6 ($\pm$30%)	0.27 ($\pm$10%)	0.30
4(1^4;22) folded		0.027 ($\pm$18%)	
4(1122;31)		0.006 ($\pm$25%)	
4(1212;11)		0.0005 ($\pm$45%)	
4(1212;14)		0.011 ($\pm$30%)	
4(2^4;33) square		0.0019 ($\pm$30%)	
6(1^6;3^6;4^3)		0.36 ($\pm$30%)	
6(1^6;523 523;417)		0.022 ($\pm$35%)	

[a]D. M. Ceperley and M. H. Kalos, in *Monte Carlo Methods in Statistical Physics*, edited by K. Binder (Springer-Verlag, New York, 1977).

determine the slope with a single calculation since the exchange takes only a finite amount of imaginary time; it is an instanton. Define the time T_e when the exchange occurs for a given path in the state P as the value in imaginary time when the path is equidistant from Z and PZ. Now Eq. (2) implies that T_e can assume $(\beta-\beta_P)$ different values; that is an extra degree of freedom the P state has that the I state lacks. Suppose that we demand that the path in the P state have its exchange time in the middle of the path: $|\beta/2-T_e|<\bar{\beta}/2$ where $\bar{\beta}$ is a constant less than $\beta-\beta_P$. Now that the exchange is confined to a known portion of imaginary time in the middle of the path T_e can assume exactly $\bar{\beta}$ different values and for this new distribution $F_P(\beta)=\tanh(\bar{\beta}J_P)$. We have checked that this method gives the same value of J_P.

We have calculated a variety of two-, three-, four-, and six-particle exchange frequencies at two densities: near melting ($v=24.12$ cm^3/mole), and near the high-pressure limit of the bcc solid ($v=20.07$ cm^3/mole). These are given in Table I. We have tested that the exchange frequencies are not sensitive to β, $\bar{\beta}$, the number of particles, and time step τ, to the 10% accuracy level of our calculations. We find that pair exchange is most frequent, but this is followed closely by planar four-atom exchange and triplet exchange. Thus, the multiple-exchange model is strongly supported. The semiempirical exchange frequencies, also in Table I, from the multiple-exchange model[1,11] are only different from the Monte Carlo (MC) frequencies by 30%. We do not expect perfect agreement since, first, many other types of exchanges are significant and, second, the semiempirical frequencies are partly determined by approximate mean-field calculations[1] on the lattice model. The ordering of exchange frequencies is correctly predicted by high-density semiclassical calculations[4] with a purely repulsive interaction, but those calculations have not yielded absolute magnitudes. Also we find that the largest exchange frequencies scale with density as the (20 ± 1)th power, in agreement with experiment.[1,2]

The specific heat and the magnetic susceptibility at zero magnetic field and at temperatures above the ordering transition can be expanded in powers of the inverse temperature:

$$C_V \propto e_2\beta^2-e_3\beta^3,\quad \chi^{-1}\propto\beta^{-1}-\theta+\beta B, \tag{4}$$

where the coefficients are products of the exchange frequencies.[1] Also the value of the magnetic field when ^{3}He becomes ferromagnetic at zero temperature is a linear combination of the exchange frequencies. The calculation of the coefficients and magnetic field has been done in two ways: first, by use of the three largest frequencies, and second, by use of all frequencies in Table I. The results, in Table II, show the importance of many types of exchange in crystal ^{3}He. Even though the frequencies of some of the exchanges are small, they are numerous. The most important frequencies are determined to better than 10%, but because of large cancellations the resulting coefficients are very inaccurate. With the present results we cannot rule out the possibility that even more types of exchanges are important. It is also possible that the interatomic potential is inaccurate. But since our results give the measured properties of ^{3}He within the errors, it is unlikely that other proposed exchange mechanisms,[2] such as ground-state vacancies or

TABLE II. High-temperature thermodynamic-expansion coefficients, as defined in Eq. (4) at melting (v=24.12 cm^3/mole) and the zero-temperature critical magnetic field for the transition into the ferromagnetic phase. The first column was obtained by our using only the three largest exchange processes (nn, t, and planar), the second column uses all calculated frequencies.

Property	Three frequencies	All frequencies	Experiment[a]
θ (mK)	0.1 ± 1.0	-2.2 ± 1.0	-1.7 ± 0.1
e_2 (mK^2)	5.0 ± 0.8	5.9 ± 1.6	5.9
e_3 (mK^3)	2.0 ± 4.0		<2.4
B (mK^2)	0.7 ± 0.8		0.0 ± 1
H_{c2} (T)	9.7 ± 2.1	19.0 ± 2.2	...

[a]Reference 11.

the coupling to phonons are relevant. The importance of so many types of exchange complicates the already difficult process of our determining from the exchange Hamiltonian such properties as the transition temperature and spin-wave velocities.

In conclusion, a first-principles method of calculating exchange frequencies in quantum crystals has been developed and tested. This calculation has verified the multiple-exchange model for ^{3}He. Three types of exchange are dominant but many others produce significant effects showing that ^{3}He is more complex than previously thought. We anticipate that the development of this accurate computational technique will lead to a much deeper understanding of exchange in quantum crystals.

We wish to thank E. L. Pollock for useful discussions, B. J. Alder for suggesting improvements on the manuscript, and the Aspen Institute for Physics (Aspen, CO), where this research was initiated. This work was partially supported by Consiglio Nazionale delle Ricerche (Italy) through the Center of Studies in Trento.

[a]Permanent address: Lawrence Livermore National Laboratory, Livermore, Ca 94550.

[1]M. Roger, J. Hetherington, and J. M. Delrieu, Rev. Mod. Phys. **55**, 1 (1983).

[2]M. C. Cross and D. S. Fisher, Rev. Mod. Phys. **57**, 881 (1985).

[3]D. J. Thouless, Proc. Phys. Soc. London **86**, 893 (1965).

[4]M. Roger, Phys. Rev. B **30**, 6432 (1984).

[5]E. L. Pollock and D. M. Ceperley, Phys. Rev. B **30**, 2555 (1984); D. M. Ceperley and E. L. Pollock, Phys. Rev. Lett. **56**, 351 (1986).

[6]R. A. Aziz, V. P. S. Nain, J. S. Carley, W. L. Taylor, and G. T. McConville, J. Chem. Phys. **70**, 4330 (1979); D. M. Ceperley and H. Partridge, J. Chem. Phys. **81**, 5833 (1984).

[7]G. Jacucci, in *Monte Carlo Methods in Quantum Problems,* edited by M. H. Kalos (Reidel, Dordrecht, The Netherlands, 1984).

[8]N. Metropolis, A. W. Rosenbluth, M. H. Rosenbluth, A. H. Teller, and E. Teller, J. Chem. Phys. **21**, 1087 (1953).

[9]C. H. Bennett, J. Comput. Phys. **22**, 245 (1976).

[10]A. K. McMahan and R. A. Guyer, Phys. Rev. A **7**, 1105 (1973).

[11]H. L. Stipdonk, G. Frossati, and J. H. Hetherington, J. Low. Temp. Phys. **61**, 185 (1985).

Quantum Simulation of Hydrogen in Metals

M. J. Gillan
Theoretical Physics Division, Atomic Energy Research Establishment, Harwell, United Kingdom Atomic Energy Authority, Oxfordshire, England
(Received 4 August 1986)

The path-integral method of quantum simulation is applied to an empirical model for hydrogen in niobium. Results for the density distribution of D, H, and the positive muon over the unit cell show the dramatic increase of quantum effects along this series. Calculations on the activation energy for diffusion confirm the importance of excited states at high temperature pointed out by Emin, Baskes, and Wilson and suggest that hydrogen diffusion is approximately classical in this regime.

PACS numbers: 61.70.Wp, 05.30.-d, 66.30.Jt

Although the study of hydrogen in metals dates back over a century,[1] our understanding of these systems is not fully satisfactory. One reason for this is the difficulty of making realistic model calculations. The purpose of this Letter is to show how the recently developed path-integral technique of quantum simulation[2] can give new insights in this field. I will present illustrative simulation results relevant to (i) the distribution of hydrogen and deuterium and the closely related positive muon[3] in the unit cell and (ii) the temperature dependence of the hydrogen diffusion coefficient in the typical bcc metal niobium.

In the bcc metals, hydrogen and its isotopes reside on the tetrahedral sites.[1] There are two local-mode frequencies ω, which for Nb:H are[4] 26.5 and 43.4 THz ($\hbar\omega/k_B = 1270$ and 2070 K). On expects quantum effects to be dominant when the temperature T is low compared with these values. For small enough T, vibrational excitations of the hydrogen will be frozen out, and its diffusion will be dominated by tunneling between vibrational ground states on neighboring sites. The well-known theory of Flynn and Stoneham[5] assumes this ground-state dominance and stresses the crucial role played by lattice distortion in the diffusion process. The theory predicts that, except at very low temperature, diffusion will be thermally activated, the activation energy being the distortion energy needed to bring the ground states on neighboring sites into coincidence. Emin, Baskes, and Wilson[6] (EBW) pointed out that excited vibrational states will become important (and Flynn-Stoneham theory will break down) for T well below $\hbar\omega/k_B$, because of the much greater transition amplitudes for these higher states, and because the frequencies ω are changed by the relevant lattice distortions. Experimentally, one finds two regimes of thermal activation in the bcc metals, with activation energies differing typically by a factor of 2; the break between the regimes is at about 250 K for Nb:H.[1] Very recent calculations of Klamt and Teichler[7] confirm the ideas of EBW and show that diffusion of hydrogen in Nb is ground-state dominated below 250 K, the break in activation energy being due to the increasing importance of excited states above this temperature.

Quantum simulation can be potentially very helpful for this kind of system, because it allows one to make model calculations without analytic approximation. This is particularly relevant in the present context, where more and more excited states come in with increasing T. The path-integral technique I use implicitly performs the thermal averaging over states, without treating all the states individually. A direct calculation of the diffusion coefficient, even with this technique, would be exceedingly difficult,[8] but I will suggest how the activation energy can be studied. The results fully support the ideas of EBW,[6] but imply in addition that the break at 250 K is closely related to the transition from quantum to nearly classical behavior. The density distribution of hydrogen and its isotopes, for which I also report results, is of interest because a comparison with diffraction measurements could give a direct test of the assumed model.

The path-integral (PI) technique of quantum simulation[2] has become increasingly popular in the last five years, and has been used to study several important problems, e.g., liquid helium[9] and the solvated electron.[10] The method is based on Feynman's PI formulation of quantum mechanics,[2] which establishes an isomorphism between the original quantum-statistical system of particles and a purely classical system consisting of cyclic chains of beads (P beads on each chain) coupled by harmonic springs. The isomorphism becomes exact only in the limit $P \rightarrow \infty$, but in practice almost exact results can be obtained with a finite P, which depends on the situation of interest. The spring constant of the chains is equal to $mP/\hbar^2\beta^2$ [m the mass of the quantum particles, β the inverse temperature $(k_BT)^{-1}$]. This implies that for small T, where quantum effects are large (and ground-state dominated for small *enough* T in the present context), the chains become relatively extended, as would be suggested by the uncertainty principle; at high T, the chains shrink to a small size and behave almost like point particles, and one approaches the classical (excited-state dominated) limit. For hydrogen in metals, the metal atoms can be treated classically from the outset, except at temperatures below about $\frac{1}{3}$ of the Debye temperature[6] (i.e., $T \lesssim 80$ K), which will not be of interest here. The classical isomorph then consists of a set of point particles representing the metal atoms and a cyclic chain representing each of the hydrogens.[10] The simulation of the classical isomorph can be done either by Monte Carlo or by molecular dynamics,[10] though the latter gives no dynamical information; either way, one is merely sampling configurations of the system with the correct statistical weight. The results to be described were obtained with the molecular-dynamics sampling procedure. The PI technique is best suited to the calculation of static averages, such as spatial distributions; the *direct* calculation of a diffusion coefficient seems to be impossible at present.

The simulations of Nb:H are based on an empirical model.[11] The metal-metal interactions are described by the recent model of Finnis and Sinclair[12] (FS). The energy U_M of the metal subsystem as a function of the positions $\mathbf{r}_{Mi}$ $(i=1,\ldots,N_M)$ of the N_M metal atoms is

$$U_M = -A\sum_i f(\rho_i) + \frac{1}{2}\sum_{i\neq j} V(|\mathbf{r}_{Mi}-\mathbf{r}_{Mj}|),$$

the "densities" ρ_i being of the pair form

$$\rho_i = \sum_{j(\neq i)} \phi(|\mathbf{r}_{Mi}-\mathbf{r}_{Mj}|).$$

The form $f(\rho)=\rho^{1/2}$ of the "embedding function" and the parametrizations of V and ϕ are those given by FS, except for two modifications. Since vibrations of the metal atoms will play an important role, it is important that the phonon spectrum be roughly correct. With the parameters of FS, the zone-boundary frequencies are too low for Nb by 25%–30%. I correct this by multiplying the energy parameters of FS by the factor 1.9. The FS model also suffers from an instability in $V(r)$ for small r, which I cure by replacing $V(r)$ by $B_0+B_1\exp(-r/d)$ inside the nearest-neighbor distance, with B_0, B_1, and d chosen to ensure continuity of the potential and its first two derivatives. For the metal-hydrogen interaction, I assume an exponential pair potential $C\exp(-r/\sigma)$. The parameters C and σ are fixed uniquely by the requirement that the two local hydrogen frequencies be reproduced; this gives $C=4.12\times10^{-15}$ J, $\sigma=1.62\times10^{-11}$ m. The calculations are all done with a lattice parameter $a_0=3.3008$ Å.

It is useful to know the fully relaxed (classical) energy of the system with a single hydrogen at various positions. Call the tetrahedral and octahedral sites T and O, and the midpoint between neighboring T sites S (Fig. 1). Relaxation calculations with the model show that T is energetically lowest, the S and O sites being higher by 0.127 and 0.156 eV, respectively. The saddle point is not exactly at S, but lies about $\frac{1}{4}$ of the way from S to O, with an energy of 0.104 eV.

I have used the potential model to make PI simulations for hydrogen ($m=1$) and deuterium ($m=2$), and also for the positive muon ($m=0.1126$). It is interesting to examine the distribution $\rho(\mathbf{r})$ of the quantum particles, i.e., the thermal equilibrium probability density of finding one at position $\mathbf{r}$ in the unit cell. Calculations of $\rho(\mathbf{r})$ were made at a temperature of 300 K with a periodically repeating system of 54 metal atoms and 5 quantum particles in the case of H and D, and a single quantum particle in the case of μ^+. For H and D, a chain number P equal to 20 is sufficient; for μ^+, I have used $P=80$, which is probably overcautious. The results for $\rho(\mathbf{r})$ are shown in Fig. 1, with the classical prediction $P=1$ for comparison. For the hydrogen isotopes, the distribution is strongly concentrated on the T sites, in accord with experiment.[1] The quantum broadening is very marked, however, and the density is quite substantial all along the "corridor" joining T sites. This implies that diffraction experiments, which measure this density, could be used to get information about the metal-hydrogen potential over a range of separations. Available diffraction results[13] for deuterium in Nb show a measurable density between the interstitial sites and could, with improvement, provide the required comparison. For the muon, $\rho(\mathbf{r})$ is spread evenly over the T–O region. The μ^+ will spend most of its time in the region of T, since there are twice as many of these sites. This is consistent with experiment,[3] but the balance may be del-

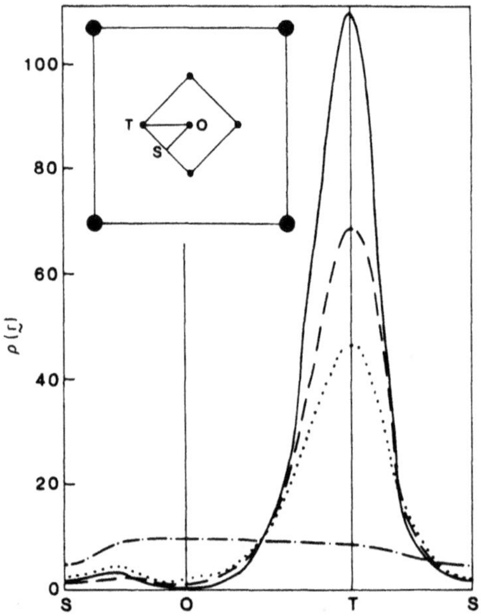

FIG. 1. The simulated density distribution $\rho(\mathbf{r})$ (normalized to the bulk density) for D (dashed line), H (dotted line), and μ^+ (dot-dashed line) in Nb at 300 K; solid line shows the classical prediction. Inset: Points T, O, and S on a face of the bcc cube.

icate, since experiment shows that in vanadium the μ^+ resides partly on the O site.[3]

I now want to argue that simulation can be used to help interpret the meaning of the break in the experimental Arrhenius slope. First, let us see what happens if one calculates the diffusion coefficient D by ordinary (classical) molecular dynamics. Results for D obtained in this way from the slope of the time-dependent mean square displacement are compared in Fig. 2 with measured values.[14] The qualitative agreement between the two at temperatures above the experimental change of slope is remarkable. The activation energy from simulation is 0.101 eV (essentially the same as our classical saddle-point energy, as expected), which is in close agreement with the experimental value of 0.106 eV. This suggests that classical mechanics provides at least a rough guide to the diffusion behavior above ~250 K (but see discussion below).

Since path-integral simulation cannot give D directly, the study of the break in Arrhenius slope requires an indirect method. The idea is that the transition rate between sites is governed by the probability of finding the center of mass (c.m.) of the quantum chain at the saddle point separating the sites. If $F(\mathbf{r})$ is the free energy of the system when the c.m. of the chain is fixed at $\mathbf{r}$, the probability distribution of the c.m. is rigorously $\rho_{\text{c.m.}}(\mathbf{r}) = \alpha\exp[-\beta F(\mathbf{r})]$. One expects that F will have its minimum value at the T site and will have a saddle point somewhere near the classical saddle point. Let ΔF be the difference of F between these two positions, and $\Delta E \equiv \partial(\beta\Delta F)\partial\beta$ be the corresponding energy. Now it is clearly correct to identify ΔE with the activation energy for diffusion at high temperature: The quantum chain is then contracted to a small size, ΔF and ΔE go to their classical values, and the identification just corresponds to classical transition-state theory. At low temperatures, the chain becomes extended, but then ΔE becomes equal to the energy difference between the ground states associated with the relaxed saddle-point and T-site configurations. But this is just the activation energy given by the Flynn-Stoneham theory[5] for diffusion in the quantum regime. Since ΔF yields the correct activation energies in both the high-temperature quasiclassical regime and the low-temperature quantum regime, it seems at least plausible that it will exhibit the transition between the two at roughly the correct temperature. The proposed method constitutes the natural generalization of transition-state theory: The transition rate is proportional to the probability of finding the system at the appropriate saddle point, but the saddle point is now in the configuration space extended to include the degrees of freedom of the quantum chain. This idea is not new, and is implicit in recent discussions of the general problem of quantum barrier crossing.[15]

To find the difference of $F(\mathbf{r})$ between any two points $\mathbf{r}$, I use the fact that $F(\mathbf{r})$ is the potential of mean force for the c.m. of the chain: If $\mathbf{f}(\mathbf{r})$ is the mean force acting on the chain when the c.m. is fixed at $\mathbf{r}$, then $\mathbf{f}(\mathbf{r}) = -\nabla F(\mathbf{r})$. Then ΔF can be expressed as $\Delta F = -\int d\mathbf{s}\cdot\mathbf{f}$, along any integration path joining the two points. I use a path going from T to S and then along the lines $S-O$ to the point where $\mathbf{f}$ vanishes. The calculation of $\mathbf{f}$ at seven points along this path allows an accurate numerical integration to obtain ΔF. I have calculated ΔF for hydrogen and deuterium at seven temperatures ranging from 68 to 763 K. The required value of P varies strongly with temperature, going from 50 to 10 over this range. The results are shown in Fig. 3 as a plot of $-\beta\Delta F$ vs β, i.e., as an Arrhenius plot of $\rho_{\text{c.m.}}$ at the saddle point divided by its value at site T. For hydrogen, a marked change of behavior is clear at ~250 K, which is close to the observed transition (cf. Fig. 2). The low-

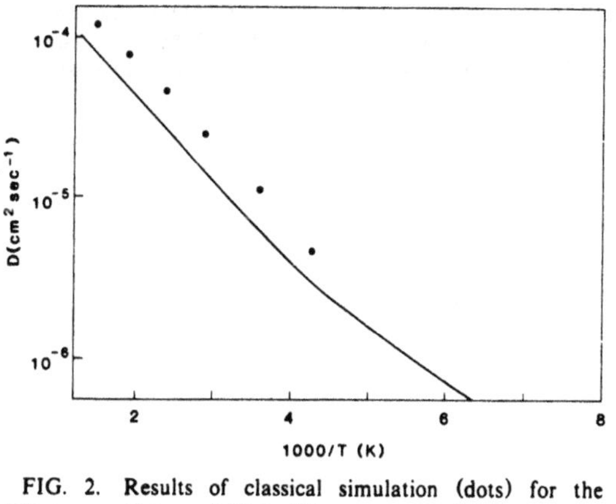

FIG. 2. Results of classical simulation (dots) for the diffusion coefficient of H in Nb compared with experiment (Ref. 11). Note the experimental break in Arrhenius slope at ~250 K.

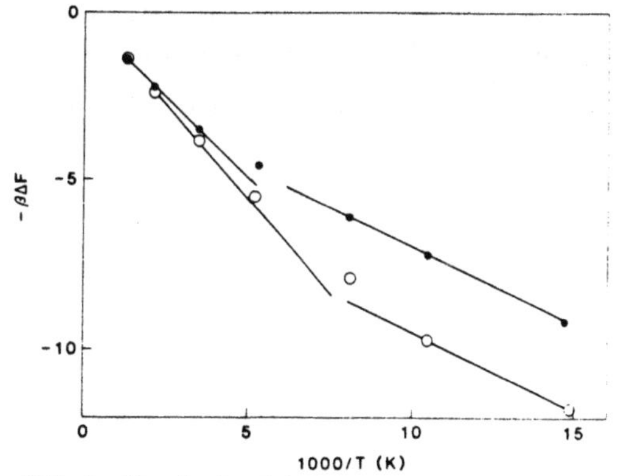

FIG. 3. The simulated free-energy difference ΔF for the center of mass of the quantum chain at the saddle point and at the T site (H: filled circles, D: open circles). Slopes of the straight lines give the activation energies at low and high temperatures.

temperature ΔE is 0.041 eV, which is in qualitative accord with (actually some 40% smaller than) the observed activation energy in this region; the high-temperature ΔE is 0.084 eV ($\sim$20% lower than experiment). As expected, the transition for deuterium is at a lower temperature ($\sim$140 K); the low-temperature ΔE is the same as for hydrogen, while the high-temperature value is 0.094 eV. This is consistent with experiments for deuterium, which so far go down only to $\sim$150 K, and find no break of slope in this range. My results agree with experiment that the high-temperature ΔE for D is greater than for H. Note that our high-temperature ΔE for both H and D lie below our classical saddle-point energy (0.104 eV). Although, as noted above, they must become equal to it for high enough T, we expect some reduction because of tunneling through (spreading of the quantum chain across) the barrier top. The experimental fact[14] that the isotope effect above the transition temperature deviates from the classical prediction shows that diffusion cannot be *fully* classical in this region. Nevertheless, my calculations show that the near equality of the high-temperature ΔE and the classical saddle-point energy is no accident, and suggest that the quantum effects in this region may be regarded as *corrections* to the classical description. I stress that the simulation results are fully consistent with the ideas of EBW[6]: The increasing dominance of excited states which causes the change of Arrhenius slope corresponds to the contraction of the quantum chains in the PI description.

The main point of this Letter is that quantum simulation opens up new ways of studying models for hydrogen in metals. One quantity that can be readily calculated is the density distribution, and the comparison of this with diffraction data could be used to test and improve models. Available data appear not to be adequate for this purpose, and it is to be hoped that new measurements will be made. My results relating to thermally activated diffusion suggest that the observed transition in bcc metals is related to the transition from quantum to nearly classical behavior.

I am grateful for illuminating discussions with Dr. A. M. Stoneham. The work described is part of the longer-term research within the Underlying Programme of the United Kingdom Atomic Energy Authority (UKAEA).

[1]For reviews, see, e.g., *Hydrogen in Metals,* edited by G. Alefeld and J. Völkl (Springer, Berlin, 1978); Y. Fukai and H. Sugimoto, Adv. Phys. **34**, 263 (1985).

[2]E.g., R. P. Feynman, *Statistical Mechanics* (Benjamin, Reading, MA, 1972); J. Barker, J. Chem. Phys. **70**, 2914 (1979); D. Chandler and P. G. Wolynes, J. Chem. Phys. **74**, 7 (1981); *Monte Carlo Methods in Quantum Problems,* edited by M. H. Kalos (Reidel, Dordrecht, 1984).

[3]A. Seeger, in *Hydrogen in Metals,* edited by G. Alefeld and J. Völkl (Springer, Berlin, 1978), Vol. 1, p. 349.

[4]T. Springer, in Ref. 3, p. 75.

[5]C. P. Flynn and A. M. Stoneham, Phys. Rev. B **1**, 3966 (1970).

[6]D. Emin, M. I. Baskes, and W. D. Wilson, Phys. Rev. Lett. **42**, 791 (1979).

[7]A. Klamt and H. Teichler, Phys. Status Solidi (b) **134**, 103, 533 (1986).

[8]E. C. Behrman and P. G. Wolynes, J. Chem. Phys. **83**, 5863 (1985).

[9]D. M. Ceperley and E. L. Pollock, Phys. Rev. Lett. **56**, 351 (1986).

[10]M. Parrinello and A. Rahman, J. Chem. Phys. **80**, 860 (1984).

[11]A previous model due to H. Sugimoto and Y. Fukai, Phys. Rev. B **22**, 670 (1980), seems unsatisfactory, because the relaxed classical saddle-point energy is only 0.061 eV, i.e., less than both the low- and high-temperature experimental activation energies (H. R. Schober and A. M. Stoneham, to be published).

[12]M. W. Finnis and J. E. Sinclair, Philos. Mag. A **50**, 45 (1984).

[13]M. J. Tew, Ph. D. thesis, Oxford University, 1973 (unpublished).

[14]J. Völkl and G. Alefeld, in Ref. 3, p. 321.

[15]P. G. Wolynes, Phys. Rev. Lett. **47**, 968 (1981); H. Grabert and U. Weiss, Phys. Rev. Lett. **53**, 1787 (1984); see also M. Sprik, R. W. Impey, and M. L. Klein, Phys. Rev. Lett. **56**, 2326 (1986).

SUBJECT INDEX

AUTHOR INDEX

PERSPECTIVES IN CONDENSED MATTER PHYSICS

Published Volumes

1. G. Margaritondo, *Electronic Structure of Semiconductor Heterojunctions*
2. A.H. MacDonald, *Quantum Hall Effect: a Perspective*
3. M. Ronchetti and G. Jacucci, *Simulation Approach to Solids*
4. W. Mönch, *Electronic Structure of Metal-Semiconductor Contacts*

Forthcoming Volumes

I. Janossy, *Optical Effects in Liquid Crystals*
H. Neddermeyer, *Scanning Tunneling Microscopy*

Executive Editor: L. Miglio

Dipartimento di Fisica dell'Università di Milano
Via Celoria, 16 I-20133 MILANO
Fax + 39/2/2366583; **Telex** 334687 INFNMI; **Tel.** + 39/2/2392.408

finito di stampare nel mese
di dicembre 1990
dalla Nuova Timec s.r.l.
Albairate (MI)

Editoriale Jaca Book spa
Via Aurelio Saffi 19, 20123 Milano

spedizione in abbonamento
postale TR editoriale
aut. D/162247/PI/3
direzione PT Milano

GPSR Compliance

The European Union's (EU) General Product Safety Regulation (GPSR) is a set of rules that requires consumer products to be safe and our obligations to ensure this.

If you have any concerns about our products, you can contact us on ProductSafety@springernature.com

In case Publisher is established outside the EU, the EU authorized representative is:

Springer Nature Customer Service Center GmbH
Europaplatz 3
69115 Heidelberg, Germany

Batch number: 10607920

Printed by Printforce, the Netherlands